Ecological
Sustainability

Understanding Complex Issues

Ecological Sustainability

Understanding Complex Issues

Robert B. Northrop
Anne N. Connor

CRC Press
Taylor & Francis Group
Boca Raton   London   New York

CRC Press is an imprint of the
Taylor & Francis Group, an **informa** business

CRC Press
Taylor & Francis Group
6000 Broken Sound Parkway NW, Suite 300
Boca Raton, FL 33487-2742

First issued in paperback 2017

ISBN-13: 978-1-4665-6512-8 (hbk)
ISBN-13: 978-1-138-07707-2 (pbk)

Library of Congress Cataloging-in-Publication Data

Northrop, Robert B.
    Ecological sustainability : understanding complex issues / Robert B. Northrop and Anne N. Connor.
    pages cm
    Includes bibliographical references and index.
    ISBN 978-1-4665-6512-8 (hardcover : acid-free paper)
    1. Sustainability. 2. Sustainable development. 3. Sustainability--Simulation methods. 4. Sustainable development--Simulation methods. 5. Human ecology. 6. Social systems. 7. Biocomplexity. I. Connor, Anne N. II. Title.

GF50.N675 2013
338.9'27--dc23
                                            2012044966

Visit the Taylor & Francis Web site at
http://www.taylorandfrancis.com

and the CRC Press Web site at
http://www.crcpress.com

We dedicate this book to our spouses, Adelaide and Michael

Contents

Preface

This book is intended for general use by readers interested in ecological sustainability and the complex issues affecting it, or by college students in a one-semester seminar or classroom course that has the purpose of acquainting scientists, including chemists, ecologists, environmental scientists, biologists, engineers (civil, mining, and sanitary), political scientists, and others, with sustainability and how we can deal with its issues. We have focused on the many complex systems that interact to challenge human sustainability; we also have introduced the reader to the basic area of complexity and complex systems. The book provides numerous case studies and examples for those wishing to go into greater depth in the study of complex systems theory. Its interdisciplinary nature makes it well suited for students of ecological systems.

Reader Background: We have assumed readers have had introductory college courses in algebra, biology, chemistry, calculus, and ordinary differential equations. Readers will also benefit from having had basic college courses in biochemistry, ecology, and perhaps cell biology. Although the book assumes a strong math background, it still can be used and understood without delving into the equations. It is the hope of the authors that students serious about acting on some of the issues presented will take it upon themselves to strengthen their math backgrounds. Effective leadership in ecological sustainability will greatly benefit from the ability to model and predict the behavior of these complex systems.

Features

Some of the Unique Features of Ecological Sustainability: Understanding Complex Issues are as follows:

1. *Sustainability* is defined in Chapter 1 and shown to be a function of the interaction of a number of complex systems, one of the more important of which is information-driven human behavior. Sustainability is seen to vary with geography and with individual countries. Accordingly, North America and Western Europe can be considered one *sustainable area* (SA). Other SAs include sub-Saharan Africa, China, Korea and Japan, India, Russia, South America, and so forth.

2. Chapter 2 treats the general topic of complexity and how humans typically react to and deal with complex situations and problems.

It describes the *law of unintended consequences* (LUC), the *social action rate sensitivity law* (SARSL), the *single-cause mentality*, and the consequences of *"not in my box" thinking*. Also introduced in Chapter 2 are directed graphs and linear signal flow graphs and how they can be used to visualize the interrelationships in the models of the complex nonlinear systems (CNLSs) associated with sustainability. Graph modules are described, and examples of modules in sustainability modeling are given.

3. Chapter 3 reviews the more important, multidimensional challenges to human sustainability, beginning with population growth and the many adverse effects of overpopulation in Section 3.2. We argue that population growth affects, either directly or indirectly (parametrically), all systems relevant to sustainability. The future of world population growth is explored. In Section 3.3, we examine the causes and effects of global warming. Freshwater is considered in Section 3.4 as a limited resource vital to robust agriculture, manufacturing, energy, production, and urban living. How greenhouse gas (GHG) emissions are affecting water resources is also covered. Cities cannot exist without water supplies. *Bee colony collapse disorder* (CCD) is described in Section 3.5 as a threat to the world's food resources. In Section 3.6, we describe the interesting phenomenon of the slow *species size reduction* caused by planetary warming and consider its effects on sustainability. In Section 3.7, we consider the effects of fossil fuel depletion and the sustainability impact and potential of underdeveloped sources of fossil fuels, such as methane hydrate, natural gas, oil sands, and oil shale.

4. Chapter 4 describes some actions we can take (and are taking) to mitigate, and even correct, certain of our impacts on our sustainability. The pros and cons of biofuels are treated in Section 4.2, including ethanol, methanol, biodiesel, hydrothermal carbonization, and syngas production in catalyzed solar thermochemical reactors. Section 4.3 describes the technologies and economics of desalination. Carbon-free (renewable) energy sources are treated in Section 4.4. These sources include wind, solar thermal, solar photovoltaic, thermophotovoltaic, thermoelectric, hydropower (including waves and tides), geothermal energy, and hydrogen, In Section 4.5, carbon-neutral energy sources are considered (wood and biomass, fuel cells, and biogenic methane). Physical means of energy storage are described in Section 4.6, including flywheels, and fusion energy sources are treated in Section 4.7. Nuclear reactors, including pebble-bed reactors, are covered in Section 4.8.1. Section 4.9 addresses the controversial topic of carbon capture and storage. The carbon cycle is described, showing natural and anthropogenic CO_2 fluxes, sources, and sinks. Section 4.10 considers water vapor as a GHG. The

important topic of designing increased energy efficiency into our lives as a means of reducing carbon emissions is treated in Section 4.11.

5. Chapter 5 details the problems inherent in nonsustainable agriculture. Current large-scale "industrial" farming practices constitute threats to individuals in the short run and to human sustainability in the long run. The cost of industrial agriculture is high and not fully reflected in prices at the grocery store. This chapter examines these threats and presents some possible mitigations and solutions. The question of whether large-scale centralized agriculture and farm animal production is necessary to "feed the masses" is also considered.

6. Chapter 6 describes food derived from four alternate sources including insects, plankton, fungi, and muscle cells grown in vitro. The dangers of overharvesting plankton are described. Edible fungi are treated in Section 6.5, including the batch-grown mycelia made into the processed food, Quorn. Fungi harmful to human survival are also described, including grain rusts and the white-nose fungal infection killing bats in the eastern United States. Section 6.6 introduces the concept of growing animal muscle cells in large-scale tissue cultures using committed stem cells and discusses some of the challenges this emerging technology will face.

7. In Chapter 7 we discuss basic economic systems and argue that to be effective in predicting economic system behavior, economists should use dynamic simulation models using agents.

8. Chapter 8 describes some effective approaches for characterizing CNLSs seen for the first time.

9. Some extrapolations, predictions, and FAQs on human sustainability are discussed in Chapter 9.

10. A comprehensive Glossary defines many of the arcane acronyms used in the disciplines covered in this text, as well as many scientific terms and abbreviations.

MATLAB® is a registered trademark of The MathWorks, Inc. For product information, please contact:

The MathWorks, Inc.
3 Apple Hill Drive
Natick, MA 01760-2098 USA
Tel: 508 647 7000
Fax: 508-647-7001
E-mail: info@mathworks.com
Web: www.mathworks.com

Authors

Robert B. Northrop, PhD, was born in White Plains, NY, in 1935. After graduating from Staples High School in Westport, CT, he majored in electrical engineering (EE) at the Massachusetts Institute of Technology (MIT), graduating with a bachelor's degree in 1956. At the University of Connecticut (UCONN), he received a master's degree in systems engineering in 1958. As the result of a long-standing interest in physiology, he entered a PhD program at UCONN in physiology, doing research on the neuromuscular physiology of molluscan catch muscles. He received his PhD in 1964.

In 1963, he rejoined the UCONN EE Department as a lecturer, and he was hired as an assistant professor of EE in 1964. In collaboration with his PhD advisor, Dr. Edward G. Boettiger, he secured a 5-year training grant in 1965 from the National Institute of General Medical Sciences (NIGMS) of the National Institutes of Health (NIH), and started one of the first interdisciplinary biomedical engineering graduate training programs in New England. UCONN currently awards MS and PhD degrees in this field of study, as well as BS degrees in engineering under the Biomedical Engineering (BME) area of concentration.

Throughout his career, Dr. Northrop's research interests have been broad and interdisciplinary and have been centered on biomedical engineering and physiology. He has done sponsored research (by the Air Force Office of Scientific Research (AFOSR)) on the neurophysiology of insect and frog vision and devised theoretical models for visual neural signal processing. He also did sponsored research on electrofishing and developed, in collaboration with Northeast Utilities, effective, working systems for fish guidance and control in hydroelectric plant waterways on the Connecticut River at Holyoke, MA, using underwater electric fields.

Still another area of his sponsored research (by NIH) has been in the design and simulation of nonlinear, adaptive, digital controllers to regulate in vivo drug concentrations or physiological parameters, such as pain, blood pressure, or blood glucose in diabetics. An outgrowth of this research led to his development of mathematical models for the dynamics of the human immune system, which were used to investigate theoretical therapies for autoimmune diseases, cancer, and HIV infection.

Biomedical instrumentation has also been an active research area for Dr. Northrop and his graduate students: an NIH grant supported studies on the use of the ocular pulse to detect obstructions in the carotid arteries. Minute pulsations of the cornea from arterial circulation in the eyeball were sensed using a no-touch, phase-locked, ultrasound technique. Ocular pulse waveforms were shown to be related to cerebral blood flow in rabbits and humans.

More recently, Dr. Northrop addressed the problem of noninvasive blood glucose measurement for diabetics. Starting with a phase I Small Business

Initiation Research (SBIR) grant, he developed a means of estimating blood glucose by reflecting a beam of polarized light off the front surface of the lens of the eye, and measuring the very small optical rotation resulting from glucose in the aqueous humor, which in turn is proportional to blood glucose. As an offshoot of techniques developed in micropolarimetry, he developed a magnetic sample chamber for glucose measurement in biotechnology applications. The water solvent was used as the Faraday optical medium.

He has written numerous papers in refereed journals, as well as 12 textbooks: *Analog Electronic Circuits* (1990); *Introduction to Instrumentation and Measurements* (1997); *Endogenous and Exogenous Regulation and Control of Physiological Systems* (2000); *Dynamic Modeling of Neuro-Sensory Systems* (2001); *Noninvasive Instrumentation and Measurements in Medical Diagnosis* (2002); *Signals and Systems Analysis in Biomedical Engineering* (2003); *Analysis and Application of Analog Electronic Circuits in Biomedical Engineering* (2004); *Introduction to Instrumentation and Measurements*, 2nd edition (2005); *Introduction to Molecular Biology, Genomics & Proteomics for Biomedical Engineers* (with Anne N. Connor) (2009); *Signals and Systems Analysis in Biomedical Engineering*, 2nd edition (2010); *Introduction to Complexity and Complex Systems* (2011); and *Analysis and Application of Analog Electronic Circuits in Biomedical Engineering*, 2nd edition (2012).

His current research interest lies in complex systems.

Dr. Northrop was on the Electrical & Computer Engineering faculty at UCONN until his retirement in June 1997. Throughout this time, he was the director of the Biomedical Engineering Graduate Program. As emeritus professor, he still teaches graduate courses in biomedical engineering, writes texts, sails, and travels. He lives in Chaplin, CT, with his wife and a smooth fox terrier.

Anne N. Connor, MA, is currently working as the director of community grants for Methodist Healthcare Ministries, a medical nonprofit organization in San Antonio, TX. Her educational background includes a bachelor's degree from Dartmouth College, where she received honor citations in chemistry and sociology. Her master's degree in communications is from the University of New Mexico at Albuquerque. She is the coauthor of the 2008 textbook, *Introduction to Molecular Biology, Genomics and Proteomics for Biomedical Engineers,* (Taylor & Francis/CRC Press, ISBN # 1420061194). She is a graduate of the Leadership Texas Class of 2006 and the Leadership America Class of 2011 and has received numerous awards for her work, most recently a humanitarian award from the health care community in San Antonio. She is a member of Phi Beta Kappa and Phi Kappa Phi.

1

Human Ecological Sustainability

1.1 Introduction

It is important to study and understand the complex relationships between the many changing factors that affect human ecological sustainability. These include, but are not limited to, human population P, freshwater supplies W, energy supplies E (including fossil fuels and renewable sources), food supplies F, natural resources N (metal ores, cement, wood, etc.), the economy [using gross domestic product (GDP) as a measure], emerging diseases D (human, animal, and plant), and pollution C (*atmospheric:* greenhouse gasses, volatile organic compounds (VOCs); *water and land:* insecticides, fertilizers, steroid hormones, bacteria, etc.). W, F, N, and C directly affect and are affected by ecosystems.

The crude relations between these factors and atmospheric CO_2 concentration $[CO_2]$, climate change, and human sustainability are shown in the simple web graph in Figure 1.1. Note that F, W, E, N, and GDP are dimensionless variables in this simple example. F, W, E, P, and N are nonnegative. SUS is a prototype human *sustainability function:* $SUS \equiv (k_f F + k_w W + k_e E + k_n N + k_{eco} GDP)/P$, where k_f, k_w, k_e, k_n, and k_{eco} are normalization constants and the population P is continually increasing. GDP/P is the *per capita* national GDP ($pGDP$).

Thus, to insure sustainability, it is necessary to keep SUS above some arbitrary threshold, φ_{sus}. In order to keep $SUS > \varphi_{sus}$, F, W, E, N, and GDP must grow, too. Therein lies the problem; F, W, E, N, and GDP have practical upper limits. Note than many other sustainability functions can be devised. In any sustainability model, food, water, energy, and wealth appear to be the most important factors for thriving; however, all four parameters affect and are affected by the ecosystems around us. In a modern, developed society, a collapsed economy can lead to governmental collapse, conflict, riots, hyperinflation, famine, mass unemployment, emigration, and so forth. $pGDP$ is not the only possible choice for an economic goodness function. However, it is generally accepted as a robust indicator of a country's economic health.

In the web graph, a path ending in an arrow shows that the net effect of the source parameter is stimulating or activating the object parameter. In our notation, paths that end in balls indicate that the source parameter is

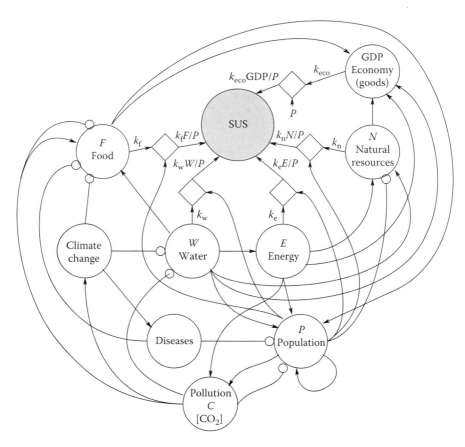

FIGURE 1.1
Simple web graph for ecological sustainability. A "crude look at the whole" (CLAW). See text for description.

inhibitory; for example, diseases, **D**, cause the rate of population growth, **P**, to decrease, as well as decrease the amount of food **F** (plant, animal) available. We have shown [CO_2] as both increasing food supply **F** (i.e., plant growth) and decreasing **F** (ocean acidification from dissolved CO_2 adversely affects fisheries).

This network diagram is a gross oversimplification, but it does illustrate the tangled web of the major variables involved in determining even a simple measure of sustainability. In the big picture, the *food, water, energy, natural resources, economy, pollution,* and *population* nodes are each in themselves composed of a number of large, complex subsystems or modules.

For example, the *food module* **F** must contain dynamic subsystems describing the growth, cultivation, harvesting, and bringing to market of many different types of food: *meat animals* (cattle, sheep, swine, goats, poultry, etc.); *grains* (rice, wheat, oats, barley, corn, etc.); edible seeds, quinoa, and *fruits* (apples, pears, plums, grapes, etc.); *vegetables* (beans, beets, eggplants, onions, peas, potatoes, turnips, etc.); and *fish* (tuna, flounder, cod, salmon, herring, etc., as well as *marine invertebrates* such as crabs, lobsters, shrimp, clams, mussels, oysters, squid, plankton, etc.). Also affecting the *food module*, but not shown, are weather (affected by climate change) and pollinators (affected by pollution). Clearly, **F** is ecosystem dependent. The dimensionless food parameter **F** itself must be a function of the kinds, quantities, and qualities of foods. Perhaps the nutritional groups (proteins, carbohydrates, lipids, etc.), total dietary calories available in a nutritional group, and retail market prices should be considered in computing an **F** for sustainability.

Another example is the *natural resources module* **N**: Natural resources include *nonrenewable resources* [fossil fuels (coal, natural gas, methane, oil), helium, metal and rare earth ores, etc.] as well as *renewable resources* (wood, biomass, and algae) and *noncarbon, natural energy sources* (wind, solar flux, water power, etc.). Certainly, one of the more complex modules is that of *energy* **E**: Energy is mostly distributed as electric power; electric power is generated mostly by the combustion of fossil fuels but increasingly from noncarbon sources. Most noncarbon sources are intermittent, which underscores the importance of buffering or storing their excess output power. Electric power from various sources and storage buffers is distributed on lossy transmission grids. Once generated, electric energy can be stored in the potential energy of pumped hydro or compressed air, in storage batteries, or even in flywheels. Clearly, there are strong connections between the **N** and **E** modules. For a general discussion of the history of modeling "the human predicament," see Hayes (2012).

The *economy module* may be the most complex module; it contains many complex subsystems affecting our sustainability, including the rate of inflation of food prices and those of other goods, a consumer confidence index, employment, taxes, government spending, private investment, and of course the *per capita* **GDP** of a country, and many other factors. It is driven by the behavior of human agents as well as natural events.

Not shown specifically in the graph is a complex *climate module*. *Climate* contains subsystems describing air and ocean currents, clouds, rain and snowfall, cyclonic storms, heat transfer by ocean currents and water vapor (WV) in the moving atmosphere, melting ice and glaciers, and so forth. Climate, of course, is affected by atmospheric greenhouse gas concentrations (e.g., $[CO_2]$ and $[CH_4]$) and particulate air pollution. Climate is a direct modulator of agricultural food production, as well as freshwater supplies and general ecosystem viability. It also affects ocean currents transporting heat and nutrients, and hence, indirectly, fisheries. Climate also directly affects energy consumption (heating and air conditioning in developed countries).

Also not addressed specifically in the web graph is a social sustainability module, including governments, legislatures, bureaucracies, courts, military, police, and so forth, at all levels (national, state, county, municipal). Failure of a central government can lead to anarchy, conflict, and the eventual evolution of new, smaller, semiautonomous social management units (e.g., tribal units, militias). See the book by Tainter (1988) for examples of the collapse of complex societies; also, Diamond (2005) wrote an interesting book on how societies choose to fail or succeed.

In the following chapters, you will be introduced to these changing factors in some detail, how they are interrelated, how mathematical modeling can be used to try to predict their future behavior, and how this knowledge might be used to mitigate or solve the challenges to our sustainability. First, the general topic of sustainability is addressed.

The transitive verb *sustain* comes to English from the Latin verb *sustinere*, to keep up, endure, withstand (from the Latin prefix *sub-*, up from under, and *tenere*, to hold, keep, occupy, possess). Sustainable is the more widely used adjective. Dictionary definitions of *sustain* include the following:

(1) To keep in existence; maintain; prolong

(2) To supply with necessities or nourishment; provide for

(3) To support from below; keep from falling or sinking; to prop

(4) To support the spirits, vitality, or resolution of; in-spirit; encourage

(5) To endure or withstand; bear up under: *sustain hardships*

(6) To experience or suffer (loss or injury)

(7) To affirm the validity or justice of

(8) To prove or corroborate; confirm

Sustainability (*n.*) thus is the ability to sustain (Morris 1973).

Sustainability, *sustainable*, and *sustain* have many meanings, which are context (and user) dependent, similar to the adjective *complex* (Bartlett 1998; Northrop 2011). Many university research centers and professional organizations have been created using the words "sustainable" and "sustainability" in their names and mission statements, illustrating that this important topic has generated considerable academic and social interest and has also attracted research on its many facets since the 1970s in sustainable ecosystems, agriculture, climatology, economics, energy resources, fisheries, and so forth (Meadows et al. 1972).

Sustainable implies a time dimension, that is, something is sustainable over time. One often hears the term *sustainable growth* used as a desirable attribute in economic systems. Applied to societies, the term sustainable growth implies continuing growth over a long period of time, and this requires a healthy economy relatively free of unemployment and inflation. From the finite sizes of our planetary ecosystems, food and freshwater supplies, energy and mineral resources, and so forth, we may infer that such "*sustainable growth*" is an oxymoron.

All biological growth rates peak (e.g., see *Hubbert curve* in the Glossary), and a biological population's number eventually reaches a steady-state, saturation value. (In a closed system, the numbers then decline.) Specifically, individual populations reach size limits set by their ecosystem's environmental resources, competition with other organisms, predators, available energy, and their genomes.

Economic systems are dependent in part on sources of goods that are in turn dependent on raw materials, which are finite and exhaustible, such as fossil fuel (FF) energy sources, ores, minerals, and freshwater. Raw materials also have economic costs set by availability, market speculation, and supply and demand. The growth rate of economic systems necessarily must slow as these materials gradually become less plentiful and, hence, more costly to extract. Technological innovations may improve the production efficiency of certain goods, but such innovations put off the inevitable decrease to economic growth. Economic systems also depend on labor to produce goods and services. Labor requires wages, coming from funds from investment capital, sales, taxes, and so forth.

Thus, human-induced growth in economic systems has limits set by finite resources and funding, just as a colony of *Escherichia coli* in a closed flask of growth medium grows to a maximum population density as it exhausts its nutrients (chemical energy sources). Thus, "sustainable growth" should be viewed to be a short-term phenomenon; it cannot be long term. What can be debated is the time scale involved. Sustainable development has been defined as development that meets the needs of the present population without compromising the ability of future generations to meet their own needs (Bartlett 1998). We predict that sustainable development of a country or region, measured, for example, by the positive first derivative of the gross domestic product *per capita* parameter (GDP) of a country, cannot really be maintained over the long term (say, 25 years), given the rates of human population growth, resource exhaustion, and ecosystem destruction.

Ecological sustainability also involves the hard-to-quantify property of *prosperity*. Prosperity is the socioeconomic condition of having financial and social success and an improved quality of life. It is a group property, associated with an increasing GDP *per capita*. However, prosperity is not solely synonymous with income or wealth; it involves health, property, family, happiness, and certain freedoms as well. Quoting Jackson (2011): "Prosperity consists in our ability to flourish as human beings—within the ecological limits of a finite planet. The challenge for our society is to create the conditions under which this is possible. It is the most urgent task of our times." Jackson also stated, "… it has been clear that something more than material security is needed for human beings to flourish. Prosperity has vital social and psychological dimensions. … an important component of prosperity is the ability to participate freely in the life of society."

Clearly, our sustainable living involves many factors; it is a complex issue because of the many interactions between the many variables involved and, above all, human behavior. Broadly, sustainable living in a region or country is seen where there is a robust economic system with adequate employ-ment and a low rate of inflation, and where there are adequate supplies of affordable food, energy, water, housing, and health care available, *per capita*, to enable "comfortable" living at a reasonable cost. Sustainable living also requires healthy ecosystems and taxable wealth to provide effective gov-ernment services including health care, public transportation systems, and communications. Wealth is also required for robust commerce, agriculture, education, health care, and housing. Robust economic planning must be pro-vided by the government to set sound, long-term policies on taxation, gov-ernment spending, market regulation, when and where government money should be invested in the private sector, and so forth.

Sustainable living is also seen to involve many complex logistic trade-offs because of population growth and finite resources. These trade-offs include, for example, the following: water for energy [to produce natural gas (by fracking) and petroleum from tar sands, oil shale]; food for energy (crop diversion—corn for ethanol, biodiesel from soy); energy for desalinating water; greenhouse gas emissions for energy and transportation; ecosystems for energy (e.g., open pit mining, fracking fluid contamination of ground-water); ecosystems for food (new arable land cleared out of forests); and so forth. Thus, our energy, water, and food needs emerge as key factors in human ecological sustainability.

Sutton (2000) argued: Sustainability is like being pregnant or alive or dead. You are or you are not. So relative sustainability is not a valid concept. Something is either sustained or sustainable or it's not. So when people say something is 'more sustainable' or 'less sustainable' they probably mean one of the following things:

- That more or fewer things can be sustained
- A system is declining slower or faster
- There is a higher or lower probability that something will be sustained
- Something [e.g., a species] that is continuing to decline is hardly sustainable!

If an ecosystem is trending toward loss of biodiversity over time and shows the collapse of certain species' populations, one can argue that the ecosystem's sustainability is in decline. But exactly when is it lost? Arbitrary, measurable criteria and thresholds can be defined and applied to test for *unsustainability*. However, the situation is not as binary as Sutton suggests; there are gray areas. A healthy, robust ecosystem can be made "sick" by human actions or climate change. If not past its tipping point, its landscape can transition to a new, less stable format, where it can continue to exist in a steady state with reduced biodiversity and diminished species, resources, and area; evidently, it is sustaining itself. Ecosystems are generally resilient. However, tipping points exist; loss of too many critical species and/or their energy sources can lead to total ecosystem collapse. This is happening on certain tropical reefs.

One objective in this book is the evaluation of the many dimensions of ecological sustainability for the living conditions of humans on Earth in the past, the present, and the future and how they interact. Also, we examine how human activities have impacted environmental, social, and economic systems on the planet and thus are affecting long-term human survival and well-being. We also address the challenging problems of generating meaningful actions to create human sustainability through technology and behavior modification through education.

1.2 Is It Possible to Model Human Ecological Sustainability?

For an in-depth analysis of human sustainability, one must necessarily examine the significant factors affecting regional and local human populations. The sustainability parameters in central China are not the same as those in the central United States, Northwestern Europe, Japan, or the Amazon basin. It is also true that the actions of humans in the United States, China, Europe, Japan, and the Amazon basin all contribute to global warming and climate change, which affect all population groups. Because of globalized trade and speed-of-light communications, economic effects related to sustainability also tend to be more globally distributed among developed and developing countries. [Consider how the recent economic problems in Greece and Spain have affected the entire European Union (EU) economy.] However, the effects of population growth on social groups and ecosystems remain substantially local.

In order to predict and plan for the untoward effects of population growth and resource consumption, it is desirable to create dynamic mathematical models of the factors affecting human sustainability. Such models deal with complex systems and thus are challenging to formulate and validate. For example, see the validated model predicting food prices by Lagi et al. (2011).

Any mathematical model used to examine aspects of human sustainability in the future must include many subsets of valid, mathematical models for sustainability factors for regional or local human populations that quantify their impacts on the Earth's resources in general, as well as on those specific populations. One widely used modeling format involves creating a *directed graph* (a digraph), from which very large sets of nonlinear ordinary differential equations (ODEs) or difference equations can be written and solved numerically. The graph's nodes or vertices are the states (or variables) identified as being relevant in ecological sustainability. The branches connecting nodes describe quantitatively how the value of the variable of "node **j**" affects the variable of "node **k**" and so forth. Thus, each parameter is interconnected

with many of the others by directed branches that describe the dynamic relationships between nodal parameters. In many graphs, some branches can
form feedback loops, either positive (often destabilizing system behavior) or
negative (often stabilizing), and transport lags (delay operations) may exist
in some branches (which can also be destabilizing). [For an introduction to
linear directed graphs (digraphs), see Section 2.3.]

An example of a nodal parameter might be the accessible volume of
water in an aquifer used by a regional population. This volume can depend
on regional climate (rainfall or snowmelt as sources, river level, etc.); it is
depleted by crop irrigation (water from the aquifer or from a nearby river or
lake charging the aquifer), mining and industrial uses, and human domestic consumption. Water quality can be degraded by agricultural, industrial,
mining (including oil and natural gas production), and domestic sources of
pollution to the point where it must be filtered, treated chemically, and/or
desalinated before human domestic consumption. For example, the Ogallala
Aquifer extends from Southeast Wyoming through Nebraska and south
through Western Kansas, Eastern Colorado, and Western Oklahoma to North
Central Texas. This aquifer, which now covers approximately 450,000 km²,
was created geologically between 2 and 6 million years ago. It yields about
30% of the nation's groundwater used for irrigation. In 2005, the United
States Geological Service (USGS) estimated that the total water reserve in
the aquifer was 3608 km³, a decline of 312 km³ (or 9%) since the early 1950s,
when substantial groundwater irrigation began. The rates of drawdown and
recharge of the Ogallala Aquifer vary with location, underscoring the complexity of modeling its reserve level. The situation in the High Plains of Texas
is particularly bad compared to Nebraska, where the recharge rate is higher
(Glantz 1989). One projection estimates that the southern Ogallala Aquifer's
volume will fall a debilitating 52% between 2010 and 2060, as corn and cotton growers continue to irrigate from it (Galbraith 2010). As the water level is
drawn down, mineral pollution rises; already, the amount of dissolved solids (mineral content) in water pumped in the southern High Plains of Texas
(around Lubbock) is approximately 1000 mg/L. There is a high fluoride content, and selenium concentrations locally are in excess of drinking water
standards. Clearly, the current rate of water use from the southern Ogallala
Aquifer is not sustainable.

Food for a human population is certainly a critical consideration for human
ecological sustainability (Keating 2011) (see Chapter 5 in this book). Food can be
further subdivided into graph modules for food source categories, that is, plant,
animal, seafood, and unconventional (cf. Chapter 6). Crops (grains, nuts, vegetables, roots, fruits, and sugar plants) require the right balance of water, temperature, soil nutrients, and sunshine; most require arable land, planting, fertilizers,
cultivation (weed and insect control), and of course, harvesting. There is a built-
in delay from the time crops are planted to the time they are harvested and
brought to market. Unpredictably, a crop can be decimated or destroyed by a
flood, drought, hail, early frost, plant viruses, or insect pests.

Continual use of arable land for the same crop (without crop rotation, fallow seasons, or soil regeneration by planting nitrogen-fixing crops such as legumes, clover, or alfalfa) leads to exhausted soil with little organic content and reduced crop yields. It appears that for any given crop (e.g., corn) and growing site, there is an optimum annual yield. The nitrates and phosphates in fertilizers used on crops are only partially absorbed; the balance of these ions percolates down into aquifers or runs off into surface ponds, lakes, and streams, thence to rivers and estuaries, where they cause eutrophication (algal blooms), which degrade fisheries. Crop planting, cultivation, and harvesting now require the burning of diesel fuel or gasoline, a nongreen activity (100 years ago, horses or farm workers were used; 200 years ago, oxen were also used). Weed control today on crops growing in over an acre is generally done by spraying chemical herbicides (e.g., Roundup™) on the crops from a helicopter, rather than by old-fashioned, on-the-ground cultivation. Economics dictate this action.

First-generation genetically modified (GM) crops have been designed that have man-made genes inserted, which, when expressed, give them immunity to Roundup. GM crops also can be given a gene for a built-in insecticidal protein (Bt). Crops also have been genetically engineered to have increased yields. First-generation GM crops, such as the FlavrSavr tomato (Northrop and Connor 2009; Section 10.2.2.1) were not without deleterious side effects. Bt-resistant insect populations have arisen from chronic exposure to Bt gene-expressing crops, which is no surprise (Northrop and Connor 2009). In 1998, the EU banned the sale of all GM foods in EU countries. In 2004, under international pressure from nations that produce genetically modified organisms (GMOs), it allowed the sale of GM sweet corn (maize). In 2006, the World Trade Organization challenged the EU's ban on other GM crops. In early 2011, the EU voted to let individual member nations decide what GM crops to ban (Dunmore 2011). Second-generation GM plants promise to not be toxic to humans and to have greater utility; one application is the insertion of a gene from sunflowers to maize and other crops, giving them drought resistance (Schenkelaars 2007; Rauf 2008).

Grain crops are used by man to make flour, breads, pastries, cereals, and so forth; to feed farm animals, cats, and dogs; and in the case of corn, to also make ethanol for fuel and oil for cooking. Part of every grain crop goes to feed insects, birds, deer, raccoons, woodchucks, rats, mice, voles, certain fungi, and so forth; this is a "normal loss" that we have some control over. We also have control over what happens to crop waste (stover); it can be turned into "green fertilizer" by composting, or the carbon energy in it can be harvested for fuel by the efficient process of *hydrothermal carbonization* (see Section 4.2.6). It can also be used to make syngas (synthesis gas), which has wide applications, and also in the production of cellulosic ethanol (Balboa 2012).

Seafood populations (fish, crustaceans, mollusks, cetaceans, plankton) are probably one of the best examples of the direct impact of human behavior on ecosystems. Again, one can assign separate, graph submodules to model the populations of each species of seafood. A classic example of marine species exhaustion is the Northwest Atlantic cod fishery. Viable (in super-abundance) from the time European settlers first reached the new world, the entire cod fishery collapsed in the late 20th century (ca. 1990), generally as a result of massive overfishing (Kurlansky 1997; Hutchings 1996). The decline in the Northwest Atlantic cod fishery began circa 1968, recovered briefly in the 1980s, and plummeted to near zero in 1989–1991. In 1992, Canada imple-mented a moratorium on cod fishing in its national waters; this was the end of 500 years of "good times" in the cod fishery. Figure 1.2 illustrates the 1992 collapse of the Canadian (Northwest Atlantic) cod fishery. An ecological tip-ping point was reached in 1990 following a peak landing in 1965 of over 300% of the average catch over 1910–1965.

The Northeast Atlantic cod fishery (in the North Sea, in the Norwegian Sea, and off the western coasts of Scotland, Denmark and Norway, etc.) also declined, but there was no precipitous decrease in cod landings as in the west. Figure 1.3 compares the cod fisheries in the Northwest and Northeast Atlantic Ocean from 1950 to 2005. All is not well in the Northeast Atlantic at present, however. The EU Fisheries Commission has mandated a mora-torium on cod fishing for 12 weeks beginning February 12, 2012 (the cod

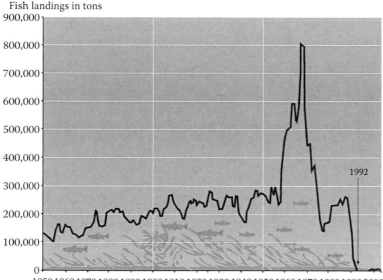

FIGURE 1.2
Fish landings in tons in the Northwest Atlantic cod fishery: Note that the peak catch of 800,000 tons in 1965 due to bottom trawlers exploiting pelagic cod stock was followed by a sharp decrease in 1978, a rebound, and then a near-total collapse in 1990. (Courtesy of the US Federal Government under the terms of Title 17, Ch. 1, Sect. 105, USC.)

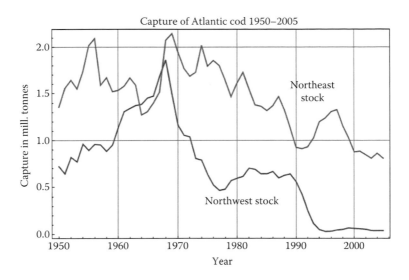

FIGURE 1.3
Plots of cod landings in the Northwest Atlantic (Canada) and Northeast Atlantic (Scandinavia, Scotland) in millions of tonnes covering 1950–2005. The Northeast cod catch exhibited a small peak over 1990–2000 as the Northwest catch collapsed. (Courtesy of the UN FAO Fishery Statistics program: www.fao.org/figis/servlet/SQServlet?file =/usr/local/tomcat/FI/5.5.23/figis/webapps/figis/temp/hqp_32285.xml&outtype = html.)

spawning season) in a 40,000 mi.^{2} (1.036 × 10^5 km^2) area north and east of Scotland and around Norway and Denmark. Only 6.4 million mature fish and approximately 48 million juvenile cod were caught in this fishery in 2010 (in 1972, 250 million adult cod were landed—300,000 tonnes!) (Dailymail 2011). The Northwestern Atlantic (Newfoundland) cod fishery has still not recovered, and this cod fishery remains closed.

According to Weldon (2006), Atlantic cod (*Gadus morhua*) have been reared in captivity for over 100 years in both Canada and Norway. At first, this was to produce yolk-sac fry to restock local wild populations. In 1977, cod were raised from eggs to mature fish in captive conditions. At present, there is an expansion of cod mariculture farms off the Shetland Islands off the western coast of Scotland and off the western coast of Norway. Fish-farming cod has its own sustainability issues. High densities of penned fish are susceptible to infection from ectoparasites such as sea lice (*Lepeophtheirus salmonis* and *Caligus* sp.), which feed on mucus, blood, and skin. These lice are particularly dangerous to juvenile cod. Endoparasites such as intestinal worms and protozoans, as well as bacteria such as *Yersinia* sp. and *Pseudomonas* sp., can also infect high-density farmed fish. As with cattle on land, the prophylactic treatment of the penned fish with broad-spectrum antibiotics will lead to eventual bacterial resistance to the antibiotics and traces of antibiotics in fish consumed by humans.

In Canada, farmed *AquaAdvantage* GM Atlantic salmon were given genes to continuously stimulate their production of growth hormone, leading to their maturity (6 to 10 lb.) in 18 months, instead of 24 to 30 months as for non-GM, farmed salmon (Northrop and Connor 2009). Similar genetic engineering can be used in cod farming to increase production rates, creating another challenge to the sustainability of the cod fishery. The potential crossbreeding of escaped, farmed, GM cod with wild fish could eventually replace the wild cod stock genome.

Scientists from Scotland's Fisheries Research Service found that farmed cod discharge 72.3 kg of nutrient nitrogen (NN) into the surrounding environment per tonne of fish production; farmed salmon discharge 48.2 kg of NN per tonne (Owen 2003). It was noted that this NN and phosphorous from fish waste can cause local eutrophication and lead to toxic algal blooms. It was estimated in 2000 by the Worldwide Fund for Nature that Scotland's established salmon farms produced the same amount of NN as the sewage of 3.2 million humans and phosphorous deposits equivalent to that of 9.4 million people (Owen 2003)!

Whaling is another example of human financial greed affecting a wild animal population. In spite of international limits and bans, the Japanese have continued to take whales, including endangered species, "for scientific research" under International Whaling Commission (IWC) rules. They are fortunate to be able to eat their "research subjects." Once plentiful before whaling (population ca. 275,000), the worldwide blue whale population declined rapidly with the advent of factory whaling ships in the 20th century, reaching an estimated low of 650 in 1964! The blue whale population then recovered (due to reduced demand for whale oil and whale meat, and international restrictions) to under 5000 in 1994 (Blue whale 2008). The present global blue whale population is estimated to lie between 3800 and 5255 animals (IWC 2007 data) (Wisteme 2011). They almost reached extinction and still could.

From the few examples cited, we see that water and food nodes are affected by climate, consumption patterns, our stewardship, and so forth. For example, it is possible to estimate relations between the level of an aquifer and its sources and sinks (including agricultural and human domestic uses). The sustainability of a fish catch (e.g., sardines) can be modeled by the fish reproduction rate, catch rate, food input rate (plankton), natural predation rate (affects the growth rate of other fish species), loss rate due to disease, and oceanographic factors, such as shifts in major ocean currents, changes in salinity and temperature, and so forth.

A particularly challenging area to model is sustainability associated with the energy nodes. There are many forms of energy being used by 21st–century humans: *nonrenewable fossil fuels* (oil, coal, peat, natural gas including methane); *nuclear energy; hydrogeneration; wind; solar thermal energy; solar photovoltaic energy; tidal and wave energy; geothermal energy; thermoelectric energy; and hydrogen and fuel cells.* Electricity can be generated from any of the preceding energy sources. "Green" electricity comes from wind, solar, geothermal, wood, and hydro sources. Hydrogen is "green" only if produced by electrolysis of water using direct current (DC) power from solar, wind, hydro, tidal, or wave sources of electricity. Fossil fuels are used for their heat energy to warm homes, to drive industrial processes, and as a source for petrochemicals used in manufacturing plastics, dyes, and explosives. Modern North Americans and Western Europeans are energy gluttons; China is rapidly becoming one. Energy can also be categorized by uses: *heating and cooling* (for human comfort); *transportation* (cars, trucks, busses, trains, planes, ships); *cooking; agricultural uses* (plowing, planting, cultivation, harvesting, making fertilizer); *industries* (steel, aluminum, metal refining, chemical production, oil refining, durable goods production, etc.); *mining; oil and gas exploration and production*, and so forth. In many of these applications, energy is still often used in a wasteful, 19th-century pattern; heat is wasted, and water is not cleaned or recycled.

The "800-pound gorilla behind the closet door" in any model of human sustainability is continuing human population growth; it is "... the engine that drives everything" (Bartlett 1998). It is one thing to examine the total world human population (see Figure 3.1 in this text), but it is perhaps more relevant to consider local population densities and consider how these regional densities affect regional ecosystems, water resources, food production, and economic and social issues. One can argue that fossil fuel energy exhaustion, rising CO_2 emissions, and global warming with its ancillary effects of drought, crop failure, coastal flooding, and so forth are all ultimately due to population growth. Local (and global) population growth must be included in any sustainability models involving energy, environmental, economic, food, and social systems. For example, it is ultimately a rising global population and its concurrent food needs that stress fisheries and causes fish population collapses (Kurlansky 1997). More people consume more energy, leading to rapidly rising atmospheric CO_2 levels, hence a higher rate of global warming with local droughts, crop failures, and so forth. There is a rich connectivity between the energy nodes, with many feedback paths. As a rule, richer, more prosperous, affluent nations consume more energy *per capita* than poorer nations—because they have the means to do so.

To answer the question of the section title: Yes, human sustainability can be modeled, but it requires many modules and subsystems; many, many states, branches, and nonlinear ODEs; the use of agents; and clever, knowledgeable modelers. It is a formidable challenge. Murray Gell-Mann's (2010) exhortation to take a "crude look at the whole (CLAW)" is a first step in embracing and attempting to understand the vast complexity of ecological sustainability issues. The "CLAW" must be gradually refined and made more complicated, however, if we are to arrive at meaningful models that have predictive value.

1.3 Why Human Sustainability Is a Complex Issue

It is clear that different human population groups (e.g., North American, Western European, Western Asian, Chinese, Indian, Micronesian, Central American, Australia–New Zealand, Equatorial African, South African, etc.) have different lifestyles and climates and hence have different requirements for energy, food, water, shelter, economic opportunity, transportation, and so forth. Obviously, the Earth's human population is not homogeneous over the planet; population groups (cities, towns) are generally concentrated along the coasts of the continents or on the borders of lakes and large rivers. They have different impacts on their local ecosystems, fisheries, water supplies, energy resources, agricultural land, the global economy, and so forth. What is a sustainable lifestyle for one population may not be so for another. It is dangerous to extrapolate sustainability issues for one population group to the whole Earth. For example, "global warming" effects are not homogeneous. Arctic ice and permafrost are melting more rapidly than Antarctic ice. Note that Arctic pack ice is warmed from below by seawater currents, while Antarctic glaciers rest on land, warmed by the Earth's core heat. (Heat transfer to and from moving water is much more rapid than through land.)

Sea temperatures are rising. Drought is more prevalent in Australia, the Sahel, the Gobi Desert, and the American Midwest. The greenhouse gas carbon dioxide is relatively well mixed in the atmosphere, but the concentration of the greenhouse gas WV tends to be denser over equatorial oceans and jungles, where solar radiation is strongest. When spatiotemporally averaged together, we can talk about global drought and global warming, but their impacts are felt acutely locally.

As an example, in considering the factors that make one author's (RBN) life in rural Connecticut, United States, sustainable, the usual dimensions of food, water, energy, and shelter are present. However, his lifestyle is one that uses energy in the profligate US manner. Electricity is used for lighting, for cooking (oven), to power the deep-well water pump, to run the oil-burning furnace (which heats hot water for domestic use as well as for home heating), and to run the washer and dryer. [Electricity in Connecticut is generated mostly from combustion of fossil fuels and is the second most expensive of any US state! It is also imported via the grid from other states and Canada (Hydro Quebec).] His January 2012 cost for electricity in Connecticut was US $0.18098/kWh; the national average price was $0.1104/kWh. Only Hawaii beats Connecticut, at $0.3454/kWh (USEIA 2011a). Natural gas [compressed natural gas (CNG)] is used in the cooking stove. Heating oil is burned in the furnace. 87-octane gasoline (E10) powers the two family automobiles (there is no public transportation; he lives in the country); gasoline runs the lawnmower and the snowblower. On the plus side, his house is superinsulated, and both automobiles get over 30 mpg. He uses high-efficiency fluorescent lighting and sets the thermostat to 60°F at night in the winter. His price for home heating oil was \$3.68/gal. (July 29, 2011); \$3.76/gal. (November 20, 2011); \$3.80/gal. (January 14, 2012); \$4.06/gal. (March 1, 2012); and \$3.82/gal. (May 18, 2012). (All with senior citizen discount.) Clearly, although the local prices of fuel oil, gasoline, and electricity are steadily increasing, his family lifestyle changes will lag these increases.

An efficient wood stove would be one answer to the high price of heating oil. However, the average price of split, seasoned hardwood is approximately \$200 per cord (May 2012) (USDOE 2012). (A cord of cut wood measures 4 × 4 × 8 ft. = 512 ft.^{3} = 14.5 m^3.) Burned at 100% efficiency, dried red oak has an energy density of nearly 15 MJ/kg. [Wood with higher moisture content releases less thermal energy (USDA 1979).] By contrast, residential heating oil has an energy density of 46.2 MJ/kg (cf. Table 4.1).

Wood is considered to be a green or renewable fuel; the CO_2 produced by its combustion is considered to be immediately recycled to grow new trees, and the ashes can be used to fertilize certain crops. The ashes contain alkaline mineral oxides as well as potash (K_2CO_3) and wood cinders (charcoal), which, ground and incorporated into the soil, form *terra preta*. Burning wood is not without environmental cost. Wood smoke contains particulates and other toxic chemicals (VOCs). Many jurisdictions require that new wood stove installations include a catalytic converter "afterburner" to raise the stack smoke to a temperature above 700°C, at which carbon monoxide and other smoke substances such as tars and *polycyclic aromatic hydrocarbons* (PAHs) will break down and oxidize, forming CO_2 and water. Many PAHs are carcinogens.

1.4 Chapter Summary

In this chapter, we have defined sustainability and introduced the reader to the many complex, interacting factors that determine sustainability, including examples of these systems.

One reason human sustainability is fascinating lies in its future uncertainty. A wide spectrum of sustainability scenarios involving food, water, energy, weather, ecosystems, and so forth exist, involving the status of our future sustainability, say, in 2050. This uncertainty is based on our inability to predict future outcomes based on our imperfect ability to model the many interacting, noisy, nonlinear, time-variable, complex systems that determine human sustainability. Adding to this uncertainty are events that may happen with high positive impact, such as future technological innovations, as well as events of low probability but high negative impact, such as floods, tsunamis, earthquakes, wars, volcanic eruptions, asteroid strikes, and so forth.

2

Review of Complexity and Complex Systems

The significant problems we face cannot be solved at the same level of thinking we were at when we created them.

Albert Einstein (Heart 2008)

2.1 Introduction to Complexity

In our fast-paced, 21st-century world, we are often challenged with having to predict how our actions (or inactions) will affect events (e.g., the outputs) in the complex systems (CSs) with which we live and interact. In particular, this includes our present and future actions intended to mitigate the adverse effects on human ecological sustainability caused by past human actions and inactions. In order not to be surprised by unexpected results, we need a systematic, comprehensive means of analyzing, modeling, and simulating CSs in order to predict nonanticipated outcomes. Sadly, comprehensive models of CSs are generally unattainable. They can be approximated, however, using well-formulated complicated models.

This chapter addresses how we can describe the important characteristics of CSs (in particular, the many CSs affecting human sustainability), model them, and use the models to try to predict unanticipated behavior. We want to be able to work competently with CSs in order to create desired results while simultaneously avoiding the pitfalls of the *law of unintended consequences* (LUC). Our mathematical models of CSs, in general, will be imperfect. Certainly, one of the major attributes of any CS is the difficulty we have in modeling it. Even so, our imperfect models, when validated, may still be able to capture the essence of the behavior of some complex nonlinear systems (CNLSs) and be used to predict their future behavior. Some attempts at modeling the factors affecting human sustainability are reviewed in a paper by Hayes (2012), including the pioneering work by Meadows et al. (1972, 2004).

We use the noun system extensively in this book. We define a system as a group of interacting, interrelated, or interdependent elements (also agents, entities, parts, states) forming or regarded as forming a collective entity. There are many definitions of system that are generally context dependent; however, this definition is very broad and generally acceptable. There are many kinds of systems; some are simple, others complicated, and many are complex. Systems often interact with other systems in varying degrees. This interaction often poses the problem of where to draw the defining boundary of a given system; that is, system boundaries are often arbitrary and fuzzy. Systems can also be classified as being linear or nonlinear (see the Glossary).

2.1.1 When Is a System Complex?

We often hear the words *complex (adj. & n.)*, *complexity (adj. & n.)*, and *complicated (adj.)* used, particularly in the media, to describe systems and situations. *Complex, complexity*, and *complicated* all come from the Latin word *complexus*, meaning twisted together or entwined. Complexity is difficult to define; hence, you will find there are many definitions given for it, depending on who is studying it, the type of system being studied, and how it is being studied. Broadly stated, we consider that *complexity* is a subjective measure of the difficulty in *describing* and *modeling* a system (thing or process) and thus being able to predict its behavior. Or we might view the complexity of a system or dynamic process as some increasing function of the degree to which its components engage in structured, organized interactions. Complexity also has been seen as a global characteristic of a system that represents the gap between component and parameter knowledge and knowledge of overall (known) behavior.

Corning (1998) suggested that complexity generally has three attributes: (1) A CS has many parts (or items, agents, units, or individuals). (2) There are many relationships/interactions among the parts. (3) The parts produce combined effects (synergies) that are not easily foreseen and may often exhibit novel or surprising (chaotic) behaviors.

Seth Lloyd (2001) compiled 45 different definitions of complexity, underscoring the plasticity of the definition problem. Quantitative measures of complexity necessarily must rely on the structure of mathematical models of CSs, most of which will be nonlinear and time variable (nonstationary).

One condition that has been proposed for a system to be called complex is that its overall behavior is destroyed or markedly altered by trying to simplify it by removing subsystems. However, a simple system can also be easily altered by removing its components or subsystems, rendering this criterion invalid. No one will dispute the complexity of the human brain. Yet it often appears to function with minimal disturbance following moderate injury, surgery, stroke, and so forth. This is because the human central nervous system (hCNS) is a *complex adaptive system* (CAS), capable of limited reorganization of its local structures and their functions following damage. It possesses the properties of *robustness and resilience*.

Bruce Edmonds (1999a, 1999b) argued that "...complexity is not a property *usefully* attributed to natural systems but only to our models of such systems." Edmonds proposed the following definition: "Complexity is that property of models which make it difficult to formulate its overall behaviour in a given language of representation, even when given almost complete information about its components and their inter-relations." The use of our models of natural CSs to characterize their complex behavior allows us to work with quantifiable formalisms such as graphs and graph theory, sets of ordinary differential equations (ODEs), and algebraic nonlinearities, and if the model can be linearized, the use of linear algebra and transfer functions is possible. For model validation, we can derive and measure *sensitivities* from models and compare them with those measured on the natural system. Note that when we consider a complex natural system (such as the human immune system [hIS]) and a mathematical model formulated from it, we might be tempted to view the model as complicated by itself. This does not mean that we should infer that the natural system can therefore be demoted to complicated status—it means only that the model is parsimonious; it sacrificed much detail while attempting to preserve the essence of one or more aspects of the natural, complex behavior.

Rosser (2008) gave a definition of *dynamic complexity* based on that of Day (1994): "...systems are dynamically complex if they [their phase-plane trajectories] fail to converge to either a point, a limit cycle, or an exponential expansion or contraction due to endogenous causes. The system generates irregular dynamic patterns of some sort, either sudden discontinuities, aperiodic chaotic dynamics subject to sensitive dependence on initial conditions, multi-stability of basins of attraction, or other such irregular patterns."

In summary, complexity must lie in the eyes of the beholder; what is complex to one observer may be complicated to another, based on observer knowledge and skills. It is safe for us to say that, in general, complexity is relative, graded, and dependent on the system and observer. You will know it when you see it.

2.1.2 Examples

There are many examples of CSs—all are characterized by having many parameters or states that are functionally interconnected, in some cases leading to nonintuitive system behavior. Some CSs that are integral to sustainability, such as economic and social systems, include the actions of *human agents* as decision makers. CSs have been subdivided into those with *fixed* (time-invariant or static) *structures, dynamic complexity, evolving complexity,* and *self-organizing* CAS (Lucas 2006). The CSs we consider in this text are generally dynamic and thus rely on the use of differential or difference equations for their model descriptions. CSs are generally *multiple-input, multiple-output* (MIMO) systems. However, others are multiple input, single output (MISO) and single input, multiple output (SIMO); a few even have a single-input, single output (SISO) architecture.

Some examples of simple systems include, but are not limited to, the following: a toaster, a mechanical wristwatch, a gyroscope, a power lawn mower, an electronic oscillator, an orbiting satellite, a pendulum, the van der Pol equation, and so forth. A simple system can be nonlinear and/or time-variable, but its behavior is always predictable and easily modeled mathematically. Its inputs, outputs, and internal branch parameters are well defined.

Some examples of complicated systems include, but are not limited to, the following: your laptop computer, a modern automobile, an automatic dishwasher, a microwave oven, an *iPad,* an income tax form, a mobile phone, a mathematical model of a CS, and so forth. Complicated systems can generally be reduced to component subsystems and successfully analyzed and modeled. They generally do not exhibit unexpected, chaotic behavior.

Some examples of CSs include, but are not limited to, the following: systems involved with ecological sustainability, which broadly include economic and social systems (these systems are dominated by the behavior of *human agents*), and ecosystems of all sorts (ecosystems are tightly coupled to relevant environmental systems). Environmental systems include climate system models. Also on the list are the following: the spread and control of epidemics (e.g., anthrax, HIV, Ebola, hantavirus, influenza, smallpox, malaria, yellow fever, Bubonic plague, "mad cow" disease, etc.); stock markets (socioeconomic systems); energy systems; meteorological systems; oceanographic systems; transportation systems; the World Wide Web; political parties; governments; multinational corporations and their management; health care systems, and so forth. A chess game played between two humans (as opposed to between a human and a computer) can be complex because of human behavior. Often, we find ecological systems grouped with economic systems as examples of CNLSs. Also included are physiological systems (especially those based on various multifunction organs, such as the liver, pancreas, hypothalamus, and kidneys) and cell biology systems, including intracellular biochemical (metabolic) pathways (in ribosomes and mitochondria) and genomic regulatory pathways (with emphasis on embryonic development).

A major challenge in understanding the behavior of any CS is how to formulate a mathematical model of that system that will allow us to predict the system's behavior for novel inputs. What are the system's relevant parameters (states), how are they functionally interconnected, what are the system's inputs and outputs, and what other systems do they interact with? Putting such information together into an effective, verified model can be quite daunting. (One approach to this model synthesis is described in Chapter 8.)

Note that the boundaries between *simple* and *complicated* and between *complicated* and *complex* system designations are *fuzzy* (Zadeh et al. 1996) and debatable, even using quantitative measures of complexity. This is because humans draw the boundaries and/or set the criteria. Try ranking the complexity of human physiological systems. A candidate for the most complex is the hCNS; the hIS or the human microbiome might be second. A greater challenge is to name the least complex human physiological system. The complexity of certain economic systems certainly rivals that of the hCNS, particularly because human agents run economic systems. Meteorological systems are complex because of scale and nonlinearity.

Generally, a well-described CS can be modeled by a *graph* composed of *nodes (vertices)* and *branches (edges)*. The nodes represent the CS's states or variables; the branches represent causal connections between the nodes. *Nonlinearity, parameter time variability (nonstationarity)*, and *noise* are generally present in the relationships (branches, pathways, edges, or arcs) interconnecting the parameters or states at the vertices or nodes of a CS's model. Once modeled by a graph, a CS's structure can be described by a variety of objective mathematical functionals used in *graph theory*, and its dynamic behavior can be studied by simulation. One can also start with a purely mathematical description of a system model and, from these ODEs, form a nonlinear signal flow graph (NLSFG) (see Section 2.3).

Agent-based simulation (ABS) (Gilbert 2007) is another more complicated, yet possibly more effective, means of modeling certain CNLSs that is receiving increased attention today. In this context, an *agent* is a mathematical computer model component that inputs information and signals and outputs behavior, signals, and information. Agents are autonomous decision-making software modules with diverse characteristics. For example, an agent can be used to model human behavior, an organization, or a biological cell. Agents are used in large groups to model interactive, autonomous, CS behavior. A network of information flow can exist between neighboring agents. Some agents also can have "memories" and are adaptive and can "learn" to copy each other's behavior. Computer models of human agents figure in *agent-based models* (ABMs) of economic and political systems, contagion, terrorism, and social CNLSs. In economic ABMs, human agent models can be programmed to produce, consume, buy, sell, bid, hoard, and so forth. They can be designed to be influenced by the behavior of other, adjacent agents. ABMs can exhibit emergent, collective behavior. Dynamic ABMs already have been used to model bacterial chemotaxis (Emonet et al. 2005) and have application in hIS modeling (immune cell trafficking). ABMs generally use many simultaneously interacting agents and hence must be simulated on large, very powerful computers.

2.1.3 Properties of CSs; Chaos and Tipping Points

As we have stressed, CSs, by definition, are hard to describe, and their input/output (I/O) relationships are difficult to understand. A large, dynamic, nonlinear CS is always a challenge to model. Generally, many, many coupled, nonlinear differential equations are required. Numerical rate constants in the equations are often poorly known or unknown and must be estimated. Often, the rate constants vary parametrically; that is, their values are functions of certain variables or states in the CS, and these parametric relationships must also be estimated.

A complex, nonlinear system may exhibit abrupt switches in its overall behavior, for example, switch from a stable, I/O behavior to a limited oscillatory one, or from one oscillatory mode to another, or to unbounded (runaway

or saturated) output(s). The study of such labile, unstable, periodic, and aperiodic behaviors in nature and in dynamic models of CNLSs is part of *chaos
theory.* These abrupt switches in CS behavior are called *tipping points*; a small
change in one or more inputs, parameters, or initial conditions can trigger a
major, I/O behavioral mode change. Such tipping-point behavior has often
been called the "butterfly effect," where ideally, a butterfly flapping its wings
in Brazil many miles away from a weather system supposedly can trigger
an abrupt change in the system, such as the formation of a tornado in Texas.
With due respect to Edward Lorenz, the butterfly effect is actually a poor
metaphor. Although all kinds of real-world CSs, including social and economic systems, can exhibit tipping points (Gladwell 2002), a certain minimum increment in the level of a signal, energy, momentum, or information
is required to trigger a rapid, global change in a real CNLS's behavior. In the
natural case of weather systems, air is a viscous, lossy medium. The gentle air
currents caused by a butterfly's wings carry negligible energy and are quickly
attenuated with increasing distance from the insect. From the same dissipative property, sound dies out with distance, and it is impossible to blow out a
candle from across a room. The so-called butterfly effect is more aptly applied
to the behavior of our mathematical models than real-world CNLSs. In fact,
it was in *weather system simulations* in 1961 that meteorologist Edward Lorenz
(1963) found that a change in the fourth significant figure of an initial condition value could trigger "tipping" and radical changes in the outputs of his
model. This change was actually $(100 \times 0.000127)/0.506127 = 0.025\%$.

Mathematical models of noise-free, deterministic, complex, nonlinear systems can exhibit *chaotic* behavior, in which output variables in the system
oscillate or jump from level to level in what appears to be a random manner.
However, on a microscale, in a noise-free CNLS, this chaotic behavior is in
fact deterministic; it involves thresholds or tipping points. Its occurrence may
at first appear random because it appears as the result of one or more minute
changes in the CNLS's inputs, initial conditions, and/or parameters. Under
certain conditions, a chaotic CNLS may exhibit bounded, steady-state, periodic *limit cycle oscillations* of its states in the *phase plane*, or even unbounded
(saturating) responses. In certain chaotic CNLSs, a combination of initial
conditions and/or inputs can trigger an abrupt transition from one nonoscillatory, steady-state, behavioral mode to oscillatory limit cycle behavior, and
then a further perturbation of an input or parameter can cause the system to
go into a new limit cycle or a new global behavioral mode.

Ecologists modeling the variation of a species' population as it inter-
acts with its environment (niche) have found that chaotic behavior can be
observed under certain circumstances that mimics data from field studies.
In some cases, the models do not show the "chaos" observed in field studies,
underscoring that the art and science of mathematically modeling complex
ecosystems is challenging (Zimmer 1999).

Much of the research on chaotic behavior has used mathematical models
consisting of systems of coupled, nonlinear ODEs. Some of the examples of

these complex mathematical systems entering chaotic behavior have been traced to round-off errors in numerical simulation of the ODEs. That is, the tipping point for the system was so sensitive that it responded to equivalent digital (quantization, round-off) noise in the computations.

One author (RBN) has found from personal experience that the choice of integration routine is very important when simulating a CNLS described by a set of stiff, nonlinear ODEs. A poor choice of integration routine was seen to lead to chaos-like noise in plots of system output variables (Northrop 2000, 2001). The noise disappeared when simple rectangular integration with a very small Δt was used.

If one plots the derivative of the kth output versus the kth output of a self-oscillating CNLS in the steady state [i.e., $x_k(t)$ vs. $x_k(t)$], the resulting 2-D *phase plane plot* will contain a closed path because $x_k(t)$ is periodic. This steady-state, closed, phase-plane trajectory is called a *simple attractor*. CNLSs may also exhibit *strange attractors* in which $x_k(t)$ is still periodic, but its SS phase trajectory contains multiple, reentrant loops, rather than a simple closed path. The Lorenz weather system is a good example of a system with a strange attractor. [See Section 3.4.3 in Northrop (2011) for a description of this three-ODE nonlinear system.]

In examining the apparently chaotic behavior of real-world CNLSs, chaos theorists have the problem of separating behavior that is intrinsic (caused by noise-free system dynamics) and behavior that is the result of random noise entering the system from within or without. Such studies rely heavily on statistical analysis of output time series and also make use of the *Lyapunov exponent* to analyze the fine structure of the trajectories of system states near their attractors in the phase plane. The Lyapunov exponent is a logarithmic measure of whether trajectories are approaching or diverging from an attractor (see the Glossary).

2.1.4 The Law of Unintended Consequences

As noted above, the nonintuitive, unexpected, chaotic behavior of natural, complex, nonlinear systems has led to the consensual establishment of the LUC. An unintended consequence occurs when a new, incremental change in an input and/or a branch parameter of a CNLS results in a behavior that is neither intended nor expected, given previous system behavior. In the real world, the unintended outcome can be adverse or fortuitous or have zero impact on sustainability.

There are several circumstances that can lead to unanticipated, unforeseen, or unintended consequences: (1) actions taken (or not taken) when there is *ignorance* of a CNLS's complete structure (part of the definition of a CNLS); (2) actions taken (or not taken) when there is a *lack of information about the system's sensitivities*; (3) actions taken (or not taken) by humans based on greed, fear, anger, faith, or other cognitive biases, rather than facts; (4) actions taken (or not taken) in systems with time lags (e.g., economic systems). An example is as follows: due to the lag, a corrective input produces no immediate effects, so the operator interprets the initial lack of response as a need for a stronger action, resulting in a delayed, excessive output, which in turn promotes another stronger, corrective input action, and so forth, leading to system-operator instability or a very long settling time.

"Actions" refers to such things as the following: the excessive administration of drugs (such as antibiotics to factory-farmed poultry and animals); chronic application of pesticides, weed killers, and fertilizers to the environment; passage of new laws attempting to regulate human behavior; application of new economic regulations; the introduction of new technologies (e.g., the World Wide Web, cameras and GPS in cell phones, public space surveillance cameras, text messaging, fish-locating sonar, etc.), raising or lowering taxes; and so forth.

A sociological corollary to the LUC is called *Campbell's law* (CL) (Gehrman 2007; Campbell 1976). Campbell stated: "The more any quantitative social indicator is used for social decision-making, the more subject it will be to corruption pressures and the more apt it will be to distort and corrupt the social processes it is intended to monitor."

Examples of CL are the recent cheating scandals associated with the US-mandated *No Child Left Behind* testing in schools. States (and years) in which school administrations have been found to "cook" testing data include the following: Maryland (2001, 2009, 2010); Utah (2002); Illinois (2002, 2003); California (2004, 2007, 2010); New Jersey (2005); Texas (2006); Ohio (2006); Washington DC (2011); Georgia (2011); and Pennsylvania (2011) (Samuels 2011; Turner 2011).

Note that the LUC and CL are not strict physical, chemical, or mathematical laws that have mathematical descriptions like the law of gravity, Newton's laws, or the laws of thermodynamics. The LUC and CL are more in the realm of Murphy's law; they are consensual laws.

Regulations applied to social and economic systems can have some unintended consequences; however, an absence of regulation can often lead to many more unintended consequences. To regulate or not to regulate, that is the question (Tabarrok 2008]) especially in economic systems; also, how to regulate optimally? [The control of CNLSs was initially considered by W.R. Ashby (1958).]

The LUC is often, but not always, manifested as part of the behavior of CNLSs. Examples of the expression of the LUC are demonstrated by certain recent ecological, political, social, and economic events: *for example,* the increase in food prices caused by the diversion of a large fraction of the US corn crop into the production of ethanol for a gasoline additive and soy beans to manufacture biodiesel fuel. (Because of system complexity, food price increases are not solely due to crop diversion.)

Introduction of xenospecies into ecosystems can have adverse, unexpected effects. For example, rabbits were introduced into Australia for sport hunting and rapidly became a crop pest, competing with sheep and cattle for forage.

Kudzu vine was introduced into the southeastern United States for erosion control and rapidly spread, choking out native species of trees and bushes. The eastern snakehead fish was introduced into several US southeastern lakes and waterways in the early 21st century. This invasive species grows rapidly—it is a hardy, voracious, apex predator that feeds on all native fish species (catfish, bass, trout, etc.) and decimates these native game-fish populations. The snakehead is an East Asian delicacy and was imported illegally to the United States as a specialty food. Unfortunately, some were released into local ponds and streams in Maryland and Virginia and have spread.

Ecosystems are complex. Multiple species' populations are held in balance by predation, competition, weather, available food, pathogens, and so forth, all forming negative feedback (NFB) loops. There were few natural predators to eat the Australian rabbits, and they had lots of food and a rapid reproductive rate. Eventually, to control their population, a lethal virus (*Myxoma*) specific for rabbits was introduced to cause myxomatosis. Myxomatosis symptoms range from skin tumors (in cottontail rabbits) to puffiness around the head and genitals and acute conjunctivitis leading to blindness; there is also loss of appetite and fever, as well as secondary infections (in European rabbits). Rabbits were effectively removed from Australian ecosystems.

Another way we have seen the LUC manifested in ecosystems is by the creation of insecticide-resistant crop pests as the result of excess application of insecticides to crops and animals. More recently, insect Bt resistance has emerged in the growing of genetically modified (GM) crops with built-in genes to continuously make Bt, a natural insecticidal protein designed to protect that particular crop. Another ecosystem example of the LUC also involves GM crops: The genes for a crop plant were modified to make the plants resistant to a specific herbicide (e.g., Roundup™). It was subsequently found that the herbicide resistance was transferred via the GM crop plants' pollen to weeds related to the crop, complicating the chemical weeding process of the crop (Northrop and Connor 2009). The LUC also occurs in complex physiological systems. For example, one side effect of taking the drug Viagra in some individuals is impaired vision. The metabolism of the neurotransmitter *nitric oxide* (NO) may be altered in the retina, as well as in the vascular tissue of the penis.

Still another example of the untoward effects of the LUC was seen in the late 1930s, when the "dust bowl" destroyed Midwestern US agriculture and drove over a million people out of the states where it occurred (Texas, Oklahoma, New Mexico, Colorado, Kansas, and Nebraska). Grassland prairie had been a stable ecosystem for millennia, resistant to floods and droughts. It was a major energy source for grazing prairie animals, including bison and antelope. In the 1930s, land speculators and the US Department of Agriculture encouraged the sale of this cheap grassland for farming money-making grain crops. The sod was "busted" by the newly available, powerful diesel tractor gang plows and grain crops planted. They did well at first because of above-average rainfall. No one bothered to ask what would happen if drought occurred. When several years of severe drought did occur in the mid-1930s, crops failed catastrophically, and the soil, unprotected by sod, suffered incredible wind-caused erosion. An area the size of Pennsylvania was turned to desert; the topsoil blew away (Condon 2008). Clearly agriculture and weather systems are closely linked. It has taken years of hard work to ecologically reclaim most of this wasted land.

The Prohibition Amendment to the US Constitution in the 1920s was intended to curb alcohol abuse and public intoxication. It also energized an entire system of illegal alcohol production, distribution, and consumption, including moonshining, smuggling, and speakeasies. Addiction proved to be a more powerful motivation to produce and consume alcohol than the governmental proscription of alcohol sales and consumption, hence the rise of organized crime in the alcohol business. Similarly, a major unintended consequence of the US "War on Drugs" is the spawning of violent criminal organizations in Mexico and Central America devoted to growing, smuggling, distribution, and sale of drugs in the US market. Aircraft, and now boats and "submarines," are used for the importation of drugs by sea, and tunnels on the United States/Mexico border have been used to bring them in by land. Again, human addiction, a very powerful motivator, was not used in the decision-making model. The lessons of the consequences of the Prohibition Amendment and its repeal were evidently not learned.

Note that the LUC can demonstrate fortuitous results, as well as adverse ones. In 1928, bacteriologist Alexander Fleming "accidentally" discovered the antibiotic penicillin by noting the bactericidal effect of unwanted *Penicillium* sp. mold growing in old cultures of *Staphylococcus* sp. in his lab. Fleming went on to isolate the active agent from the mold and named it penicillin. He found that it was bactericidal for a number of gram-positive bacterial pathogens, for example, those causing scarlet fever, pneumonia, gonorrhea, meningitis, and diphtheria, but not those causing typhoid or paratyphoid. His discovery led to the eventual development of many important antibiotics, some natural and others synthetic. Fleming shared a Nobel prize for his discovery in 1945.

Ironically, overmedication with penicillin and related antibiotics has also unleashed the dark side of the LUC, that is, through the evolution of antibiotic-resistant bacterial strains, particularly in hospitals, leading to life-threatening nosocomial infections including methicillin-resistant *Staphococcus aureus* (MRSA). The chronic, prophylactic use of penicillins and tetracyclines in cattle feed in factory farms raising hogs, cattle, and chickens has also led to the emergence of MRSA in the farms and in about 40% of hog farm workers (Wallinga and Mellon 2008).

Another example of the good side of the LUC involves common aspirin, long used for mitigation of headaches and joint pain. It was soon discovered that aspirin has the beneficial effect of inhibiting blood clotting, an unintended consequence leading to the use of low-dose aspirin as a prophylactic against coronary heart attacks and embolic strokes. An overdose of aspirin, however, can lead to hemorrhages and hemorrhagic stroke (again, the dark side).

To satisfy the insatiable world demand for gasoline, ethanol distilled from fermented sugar cane and corn is used as a fuel additive in concentrations ranging from 10% to 85% in order to stretch the petroleum-based gasoline supply. This ethanol is called a *carbon-neutral* (green) *biofuel* because it comes from annual plant crops that use CO_2 to grow. The fermentation process that produces ethanol releases CO_2 into the atmosphere (one CO_2 molecule is released when one pyruvate molecule is enzymatically decarboxylated to make acetaldehyde, which is enzymatically converted to ethanol) (Northrop and Connor 2009; Figure 4.1).

In the United States, the demand for ethanol has resulted in farmers planting more corn (instead of, say, soybeans or vegetables), thus using more diesel fuel, fertilizer, herbicides, and insecticides. Sadly, increased carbon-neutral ethanol production from corn has resulted in reduced production of corn for food use, the reduction of arable land available to grow other food crops, as well as an increase in the price of corn/bu. An obvious unintended consequence of this shift in corn usage has been increased food prices, especially beef, pork, and chicken, traditionally fed on corn. (The residue of the corn fermentation process is fed to animals, but most of the light carbohydrate energy in the residue has been lost to the fermentation process for making ethanol.) Still another unintended consequence of biofuel production is the consumption of significant amounts of freshwater for irrigation, depleting public water supplies in areas already having low rainfall.

Another carbon-neutral fuel substitute is biodiesel made from natural plant oils. New, arable land is being reclaimed from forests in order to grow soybeans for both food and oil for biodiesel in Indonesia, Brazil, India, and China. Biodiesel is also derived from oil seed crops such as canola, corn, cottonseed, crambe, flaxseed, mustard seed, oil palm, peanuts, rapeseed, and safflower seed, to name some sources of plant oils (Faupel and Kurki 2002; Ryan 2004). Biodiesel may eventually be derived from lipids from GM bacteria or fungi.

Where the LUC comes into play in the production of "green" alternate fuels has to do with CO_2 emissions that affect the *planetary weather system*, in particular, global warming. In their paper, Searchinger et al. (2008) pointed out that the extensive, new, arable land reclamation required to grow carbon-neutral fuel crops is not without cost. The heavy machinery used for land clearing, tillage, planting, and harvesting all burn fossil fuels (FFs) and emit CO_2 over and above the CO_2 that will be formed when "green" ethanol and biodiesel are burned. The industrial production of fertilizers and herbicides for these new crops also adds to the atmospheric CO_2 burden. Searchinger et al. have calculated the time required for a new, carbon-neutral fuel crop to achieve a net reduction in CO_2 emissions, considering the carbon cost of crop production. Net reduction times were found to range from 167 to 423 years, depending on the fuel crop and its location.

One way to mitigate the unintended consequence of this excess CO_2 emission is to derive ethanol fuel from crop waste, compost, and garbage, and switch grass crops that require minimum energy for tillage and processing. GM bacteria and yeasts will be required to improve fermentation efficiency. Another way to reduce CO_2 emissions is very obvious; *burn less fuel by mandating increased fuel efficiency of vehicles, the use of more public mass transportation, implementation of improved insulation efficiency of buildings, and more reliance on noncarbon energy sources (solar, wind, hydro, tidal, nuclear).* The United States is lagging in these efforts compared with Western Europe.

2.1.5 Complex Adaptive Systems

A CAS has the capacity to change its parameters (nodes, branch connections) in order to optimize its performance, a form of learning. The term *complex adaptive system* was coined at the Santa Fe Institute by J.H. Holland, M. Gell-Mann, and others. Examples of *CASs* include the human brain, the hIS, certain economic systems, ant colonies, manufacturing businesses, adaptive artificial neural networks, and so forth. CASs generally develop *resilience* as they adapt. Holland's (1995) clear definition of a CAS was as follows: "A Complex Adaptive System (CAS) is a dynamic network of many agents (which represent cells, species, individuals, firms, nations) acting in parallel, constantly acting and reacting to what the other agents are doing. The control of a CAS tends to be highly dispersed and decentralized. If there is to be any coherent behavior in the system, it has to arise from competition and cooperation among the agents themselves. The overall behavior of the system is the result of a huge number of decisions made every moment by many individual agents."

A key feature of all CASs is that their behavior patterns as a whole are not determined by centralized authorities but by the collective results of interactions among independent entities or agents. CAS systems are generally *robust* as well as *resilient*.

An important feature that characterizes nearly all CASs is that their general behavior patterns are not determined by centralized decision makers ("deciders"), but rather are determined by the net results of interactions between a number of independent entities (agents or modules). It is these network interactions that contribute to a CAS's complexity. Each individual entity (and class of entities) acts on the CAS with a built-in, basic set of behavioral rules. (One of the rules must be that the elements of the CAS act together.) We see this "whole is greater than the sum of the parts" property in the agent-driven behavior of bees and termites, migrating flocks of birds, schooling fish, and the stock market. The behavioral "rules" may be viewed as a set of subsystem I/O relations. That is, a CAS is in fact made up of many subsystems. It is tempting to think that if we learn "the rules" governing subsystem behavior, we can model certain CASs with confidence and thus be able to predict their group behavior to novel inputs. Using ABM, we can also see the effects of "tinkering" with innate rules on overall CAS model behavior. The interested reader should read H.G. Schuster's (2005) comprehensive, introductory text on CASs.

2.2 Human Responses to Complexity

2.2.1 Introduction

In general, human thinking has not evolved to deal effectively with complexity and CSs. Modern humans have descended from short-lived hunter–gatherers whose greatest threats were sudden physical attacks by apex predators, other humans, famine, and diseases. These threats were generally simple in their origins. Today, the threats to the sustainability of the human race include a spectrum of relatively slowly happening adverse conditions involving the interrelated, CSs that now surround us, including the following:

(1) A slow, steady increase in the world's population.
(2) Climate change, strongly correlated with rapidly rising, anthropogenic greenhouse gas concentrations (leading to global climate change, storms, ice cap and glacier melting, sea rise, floods, droughts, and food shortages).
(3) Loss of arable land causing decreased areas for crops.
(4) A slow decline in the availability of agricultural, industrial, and potable water due to climate change and pollution from agriculture, industry, and cities.
(5) Slow depletion of petroleum, coal, and natural gas (FF) resources while demand for energy is rapidly growing.
(6) The relatively rapid evolution of antibiotic-resistant bacteria, pesticide-resistant insect pests, and herbicide-resistant weeds. This evolution is accelerated by our chronic use of herbicides, pesticides, and antibiotics in "factory farms."
(7) The slow decay of metropolitan infrastructures, including highway bridges, metropolitan underground water distribution systems, electric power distribution systems, sewer systems, and so forth.
(8) The overloaded capacity of electrical power grids and the rising need for more electrical power. Electric power is replacing FFs in transportation systems and industries because it is perceived as more "green." Clearly, it is green only if noncarbon resources (e.g., hydro, wind, solar, tide) have led to its production.
(9) Economic effects such as stagflation, caused by all of the above-cited adverse conditions.

2.2.2 The Social Action Rate Sensitivity Law

In general, when something bad for a human population happens slowly enough, it fails to trigger a meaningful, organized, social response intended to prevent, combat, and correct it until martyrs are made. These bad things include infrastructure faults such as water main breaks, bridge collapses, dam and levee failures, and so forth, and natural disasters such as tsunamis, hurricanes, floods, tornadoes, earthquakes, and volcanic eruptions that happen infrequently and fairly randomly. The martyrs, in the usual sense, can be human lives, but humans also can be affected adversely by extreme local damage to their environment, food sources, water supplies, or to an economic or social system. This human behavioral inertia is characterized by what we call the *Social Action Rate Sensitivity Law* (SARSL).

It appears that the human brain has evolved the ability to remember and predict the timing and location of dangers to individuals before they occur. However, we are apparently handicapped by the high threshold in our ability to detect and act on the low rate of change of the cues that require our meaningful, *cooperative,* corrective responses. A sociodynamic tipping point is involved. For example, it appears that we need to have an interstate highway bridge collapse and lives lost before our attention is drawn by the media and experts to the need for costly prophylactic bridge repairs and regularly scheduled maintenance operations. Failing bridge supports are generally "out of sight, out of mind" for the public. [See the article on the slowly degrading infrastructure in the United States by Petroski (2009).] Strengthening dams, dikes, seawalls, and levees is usually done *following their failure, not in anticipation of a potential failure* (remember Hurricane Katrina, New Orleans, levees, and the Army Corps of Engineers?).

Figure 2.1 illustrates the hysteresis and tipping points in a generic SARSL process (curve **A**). The socially perceived seriousness of a problem is a nonnegative quantity, S. Active, cooperative public attitude is in response not only to S but also to its rate of change $\dot{S}$. We assume that what drives the change in public attitude is a *graveness function*, $G = S + k\dot{S}$. In other words, the rate of change of a problem figures heavily in determining when public attitude shifts, not just S as suggested by Scheffer (2009). Note that in the case of discrete S events, such as terrorist attacks or tornadoes, $\dot{S}$ can be taken as the mean frequency of these untoward events, or events/time. k is a context-dependent, positive constant. Scheffer noted that a society or group that has a great deal of peer pressure (and interagent communication) will generate a response curve with sharp tipping points and hysteresis (curve **A**). If the peer pressure is weak, individual agents slowly arrive at their own, individual opinions, and a smooth, sigmoid response curve is generated (curve **B**) with no tipping points.

Our innate ability to deal with CSs and our predictive and correlative abilities as a species are limited, especially in detecting and correcting slow-onset disasters including situations affecting human sustainability. We need specific training in the area of how to deal with CSs.

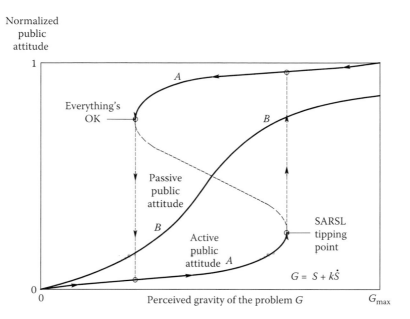

FIGURE 2.1

Illustration of the SARSL. Curve *A* illustrates behavior of a society with high peer pressure (high social connectivity). Note the behavioral hysteresis. Curve *B* is for societies with passive public attitudes.

2.2.3 Single-Cause Mentality

We humans generally tend to oversimplify our analyses of CSs and their models. People tend to use intuition instead of scientific method and verifiable data and to apply *reductionism*, often assigning a *single cause (or cure)* for any unwanted output or condition (e.g., for the steadily increasing price of gasoline and heating fuel oil and for the rate of unemployment). We call this the *single-cause mentality*. Such single-cause thinking is generally based on hunches, rumors, and poor extrapolations rather than on observed facts and detailed, quantitative models. It can lead to fixing the blame on a single individual (a scapegoat), such as a high-profile political leader, when something goes wrong in a complex economic system (i.e., high unemployment, inflation, high gasoline prices, debt) or when governmental corrective actions produce delayed, unanticipated, and/or weak results.

We often rely on the manipulation of a single input to try to attain the desired effect in a CS (e.g., adjusting the prime lending rate to stimulate the national economy). Complexity science has indicated that *it is often more effective to try multiple inputs and select those that prove more effective in obtaining the desired result.* In other words, we need to find estimates for the CNLS's *gain sensitivities* for the output states under attention. Gain sensitivities can be estimated by simulation and, of course, may be measured on the actual

CNLS whenever possible. However, such measurements are valid on a CNLS only around a stable, steady-state *operating point*. That is, the system is not oscillating or responding to a rapidly changing, time-varying input.

To further illustrate the single-cause mentality—the inability of otherwise thoughtful persons to handle CS thinking—consider the matter of *Lyme disease* and how it is spread. The Lyme disease pathogen (*Borrelia burgdorferi*) is a spirochete not unlike that which causes syphilis. It is spread by the deer tick *Ixodes scapularis*, which can feed on humans, small mammals, and birds. Infected *Ixodes* ticks were initially found in South Central Connecticut and are now found in rural areas throughout the state. A heavy density of Lyme disease cases has now been recorded through southern Maine, southern Massachusetts, Cape Cod and the islands, Long Island, Vermont, Connecticut, southeastern New York, New Jersey, Maryland, southeastern Pennsylvania, Texas, and strangely, western Wisconsin, eastern Minnesota, and California (Incidence 2011; Lyme Cases 2009). Lyme disease cases have been noted in all 50 US states. The US Centers for Disease Control and Prevention reported cumulative Lyme disease cases by state for the period 1990–2008. The top 10 were as follows: New York (87,192); Pennsylvania (51,266); Connecticut (45,938); New Jersey (39,387); Massachusetts (21,818); Wisconsin (15,463); Maryland (15,070); Minnesota (9600); Rhode Island (7673); and Delaware (4856). The fact that Colorado had nine cases and Hawaii six might be attributable to human travel from Colorado and Hawaii to tick "hot spots" like New England or Wisconsin and infected ticks "hitchhiking" to Colorado and Hawaii on these tourists, their clothing, and luggage.

Reservoirs for the *B. burgdorferi* pathogen include small ground-dwelling mammals including voles, mice, shrews, moles, rabbits, chipmunks, and perhaps squirrels, as well as whitetail deer. Ground-feeding migratory birds can also be infected *Ixodes* hosts and provide rapid, long-range transit for the ticks to spread the disease (Gylfe et al. 2000). When infected *Ixodes* adults (and their nymphs) feed on human blood, they infect their hosts with Lyme disease. The treatment of chronic Lyme disease is now controversial; one line of treatment that uses a very long-term administration of antibiotics is apparently not based on objective clinical or laboratory evidence of infection as a diagnostic criterion (Federer et al. 2007).

On March 23, 2008, an editorial appeared in the *Hartford Courant* (newspaper) entitled "Taking Aim at Lyme Disease." The editorial writer advocated severely reducing whitetail deer populations in the tick-infested regions as a means of Lyme disease epidemic remediation. Hunting by man presumably would be the method employed. This editorial is an excellent example of the single-cause mentality inappropriately applied to a complex ecosystem. N. Zyla responded on March 25, 2008 to this editorial with a letter in the *Courant* entitled "Are Deer the Biggest Culprit?" Zyla cited an article in the January 22, 2008 issue of *New England Journal of Medicine (NEJM)* in which the point was made that Lyme disease... "infects more than a dozen vertebrate species, any one of which could transmit the pathogen [*Borrelia*] to feeding ticks and increase the density of infected ticks and Lyme disease risk." The article stated that deer, in particular, "are poor reservoirs" for *Borrelia*. The *NEJM* article correctly commented: "Because several important host species [mice, voles, chipmunks & shrews] influence Lyme disease risk, interventions directed at multiple host species will be required to control this epidemic." Yes, deer are mobile hosts in terms of spreading the ticks, but eliminating them will not eliminate the reservoirs of *Borrelia*-infected small mammals and birds, or the ticks. Zyla correctly urged looking "at the bigger picture."

Another letter to the editors in the March 25, 2008 issue of *Hartford Courant* from S. Jakuba observed that the ticks not on animal hosts have predators in the form of ground-nesting wild birds (e.g., pheasants, partridges, wild turkeys, quail, etc.). (But to what extent?). He also observed that coyotes prey on these birds (and their eggs) (so do foxes, skunks, weasels, opossums, and hawks). Jakuba advocated getting rid of the coyotes as well as reducing the deer population. We note that coyotes also feed on mice, voles, moles, shrews, rabbits, and chipmunks, so eliminating an apex predator such as coyotes in the Lyme zones could activate the LUC by contributing to a population increase in these *Borrelia* reservoir rodents.

We note in closing this section that *Ixodes* ticks climb up onto the leaves of bushes and tall grasses where they are relatively safe from predation by ground birds, where they can easily be brushed onto passing humans, dogs, and deer. Ecosystems are complex; there are no simple fixes for the spread of Lyme disease.

2.2.4 The "Not in My Box" Mentality

Another shortcoming of humans in interacting with coupled, complex, non-linear system modules is the *not in my box mentality* (Smith 2004). This is a myopic approach to manipulating a CS in which the manipulator has been restricted to dealing with a circumscribed part of a CNLS (e.g., a module), neglecting the effects his/her manipulations have on the balance of the system ("It's not my responsibility"). The module boundary may have been set arbitrarily by the manipulator or by poor bureaucratic planning and external rules. The *not in my box mentality* provides a shortcut to the results of the *LUC*.

An illustration of *not in my box* behavior was seen in the attempted downing of a Detroit-bound airliner with an explosive pentaerythritol tetranitrate (PETN) on December 25, 2009, by Nigerian student terrorist, Umar Farouk Abdulmutallab (the "underwear bomber"). Abdulmutallab's father, concerned by his son's association with religious radicals in Yemen, visited the US Embassy in Abuja, Nigeria, on November 19, 2009, and reported his son's behavior. The embassy staff, following protocol under a security program called *Visa Viper*, promptly sent information on to the National Counterterrorism Center in Washington for entry into their database. According to Johnson (2009), "...neither the State Department nor the NCTC ever checked to see if Abdulmutallab had ever entered the US or had a valid entry visa; information readily available in separate consular and immigration databases. 'It's not for us to review that,' a State

Department Official said" (our underline). Such checking was evidently *outside their box*. According to the US Homeland Security Department, Abdulmutallab had twice obtained US visas, had actually twice before visited the United States (in 2005 and 2008), and had once been denied a UK visa. Obviously, in 2009, the complex communications network involved in US counterterrorism efforts still needed "tuning." Their "boxes" needed to be interconnected.

2.2.5 Complexity and Human Thinking

We see many examples every day of *reductionism, single cause, SARSL,* and *not in my box behavior* in the way people deal with complexity. For example, consider the US economy, including energy costs, recession, and inflation. We were told by economists that manipulation of the *federal funds lending rate* by the Federal Reserve System would dampen inflation and the symptoms of recession. The theory had merit, but certain effects of this lending rate reduction are felt only after a destabilizing *time delay* of approximately 12 months. The recession of 2008–2011 occurred anyway.

The US government also tried to stimulate the economy by modest tax refund payments in May 2008. There was uncertainty whether this cash injection would be effective in reducing the symptoms of recession. Economists debated whether this cash would be used to pay for local goods and services including gasoline, spent to pay off credit card and mortgage debts, or saved. Clearly, little research on human economic behavior had been done on how the refunds would be used before the decision to provide them was made. In retrospect, the United States officially entered a recession (as of November 2008), in spite of the refunds. Were these government actions too little, too late, the wrong input, or all of the above?

Consider the complex issues in the corn-for-ethanol debate. In an op-ed article, "Food-To-Fuel Failure" by Lester Brown and Jonathan Lewis, that appeared in the *Hartford Courant* on April 24, 2008, some of the factors affecting this energy/economic system were discussed. This article rightly lamented the failure of the congressional "food-to-fuel" mandate. The authors pointed out that although approximately one quarter of the US corn crop is now used for the production of ethanol to add to automotive gasoline, there has been only a 1% reduction in the country's oil consumption. [Most gasoline sold in the United States now contains 10% ethanol (E10); in Brazil, 85% ethanol fuel is common.] One of the goals of adding ethanol to gasoline was to increase the available fuel supply, so only a 1% reduction in oil consumption is not surprising; in fact, this small reduction may have been due to the monotonically rising cost of gasoline through October 2008 and beyond (see Figure 7.6).

Also, according to Brown and Lewis, the food-to-fuel mandates have caused "in large part" the current escalation in food prices. However, riots that have taken place in Haiti and Egypt are because of local food shortages and high prices and are not directly linked to US ethanol production. Food is a finite resource, and global food shortages are caused by a number of factors, including but not limited to bad weather (drought in Australia wheat farms in 2007), insect pests, as well as excessive demand linked to steady world population growth and standard-of-living increases in China and India. Yes, there is a close causality between higher US food prices and corn crop diversion from food use. However, the high prices of diesel fuel and gasoline must also be factored in to the prices of foods that are transported to US markets, some from Florida, Southern California, Mexico, and Central and South American countries in the winter. Farmers must pay more to run their tractors for plowing, cultivation, and harvesting, adding to food costs. Americans expect their foods to be neatly wrapped in plastic on Styrofoam trays and generally take their food purchases home in plastic bags, all made from expensive petroleum derivatives. Also, expensive FF energy is required to produce the fertilizers, herbicides, and insecticides used in modern agriculture. Diversion of arable land to grow more corn for ethanol has caused farmers to plant less wheat. Projected shortages of wheat and corn have led to a spate of speculative buying of these grain futures, further driving up their cost and, consequently, the cost of foods made from, or fed with, corn and wheat. These foods include not only those for human consumption but also those consumed as cattle and poultry foods. In a kind of chain reaction, as shortages appear in corn and wheat, prices of alternate grains (including rice, oats, barley, etc.) are also driven up by international investors, speculators in grain futures, and market demand. We expect that higher-yield, GM grain crops will be planted and consumed, regardless of any ecological and health consequences. (In modeling the CNLS of the US food supply, agents should be used to try to model consumer and producer confidence, market speculators, buyer and seller greed, and paranoia-including commodity hoarding.)

A quantitative, mathematical modeling study of the economic effects of US government ethanol subsidies was done by Taheripour and Tyner (2007). They traced the money flow from subsidies to the distillers through to rising real estate prices for farmland used to grow corn.

2.3 Signal Flow Graphs and Mason's Rule

2.3.1 Introduction

Using *mathematical graph theory*, many large, complex, nonlinear systems such as found in sustainability analyses, social networks, ecosystems, economic systems, and so forth can be modeled by graphs (also called webs by ecologists) characterized by *nodes* [also known as (aka) *vertices*] and *branches* (aka *edges*). Such graphs can be analyzed for their quantitative features including *topology, node degree, degree distribution, cyclomatic number, signal processing ability, sensitivities,* and other metrics such as their *Shannon index* and *Weiner index*. Ecologists have used the node/edge graph format to qualitatively describe *food webs* in ecosystems (Allesina et al. 2009). Their 3-D graphs' nodes often use size and color to code for "importance" and size of variables, and edges that are tapered tubes (rather than simple lines) to qualitatively illustrate the degree of influence one node has on another (Yoon et al. 2004; Melián and Bascompte 2004; FoodWeb3D 2004).

Economists are also beginning to use web graphs to describe complex economic subsystems or modules. As with ecosystems, the graphs are qualitative: clusters of colored nodes connected by straight edges. They show the relationships between various goods markets, producers, and consumers (Economic Complexity Observatory 2011).

Molecular biologists have also used the large, node/edge graph format to describe protein–protein (P–P) interactions. The graph in Figure 5A in the work of Stelzl et al. (2005) shows 911 interactions between 401 human proteins! Stelzl et al. also color-coded their protein nodes and used a two-bit color scale on the edges: green = 3 quality points (qps) (denoting P–P interaction confidence level), blue = 4 qps, red = 5 qps, and purple = 6 qps [also see Stelzl and Wanker (2006)]. (Note that Vester also used a 2-bit scale [0,1,2,3] to characterize the nodal interactions of his graphs; see Section 8.3 in this text.) A field where this qualitative/analytical approach would be very useful is in complex economic systems.

Below, we introduce the use of quantitative, directed graphs (digraphs) with a description of linear *signal flow graphs* (SFGs). SFGs are a specialized subset of *mathematical graphs* that were developed specifically to do pencil and paper analysis of large, *linear*, dynamic systems characterized by sets of coupled, linear, ordinary, differential (state) equations (ODEs) (Mason 1953, 1956; Truxal 1955; Northrop 2011). In the vocabulary of *graph theory*, SFGs are graphs with *signal nodes* (vertices) and *directed branches* (edges, arcs). Furthermore, SFGs belong to the class of directed graphs or digraphs, because their *edges* (branches or arcs) denote *unidirectional information (signal) flow*. In graph theory, the "order" of a graph is the number of vertices (nodes), and its "size" is the number of branches (edges). Both increasing order and size of a graph are rough indications of its increasing complexity. Below, we consider linear SFGs as a special case of mathematical graphs.

Note that SFGs were initially developed for easy analysis of *linear* dynamic systems, described by sets of *linear* ODEs. However, in modeling CNLSs with digraphs, we lose the ability to use Mason's gain formula to calculate linear I/O transfer functions between certain nodes. (A transfer function is applicable only to linear, dynamic systems.) However, an SFG with nonlinear branch gains still offers us insight into the CNLS's topology and its complex organization, including any modules, and how its overall structure can be described mathematically.

2.3.2 Signal Flow Graphs

SFGs are a venerable, linear, time-invariant (LTI) systems tool that was developed by Prof. S.J. Mason at the Massachusetts Institute of Technology in 1953, to enable easy, pencil-and-paper analysis of large, LTI dynamic systems. Mason published his SFG gain formula in 1956. SFGs enable one to take a large set of simultaneous, linear ODEs and linear algebraic equations describing a dynamic, linear, time-invariant system (LTIS) and put them in graphical form having *nodes* and *directed branches* (signal conditioning paths). Then using *Mason's gain formula*, one can easily find the I/O signal transfer function(s) for the system. The SFG approach is ideally suited for pencil-and-paper work because it replaces tedious, error-prone matrix inversion algebra as a necessary step to finding a state system's transfer function. Note that SFGs were developed more than a quarter century before PCs running linear systems simulation applications such as CSMP, MATLAB®, Simulink®, and Simnon appeared, yet they still have utility in helping us obtain a "crude look at the whole" (CLAW) in CNLSs.

SFGs have found application in describing and analyzing linear analog electronic circuits, linear control systems (both discrete and analog), linear dynamic mechanical systems, linear pharmacokinetic systems, and linearized physiological and biochemical systems (fortunately, some physiological and biochemical systems can be approximated by linear ODEs). All signals in an SFG are assumed to be in the frequency domain (i.e., are represented by Laplace, Fourier, or z-transform variables).

In this text, we use SFG digraph architecture to describe certain CNLSs; however, the nonlinearities prevent the realization of transfer functions by use of Mason's formula. The nonlinear SFG architecture does give us insight to nodal connectivity, feedback paths, and modules, however.

An SFG has two components: *unidirectional branches* that condition signals and *signal nodes*. Signals from incoming branches to a node sum algebraically at the node. The input signals that are summed at a node p, x_p, are multiplied by the *transmission* or *gain* of any branch leaving node p. The input to another node q from a branch pq is the product of the branch's transmission T_{pq} times the source (p) node's signal, x_p, that is, $x_p T_{pq}$. The conditioned signal, $x_p T_{pq}$, does not directly affect the signal at source node p (i.e., it is not subtracted from x_p, only added to x_q).

Some definitions relevant to SFGs are as follows:

> *Branch:* A *directed gain function* that operates on the signal at a *source node* and adds the result to the signal at the *sink node* onto which the branch terminates. (A branch is also called an *edge*.) In SFGs, branch gains are *linear* operations or functions; in nonlinear SFGs, branch gains can be nonlinear operations. In nonlinear SFGs, Mason's gain formula does not apply.

Path: Any collection of a succession or concatenation of *branches* between nodes traversed in the same direction. The definition of path is general because it does not prevent any node from being traversed more than once (Kuo 1982).

Forward path: A path that starts at an *input* or *source node* and ends at an *output node,* along which no node in the SFG is traversed more than once. Many CSs' SFGs have more than one forward path.

Forward path gain: the net gain of a forward path.

Loop: A path that originates from a certain node *i* and also terminates on node *i*. (This is also known as a *cycle* in graph speak.) The loop path encounters no other node more than once. Loops with parallel paths can be reduced by adding their gains, forming an *independent loop*: an independent loop cannot be made up by simple addition from other loops (Truxal 1955).

Loop gain: The net path gain around a loop.

Node: Summing points where signals from all input branches are added algebraically. Branches leaving a node do not affect the node's signal. (Nodes are also called *vertices*.)

Nontouching loops: these loops share no nodes in common.

2.3.3 Examples of Linear SFG Reduction

As an introductory example of signal manipulation in SFGs, consider the simple SFG in Figure 2.2a. Here, the signals are as follows: $y_1 = T_1 x_1$, $y_2 = T_2 x_1$, and $y_3 = T_3 x_1$. In Figure 2.2b, $y_1 = x_1 T_1 + x_2 T_2 + x_3 T_3$. When several branches are in series forming a path as in Figure 2.2c, $y_4 = T_1 T_2 T_3 x_1$. When they are in parallel as in Figure 2.2d, $y_1 = x_1(T_1 + T_2 + T_3)$. The T_x are frequency domain transfer functions.

The *systematic gain formula* for SFGs developed by S.J. Mason in 1956 is deceptively simple. However, it does require some interpretation. Mason's gain formula is written as

$$\frac{V_{ok}}{V_{ij}} = H_{jk} = \frac{\displaystyle\sum_{n=1}^{N} F_n \Delta_n}{\Delta_D} \tag{2.1}$$

where

V_{ok} is the frequency domain output signal at node *k*, and V_{ij} is the frequency domain input signal at node *j*.

H_{jk} is the net transmission (transfer function) from the *j*th (input or source) node to the *k*th (output or sink) node.

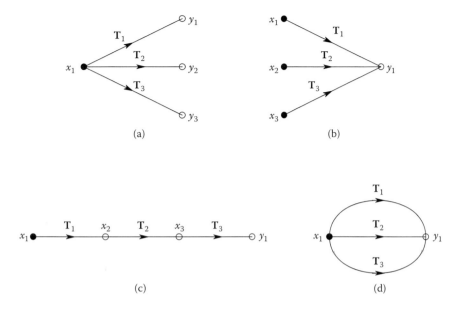

FIGURE 2.2
Four basic, directed SFGs. See the text for analysis. (From Northrop, R.B., *Introduction to Complexity and Complex Systems*, 2011, CRC Press, Boca Raton, FL. With permission.)

F_n is the transmission of the *n*th *forward path*. The *n*th forward path is a connected path of branches beginning on the *j*th (input) node and ending on the *k*th (output) node along which no node is passed through more than once. An SFG can have several forward paths, which can share common nodes.

Δ_D is the SFG gain *denominator*: $\Delta_D \equiv 1 -$ [sum of all individual loop gains] + [sum of products of pairs of all nontouching loop gains] − [sum of products of nontouching loop gains taken 3 at a time] + ...

Δ_n is the *cofactor* for the *n*th forward path. $\Delta_n \equiv \Delta_D$ evaluated for nodes *that do not touch* the *n*th forward path (see examples below). $n = 1, 2, 3,....$

N is the total number of forward paths in the SFG.

The best way to learn how to use Mason's formula on systems' SFGs is by example. In our *first example*, Figure 2.3 illustrates a simple SISO, NFB system. There is only one forward path. The loop touches two of its nodes, so $\Delta_1 = 1$. F_1 is seen to be $1 \times K_v/(s + a)$, and

$$\Delta_D = 1 - \left[\frac{-K_v \beta}{(s+a)s} \right]. \tag{2.2}$$

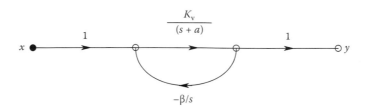

FIGURE 2.3
SFG for the simple, single-loop, NFB system of Example 1. (From Northrop, R.B., *Introduction to Complexity and Complex Systems*, 2011, CRC Press, Boca Raton, FL. With permission.)

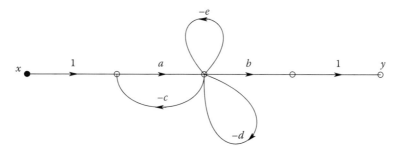

FIGURE 2.4
Three-loop, NFB system SFG for the second example. (From Northrop, R.B., *Introduction to Complexity and Complex Systems*, 2011, CRC Press, Boca Raton, FL. With permission.)

The overall Laplace transfer function is put in *time-constant form*. Note that the gain's denominator is of the second order:

$$\frac{Y}{X} = \frac{s/\beta}{s^2/K_v\beta + sa/K_v\beta + 1}. \tag{2.3}$$

In the *second example,* the SFG is shown in Figure 2.4. Now, $n = 1$, $F_1 = ab$, $\Delta_D = 1 - [- ac - e - d] + 0$, and $\Delta_1 = 1$. From this, we obtain the SISO gain in simple, rational algebraic form:

$$\frac{Y}{X} = \frac{ab}{1 + ac + e + d}. \tag{2.4}$$

In the *third example,* shown in Figure 2.5, the SFG is more complicated; there are three forward paths, so $n = 3$, $F_1 = abc$, $F_2 = def$, $F_3 = -akf$, $\Delta_D = 1 - [-bh - eg]$

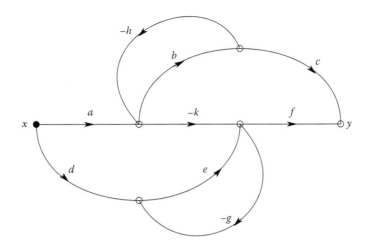

FIGURE 2.5
SFG for the system of the third example has two NFB loops and three forward paths. See text for analysis. (From Northrop, R.B., *Introduction to Complexity and Complex Systems*, 2011, CRC Press, Boca Raton, FL. With permission.)

+ $[(-bh)(-eg)] - 0$, $\Delta_1 = 1 + eg$, $\Delta_2 = 1 + bh$, and $\Delta_3 = 1$. The transfer function is thus

$$\frac{Y}{X} = \frac{abc(1 + ge) + def(1 + bh) - akf}{1 + bh + ge + bhge}. \tag{2.5}$$

In the *fourth example*, Figure 2.6 illustrates a state-variable form SFG: here, $n = 4$, $F_1 = b_3$, $F_2 = b_2/s$, $F_3 = b_1/s^2$, $F_4 = b_0/s^3$, all $\Delta_k = 1$, and $\Delta_D = 1 - [-a_1/s - a_2/s^2 - a_3/s^3] + 0$. The SFG's Laplace transfer function is easily seen to have a cubic polynomial in the numerator and denominator:

$$\frac{X}{Y} = \frac{b_3 s^3 + b_2 s^2 + b_1 s^1 + b_0}{s^3 + a_1 s^2 + a_2 s^1 + a_3}. \tag{2.6}$$

In the *fifth example*, a molecule, C, is synthesized in the mitochondria of a cell at a rate Q_C. Its concentration is $C_m \, \mu g/L$ in the mitochondria. It diffuses out of the mitochondria into the cytoplasm, where its concentration is C_c. It next diffuses through the cell membrane to what is basically zero concentration outside of the cell. The two compartments are (1) the mitochondria and (2) the cytoplasm around the mitochondria. The compartmental state equations are based on simple diffusion (Fick's first law). V_m and V_c are compartment volumes.

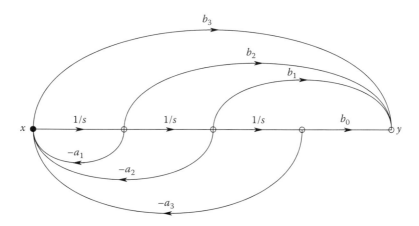

FIGURE 2.6
SFG in state-variable format for the fourth example. There are three feedback loops and four forward paths. See text for analysis. (From Northrop, R.B., *Introduction to Complexity and Complex Systems*, 2011, CRC Press, Boca Raton, FL. With permission.)

$$V_m \dot{C}_m = \dot{Q}_0 - K_{12}(C_m - C_c)\,\mu g/\min \qquad (2.7a)$$

$$V_c \dot{C}_c = K_{12}(C_m - C_c)K_2 C_c\,\mu g/\min \qquad (2.7b)$$

Written in state form, we have

$$\dot{C}_m = -C_m(K_{12}/V_m) + C_c(K_{12}/V_m) + \dot{Q}_0/V_m\,\mu g/(L\min) \qquad (2.8a)$$

$$\dot{C}_c = C_m(K_{12}/V_c) - C_c(K_2 + K_{12})/V_c\,\mu g/(L\min) \qquad (2.8b)$$

Note that mass diffusion rates depend on *concentrations* or mass/volume, so the diffusion rate constants, K_{12} and K_2, must have the dimensions of liters per minute.

From the ODEs Equations 2.8a and 2.8b, we see that the system is linear and can be described by an SFG as shown in Figure 2.7. The SFG can easily be reduced by Mason's rule to find the transfer function, $C_c/Q_0(s)$: In this example: $n = 1$, $F_1 = (1/V_m)(1/s)(K_{12}/V_c)(1/s)$, $\Delta_1 = 1$, and $\Delta_D = 1 - [(-K_{12}/sV_m) + (-(K_2 + K_{12})/sV_c)] + (K_{12}^2/s^2 V_m V_c) + [(-K_{12}/sV_m)(-(K_2 + K_{12})/sV_c)] - 0$.

This somewhat involved denominator turns out to be only a quadratic in s with two real roots. After some algebra, we can write

$$\frac{\dot{C}_c}{\dot{Q}_c} = \frac{K_{12}/V_m V_c}{s^2 + s[K_{12}/V_m + (K_{12} + K_2)/V_c] + K_2 K_{12}/V_m V_c}$$

$$= \frac{K}{(s+\mathbf{a})(s+\mathbf{b})} \tag{2.9}$$

Thus, the two-compartment system governed by diffusion is seen to have linear, second-order dynamics with two real poles at $s = -\mathbf{a}$ and $s = -\mathbf{b}$ after factoring. MATLAB's *Roots* utility can be used to numerically factor the denominator to find the $\mathbf{a}$ and $\mathbf{b}$ natural frequency values.

In a sixth and final example, we consider the SFG of a *nonlinear* compartmental system described by Godfrey (1983). This is a two-compartment, pharmacokinetic system having six nodes in its SFG. It also has three loops and two nontouching loops. The ODEs describing the system are written below:

$$\dot{\mathbf{x}}_1 = -k_1 \mathbf{x}_1^2 - \frac{V_m \mathbf{x}_1}{K_m + \mathbf{x}_1} + k_{12}\mathbf{x}_2 + \mathbf{u}_1 \tag{2.10a}$$

$$\dot{\mathbf{x}}_2 = \frac{V_m \mathbf{x}_1}{K_m + \mathbf{x}_1} - k_{12}\mathbf{x}_2 - \mathbf{k}_{02} \tag{2.10b}$$

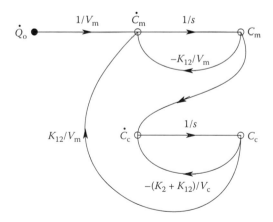

FIGURE 2.7
Two-state, linear SFG representing the ODEs of Equations 2.8a and 2.8b in the fifth example. See text for analysis. (From Northrop, R.B., *Introduction to Complexity and Complex Systems*, 2011, CRC Press, Boca Raton, FL. With permission.)

Now we define $f_1(\mathbf{x}_1) \equiv \dfrac{V_m \mathbf{x}_1}{K_m + \mathbf{x}_1}$ and $f_2(\mathbf{x}_1) \equiv k_1 \mathbf{x}_1^2$.

Sadly, because this system is nonlinear, Mason's formula cannot be used to find the gain or transfer function $\mathbf{x}_2/\mathbf{u}_1$. The system's SFG is used only to illustrate the system's signal architecture (where positive and negative feedback loops are located). This nonlinear SFG, shown in Figure 2.8, has 6 nodes, 2 integrators, 1 squarer, a nonlinear function block, and a total of 10 directed branches. We can use the SFG to illustrate some properties of directed graphs, however. First, let us compile a table of the system's vertex (nodal) degrees (Table 2.1).

The *degree sum formula* in graph theory states that for a given graph, $\mathbf{G} = (\mathbf{V},\mathbf{E})$,

$$\sum_{v \in V} \mathbf{deg}(v) = 2|\mathbf{E}|, \tag{2.11}$$

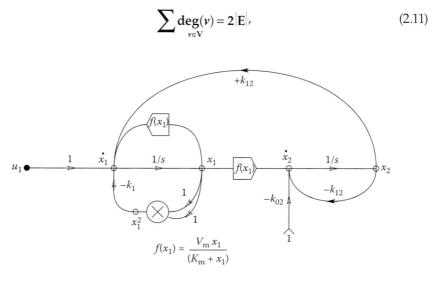

FIGURE 2.8
Nonlinear pharmacokinetic system of the sixth example is described by this SFG. See text for analysis. (From Northrop, R.B., *Introduction to Complexity and Complex Systems*, 2011, CRC Press, Boca Raton, FL. With permission.)

TABLE 2.1

Tabulation of Vertex (Nodal) Degrees for the Nonlinear System of Equations 2.10a and 2.10b

Vertex (Node)	Indegree	Outdegree
u_1	0	1
$\dot{x}_1$	4	1
x_1	1	3
$\dot{x}_2$	3	1
x_2	1	2
k_{02}	0	1

since each edge (path) is incident on two vertices (nodes). This relation implies that in any graph, the number of nodes with an odd degree is even. For the example system, $G = (6,9)$, and $\sum \deg(v) = 2(9) = 18$. In Table 2.1, note that $\sum$ indegrees $+ \sum$ outdegrees $= 18$.

2.3.4 Measures of SFG Complexity

The complexity of a linear SFG generated from a set of linear, simultaneous algebraic equations or ODEs is largely in the eye of the beholder. Certainly we can agree on some objective criteria, such as the total number of nodes, $\mathbf{N}$, the total number of directed branches, $\mathbf{B}$, and the total number of independent feedback loops, $\mathbf{L}$. If we create some monotonically increasing complexity index, $\mathbf{C_I} = f(\mathbf{N}, \mathbf{B}, \mathbf{L})$, then we are still faced with establishing a criterion for the numerical functional, $\mathbf{C_I}$, above which the system/SFG is considered complex. Intuitively, we see in the case of a *complete graph,* where every node is connected to every other node by a direct branch, that the number of branches rises with the square of the number of nodes. For example, a 100-node complete graph has 4950 branches, while a 1000-node complete graph has 4.9950E+5 branches! A dynamic, nonlinear system with 100 nodes forming a complete graph may be described by over 4950 ODEs.

If we go back to Mason's Equation 2.1 for the SISO gain of a linear system, we see that generally, it will yield a transfer function that is a rational polynomial in the Laplace variable, $\mathbf{s}$. The denominator will be of order $\mathbf{n}$, the numerator is of an order $\mathbf{m}$, $0 \leq \mathbf{m} \leq \mathbf{n}$. $\mathbf{m}$ and $\mathbf{n}$ and the system's polynomial coefficients determine the number of the systems poles and zeros and its pole and zero positions in the complex $\mathbf{s}$-plane. The pole and zero positions determine the system's dynamic, I/O behavior for an arbitrary input, as well as its steady-state, sinusoidal frequency response. The poles also determine the linear system's stability.

As we have seen above, an SFG can also be used to describe CNLSs. However, Mason's rule can no longer be used to determine I/O gains for the nonlinear system. The SFG in this case "two-dimensionalizes" an abstract set of nonlinear ODEs and aids in visualizing functional relationships. Such a nonlinear SFG will, in general, have linear gains in some of the branches, and some branches will be nonlinear functions of the signal value at the originating node [e.g., Hill functions, algebraic functions, e.g., $x_2 = a + bx_1 + cx_1^3$, or $x_2 = k \tanh(a\, x_1)$] or be parametric gain functions determined by another node variable, x_j. Obviously, such large sets of nonlinear ODEs are solved on computers. The index, $\mathbf{C_I}$, will no longer be a valid measure of system complexity because large, internally interconnected nonlinear systems often have unexpected and surprising behavior (chaos), dependent not only on their initial conditions but also on the amplitude and rate of change of their inputs. [See Sections 1.4.4 and 1.5 in Northrop (2000).]

A comprehensive treatment of complex networks and graph theory may be found in the extensive review by Boccaletti et al. (2006). Besides reviewing graph theory, this paper considered social network phenomena, including dynamic effects (modeling cascading failures and congestion in communication networks); spreading processes (epidemics, rumors); the synchronization of coupled oscillators; applications to social networks, the Internet, and the World Wide Web; metabolic, protein, and genetic networks; CNS networks; and so forth.

2.4 Modularity

2.4.1 Introduction

Often, when viewing a large, 2- or 3-D plot of a CS's graph, the eye can pick out clusters of nodes that apparently "talk more amongst themselves" than to other peripheral nodes. The local density of the edges is the giveaway in this subjective identification of a *structural module*, or subsystem graphs with modularity are not *complete*. Modularity is an abstract concept that can refer to the patterns of mathematical interactions between a graph's nodes and in the organization of graphs modeling living, economic, and social processes. However, there are objective, mathematical criteria based on topology and thresholds for identifying what collections of nodes comprise a module in a large graphical network. Some of these metrics are described below.

We are familiar with modules that occur in man-made systems. Modules in the form of software subroutines can be linked into different, complex computer programs. Modular design of interchangeable, replaceable parts in mechanical systems leads to easier design and maintenance, as well as robust operation. In analog electronic systems, the operational amplifier (op amp) often serves as a vital part of a modular gain element or signal filter. Other integrated circuits (ICs) serve as analog radio frequency (RF) and power amplifiers, digital ICs store data (random access memories [RAMs] and read only memories [ROMs]), and programmable logic array (PLA) modules are used in digital controllers.

Modularity is widely encountered in all biological systems, regardless of scale. Alon (2003) reminded us, "Biology displays the same [modular] principle, using key wiring patterns again and again throughout a network." He remarked that many metabolic networks use biochemical regulatory architectures such as feedback inhibition. "The [gene] transcriptional network of *E. coli* has been shown to display a small set of recurring circuit elements termed 'network motifs.'" The *network motifs* can perform tasks such as filtering out random input fluctuations, generating oscillating gene expression patterns, and upregulating and downregulating gene expression. The same motifs were also found in the transcriptional networks of yeast. Alon was of the opinion that nature appears to converge on these functional biochemical circuit patterns over and over again in different, nonhomologous systems.

They are evidently part of a universal, functional architecture found in all biochemical systems. Research needs to be done to characterize the modules inherent in economic and social systems.

It appears that modularity can allow the creation of robust CSs from smaller, complex subsystems. Lipson et al. (2002) argued, "It has long been recognized that architectures that exhibit functional separation into modules are more robust and amenable to design and adaptation." These authors pointed out that modularity creates a functional separation (or isolation) that reduces the amount of coupling between internal (i.e., inside a module) and external changes. Thus, evolutionary pressures may more easily rearrange the inputs to modules without altering their intrinsic behaviors. Evidence for this preservation of modular function has been seen in certain developmental experiments with *Drosophila* larvae (Halder et al. 1995). Albert (2005) observed that "...modules should not be understood as disconnected components but rather as components that have dense intracomponent connectivity but sparse intercomponent connectivity." The challenge in identifying functional modules is that a module is not always a clear-cut subnetwork linked in a clearly defined manner, "...but there is a high degree of overlap and crosstalk between modules" (Albert 2005). Thus, module description again involves fuzzy boundaries.

We see modularity allowing an evolutionary change in one module causing minimum disturbances in the functioning of other modules in the CNLS, as long as the paths between modules are held relatively constant.

2.4.2 Measures of Modularity

A challenge exists in finding objective algorithms for the identification of modules in a large graph or network. Hallinan (2003, 2004a), Hallinan and Smith (2002), and Hallinan and Wiles (2004) have reviewed some of the approaches taken for module detection and have done research on quantifying modularity in graphs. Hallinan developed a criterion called the *iterative vector diffusion algorithm* (IVDA) and applied it to randomly generated networks, social networks derived from *Internet relay chat*, and protein regulatory networks in brewer's yeast (*Saccharomyces cerevisiae*).

Hallinan (2004b) defined an *average cluster coefficient C* as a measure of the extent to which the neighbors of nodes are linked to each other in a graph:

$$C = (1/N) \sum_{i=1}^{N} \frac{c_i}{n_i(n_i - 1)/2} \qquad (2.12)$$

where N is the total number of nodes in the network, c_i is the number of connections between neighbors of root node i, and n_i is the number of neighbors of root node i. (A neighbor node k of a root node i is connected to it by an $i \leftrightarrow k$ edge.)

Hallinan also devised a *modular coherence algorithm*, which is applied to the nodes of a previously identified module and measures the relative proportions of intramodule and intermodule links (edges). It assigns a *cluster coherence value*, χ, given by

$$\chi = \frac{2k_i}{N(N-1)} - (1/N)\sum_{j=1}^{N} \frac{k_{ji}}{(k_{jo} + k_{ji})} \qquad (2.13)$$

where N is the total number of nodes in the entire network, $k_i =$ the total number of edges between all nodes *in the module*, $k_{ji} =$ the number of edges between node j and other nodes within the module, and $k_{jo} =$ the number of edges between node j and other nodes *outside the module*.

Cluster coherence is a measure of the relative proportion of edges between nodes within and outside of a previously identified module. χ ranges from −1 (no coherence) to +1 (a fully connected, stand-alone module or subgraph).

The first term in the cluster coherence equation is simply the proportion of possible links between the nodes of the graph. The summation term is the average proportion of edges per node, which are internal to the module. Thus, a highly connected module with few external edges will have a lower χ value than a highly connected node with many external edges. Note that the formal condition for a *strongly connected component* (SCC) (or module) in a *directed graph* is where there is a path from each node in the SCC to every other node in the SCC. In particular, this means paths in each direction: a path from node **p** to **q**, and also a path from **q** to **p**, generating simple feedback.

2.4.3 Examples of Modules in Sustainability Models

2.4.3.1 Introduction

Modules exist resulting from our formulations of system dynamics required to model complex ecological sustainability. Such modules exist as the result of humans interfacing with social systems, economic systems, and ecosystems. These modules allow us to attack a very complex problem more easily by separately analyzing its component modules. Such an approach is often counterproductive when dealing with a large, tightly coupled CNLS and must be used cautiously. Essential aspects of overall CNLS behavior (such as tipping points) may be lost by not dealing with the whole graph with all of its modules *in situ*.

2.4.3.2 Cod Fishery

One example of a module in ecological sustainability involves food, specifically fisheries, more specifically, the North Atlantic cod fishery. Any ecosystem is complex due to the many interactions between its species, their food sources, the oceanographic parameters, and so forth. We are fortunate that to model the complex cod fishery in a region, a simple, two-ODE, Lotka–Volterra type of ecosystem model can be used. The model below was created by one author (RBN) to illustrate the dynamics of managed, sustainable fishing. It uses only two, coupled, nonlinear ODEs. Coupling to other ecosystem nodes is assumed to be done parametrically. The model's nonlinear state equations are as follows:

$$\dot{x}_1 = fd x_1(t - D_1) - \alpha_1 x_1 x_2 - \beta x_1^2 - L_h x_1 \quad \text{"herrings" } (1000\,\text{fish/km}^2)/\text{month}$$

(2.14a)

$$\dot{x}_2 = \alpha_2 (x_1 x_2)(t - D_2) - L_c x_2 - b_o x_2 \quad \text{"cod" } (1000\,\text{fish/km}^2)/\text{month} \quad (2.14b)$$

$$\dot{cc} = L_c x_2 n_{uc} \quad \text{"cumulative cod catch" } cc \text{ in thousands of fish} \quad (2.14c)$$

$$\dot{ch} = L_h x_1 n_{uh} \quad \text{"cumulative herring catch" } ch \text{ in thousands of fish} \quad (2.14d)$$

$$fd = f_0 + f_1 \sin(2\pi t/P) \quad \text{"cyclic density of Herring's food"} \quad (2.14e)$$

$$\textbf{cumcod} = 0.01 cc \quad \text{"cod catch scaling factor"} \quad (2.14f)$$

where x_1 is the herring (the prey species) density in thousands of fish/km² sea surface; x_2 is the cod (predator fish) species density in thousands of fish/km²; cc is the cumulative (integrated) cod catch over time (a quantity to be maximized); ch is the cumulative herring catch; n_{uc} is the catch-to-market efficiency for cod; n_{uh} is the catch-to-market efficiency for herring; and fd is the herrings' food (e.g., plankton), which varies cyclically with season, sic: $fd = f_0 + f_1 \sin(2\pi t/P)$, where $P = 12$, a 12-month period, and $f_1 \geq f_0$; t is in months. D_1 is the delay in months for herring growth to maturity, D_2 is the delay in months for cod growth to maturity after feeding (i.e., the product $x_1 x_2$ is delayed D_2 months), and α_1 is the cod predation rate constant. b_o is the cod loss rate from "natural causes," for example, predation by bigger fish. β is the competition loss constant for herrings. L_c is the cod fishing law, for example, $L_c = L_{c0}$ if $x_2 > \varphi_c$, else 0 (i.e., cod fishing is prohibited until the mature cod density exceeds a threshold value, φ_c). A similar fishing law L_h exists for herring: $L_h = L_{h0}$ if $x_1 > \varphi_h$, else 0. We used the initial values (IVs): $x_1(0) = x_2(0) = 3$. A select region of the North Atlantic is considered.

Figure 2.9 illustrates a nonlinear SFG describing the nonlinear simulation equations. Note that human population density (PD) does not appear explicitly; rather, it affects the fishing law parameters L_{c0}, φ_c, L_{h0}, and φ_h.

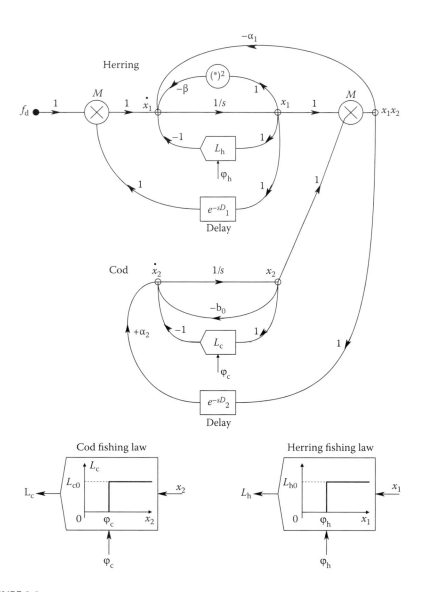

FIGURE 2.9
Nonlinear SFG illustrating the model for the cod–herring fishery described by coupled, Lotka–Volterra–type ODE dynamics. Two states are used: cod and herring PDs. The fishing laws illustrated allow taking of fish only when the fish PDs exceed preset thresholds, φ. The rate of catching cod is proportional to the cod PD, x_2, only when $x_2 > \varphi_c$. See text for analysis.

Similarly, climate and ecological conditions affect the cod growth rate, α_2, and herring food growth rates, f_0 and f_1, by affecting plankton densities, which affect **fd**.

Figure 2.10 illustrates the system's behavior when only cod are fished. The following parameters were used: IVs $x_1(0) = x_2(0) = 3$; $\alpha_1 = 0.5$, $\alpha_2 = 1.0$, $\beta = 0.1$,

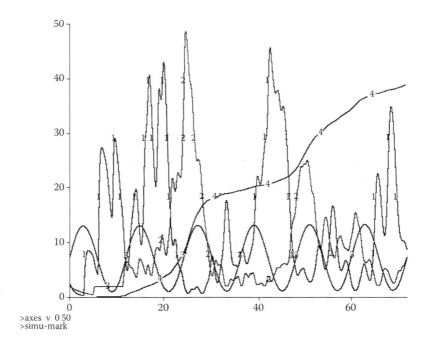

>axes v 0 50
>simu-mark

FIGURE 2.10
Simnon simulation showing the cod–herring fisheries model behavior when only cod are fished. $\varphi_c = 0.75$. See text for details. Traces are as follows: $1 = x_1$ (herring PD), $2 = x_2$ (cod PD), $3 = fd$, $4 = $ cumcod. Time in months.

$b_0 = 0.3$, $\mathbf{D}_1 = 3$ months, $\mathbf{D}_2 = 5$ months, $\varphi_c = 2$, $\varphi_h = 1000$, $f_0 = 7$, $f_1 = 7$, $n_{uc} = 0.85$, $L_{h0} = 5$, $L_{c0} = 8$, $P = 12$ months, and t is the simulation time in months.

Table 2.2 illustrates the variation of the 72-month cumulative cod catch with φ_c; herring are not fished [φ_h is made very large (1000)].

In Figure 2.11, the herrings are now also fished according to a threshold law. Parameters are as above but include $\varphi_h = 4$, $L_{h0} = 5$, and $n_{uh} = 0.9$. Note that the thresholds φ_c and φ_h can be manipulated *by* trial and error to maximize *both* cc and ch at $t = 72$ months. Note that every herring caught means one less to feed the cod, and every cod caught means more herrings swim on to be caught by man.

While the preceding simple model may give a good approximation for the prey species' population dynamics, the herring are enmeshed in their own food web: They eat zooplankton and smaller fish and, in turn, are preyed upon by not only cod but also other apex predators such as tuna, dolphins, baleen whales, whale sharks, and so forth. The model illustrates that managed fisheries (cod, herring) can have maximum long-term yields if the fishing rates are optimized. Also, as we have seen in Chapter 1, overfishing causes the populations to crash.

TABLE 2.2

Cumulative Cod Catch after 72 Months Fishing
Only Cod, Varying the Fishing Law Threshold φ_c

φ_c	cc
0.25	3960
0.50	4056
0.65	4099
0.75	4109
1.00	4089
2.0	3854
3.0	3612
4.0	3589
5.0	3540
10.0	29.60

Note: A maximum yield is seen to occur for $\varphi_c = 0.75$. Ten
simulations were run. (**cc** is in thousands of fish.)

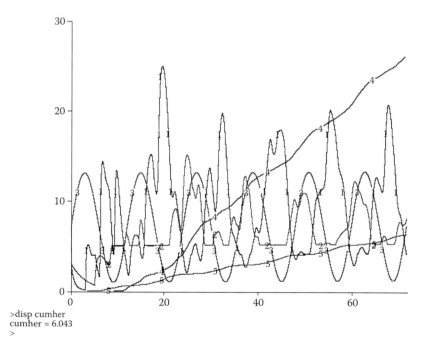

>disp cumher
cumher = 6.043
>

FIGURE 2.11

Results of a simulation of the cod–herring fishery when both cod and herring are fished according to threshold algorithms. By trial and error, it is possible to adjust both φ_c and φ_h to maximize
both cc and ch over 72 months. $\varphi_h = 4$, $L_{h0} = 5$, $n_{uh} = 0.9$, $\varphi_c = 0.75$, $L_{c0} = 8$, $n_{uc} = 0.85$. Traces are as
follows: $1 = x_1$, $2 = x_2$, $3 = fd$, $4 = $ cumcod, $5 = $ cumher. At $t = 72$ months, cumher = 6.043. (Multiply
the cumher and cumcod catches by 100 to get true 72-month cumulative catches in thousands
of fish.)

2.4.3.3 Aquifer

As a second example of an important module in sustainable food production, the fluxes in freshwater in a major aquifer are examined. An aquifer is a regional underwater reservoir, formed by local geology. Its maximum water content can be Q_{max} m³. It is replenished by regional rainfall and snowmelt percolating down through the soil. The average rainfall per day in the aquifer-charging region varies and can be described in cubic meters per day. Call the rainfall rate $\dot{R}$ m³/day. The aquifer's volume Q in cubic meters is depleted by deep-well pumping for regional domestic use, agriculture, and various industrial purposes such as gas and oil production. These depletion rates are $\dot{L}_D$, $\dot{L}_A$, and $\dot{L}_I$ m³/day, respectively. Assume that when $Q < 0.3Q_{max}$, salt begins entering the aquifer at a rate $\dot{S}$ kg/day. Thus, the water in the depleted aquifer gradually turns saline, and at a certain threshold, salinity, φ_s kg/m³, has too high a salt content for domestic, agricultural, and industrial use without an energy-intensive desalination process. $\dot{L}_D$, $\dot{L}_A$, and $\dot{L}_I$ are set to zero, and water is then pumped to the desalination plant at rate $\dot{L}_{DS}$ m³/day. Two nonlinear ODEs can be written to describe the water fluxes in the aquifer:

$$\dot{Q} = K_R\dot{R} - \dot{L}_D - \dot{L}_A - \dot{L}_I - \dot{L}_{DS} \tag{2.15a}$$

$$\dot{S} = f(Q) = \dot{S}_m - M_S Q, \quad M_S = (\dot{S}_m/0.3Q_{max}), \quad \dot{S} = 0 \text{ for } Q > 0.3Q_{max} \tag{2.15b}$$

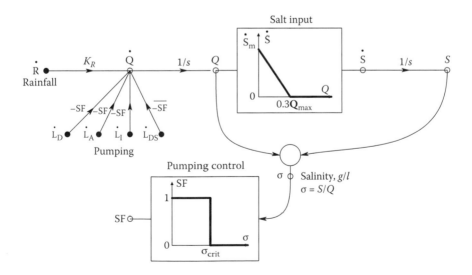

FIGURE 2.12
Nonlinear SFG for the aquifer example. Note that the rate of pumping (SF) is turned off when the aquifer's salinity exceeds σ_{crit}.

$\sigma = S/Q$. If $\sigma > \sigma_{crit}$, $\dot{L}_O, \dot{L}_A, \dot{L}_I \equiv 0$ and $\dot{L}_{DS} > 0$, else $\dot{L}_O, \dot{L}_A, \dot{L}_I > 0$ and $\dot{L}_{DS} \equiv 0$.
S and **Q** are nonnegative.

Figure 2.12 illustrates a nonlinear SFG for this simple aquifer system.

2.4.3.4 Epidemic

Another important module affecting human sustainability is the propagation of an infectious disease in a closed, dense, urban population. The system described below is based on the Kermack–McKendrick [aka susceptible, infected, and recovered (SIR)] infectious disease models (Mathworld 2011; Li and Zou 2009; Brauer and Castillo-Chávez 2001). In this example, we include deaths, so it is in fact a susceptible, infected, recovered, and dead (SIRD) model. This model is set up as an initial value problem (IVP) and is described by the following ODEs:

> **N** = **x** + **y** + **z** + **m** Closed (quarantined) system. **N** = total alive humans at start. Initially, **N** = $\mathbf{x_o}$ + $\mathbf{y_o}$.
>
> $\dot{x} = -\beta xy$ Rate of uninfected persons. **x** = number of uninfected persons; $\mathbf{x_o}$ initially.
>
> $\dot{y} = \beta xy - \gamma y - \delta y$ Rate of flu infection. $\mathbf{y_o}$ = number of patients initially with flu.
>
> $\dot{z} = \gamma y$ Rate of recovered (now immune) flu victims.
>
> $\dot{m} = \delta y$ Death rate from flu infection.

where **x** = number of uninfected, susceptible (*S*) persons in the city; **y** = number of living flu victims (*I*); **z** = number of recovered immune persons (*R*); **m** = number of persons killed by the virus (cannot exceed N_o); **N** = total population of the city at time $t = 10^4$; **x**(0) = initial population of uninfected persons = 1.E+4; **y**(0) = $\mathbf{y_o}$ initial number of infected persons = 5 (came in on Flight 293 from Hong Kong.); **z**(0) = **m**(0) = 0; and time t in weeks.

Figure 2.13 illustrates the nonlinear SFG describing the infection model. Note that there are four integrators, nine nodes, and one multiplier. Note that no one in the initial population x_o is immune. Figure 2.14 illustrates a Simnon simulation of the SIRD model given above. (Trace 1 = **x**, Trace 2 = **y**, Trace 3 = **z**, and Trace 4 = **m**; time t in weeks, $\beta = 1.\text{E-}3$, $\gamma = 0.5$, $\delta = 0.05$.) Note the approximately 1-week "incubation period" inherent in the simulation results (there are no actual delay operators in the model). The "incubation period" is the result of the nonlinear system dynamics; there is no delay operator in the model equations.

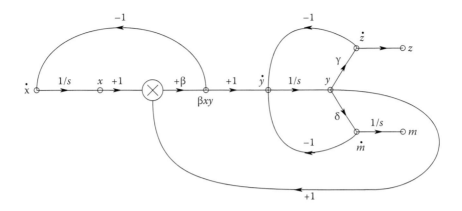

FIGURE 2.13
Four-state, nonlinear SFG describing a modified Kermack–McKendrick–based model for an infectious disease: SIRD. See text for details.

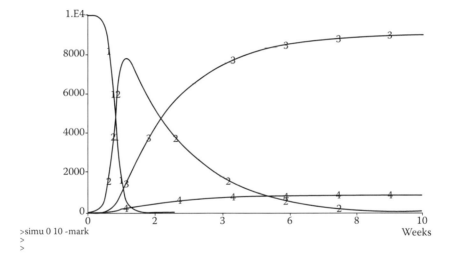

>simu 0 10 -mark
>
>

FIGURE 2.14
Simnon simulation of the SIRD system as an IVP. Note that there is approximately a 1-week "incubation period" inherent in the model before the number of infected, living flu victims (y) begins to rise steeply. There is also an approximately 1.5-week delay before the immune systems of the infected persons begin to conquer the virus. Because of the conferred immunity of the recovered patients, the steady-state death toll reaches a plateau of approximately 10% of N_o. Trace 1 = x (uninfected persons); Trace 2 = y (living flu victims); Trace 3 = z (recovered, now-immune persons); Trace 4 = m (persons killed by the flu); time t in weeks; β = 1.E-3, γ = 0.5, δ = 0.05, N_o = 104, y_o = 5 (initially infected persons on the uninfected population at t = 0.) Time in weeks.

2.5 Chapter Summary

In this chapter, we have introduced the concepts of complexity and CSs and illustrated how human sustainability is dependent on many interrelated CSs. Much of the difficulty we humans have with working with CSs comes from the fact that the behavior of certain CSs depends on information-driven human behavior. The general properties of CSs were described, including chaos and tipping points. The distinction between complex and complicated systems is often arbitrary and fuzzy. CASs were introduced. We described the well-known LUC, as well as the SARSL, the "single-cause" mentality, and the "not in my box" mentality.

In order to model CNLSs, the concept of directed graphs was introduced. Large linear systems can easily be modeled by SFGs using Mason's rule. The organization and topology of nonlinear systems can also be clarified using the directed graph format, although Mason's rule cannot be used to solve for their I/O characteristics; computer simulation must be used.

Modularity in CSs was introduced, and it was shown how the analysis of graph modules can often expedite our understanding of the CSs found in sustainability studies. Our understanding of CSs depends heavily on our models of them, which are often imperfect and approximate. The unique behavior of a CS model may be lost if too many shortcuts, approximations, or simplifications are made to streamline simulations.

3

Multidimensional Challenges
to Human Sustainability

3.1 Introduction

One hears debates today on what is the most serious, current challenge to human sustainability. Actually, there are many important factors affecting human *un*sustainability, and it is difficult to rank them by their gravity. Most adverse factors are slowly acting; their effects at present are felt at a rate that it easy for people to adapt to or even ignore. These factors also interact, which makes it hard to separate their effects on unsustainability. They include, but are not limited to the following: (1) population growth; (2) loss of freshwater due to climate change, overuse, and pollution; (3) famine, resulting from decreased food production due to water shortages, insects, crop diseases, and so forth; (4) economic collapse following inflation and mass unemployment, leading to governmental crisis; (5) exhaustion of fossil fuel (FF) energy sources leading to inflation; (6) failure to develop alternate energy sources in a timely manner; (7) loss of arable land due to overfarming and drought; (8) profligate, anthropogenic greenhouse gas (GHG) emissions accelerating global climate change; (9) human migrations; (10) conflict; and (11) pestilence (affecting humans, domestic animals, ecosystems, and plants). Pestilence and death also lead to decreased food production, starvation, and economic collapse, a positive-feedback loop.

We have ranked human population growth as the *primary challenge* to human sustainability: it is the slow, root cause of most factors that are now threatening human sustainability. The *number two challenge* may be freshwater shortages. Water is very important for human consumption, human waste disposal, agriculture, manufacturing, electric power generation, mining, and so forth. City life would be impossible without it. Famine could be the *number three challenge*. Famine results from the food supply being inadequate for the nutritional demands of the local population (both in calories and nutrition). An inadequate food supply can be caused by crop failures (even genetically-modified (GM) crops), which, in turn, can be caused by drought, heat, cold, insect or rodent infestation, plant viruses and fungi, river flooding, coastal flooding, storms, hail, and so forth, as well as loss of arable land to desertification, soil exhaustion, and soil erosion. Loss of fisheries through overfishing, pollution, and water temperature and pH changes also affects available food. An inadequate food supply causes inflation of food prices and, eventually, mass migrations and conflict.

For an in-depth, introductory overview of the interconnectivity and many feedback loops affecting ecological and environmental systems, the interested reader should see Chapter 6 in the text by Brennan and Withgott (2007).

3.2 The Challenge of Population Growth

3.2.1 Introduction

The world's population has grown from approximately 1 million around 8000 BCE at the dawn of agriculture to a plateau in 1 BCE of approximately 200 million. Due to pestilence and disease, the world population dropped approximately 50% between 541 and 800 CE. It grew again erratically after 800 CE because of epidemics. The population was approximately 350–375 million by 1400 CE and fluctuated between 1300 CE and 1600 CE because of bubonic plague and cholera epidemics. China's population at the founding of the Ming dynasty in 1368 was approximately 60 million; it approached 150 million by the end of the dynasty in 1644 CE.

In 1900, regional populations were as follows: Africa, 133 million; Asia, 946 million; Europe, 408 million; Latin America and Caribbean, 74 million; North America, 82 million; total, 1.643 billion. The 2010 world population was approximately 6.884 billion. Sometime in March 2011, the world's population exceeded 7 billion humans (US Census Bureau 2012).

Figure 3.1 illustrates the growth of the world's population since 10,000 BC. Note the log-linear (power-law) growth beginning *circa* 4000 BC. The small oscillations superimposed on the curve between 1300 and 1700 AD are from the massive pestilence outbreaks (e.g., bubonic plague and cholera) in crowded, unsanitary European and Asian cities. Note that the overall curve appears to have several log-linear segments. The so-called world "population explosion" is clearly shown on the logarithmic scale, beginning about 1950.

Population growth is the engine that today directly or indirectly drives most of the challenges to human sustainability. The rate of population growth varies widely over the globe, generally being larger in undeveloped, underdeveloped, and developing regions and in countries where the effects of high population densities challenge ecosystems, economic systems, and social order. The population growth rate is the difference between the birth rate and the death rate at a given time (Bloom 2011). Figure 3.2 illustrates the fertility rates of countries averaged over the period 2005–2010 (CIA 2012).

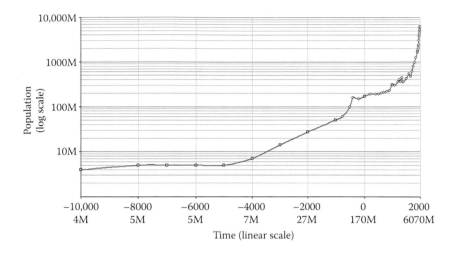

FIGURE 3.1
World population growth since 10,000 BC. Note: vertical scale is logarithmic. (Courtesy of Waldir, Wikimedia Commons. http://en.wikipedia.org/wiki/File:World_population_growth_ (lin-log_scale).png.)

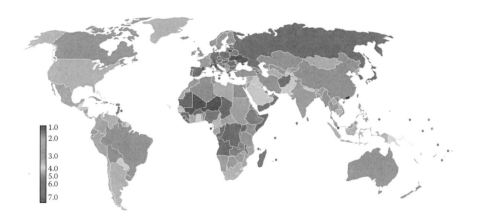

FIGURE 3.2
(**See color insert.**) Map of fertility rates of countries of the world (2005–2010). (Courtesy of Supaman89, Wikimedia Commons. http://en.wikipedia.org/wiki/File:Countriesbyfertilityrate. svg.)

Table 3.1, based on United Nations (UN) data, shows how world total fertility rate (TFR) has varied over 5-year intervals from 1950 to 2010. TFR is extrapolated through 2100.

This table shows the statistical fertility of an imaginary, typical woman who passes through her reproductive life subject to all the age-specific fertility rates for ages 15–49 that were recorded for a given population in a given year. The TFR represents the mean number of children a "typical" woman

TABLE 3.1

World Historical and Predicted TFRs
(1950–2100)

Years	TFR
1950–1955	4.95
1955–1960	4.89
1960–1965	4.91
1965–1970	4.85
1970–1975	4.45
1975–1980	3.84
1980–1985	3.59
1985–1990	3.39
1990–1995	3.04
1995–2000	2.79
2000–2005	2.62
2005–2010	2.56
2010–2015	2.43
2015–2020	2.19
2020–2025	2.10
2025–2030	2.02
2030–2035	1.95
2035–2040	1.82
2040–2045	1.74
2045–2050	1.67
2050–2055	1.61
2055–2060	1.54
2060–2065	1.47
2065–2070	1.38
2070–2075	1.32
2075–2080	1.27
2080–2085	1.25
2085–2090	1.24
2090–2095	1.24
2095–2100	1.23

Source: UN, 2010, UN Data: Total fertility rate (children per woman). Available at: http://esa.un.org/unpd/wpp/p2k0data.asp/, accessed November 14, 2011. Medium variant.

would have were she to fast-forward through all the childbearing years in a single year, under all of the age-specific fertility rates for that year. It may also be viewed as the number of children a woman would have if she were subject to prevailing fertility rates at all ages from a single given year, assuming she survives through all of her childbearing years. (Note: the US TFR was 2.1 in 1970–1974, the same as in 1930–1934. The highest regional TFR in 2000 was in Mali, at 6.89.)

Note that the TFR varies from country to country and region to region. Many factors affect fertility. Figure 3.3 illustrates how world TFR varies with gross domestic product (GDP) *per capita* (in 2004 US $), computed by the US Central Intelligence Agency (CIA). Note the steep decrease in TFR as GDP increases, generally reaching a level of approximately 1.25 for country GDP > $15,000.

The UN projected that the world population will stabilize at 9.3 billion in 2050 (UN 2010). This prediction assumed a decline from the 2005–2010 estimated world total fertility rate of 2.56 children per woman to 2.02 children per woman in the years between 2045 and 2050 (Gaia 2010; Bloom 2011; Lee 2011). If mothers average 0.5 child more after 2045, the world population may peak at 10.6 billion in 2050. If there is 0.5 child less after 2045, the world population will stabilize at approximately 8.1 billion in 2050. There are lots of "ifs" in the UN models. The current world TFR rate averages 2.56 today; it was 4.95 in 1950! China, the country with the world's largest population, recently released data from its 2010 national census. The total population in Mainland China is approximately 1.34 billion. The population growth rate was down to 0.57% in 2000–2010, compared to the rate of 1.07% in 1990–2000. The estimated current Chinese TFR rate is 1.4, well below the 2.1 considered to be the "replacement rate." It was 5.8 in 1950.

The slower Chinese population growth is generally matched by a shift in the population age distribution; people over 60 now represent 13.3% of the

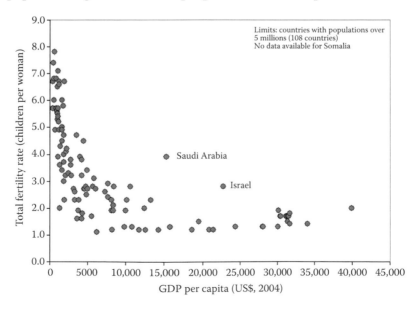

FIGURE 3.3
Graph of total fertility rate versus GDP per capita. Note that as GDP per capita rises, TFR falls. The outliers are Saudi Arabia and Israel. (Courtesy of the US CIA World Factbook.)

population, up from 10.3% in 2000. In the same period, the percentage of persons under 14 declined from 23% to 17%. Among newborns, there are now only 100 girls for every 118 boys. Male child preference, coupled with the one-child policy and the availability of prenatal gender determination by ultrasound, encourages sex-selective abortions (Economist 2011d).

What has happened to the Chinese population age distribution is the phenomenon of *global aging*. It is the consequence of countries reducing their TFRs. Global aging is a shift in the world population's age histogram toward the high-age end. Specifically, it can be seen as the proportional rise in the over-65 population with time. This preponderance of elderly citizens is made possible by global improvements in health care (decreasing the elderly mortality rate) as well as a sharp decrease in birth rate, and it translates into a decrease in the proportion of working-age population. The global aging phenomenon appears to be more advanced in developed countries, while many less developed countries maintain a higher TFR. Proportionally fewer workers will impact the economic sustainability of countries. Just how much the proportional decrease in the working-age population of various developed countries will affect their economies and social well-being remains to be seen (Hayutin 2007). Some of the countries that will be more affected in time by the global aging phenomenon are the following: the European Union (EU) countries, Russia, China, South Korea, Japan, and to a lesser degree, Canada and the United States. Hayutin argues, "The potential for unrest is greatest in Africa and the Middle East, where the economies are unable to support predominantly young populations."

Figure 3.4 shows the population density (PD; number of persons per square kilometer) of countries in the world as of 2007. Note that India's PD exceeds China's PD.

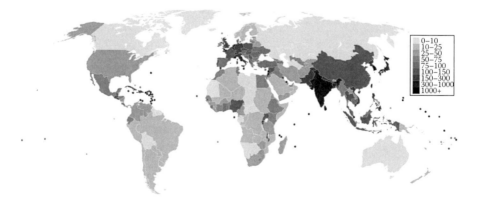

FIGURE 3.4
Map of countries by PD (as of 2007). (Courtesy of Contreras, M., Wikimedia Commons. http://en.wikipedia.org/wiki/File:Countries_by_population_density.svg.)

Many early thinkers and writers were aware of the adverse effects of over-population, including Plato (427–347 BC), Aristotle (348–322 BC), Confucius (551–478 BC), Tertullian (160–220 AD), Giovanni Botero (1540–1615 AD), Richard Hakluyt (1527–1616), and of course, the best known, Thomas R. Malthus (1766–1834) (see Simpkins 2012 for a biography and bibliography of T.R.M.).

Thomas R. Malthus is widely known for his theories concerning population and the factors determining its increase or decrease. In his famous *An Essay on the Principle of Population*, Malthus observed that sooner or later, a population gets checked by *conflict (wars), famine, disease*, and *widespread mortality (death)* related to poverty. These checks are, in fact, the modern Four Horsemen of the Apocalypse. Malthus observed that there are two types of checks that hold populations within resource limits (carrying capacity): *positive checks*, which raise the death rate (i.e., the Four Horsemen, with poverty), and *preventative checks*, which lower the birth rate (Gaia 2010). The short list of preventative checks included the following: birth control, abortion, prostitution,* celibacy until marriage, and strangely, raising the standard of living (income, education), which, in addition to decreasing the birth rate, also decreases infant mortality.

Some essays on the effects of overpopulation on human sustainability have been written by Bartlett (1997), Hardin (1968, 1972), Ehrlich and Holdren (1971), Ehrlich (1995, 2010), and Ehrlich and Ehrlich (2009, 2010).

3.2.2 Effects of Overpopulation

A succinct list of the direct and indirect effects of human overpopulation necessarily must include, but is not limited to, the following:

- *Inadequate freshwater:* Demand exceeds supply for drinking, sanitation, agriculture, industry, and so forth (Shiklomanov 2000).
- *Depletion of natural resources:* Natural deposits of FFs, metal ores, minerals, and so forth are becoming exhausted and more expensive to extract, fuelling inflation. The cost of developing new sources of FFs is rising (Hubbert 1982).
- *Increased environmental pollution:* Of air, water (including aquifers, rivers, lakes, and seas), and soil. Pollution is from mining, industry, manufacturing, fertilizers, pesticides, and animal and human waste.
- *Deforestation:* Wood is required for building and fuel. Cleared land is used to grow crops. Deforestation leads to loss of ecosystems, species, and soil erosion, as well as reduced CO_2 uptake (Wilson 2002).

---

* Prostitution can be a mechanism for spreading STDs, including HIV. Thus, one might consider it a positive check, as well.

- *Increase in GHGs and associated climate effects:* From burning FFs to make electric power (energy demands grow with population); also from burning forests for swidden agriculture (Tainter 1988).
- *Loss of arable land:* From overfarming, climate change, erosion, desertification (UNEP 2012).
- *Mass species extinctions:* From overfishing, bad forestry practices, climate change, and habitat destruction (Leakey and Lewin 1996; Pimm et al. 1995).
- *Poverty:* From unemployment, inflated food and energy prices, loss of investments; further rise in wealth inequality.
- *Malnutrition:* Malnutrition is a direct result of population-driven poverty and food shortages.
- *Inflation:* Driven by the demand for dwindling food supplies and energy sources.
- *High infant and child mortality:* Correlated with poverty, malnutrition, and lack of water.
- *Raised death rate in countries with the highest population densities:* A lack of freshwater leads to epidemics and pandemics.
- *Elevated crime rate:* People stealing to survive, drug cartel wars. Rise of petty crimes, including the thefts of agricultural products from farms for food or resale (Semuels 2011), and the thefts of valuable metals (e.g., copper) from utilities and public and private places.
- *More restrictive laws, loss of personal freedom:* Reaction to elevated crime rate, civil unrest.
- *Wars over scarce resources:* Over the remaining FF deposits, water, pipeline routes, cropland.

This list can be expanded and elaborated on. The point to remember is that there is a dense, parametric connectivity between the local PD nodes and all aspects of human sustainability.

Paul Ehrlich's central argument on overpopulation (from his 1968 book, *The Population Bomb*) is as follows:

> "A cancer is an uncontrolled multiplication of cells; the population explosion is an uncontrolled multiplication of people. Treating only the symptoms of cancer may make the victim more comfortable at first, but eventually he dies—often horribly. A similar fate awaits a world with a population explosion if only the symptoms are treated. We must shift our efforts from treatment of the symptoms to the cutting out of the cancer. The operation will demand many apparent brutal and heartless decisions. The pain may be intense. But the disease is so far advanced that only with radical surgery does the patient have a chance to survive."

In his final chapter, Ehrlich concludes that a partial solution to the "population bomb" would be "compulsory birth regulation... [through] the addition of temporary sterilants to water supplies or staple food. Doses of antidote would be carefully rationed by the government to produce the desired family size."

Such a draconian, "brave new world" solution imposed by a government would never be accepted at present in democracies; the "temporary sterilants" would probably be synthetic steroid hormones with many untoward side effects on humans as well as on domestic and wild animals that accidentally ingest it. They are the sorts of chemicals we now pay extra money in a supermarket to avoid having in our food.

Another effect of overpopulation seen at the time of this writing, particularly in India and other Asian and African countries, will continue and worsen as the world's population continues to grow. This effect is inflation in food prices, an obvious result of limited supply and increased demand. If one factors in food shortages caused by weather (droughts, floods, etc.), pests (plant viruses, insects particularly those resistant to insecticides, rodents, etc.), and loss of pollinators, we will see further decreases in food supplies per capita, driving more food price inflation. See the interesting food price modeling papers by Lagi et al. (2011, 2012).

Some economists [Thomas Sowell (Hoover 2010) and Walter E. Williams (Econfaculty 2010)] have argued that poverty and famine are in fact caused by bad government economic policies, not by overpopulation. Poverty and famine are complex social phenomena with many nodes and branches. Certainly, bad government economic policies contribute to poverty and famine, but there are a plethora of other causes, with population growth being the most important in our view. Government policies and laws are made by fallible politicians who are ill-equipped to deal with complex issues and often focus on short-term economic goals and status-preserving decisions.

Economist J.L. Simon (1998), in his book, *The Ultimate Resource 2*, argued that a *higher population leads to more specialization and technological innovation, which in turn leads to a higher standard of living*. This argument can be refuted by consideration of the *law of diminishing marginal returns* (Tainter 1988). For example, if we plot "standard of living" and population versus time for a country, state, region, or city, we see that at first, the standard of living rises, reaches a peak, and then begins to fall as the population increases further. This is because, in effect, the available technology for sustainable living saturates—a limit is reached where technical innovation, financial investments, and social and energy inputs by the increasing population can no longer produce effects that can correct for the unsustainable effects of overpopulation. Increasing inflation and unemployment generally follow (i.e., stagflation), and then loss of tax revenues, leading to economic and social collapse.

3.2.3 Mitigation Measures for Human Overpopulation

> The paradox imbedded in our future is that the fastest way to slow our population growth is to reduce poverty, yet the fastest way to run out of resources is to increase wealth (Gaia 2010).

We have argued that overpopulation is one major threat to human sustainability on the planet. It is important to address solutions for this problem. One obvious means is to reduce the birth rate. The other is to increase the death rate. We will address the latter action first.

Clearly, as a moral and ethical imperative, we as humans cannot and must not actively increase the human death rate in order to check population growth. However, there is a sad history of human infanticide that stretches from prehistoric Paleolithic and Neolithic times through the golden age of Greece to the early Persian Empire, and even to the present epoch, the Anthropocene time. Infanticide has occurred in nearly all ancient societies studied, including, but not limited to, Arabia, Babylonia, Carthage, China, Egypt, Greece, Hawaii, the Inca Empire, India, Innuit, Japan, Native American tribes, Rome, Russia, and Syria. It even occurred in early Colonial America. Recently, it has been documented in Papua New Guinea, the Solomon Islands, Benin (West Africa), rural India, China, and so forth (Milner 1998; Karabin 2007; Mohanty 2012; IRIN 2005; Papua 2008). Milner stated: "While there are many diverse reasons for this wanton destruction, two of the most statistically important are poverty and population control. Since prehistoric times, the supply of food has been a constant check on human population growth. One way to control the lethal effects of starvation was to restrict the number of children allowed to survive to adulthood." In addition to the obvious check on population growth versus resources for a society, it appears that in pre-20th-century Hawaii, infanticide was practiced on babies with obvious birth defects or born by breech delivery (Elkin 1902). In 21st century Northern Benin, unless a baby (of either sex) is born head first and face upward (normal birth), many communities believe that the child is a witch or sorcerer. Tradition demands that the baby must be killed. Minor developmental anomalies also can condemn a Benin child, for example, if its first tooth appears in the upper jaw (IRIN 2005).

Wars have always have had a dampening effect on population growth. For example, consider the massive death tolls in the First World War (Strachan 2003). Population numbers have always rebounded after major wars, however. Wars are evil and should be avoided, and we earthlings are hopefully progressing beyond the use of weapons of mass destruction (WMDs) and the eras and genocidal acts of Hitler (Nazi Germany), Stalin [Union of Soviet Socialist Republics (USSR)], Tojo (Japan), Pol Pot (Cambodia), Jean Kambanda (Rwanda), and Saddam Hussein (Iraq).

It is a certainty that an increase in the global death rate will occur naturally as the result of overpopulation. Famine, water shortages, and decreased standard of living lead to physiological stress, starvation, weakened immune systems, and a rise in contagious pestilence [e.g., bubonic plague, cholera, yellow fever, Ebola virus, smallpox virus, Marburg virus, and STDs (syphilis, HIV/AIDS), etc.]. Bacteria and viruses in particular have the propensity to mutate relatively rapidly and thus avoid the adaptive human immune system and the toxic effects of the antibiotic *du jour*. Consider the origin of MRSA. The *Four Horsemen of the Apocalypse* (cf. Glossary) are the natural "population control" for humanity. In our view, the modern horsemen represent wars, famine, pestilence, and death.

Population control by decreasing the birth rate has been controversial, generally on religious, faith-based grounds. It is, however, something we can address proactively. The major opposition to population control by abortion, sterilization, and contraception is made by the Roman Catholic Church. Pope Benedict XVI stated: "The extermination of millions of unborn children, in the name of the fight against poverty, actually constitutes the destruction of the poorest of all human beings" (Vatican 2009). However, people living on the edge of poverty do not need extra mouths to feed and the expenses required to properly raise and educate more children. If a child has not yet been conceived, he or she cannot be "exterminated." The education of women in poor countries about the basics of human reproductive physiology and the use of basic contraceptive measures (abstinence, condoms, pessaries, IUDs, and "the pill") are strategies that must continue if population growth is to be slowed.

Male sterilization by vasectomy (tying off or cutting the two *vas deferens* sperm ducts) is a more expensive, invasive approach to birth control that is unlikely to be seen in underdeveloped countries. A forced-sterilization program was implemented by Indian Prime Minister Indira Gandhi in the 1970s. The Indian government required that men with two or more children had to submit to sterilization (vasectomy) (Times 2011; Guardian 2012). By 1973, over 7 million vasectomies were performed in India due to propaganda and cash initiative schemes (Vasectomy 2008). If contraception is effective, there should be no need for abortion, the most controversial birth control procedure.

Worldwide, it has been observed that countries with the higher standards of living have lower birth rates than do poor, underdeveloped countries. One reason for this inequality may be the fact that education about reproductive physiology and the use of contraceptives is not readily available in underdeveloped countries for practical and/or faith-based reasons. Such education should be compulsory for *all middle-school students* (and their parents) in *all countries* if we are to mitigate the rate of world population growth. Quoting Hardin (1968):

> "... it is the role of education to reveal to all the necessity of abandoning the freedom to breed. Only so, can we put an end to this aspect of the tragedy of the commons."

A third means of downsizing a regional population is by the emigration of refugees. When the carrying capacity of a region is exceeded, as, for example, by crop failure and famine, people can embark for "greener pastures." A classic example is the great Irish potato blight and famine of 1845–1852. Many Irish farmers, tradesmen, and workers pulled up stakes and sailed for the eastern US cities of Boston and New York, where they could find employment and food. Over 1 million people died in Ireland as a direct or indirect result of this famine, which had roots in a potato monocrop plant pestilence as well as British government policies (Donnelly 2011; History Learning 2012; Kinealy 1995). In 2011, hundreds of thousands of refugees in Sudan were fleeing famine from drought-caused crop failure and conflict and were marching to South Sudan, Ethiopia, and elsewhere in search of food and water in refugee camps (Reuters 2012).

3.2.4 Population Growth in Ecosystems

Healthy ecosystems are necessary for global sustainability. Nature has devised a number of intrinsic, feedback mechanisms to balance animal and plant populations in ecosystems at sustainable levels. These negative-feedback mechanisms can be broadly divided into two categories: density-independent checks and density-dependent checks on population growth (Kimball 2011).

Density-independent checks on population growth include mortality due to weather anomalies (droughts, freezes, hurricanes, floods, and tsunamis), fires, anthropogenic habitat loss, and so forth. Habitat losses affect available food and living space; they come from human encroachment into a habitat or ecosystem, such as deforestation and filling in or polluting marshes and swamps.

Density-dependent checks on population growth can be due to several factors, including the following:

- *Intraspecies competition:* High population densities of a species can lead to depletion of its food (energy) source(s). This in turn can lead to famine, premature mortality, and loss of reproductive ability. The population crashes, to slowly recover, as does the food source(s).

- *Interspecies competition:* Two species that share the same finite food sources, nesting sites, place in the sun (plants), and so forth will limit each other's population growth to levels lower than that of either species alone in the same ecosystem.

- *Reproductive competition:* Crowded living conditions can cause the reproduction rate of a species to decline. Examples are egg-laying reduction in fruit flies, increased infant mortality in rats due to decreased maternal care and cannibalism, and honeybee queens reducing their egg-laying rate when food (flower pollen and nectar) becomes scarce.

- *Migration:* Some species will emigrate from a high-PD habitat to one with lower PD, hence more food.
- *Predation:* As a PD grows, predators are able to harvest it more easily. Mathematically, this can be stated by the simple ODE: $\dot{\mathbf{a}} = -k_1\mathbf{ap} + k_r\mathbf{a}$, where $\mathbf{a}$ = animal population, $-k_1\mathbf{ap}$ = rate of animal loss due to predator density ($\mathbf{p}$), and $+k_r\mathbf{a}$ = rate of animal reproduction ($\mathbf{p}$ assumed constant).
- *Parasitism and epidemics:* The same ODE as above applies, except $\mathbf{p}$ is parasite density whose rate of growth is simply $\dot{\mathbf{p}} = +k_r\mathbf{ap} - k_1\mathbf{p}$. Parasites and infectious bacteria are able to propagate more rapidly under high host PD conditions. This is seen for human populations (e.g., plague, cholera, dysentery, flu) as well as animals and plants. Fish-farmed salmon and cod are raised in high-density pens and are liable to have fish louse infestations and certain endoparasite infections.

All ecosystems/environments have an equilibrium *carrying capacity* for a species. This is easily illustrated by an experiment where a small number of bacteria are inoculated into a flask of nutrient broth, from which a PD versus time plot is created. The result is invariably a sigmoid or S-shaped curve (the logistic growth curve) that can be simply modeled by the single nonlinear ODE:

$$\dot{\mathbf{b}} = r\mathbf{b} - r\mathbf{b}^2/K \tag{3.1}$$

where $\mathbf{b}$ = bacterial density, and r and K are nonnegative constants, given $\mathbf{b}(0) = \mathbf{b}_o$.

Zero population growth rate is reached when $\mathbf{b} = K$. If K is then sharply reduced, $\mathbf{b}(t)$ will decline. Simple calculus tells us that $d\mathbf{b}/dt$ is maximum when $\mathbf{b}(t) = K/2$. $K/2$ is called the *maximum sustainable yield.*

Kimball (2011) calls organisms that have a high-growth-rate strategy in order to fill their habitat and avoid competition *r*-strategists. *r*-strategy also can compensate for high egg, seed, and infant mortality in a species. Kimball stated: "When a habitat becomes filled with a diverse collection of creatures competing with one another for the necessities of life, the advantage shifts to *K*-strategists. *K*-strategists have stable populations that are close to K [in Equation 3.1 above]. There is nothing to be gained from a high *r*. The species will benefit most by a close adaptation to the conditions of its environment."

Certain animal populations go through steady-state, predator–prey population limit cycles. The best-known examples of these oscillations are the approximately 10-year period cycles in the populations of lynxes and snowshoe hares in northern Canada. The peaks in the hare cycles lead the peaks in the lynx cycles. Other animals whose populations cycle are voles and their

predators in Finland (3-year cycles); red grouse and their nematode parasites in Scotland (4- to 8-year cycles); and lemmings and their predators in
northeast Greenland (4-year cycles). The causes for population cycling are
complex, but delays in reproduction or reestablishment of prey food sources
can be involved (cf. Northrop 2011, Sections 3.4.3 and 3.4.4).

3.3 Global Warming

Probably no other area of earth science has received so much attention in the
past 20 years as the study of global warming (GW). GW is not just about a
global, 2°C or 3°C mean temperature rise but also about the myriad of effects
this warming will have on planetary weather, ecosystems, the environment,
and food and water supplies, and hence human sustainability. Droughts,
severe storms, ocean rise from melting polar land ice caps and glaciers, ocean
ecosystem degradation due to falling pH and rising water temperatures,
freshwater shortages, decreases in food production rates, and so forth are
all expected. In 2011, the UN World Meteorological Organization's (WMO)
deputy director, R.D.J. Lengoasa, presented data in climate talks in Durban,
South Africa, that underscored the warming phenomena. He said that the
warmest 13 years of average global temperatures have all occurred in the
15 years since 1997. The year 2011 has been one of extreme weather; drought
in East Africa has left tens of thousands dead, and there have been deadly
floods in Asia and 14 weather catastrophes in the United States with damage
topping $1 billion each. The WMO said that Arctic sea ice volume in 2011 was
the lowest in recorded history, and its coverage was second lowest on record.

The largest temperature departure from the norm in 2010 occurred in
northern Russia, where temperatures reached an average of 7.2°C above
average in some locations, and some stations reported temperatures 16°C
above normal in the spring (NOAA 2011b). The National Oceanic and
Atmospheric Administration (NOAA) attributed this heat event to "atmospheric natural variability." Statistically, the Russian heat event may be one
of the first such events in a future climate pattern being established globally.
Taken alone, it is not part of such a pattern, hence the NOAA attribution.
The Russian heat wave arrived with little precipitation, low wind speed, and
serious forest and peat bog fires. The fires put a high concentration of carbon
monoxide and smoke in the air, contributing to the deaths of approximately
56,000 persons (Friedrich and Bissoli 2011). A detailed analysis of this 2010
Eastern European heat event can be found in the detailed online paper by
Barriopedro et al. (2011). These authors show that there are many climato-
logical factors that contributed to the heat wave, including decreases in soil
moisture, reduced relative humidity, and "enhanced variability of surface
net radiation."

GW is real; physical measurements of air, ground, and ocean temperatures; the use of satellites carrying infrared (IR) sensors to measure land and air temperatures; and satellite surveillance (visible and IR) of melting glaciers and shrinking polar ice caps have all contributed to indisputable knowledge of the fact that the mean temperature of the Earth has been slowly rising over the last 110 years. A graph of "reconstructed mean Earth temperatures" synthesized from 10 different published reconstructions of mean temperature changes over the last 800,000 years illustrates that the Earth has shown major, natural fluctuations in mean temperature (ice age to warm period) about twice per 100,000 years. Figure 3.5a and b illustrates these reconstructed temperature trends. Figure 3.5a illustrates reconstructed mean Earth temperatures for the past 800,000 years. Figure 3.5b shows more recent mean temperatures from 500 AD to 2008 AD that include our ongoing GW trend. The darker temperature record that shows the present GW trend is the result of physical measurements rather than reconstructions. There

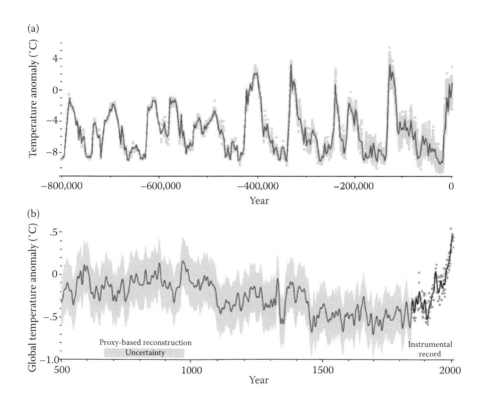

FIGURE 3.5

(a) Reconstructed mean Earth temperatures, covering 800,000 BP to 2008. (b) Reconstructed mean Earth temperatures, covering 500 BP to 2008. The dark trace denotes temperatures measured by IR sensors from space and/or surface thermometers. (Courtesy of NASA. http://earthobservatory.nasa.gov/Features/GlobalWarming/page3.php.)

was a so-called Medieval Warm Period from about 900 to 1200 AD. Fairly normal temperature variations followed for about 200 years, and then there was a "Little Ice Age" (LIA) from about 1400 to 1800 AD. After about 1900 AD, the mean planetary temperature has risen steadily.

The reason the Earth's mean temperature is currently rising is that the net planetary heat balance is positive; the Earth now is absorbing more heat than it can radiate into space, hence the late 20th- and 21st-century average temperature rise. The mean planetary temperature rises until at some higher temperature, the blackbody (BB) radiation of long-wave infrared radiation (LIR) energy into space from the planet's surface balances the net input radiation, and an equilibrium temperature (T_2) is reached.

There are two major sources of heat energy on the Earth's surface: heat from the molten core of the Earth and heat from incident solar radiation (insolation). Some heat also comes from the combustion of fuels and the decay of vegetation (oxidation). Heat on Earth is stored in (and released from) gasses in the atmosphere [including water vapor (WV)], the oceans (70% of the Earth's surface) and other bodies of water, and the landmasses. Planetary heat is lost only through radiation into space. The National Aeronautics and Space Administration (NASA) claimed that over the time interval 1979–1999, satellite measurements showed that the insolation varied no more than 0.2% [around 1380 J/(s m²)] (Sun 2004).

By the Stefan–Boltzmann law, the total heat power per square meter radiated into space in a broadband photon spectrum from an ideal BB at T_e Kelvin is

$$W_{bb} = \varepsilon\sigma\left(T_e^4 - T_s^4\right) W/m^2$$

where $\sigma = 5.672 \times 10^{-8}$, ε is the _emissivity_ of the radiating surface ($0 < \varepsilon \leq 1$; $\varepsilon = 1$ for an ideal BB), and T_s is the equivalent Kelvin temperature of space.

The Earth's average emissivity depends on the degree of cloud cover, the area of high-albedo polar ice, and the concentrations of all GHGs in the atmosphere, including the ubiquitous WV. Note that the emissivity of the spherical Earth is a function of latitude θ and longitude φ, and thus $\varepsilon = \varepsilon(\theta, \varphi)$. T_e is the mean effective BB temperature of the Earth; also, $T_e(\theta, \varphi)$. T_s is the mean average temperature of the space surrounding Earth. At night, $T_e \gg T_s$; in the day, $T_e < T_s$. Thus, the positive incremental increase in the heat input to a BB in the day will result in a corresponding positive increment in the Kelvin temperature T_e of the BB, increasing the radiation power density lost into space at night, establishing a new equilibrium for temperature where heat input equals heat lost. This process is slow for the Earth because of the incredible thermal mass of the oceans and land and phase changes by moving water and WV. (Melting ice absorbs heat; however, the condensation of WV to rain or snow releases heat.) The Earth's rotation on its axis further complicates the analogy of the Earth as a BB engaged in the input/output of radiant energy. The dark (night) side of the Earth is the radiator; in the day, the Earth receives energy from insolation.

What is currently being debated and politicized are the causes of GW and how fast and how much will occur. (Which mathematical model for climate change do we believe to be more valid?) One cause for GW is the anthropogenic and natural emission of GHGs into the atmosphere, the arch villain of which (by concentration) is carbon dioxide, followed by methane, nitrous oxide, ozone, chlorofluorocarbons, and so forth. Omnipresent in large quantity is the IR-absorbing gas, atmospheric WV (AWV). AWV has always been present since the planet first developed oceans. Past ice ages and GW periods have occurred in spite of the omnipresent AWV. All of the GHGs absorb incoming LIR in unique, discrete wavelength bands. This absorbed LIR energy heats the GHGs, and they reradiate LIR energy both up into space and down to the Earth's surface, which in turn radiates into space, albeit at a lower BB temperature, T_e, at night.

The measurable rise in planetary temperature is highly correlated with the atmospheric CO_2 concentration ([CO_2]), a good percentage of which now comes from the combustion of FFs [coal, natural gas (NG), oil] used to make electric power and heat for industry and domestic uses. For the past 800,000 years, planetary atmospheric [CO_2] has undergone cycles with lows of approximately 180 ppm and peaks of around 275 ppm. The period of these natural oscillations was roughly 75,000 years (Yergin 2011). Starting in the late 19th century, global atmospheric [CO_2] began to rise (the industrial revolution was well underway) and shot up sharply beginning in the mid-20th century to its April 2012 value of 396.18 ppm at the Mauna Loa sampling station (NOAA 2012). This is approximately a 44% increase from the pre–industrial revolution (early 19th century) peak level of about 275 ppm.

Atmospheric methane is the second most important anthropogenic (and natural) GHG next to CO_2. Methane gas is about 21–23 times as effective a GHG in absorbing and retaining heat as is carbon dioxide (Boucher et al. 2009). Atmospheric methane concentration began to rise from a pre–industrial revolution (1700–1800) global average concentration of approximately 0.7 ppm (Forster et al. 2007) to the 2007 North Pole value of approximately 1.875 ppm (NOAA graph 2009). This is an increase by a factor of 2.68. (The southern hemisphere methane concentration [CH_4] is currently about 0.2 ppm lower than in the northern hemisphere. This may be due to the fact that the Arctic is warming faster than the Antarctic, releasing CH_4 from permafrost and coastal sea bottom methane clathrate deposits.) The local atmospheric methane concentration in Western Greenland on August 15, 2011, reached approximately 2.2 ppm (Webster et al. 2012).

The demonization of anthropogenic CO_2 emissions as the sole cause of GW is a good example of the "single-cause mentality" applied to a complex system. GW is not a simple process; there are other GHGs; natural heat storage and transport mechanisms in the atmosphere and oceans; and natural variations in the solar heat input to the atmosphere, land, and oceans. These variations come from reflecting particles in the upper atmosphere (e.g., from pollution, clouds, contrails), absorption and reradiation of LIR energy by GHGs, small increases or decreases in the sun's output energy flux, and changes in the Earth's orbital distance from the sun caused by its axis of rotation being inclined with respect to the plane of its orbit, which is not exactly circular.

Solar irradiance measured outside the Earth's atmosphere has a nearly pure BB Planck spectral curve. However, the direct solar irradiance spectrum on Earth at sea level is an attenuated BB curve with a number of "notches" taken out of it at various wavelengths by energy absorption by the atmospheric GHG molecules that the solar photons must pass through. These include significant absorption notches from AWV, CO_2, CH_4, O_2, and O_3. This energy absorption by these gasses heats the gasses and the atmosphere, which, through natural atmospheric movements, transports the heat to cooler parts of the globe. The warm GHGs also radiate long-wave IR back into space, and also to the ground and oceans, where some heat is stored. [A detailed but readable summary description of this complex, heat exchange process can be found in scienceofdoom (2011).] Heat is also transferred by ocean currents from warmer, equatorial regions to cooler northern and southern (polar) regions.

Unfortunately, the GW phenomenon is made more complex by the fact that the ubiquitous AWV can undergo phase changes, releasing heat when and where it condenses to water (as clouds, rain, fog) and absorbing heat when liquid water changes phase to AWV. When snow and ice melt to water, or sublime to AWV, they also absorb heat. Clouds have albedo, that is, they reflect incoming and outgoing radiant energy, as do microparticles of certain pollutants. As glaciers, snow packs, and polar ice flows melt, they absorb the heat of fusion of water and cool the air and water in their vicinity. The atmosphere and the oceans have currents that transport heat from equatorial to polar regions. The northern hemisphere jet stream flows from west to east. The factors determining the flow of these winds and currents (magnitude and direction) are complex. Recall the *La Niña–El Niño* current oscillations in the Pacific and their effects on continental climates and rainfall. A particularly strong *La Niña* event occurred in 2011. One scenario related to GW predicts that the influx of cold freshwater from melting Arctic ice flows will stop the Gulf Stream from carrying equatorial heat to Western Europe and, paradoxically, cause colder weather in the United Kingdom and Scandinavia. (This is a negative feedback for local GW.) Arctic warming is causing the release of methane from melting permafrost in tundra, swamps, and marshes, as well as from the sea bottom of the East Siberian Arctic Shelf. Methane is about 21–23 times as effective a GHG as is CO_2, so this process is a positive feedback for the GW process. As glaciers and ice caps melt, the seas will slowly rise, flooding low-lying coastal cities and islands. An extensive, scientific measurement-based literature on the causes, effects, and mitigations of GW can be found in four reports issued by the Intergovernmental Panel of Climate Change (IPCC) (2007a, 2007b, 2007c, 2007d).

The problem of what man can do about GW, its causes, and its effects on sustainability has been treated in the interesting book by David G. Victor (2011), *Global Warming Gridlock*. Victor discusses three major challenges to the mitigation of GW effects:

(1) The first is by cutting GHG emissions. A major way to effect this goal is to burn less FFs. This will be hard in the face of growth in global industrialization and population and the exploitation of newly discovered FF reserves. Economic growth and prosperity are intimately linked to the amount of FF energy a country burns. Fuel efficiency must be mandatory or it will not be implemented in the face of these other factors. Another way to cut GHG emissions is by using a significant fraction of our energy from noncarbon sources (solar, wind, wave, tidal, falling water, nuclear fission, and fusion).

(2) Technological innovation is required for fuel efficiency. Some of these engineering changes, such as automobile Corporate Automotive Fuel Efficiency (CAFE) standards, will come through government regulation; others will result from motivation inspired by empty consumer and industrial pocketbooks.

(3) Individual countries, states, and cities will have to adapt to the effects of GW; their taxpayers will carry the burdens. For example, certain coastal cities will have to build dikes, floodgates, and pumps to fend off rising oceans. The Netherlands has used dikes, floodgates, and pumps for many years. The city of New Orleans already has these in the United States. Other coastal US cities such as Boston, Providence, Miami, and New York City will need to act in the latter 21st century to gain protection from rising sea levels and storm surges.

Figure 3.6 illustrates the correlation between the global atmospheric CO_2 concentration (top) and the global mean temperature anomaly (increase from datum in 1958; bottom graph). The 52 years between 1958 and 2010 are considered.

Figure 3.7 shows that the global mean sea level change, considered from 1880 to 2006, is currently rising at a rate of approximately 3.2 mm/year, or 3.2 cm/decade. This rise is due to melting landmass ice, not ice floes. This melting will accelerate as melting glaciers expose more dark earth land surface, which absorbs LIR more effectively than snow or ice, and the rising temperatures will release the GHG methane, trapped in clathrate crystals in tundra and on the polar sea bottoms. One estimate of the total mass of methane trapped in Arctic permafrost as clathrate hydrates is 8×10^{11} tonnes (Economist 2012h). We can only guesstimate how much of this effective GHG will be released into the atmosphere as the Arctic warms and land ice cover melts. Methane is also being released from the coastal Arctic oceans' bottoms due to warmer sea currents. The current atmospheric methane concentration in the northern hemisphere is approximately 1.8 ppm and rising.

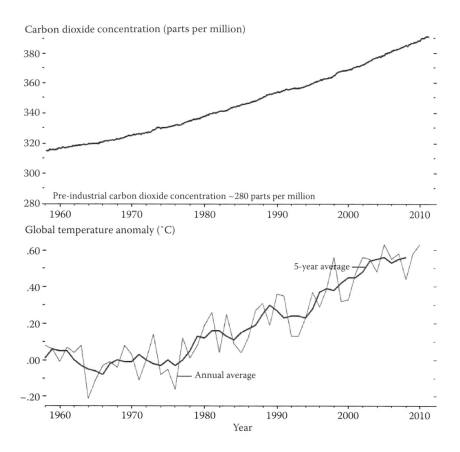

FIGURE 3.6
Apparent correlation between the rise in mean atmospheric CO_2 concentration (top graph) and the increase in the global temperature anomaly (0.00°C in 1958 to about 0.60°C in 2010). (Courtesy of Riebeek, H., The carbon cycle, NASA, 2011. http://earthobservatory.nasa.gov/Features/CarbonCycle/printall.php.)

By 2100, the mean sea level may rise as much as 1 m. Most of the sea volume increase will be from landmass (glacier) and ice melt, and a small part will be from water thermal expansion (e.g., the volume of 1 g of water at 0°C is 1.00000 cm³ and expands to 1.00177 cm³ at 20°C). The influx of cold, fresh, meltwater may disrupt heat-carrying ocean conveyor currents and alter ocean ecosystems. Arctic polar ice is melting at an unexpectedly high rate. Greenland's land ice sheet melting rate is estimated to currently be approximately 200 gigatonnes (Gt) per year (Economist 2012h). This melting represents a fourfold increase over the melt rate a decade ago. This melting process is autocatalytic, that is, as the ice melts, it exposes the dark ground beneath it, which is more effective at absorbing LIR radiation, which warms the nearby ice, causing accelerated melting. The mean temperature in Greenland has gone up by 1.5°C, compared to a 0.7°C increase globally. One reason for this

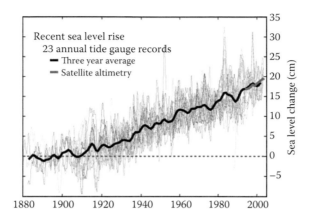

FIGURE 3.7
(See color insert.) Recent global mean sea level rise (1880–2006). (Courtesy of Waldir, Wikimedia Commons. http://en.wikipedia.org/wiki/File:World_population_growth_(lin-log_scale).png.)

anomalous rise in Arctic temperature may lie in the ocean conveyor currents carrying heat north through the Bering Strait from the Pacific Ocean and the currents through the Greenland and Barents Seas from the Atlantic.

In agricultural areas undergoing GW-induced drought, crops will be shifted from irrigation-requiring crops like maize and soybeans to xerophytes such as *Miscanthus* sp. and switchgrass for cellulosic ethanol (fuel) production and *Crambe abyssinica* as a biodiesel source. Genetic engineering is being used to make common crops drought resistant, such as corn (maize) (Schenkelaars 2007). More efficient use of crop water must be made, or the effects of drought will be compounded; no more spraying it into the air in the daytime.

3.4 Water and Sustainability

3.4.1 Introduction

Next to human population growth, freshwater resources have an enormous effect on sustainability. As the ancient Romans showed us, cities cannot exist without water. In addition, much of agriculture requires irrigation, industries require water (for steam generation, cooling, condensing, chemical reactions, etc.). Water is also required for the production of petroleum from oil shale and oil sands, production of natural gas (fracking), hydropower generation, fish farming, sewage treatment, etc., to mention some of its more important uses in modern society. Our freshwater comes from a variety of sources: lakes, rivers, shallow wells, deep wells tapping aquifers such as the Edwards and the Ogallala Aquifers in Texas, direct rainfall collection, and the desalination of brackish and seawater. Ultimately, all freshwater sources are dependent on rainfall, snow, and ice melt.

Just as there is a global carbon cycle, there is also a solar energy–driven water cycle (solar energy drives the whole water cycle). The total water volume on earth (ice, liquid, vapor reduced to water) is estimated to be 326 million mi.^{3} (1.36×10^9 km^3). Of this 1.36×10^9 km^3 total, approximately 1.29×10^4 km^3 is found in the atmosphere. The eight major "compartments" (nodes) in the Earth's water cycle include the following: (1) the atmosphere (as vapor, clouds); (2) the oceans (saltwater; about 1.338×10^9 km^3); (3) lakes (freshwater); (4) rivers (freshwater); (5) glaciers and ice caps (freshwater ice); (6) aquifers (freshwater); (7) ice flows (floating on polar oceans); and (8) the Earth's land surface (including plants and soil). Fluxes *into the AWV compartment* are from ocean evaporation; plant transpiration; evaporation from lakes, swamps, marshes, and surface soil; and WV from anthropogenic sources (burning fuels, etc.). We note that plants generally transpire much more water than they use in photosynthesis, in some cases, up to 98% more. Plants regulate their stomatal openings to balance their need for admitting CO_2 for photosynthesis and retaining water for necessary cell turgor. Higher atmospheric $[CO_2]$ leads to reduced stomatal opening, hence reduced water loss (McLaughlin 1988). The total estimated plant evapotranspiration is 5.05×10^5 km^3/year. Water flux *from the atmosphere* is from precipitation (rain, snow, sleet, hail, graupel, fog) on land and seas. Approximately 5.05×10^5 km^3 falls as precipitation each year (Water Cycle 2011). Flux into the oceans includes groundwater flow, river flow, and surface runoff. [For example, the Mississippi River puts, on the average, 5.18×10^2 km^3 of freshwater per year into the Gulf of Mexico (Corps of Engineers 2011)]. There is also freshwater input to the oceans when glaciers and ice flows melt. Figure 3.8 illustrates the salient, qualitative features of the Earth's water cycle, including sources and sinks.

The residence time (RT) for a water molecule is the mean time it will spend in a "storage compartment" (node). Some approximate RTs are as follows: atmosphere, 9 days; rivers, 2–6 months; lakes, 50–100 years; deep aquifers, 10^4 years; shallow groundwater, 100–200 years; oceans, approximately 3200 years; Antarctic ice cap, 20,000 years (some Antarctic ice is over 800,000 years old); glaciers, 20–100 years; soil moisture, (variable) 1–2 months; seasonal snow cover, 2–6 months (Pidwirny 2011).

The Earth's population is faced by a gradual, *per capita* shortage of water. This shortage appears to be driven by two factors: (1) a shortage of recharge water from rainfall and snowmelt in certain geographical areas related to planetary climate change and (2) an increase in all forms of water consumption that is driven by factors dependent on population growth, expanding agriculture, mining, and industrialization. We will address the first factor below.

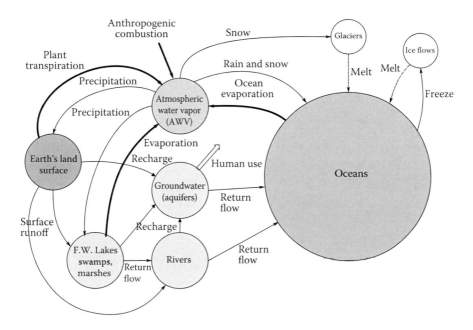

FIGURE 3.8
Global water cycle showing the various compartments, which, at any time, store water as vapor, liquid, or ice.

3.4.2 Drought and GW

While the extent of GW caused by anthropogenic sources is model dependent and debatable, the fact that it is happening is a measurable certainty (NOAA 2008a). Already, in the summer of 2012, a severe agricultural drought has devastated corn crops in the central United States. This will impact the prices of ethanol and human and animal food in general.

Drought is defined as "...a recurring extreme climate event over land characterized by below-normal precipitation over a period of months to years. Drought is a temporary dry period, in contrast to permanent aridity in arid areas" (Dai 2010). Dai classified drought into three types:

(1) *Meteorological drought* is below-normal precipitation that persists for months to years. It is often accompanied by above-average temperatures and precedes and causes other types of droughts. The causes of meteorological drought are complex and can involve shifts in ocean currents and sea surface temperatures (SSTs) that are the direct or indirect results of GW.

(2) *Agricultural drought* is a period with dry soils resulting from below-average rainfall and above-average evaporation. Agricultural drought leads to reduced crop yields, topsoil loss from winds, and the need to irrigate.

(3) *Hydrological drought* occurs when lake levels, river streamflow, and aquifer water storage all fall below long-term average levels. Water is depleted by evaporation and human use and is not replaced by precipitation.

To quantify drought in a region, various drought indices have been developed. These include the following: the *Palmer Drought Severity Index* (PDSI), *Standardized Precipitation Index* (SPI), *Rainfall Deciles* (RD), *Computed Soil Moisture* (CSM), *Palmer Moisture Anomaly Index* (Z-index), *Total Water Deficit* (S), *Palmer Hydrological Drought Index* (PHDI), *Surface Water Supply Index* (SWSI), *Drought Area Index* (DAI), and the *Drought Severity Index* (DSI). Dai (2010) gave a detailed description of these indices and their applications. Quoting Dai: "Recent studies revealed that persistent dry periods lasting for multiple years to several decades have occurred many times during the last 500–1000 years over North America, West Africa, and East Asia." Dai went on to say that these natural drought periods are linked to tropical SST variations in which La Niña–like SST events in the South Pacific often lead to widespread drought in North America, and El Niño–like SST warming in the Pacific caused droughts in East China. The recent Sahel droughts in West Africa were linked to the southward shift of the warmest SSTs in the tropical Atlantic and warming in the Indian Ocean.

Computer modeling of future worldwide drought conditions has led to a number of scenarios. While differing in some details, these models predict a worldwide, general loss of precipitation over the rest of the 21st century (UCAR 2010). Figure 3.9a through d illustrates the results of predictive global drought simulations based on 22 computer climate models. Dai's National Center for Atmospheric Research (NCAR)-sponsored studies predicted that most of the western two-thirds of the United States, including Texas, Oklahoma, Kansas, Iowa, and Nebraska, will be significantly drier by 2030. This will be bad for irrigated agriculture; food and ethanol production will decrease in this region in the face of a growing population and dwindling water supplies. Drought in this region will also directly limit NG and shale oil production. Other countries and regions that could see significant drought include the following:

- "Much of Latin America, including large sections of Mexico and Brazil
- Regions bordering the Mediterranean Sea, which could become especially dry
- Large parts of Southwest Asia
- Most of Africa and Australia, with particularly dry conditions in regions of Africa
- Southeast Asia, including parts of China and neighboring countries" (UCAR 2010)

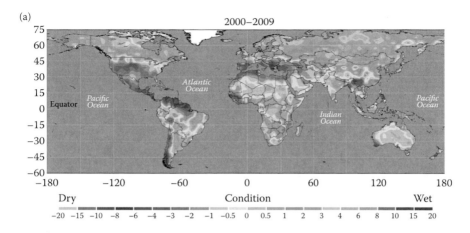

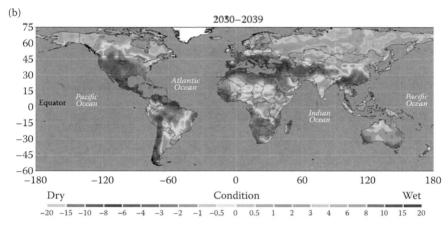

FIGURE 3.9
(See color insert.) Color maps of computer-generated model drought predictions. (a) 2000–2009 (present: model verification). (b) 2030–2039. (c) 2060–2069. (d) 2090–2099. Note that the American Heartland and all of Europe surrounding the Mediterranean Sea are predicted to have severe drought. (From Dai, A., NCAR/CGD, University Corporation for Atmospheric Research. With permission.)

The northeast United States (New England), the Canadian Maritime Provinces, and the northern Canadian provinces appear to be affected less by drought in the NCAR model's 2090–2099 scenario. Other areas expected to experience lesser drought include much of Northern Europe, Russia, and Alaska.

Cook et al. (2009) have pointed out that severe megadroughts have occurred in North America during the last thousand years, principally during the Medieval Climate Anomaly (MCA) period and into the early part of the LIA. The MCA was a significant climatic warming in the northern hemisphere that occurred between 900 and 1200 CE, followed by the LIA

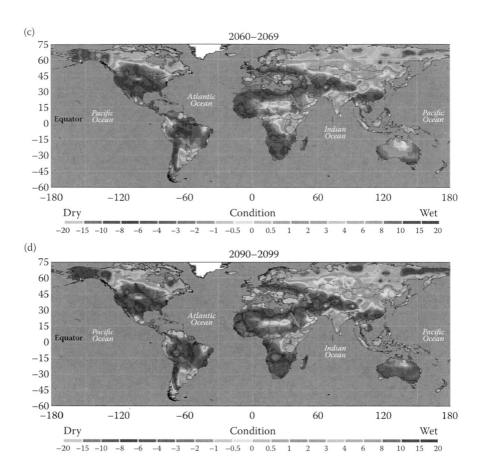

FIGURE 3.9

(Continued) (See color insert.) Color maps of computer-generated model drought predictions. (a) 2000–2009 (present: model verification). (b) 2030–2039. (c) 2060–2069. (d) 2090–2099. Note that the American Heartland and all of Europe surrounding the Mediterranean Sea are predicted to have severe drought. (From Dai, A., NCAR/CGD, University Corporation for Atmospheric Research. With permission.)

between 1400 and 1850 CE. Figure 3.5 illustrates a composite reconstruction of northern hemisphere temperature shifts from a number of proxy sources (e.g., tree rings, pollen, sediments, ice cores, etc.). Note that from the figure, the Medieval Warm Period was followed in approximately 200 years by the LIA, which has now turned into a GW event.

The megadrought events that occurred prior to the strong warming that began with the industrial revolution and the burning of FFs demonstrated that our planet has the capability of generating drought events on its own, which can last for several decades to hundreds of years or more. Well correlated to these events are the cool *La Niña*–like SSTs in the tropical Pacific *El Niño*–Southern Oscillation (ENSO) region. Cook et al. (2009) stated: "There is a strong indication that inter-annual drought and wetness in the Mississippi Valley are associated with the state of the *North Atlantic Oscillation* (NAO)."

In our view, factors governing entry into a megadrought event are complex and involve climatic tipping points. When the extra trapped thermal energy from GHG emissions was considered in the IPCC model projections, the occurrence of increasing droughts in the 21st century appears to be a valid prediction. Still, there always will be doubters; Cook et al. (2009) commented: "While there is no guarantee that the response of the climate system to greenhouse gas forcing will result in megadroughts of the kind experienced by North America in the past, the IPCC model projections are not comforting." We must wait and see how GHG forcing will affect the ENSO and the NAO, which are strong drivers of wet/dry and warm/cold conditions.

The predicted global drought scenarios will happen slowly and probably will be modulated by natural climate variability. With a warmer climate, droughts, storms, hurricanes, and floods could become more frequent and last longer (Hayes 2010). There will be more stored thermal energy in the oceans (SSTs will be higher), landmasses, and atmosphere. Existing long-range ocean conveyor currents will be disrupted by freshwater from melting polar ice caps. All models of future climate that we have seen predict that regions of the globe will be dryer; hence, water supplies, agriculture, and thus food supplies will be affected, challenging human sustainability. It is the model-based severity of these droughts that is open to debate.

The Texas Water Development Board (TWDB) predicted that by 2060, statewide demand is projected to rise to 22 million acre-feet (Maf) (2.7×10^{10} m³) per year. The availability supply is expected to decline from 17 Maf (2.1×10^{10} m³)/year to about 15.3 Maf (1.89×10^{10} m³)/year in 2060, as some aquifers become depleted, and areas of the state will come under new water conservation regulations. The TWDB forecasts a total statewide water shortfall of about 8.3 Maf (1.02×10^{10} m³)/year by 2060. This predicted water crisis could result in approximately \$115.7 billion in state economic losses per year by 2060 (Economist 2011g). Texas will really have to scramble to head off this water crisis; new resources will have to be developed (e.g. desalination-turning oil and NG energy back into water), and new, effective conservation methods will have to be developed.

Shrinking glaciers in the high mountains of Pakistan, Kashmir, Tibet, Nepal, and Bhutan are threatening the future supply of water in rivers such as the Indus (in Pakistan), the Ganges in northern India, and the Brahmaputra in Assam and Bangladesh. These rivers and their tributaries supply water for agriculture, domestic use and hydropower generation to hundreds of millions of people, populations that are increasing. Decreased monsoon rainfall, shrinking glaciers and reduced snowfall in the mountains caused by GW-induced weather changes, coupled with population growth and industrialization in India and China, will exacerbate low river flows and are leading to regional competition for these limited water resources (Economist 2011h).

What can be done about drought? Use water more efficiently: if one irrigates, use drip irrigation on plant roots at night when it is cool, rather than spray water in the air in the sun where a large fraction will evaporate. Conserve water at home: Use low-volume-flush toilets, wash clothes less frequently, do not water lawns, landscape with dry-climate plants (xerophytes), catch rain and use it for gardening, and recycle treated *gray water* for gardening and sanitary use. Desalination is an energy-intensive source of potable water (cf. Section 4.3) that helps coastal cities to exist in arid regions. Accelerate development of GM crop plants (sunflowers, corn, cotton, soybeans, etc.) that are drought tolerant.

See the paper by Nus (1993) for a review of the physiology and morphology of drought-resistant plants. Nus commented that desert plants such as cacti and spurges have the adaptation of keeping their stomates closed during the heat of the day and opening them at night when evaporative demand is lower. (Stomates are the many small pores in the leaves of all higher plants that allow them to regulate the intake of CO_2 for photosynthesis, exhale the O_2 produced, and transpire the WV not used in the process of photosynthesis.) The stomates of all other higher plants are open during the day and closed at night. Other morphological adaptations for dry climates are deep root systems and small, thick, fleshy leaves.

3.5 Bees, Pollination, and Food Crops

3.5.1 Introduction

In this section, we describe a threat to agricultural sustainability posed by the worldwide epidemic of honeybee colony collapse disorder (CCD) as well as the rapid decline of other important insect pollinator species including bumblebees (Cameron et al. 2010; Mader et al. 2010; Potts et al. 2010).

Certain crops such as almonds, blueberries, peaches, and apples are heavily dependent on bee pollination, while others such as oranges, cotton, and soybeans are less dependent (Beespotter 2007).

There are over 200 species of bumblebees worldwide (Goulson 2003). There are approximately 45 bumblebee species pollinating in the United States (Mader et al. 2010). Cameron et al. (2010) have shown that the relative abundances of four species of North American bumblebees (*Bombus occidentalis, B. pensylvanicus, B. affinus,* and *B. terricola*) have declined by as much as 96% in certain locations and that their surveyed graphical ranges have contracted by 23%–87%, some within the last 20 years! The causes for this decline probably involve insecticide exposure, but may involve some other things that affect the bees' immune systems (e.g., viruses, fungi). It is not known whether the bumblebee decline is related to honeybee CCD.

In North America, it is believed that over 30% of the food for human consumption originates from plants pollinated by bees. Bumblebees and honeybees are pollinators for the following important crops: alfalfa, almonds, apples, apricot, avocados, blackberries, blueberries, Brazil nuts, broad beans, buckwheat, cantaloupe, cashews, celery, cherry, clovers, coriander, cotton, cranberries, cucumber, currants, eggplant, fennel, gherkin, gooseberry, gourds, grapefruit, kiwi fruit, lemon, Lima beans, lupin, mangoes, marrow, melons, mustard, oilseed rape, oranges, peaches, pears, peppers, plums, pumpkins, raspberries, runner beans, soya beans, squashes, strawberries, string beans, sunflowers, tomatoes, turnips, vetches, watermelons, and zucchini (Bumblebee 2011). Bumblebees pollinate by brushing pollen onto their bodies and then transferring it to the flower's pistil as they rummage for pollen and nectar in a blossom. They also "buzz pollinate" (at about 400 Hz) by vibrating their wings while inside a flower, causing the flower's pollen to become airborne and reach the pistil.

Some plants are *self-fertile* and can pollinate themselves; other plants have physical or chemical barriers to self-pollination. Peach flowers are self-fertile, but insects can also act as pollinators. Some common self-pollinators include, but are not limited to, the following: barley, beans, beets, corn, cowpeas, eggplant, endive, lettuce, oats, peas, peppers, tomatoes, and wheat. Wind can help in self-pollination.

An awesome number of honeybees are used to pollinate large monocrops in the United States: for example, in the almond orchards in California, where approximately 1.5 million strong honeybee hives are trucked in the spring (Flottum 2010). [In the summer, each mature beehive or colony may have about 60,000 bees (Lovgren 2007), so the total number of bees pollinating (and making honey) in California almond orchards is approximately 6×10^{10} (60 gigabees!).] New York's apple crops require approximately 30,000 hives; Maine's blueberry crop uses approximately 50,000 hives/year. Monocrops require large concentrations of pollinator bees at bloom times and none before or after. The US solution so far has been for commercial beekeepers to truck their hives from south to north (Texas and Florida to Canada), following the blooms of a crop. The loss of entire honeybee hives from CCD has meant a reduction in the number of pollinators, hence crop yields. The slack is being taken up in part by the commercial use of domestic bumblebee hives.

Bumblebees are already used in greenhouses to pollinate tomatoes, eggplants, and peppers. They are more efficient pollinators than honeybees; they seldom sting, they fly faster and at lower temperatures (10°C) than honeybees, and they have longer tongues and can reach deeper into certain flowers that honeybees avoid. They are truly random foragers; they do not do the bee dance that honeybees do to notify their peers where there is a source of nectar and pollen. Their random search strategy makes them more effective in covering a 360° circle around their hive for food. Bumblebees' nests are generally underground but can be in old wood. Their nests are generally not large, and they make little honey, just enough for their larvae and immediate needs. Their value to humans is as pollinators. A bumblebee nest contains a single egg-laying queen and 150–400 workers and males (drones) (Biobest 2011). It was estimated that wild pollinators in California agriculture account for $937 million to $2.4 billion in crop value (Guy 2011). In 2000, the total US crop value attributable to honeybee pollination was estimated to be approximately $14.56 billion (Mburu et al. 2006; Morse and Calderone 2000).

In 2008, approximately US $217 billion worth of agricultural products was pollinated by all bees, worldwide (ScienceDaily 2008). A table by UNEP (2010) based on 2009 data showed that all pollinators contributed approximately €153 billion to the total value of human food crops worldwide, which is about 9.5% of the total value of global food production at that time. In 2011, honeybees were estimated to pollinate approximately $15 billion of food crops per year in the United States, approximately US $1.14 billion in Canada, approximately US $3 billion in the European Community (EC) and US $2.3 billion in New Zealand (ICIMOD 2012), and between £120 and 200 million per year in the United Kingdom (Sigma Scan 2011).

Other insects also pollinate; these include certain wasps, ants, beetles, moths, butterflies, flies, and solitary bees. Animals also pollinate; these include certain bats and birds (e.g., hummingbirds, sunbirds, spider hunters, and honeyeaters) that pollinate the durian (*Durio zibethinus*), for example. Flower petals that attract insects are often brightly colored, have a strong scent, and produce nectar. The brightly colored flowers attract insects, as does the strong scent. Flower petals that attract birds often are tube shaped and red or yellow in color and have no scent. Flower petals that attract mammals, such as bats, often have a fruity scent, white petals, and flowers that only open at night (Mauseth 2008).

3.5.2 CCD and Its Possible Causes

Honeybees are social insects that live under high-density conditions in their hives. Such high densities ordinarily would make them particularly susceptible to bacterial, fungal, and parasitic infections. However, bees have evolved behavioral defenses as well as individual, innate, immune protection. Among the behaviors that protect them are mutual grooming and the relentless maintenance of a clean, sheltered environment. They construct their nests from antimicrobial materials and raise their offspring in individual, sterile nurseries. They also secrete antimicrobial substances that reduce the viability and growth of pathogens in their hives, honeycombs, and honey. "Nurse" workers identify and remove any infected larvae from the hive. Also, sick workers fly away from the hive, protecting their fellows from infection (Underhill 2009; Evans et al. 2006).

Bee immune systems are *innate,* that is, they respond directly to threats to the animal by bacteria, fungi, and parasites using their circulating antimicrobial peptide (AMP) molecules that attach to pathogens and either render them harmless or tag them for destruction by hemocyte/phagocytes circulating in the animal's hemolymph (Northrop 2011, Section 5.4). Hemocyte/phagocytes also can directly attack certain pathogens by encapsulation or phagocytosis. For example, the Asiatic honeybee, *Apis cerana*, has a number of known AMPs: 7 different *defensin* peptides, 2 different *abaecin* peptides, 4 *apidaecin* peptides, and 13 different *hymenoptaecin* peptides (Antúnez et al. 2009; Xu et al. 2009). Bee hemolymph also contains *lysozyme (N-acetylmuramylhydrolase)*, which attacks primarily Gram-positive (G+) bacteria as well as some varieties of *Escherichia coli* (G−) bacteria.

Bees must also combat fungal infections. The known bee AMPs, including certain lysozymes, are generally ineffective against bee fungal and yeast invaders (mycelia and spores), which include the following: *Ascosphaera apis, Aspergillus* sp., *Aureobasidium pullulans, Trichoderma lignorum, Mucor hiemalis, Rhizopus* sp., and *Torulopsis* sp. (Gliński and Buczek 2003). Bee blood cells (hemocytes) are the main active defense again fungal infections. When the body cavity is exposed to small numbers of bacteria or fungal spores, the hemocytes enclose them in phagosomes by the process of phagocytosis. In the final stage of the phagocytic process, the engulfed spores or fragments of fungal mycelia are broken down by enzymes in a phagolysosome that fuses with the phagosome inside the hemocyte. The other process implemented against invading fungi is encapsulation by hemocytes. Chemical signals from the hemocytes cause other blood cells, for example, granular cells and plasmatocytes, to form a capsule around the fungus, inactivating it.

Honeybee CCD is characterized by the sudden disappearance of nearly all the bees in a hive (van Engelsdorp et al. 2009; Steering Committee 2009, 2010). One morning, a beekeeper takes the cover off of a honeybee hive, and all the workers are gone! There are no dead worker bees in the hive or on the ground around it. Bee behavior is such that a sick worker bee will fly from the hive; this prevents it from infecting its fellow bees (Underhill 2009). CCD is occurring worldwide: all over the United States and Western Europe, and also Taiwan. In the 2010/2011 winter, the US Department of Agriculture (USDA) reported that there was an estimated annual honeybee loss of approximately 30% *from all causes*, which was statistically similar to losses reported in 2007, 2008, and 2009. Before the 21st century, large-scale honeybee colony losses have also been documented: "Since 1869, there have been at least 18 discrete episodes of unusually high colony mortality documented internationally" (van Engelsdorp et al. 2009).

The exact causes of CCD are still unknown, but research to date points to several causative factors that probably act synergistically, including systemic insecticides sprayed on crops near hives that may compromise the bees' immune systems. These insecticides can end up in a plant's pollen, which bees use for food. In addition, the miticides *Coumaphos* and *Fluvinate* (used to combat the *Varroa* sp. tracheal mites that parasitize bees) may act individually or synergistically with other bee pathogens to compromise the bees' immune systems (Steering Committee 2010).

Infection by the microsporidian parasite fungus *Nosema ceranae* has also been found to degrade honeybees' immune systems. Found in *all* dead bees from CCD hives were both *N. ceranae and* a newly identified DNA virus, an invertebrate iridescent virus, IIV6. Infection by the combination is always fatal (Leal et al. 2010; Bromenshenk et al. 2010). Just when the root causes of CCD seemed to be emerging, Foster (2011) presented data that suggested that there is "insufficient evidence to conclude that bees are a natural host for IIV6, let alone that the virus is linked to CCD."

Other factors that may be contributory to CCD are stresses caused by hive malnutrition, transporting hives, and infection of bees by the tracheal mite *Varroa destructor*. These mites carry a number of viruses, but the Israel acute paralysis virus (IAPV) shows a significant association with CCD; the IAPV has been found in 25 of 30 tested CCD colonies but only in one of the normal colonies. The problem in analyzing a syndrome like CCD that presumably involves immunosuppression is that an immunocompromised bee can be infected by many potentially lethal viruses, fungi, or parasites, many of which can further weaken its immune system, and *it is very difficult to separate cause from effect in CCD*.

Clearly, any combination of factors that weakens the bees' immune systems makes them susceptible to bacterial and fungal infections and could contribute to honeybee CCD. An analogous situation is the effect of the various immune-deficiency viruses on the complex immune systems of mammals, the most notable case of which is HIV, which gives rise to AIDS by infecting key human immune system cells (Northrop 2010).

In the case of bees with CCD, the ability of their immune systems to synthesize AMPs may be compromised lowering their ability to fight invading pathogens and/or their ability to produce normal hemocyte/phagosomes. Because the key to understanding bee CCD appears to lie in their immune systems, research should be directed toward differential analysis of normal versus CCD bee AMP titers and a study of what happens to CCD bees' phagosomes.

3.5.3 The Impact of CCD on Our Food Supply

One indicator of the decline in natural (and domestic) insect pollinators is decreasing crop yields and quality, despite the necessary, expensive, energy-dependent agronomic inputs of water, fertilizers, herbicides, pesticides, and cultivation. An extreme example of the effect of a total loss of pollinators can be seen in *Maoxian* County in the *Hengduan* Mountains of China, where orchards of pears and apples are hand-pollinated by humans. Bees cannot be used because the Chinese farmers make excessive use of pesticides, even during flowering season (Partap 2002; Tang et al. 2002). Partap commented: "Hand pollination is an interesting method of pollinating crops and provides employment and income generating opportunities to many people during apple [and pear] flowering season. But at the same time it is an expensive, time-consuming and highly unsustainable proposition of crop pollination owing to the increased labour scarcity and costs. Moreover, a large part of farmer's income is used in managing pollination of their crops."

Interestingly, in spite of all of the attention and research paid to CCD and bumblebee decline, there is no good quantitative estimate on how these conditions will affect US and world food availability and prices; the only consensus is that they will decrease the availability of certain food crops and consequently increase their prices.

In order to quantify the effects of CCD, what is needed is a good numerical estimate of the sensitivity $\left(S_B^F\right)$ between food prices (F_p) in a region and the number of pollinators (**B**) in that region. The fractional loss of pollinators due to honeybee CCD (and similar afflictions to bumblebee colonies) is **ΔB/B**. **ΔF$_p$** is the increase in food prices due to **ΔB/B**. The sensitivity S_B^F is given by

$$S_B^F \equiv \frac{\Delta F_p/F_p}{\Delta B/B} = \frac{\Delta F_p}{\Delta B}\left(\frac{B}{F_p}\right) = \frac{\partial F_p}{\partial B}\frac{B}{F_p} \text{ or } \Delta F_p/F_p = S_B^F(\Delta B/B) \qquad (3.2)$$

In general, **ΔF$_p$/F$_p$** will increase with increasing |**ΔB/B**|. Ideally, we want S_B^F to be as small as possible.

One simple formula used to calculate the benefits to the farmer, *W*, of using bees to pollinate a coffee crop was

$$W = S \times \Delta q(p - c)$$

where *W* = dollars, *S* = crop area (ha), Δq = fractional increase in production per hectare as a consequence of pollinating this crop, *p* = farm gate coffee price, and *c* = variable costs related to coffee harvest.

Mburu et al. (2006) gave an example of applying this formula to a coffee plantation in Costa Rica. They showed that pollination added 7% to the plantation's income.

The food price increases due to bee losses will probably be less in countries that rely heavily on wind-pollinated grain crops such as rice. In the United States, honeybees alone are estimated to pollinate approximately $15 billion in US crops, and bees in general pollinate some 90% of the world's commercial plants (Jha 2011).

Another approach to evaluating the impact of pollinator decline (wild and domestic) was described in a paper by Allsopp et al. (2008). These authors did an economic study of the cost of replacing lost natural pollinator services using human pollen dusting and pollination by hand in the Western Cape (South African) deciduous fruit industry (e.g., apples, apricots, plums, pears, etc.). The bottom line: replacement pollination by man is tedious and expensive.

It is expected that farmers worldwide will make more use of alternate pollinators (Mader et al. 2010) if the rates of honeybee CCD infection and bumblebee die-off continue unabated. See the interesting paper by Bauer and Wing (2010) for their predictions of the quantitative details on the global consequences of pollinator declines.

3.6 Species Size Reduction Due to Habitat Warming: Another Challenge to Our Food Supply

It has long been known that the size of the ubiquitous North American species, the white-tailed deer (*Odocoileus virginianus*), is much larger in the colder, northern limits of its US range (Maine, Vermont, Wisconsin, Minnesota, Montana, etc.) than in its warmer southern habitats (Florida, Texas, Alabama, Georgia, etc.). Black bears also follow this rule. This increase in mammal size (R) with northern (colder) habitats is related to the animal's area/volume $\propto R^{2/3}$ ratio. Body heat loss in a cold climate is proportionally less for a large R animal, giving large mammals a survival edge. This phenomenon can also be viewed as a reduction of natural size with *increased* average habitat temperature.

What was not known until recently is that many animal species' sizes in a given habitat can shrink over a relatively short period of time (30–10 years) given environmental warming of that habitat (Gardner et al. 2011). This size reduction with increasing habitat temperature evidently applies to plants as well as animals (Sheridan and Bickford 2011). "[I]ndividuals reared at lower temperatures grow more slowly, but are larger as adults than individuals reared at warmer temperatures" (Forster et al. 2011). Forster et al. observed that "at warmer temperatures a species grows faster but matures even faster still, resulting in them achieving a smaller adult size." The reason for this size reduction with temperature may also be related to the fact that the ratio of skin area to body volume (mass) scales as $R^{2/3}$, where R is the size metric. A smaller R means there is less skin area to lose (or gain) heat (or water) per body mass. A smaller animal also requires less food energy per day to thrive. In vertebrates, growth hormone levels [GH] are one determinant of adult body size and so are sex hormones (androgens, estrogens). Animal development is an extremely complex process involving the sequential expression of many genes, including the homeobox (**HOX**) genes, and the regulated expression of hormones, enzymes, and other genomic regulatory molecules such as micro-RNAs. All biochemical reactions are increasing functions of temperature (Q_{10}) over a wide range; however, enzymatic reactions generally have an optimum temperature range, T_U–T_L. Above T_U, Q_{10} drops off sharply.

Yet another potential challenge to human sustainability in terms of our food supply may come from this documented reduction in the size (and mass) of certain amphibians, fish, animals, and plants that is correlated to the GW-caused rise in the mean temperature of their habitats. For example, the average size of the Scottish Highland Soay sheep has now decreased by 5% compared to its size in 1985. Other species showing measurable size reduction with temperature include the following: cotton, corn, strawberries, bay scallops, shrimp, crayfish, carp, Atlantic salmon, frogs, toads, iguanas, hooded robins, red-billed gulls, California squirrels, and wood rats. The paper by Sheridan and Bickford reported that each degree of habitat warming has been shown to decrease the size of marine invertebrates by up to 4%, salamanders up to 14%, and fish from 6% to 22%. "Over the past century, various plant species have shown significant negative correlations between growth and temperature…resulting in smaller grasses, annual plants and trees in areas that are getting warmer and drier." Their paper cites the results of their experiments manipulating temperature, showing that the biomass in some grass, grain, and fruit plants was 3%–17% smaller for every degree Celsius warming.

To quantify the impact of warming on the human food web, one must look at the dietary spectrum consumed by a given population. If a certain fish regularly consumed is, on the average, 16.7% smaller due to ocean warming, a person (or predator) would have to eat six of the smaller fish to get the same nourishment as five normal-sized fish. Thus, catch size would have to increase by 120% to meet this need. Could it? Some of our fisheries are already severely depleted (e.g. wild Atlantic cod). Thus, in general, shrinking food size means catching (or growing) more food to meet the same nutritional level for a given population. Clearly, further research must be done to quantify the impact of shrinking food (plants, animals) size on the ecosystems that provide human food and, ultimately, human sustainability.

3.7 FF Energy and Sustainability

3.7.1 Introduction

The past 150 years have been an epoch of FF discovery, extraction, and utilization. One of the first FFs to be used for home heating was peat mined from bogs. Coal mined from underground proved to be a more efficient energy source and was used to heat homes as well as provide energy for industry and rail transportation. With the discovery of in-ground deposits of crude oil, the whole energy landscape changed. Here was an energy source that could be refined and used to heat homes, power the ubiquitous internal combustion engine, and also power turbines that powered aircraft and ships. A fourth, important, in-ground FF energy source to be recently exploited is NG. Coal, oil, and NG fueled the age of development—industry, electric power generation, transportation, petrochemical production, and so forth—in the latter half of the 19th and in the 20th century. The increasing demand for these energy sources, coupled with world population growth, soon led to the realization that they had finite volumes on the planet, and someday, they would become less available and hence more expensive. In this section, we discuss these energy sources, their sources, and their reserves. A detailed history of the worldwide development of petroleum resources in the 19th,

20th, and 21st centuries may be found in the interesting book by Yergin (2011) *The Quest.*

3.7.2 Natural Gas

NG is created by two mechanisms: contemporary production is from methano-genic *archaea* bacteria living in marshes, bogs, landfills, and shallow sediments; it is mostly methane + CO_2. Gas from landfills is most easily acquired. Fossil NG is trapped deep in the Earth and is the result of the decomposition of buried organic material under high pressure and temperature over millions of years. Most of the NG produced in the world is from this thermogenic, fossil source. Thermogenic NG is actually a mixture of gasses: typical fossil NG contains approximately 95% methane (CH_4); 2.5% ethane; 0.2% propane; 0.03% isobu-tane; 0.03% *n*-butane; 0.01% each of *n*-pentane, isopentane, and hexanes; 1.3% N_2; 0%–8% CO_2; 0.02% O_2; 0%–5% H_2S; and traces of H_2 and helium (Uniongas 2011; NG 2011). Clearly, compositions vary with deposit.

Syngas, also known as *synthesis gas* or *town gas,* is a synthetically produced mixture of methane, carbon monoxide, hydrogen, and other trace gasses. It can be produced by treating coal chemically or in the process of making coke. Recently, a solar-powered process, currently under development, has been devised to form syngas directly from CO_2 and water (see Section 4.2.6).

NG is associated with oil deposits that are approximately 1 to 2 mi. below the Earth's surface; deeper deposits, far underground, generally contain mostly NG. Methane is often found in deep coal mines as well.

An unusual and, at present, untapped source of methane is in crystallized methane hydrates (MHs) trapped in deep marine bottom sediments and also in permafrost on land. As of 2012, several prototype systems are being evalu-ated to harvest methane from hydrates (Japan Times 2012; Magiawala et al. 2007; ORNL 2002).

Table 3.2 lists the world's proven NG reserves and those for the 25 top gas-producing countries.

Recent claims for the size of NG reserves under the South China Sea vary widely. These reserves are in deepwater and around the Paracel and Spratly Islands, and six countries contest the ownership of at least part of these reserves: China, the Philippines, Vietnam, Malaysia, Taiwan, and Brunei. Several Chinese studies estimated the total South China Sea NG reserves to be around 2×10^{15} ft.^{3} (5.67×10^{13} m^3). A Husky Energy/Chinese National Offshore Oil Corporation study in April 2006 announced that proven NG reserves near the Spratly Islands ranged from 4×10^9 to 6×10^{12} ft.^{3} (1.13×10^8 to 1.7×10^{11} m^3) (Lopez 2011; Mellgard 2010).

Figure 3.10 illustrates the almost exponential rise in US NG production from 1900 to 1970. Note that production rose almost exponentially from less than 1 trillion ft.^{3} (tft^3) (2.832×10^{10} m^3) in 1900 to a peak of approximately 22.5 tft^3 (6.37×10^{11} m^3) in 1970, dropped off to approximately 17 tft^3 (4.81×10^{11} m^3) in about 1985, and then slowly rose again to approximately 20 tft^3

TABLE 3.2

Top 25 NG-Producing Countries, plus the EU and the World

Rank	Country/Region (Date)	NG Proven Reserves (m^3)
1	Russia (2010)	5.50×10^{13}
2	Iran (1/2011)	3.35×10^{13}
3	Turkmenistan (1/2012)	2.62×10^{13}
4	Qatar (1/2010)	2.547×10^{13}
5	United States (1/2009)	7.716×10^{12}
6	Saudi Arabia (1/2010)	7.461×10^{12}
7	Azerbaijan (1/2010)	6.071×10^{12}
8	Venezuela (7/2011)	5.525×10^{12}
9	Nigeria (1/2010)	5.246×10^{12}
10	Algeria (1/2010)	4.502×10^{12}
11	Iraq (1/2010)	3.170×10^{12}
12	Australia (1/2010)	3.115×10^{12}
13	Indonesia (1/2010)	3.001×10^{12}
14	Kazakhstan (1/2010)	2.407×10^{12}
15	Malaysia (1/2010)	2.350×10^{12}
16	Norway (1/2010)	2.313×10^{12}
17	Uzbekistan (1/2010)	1.841×10^{12}
18	Kuwait (1/2010)	1.798×10^{12}
19	Canada (1/2010)	1.754×10^{12}
20	Egypt (1/2010)	1.656×10^{12}
21	Libya (1/2010)	1.539×10^{12}
22	Netherlands (1/2010)	1.416×10^{12}
23	Ukraine (1/2010)	1.104×10^{12}
24	India (1/2010)	1.075×10^{12}
25	Oman (1/2010)	8.495×10^{11}
–	*The EU* (1/2010)	$\mathbf{2.25 \times 10^{12}}$
–	*The World* (1/2012)	$\mathbf{3 \times 10^{14}}$

Source: Natural gas, 2012, *Natural Gas Gross Withdrawals and Production,* US EIA data. http://205.245.135.7/dnav/ng/ ng_prod_sum_dcu_NUS_a.htm.

(5.66×10^{11} m^3) in 2005 (USEIA 2011b). Further USEIA data show that US NG gross withdrawals rose steadily: 23.5 tft³ (2006), 24.7 tft³ (2007), 25.6 tft³ (2008), 26.1 tft³ (2009), 26.8 tft³ (2010), and 28.6 tft³ (2011) (Natural gas 2012).

NG is generally processed to remove impurities and harvest the higher-molecular-weight aliphatic hydrocarbons such as ethane, propane, butanes, and pentanes. Impurities such as H_2S, Hg vapor, H_2O, and N_2 are also removed. If present in sufficient concentration, the valuable gas helium is also extracted. Because of the explosive nature on NG in air, commercial NG

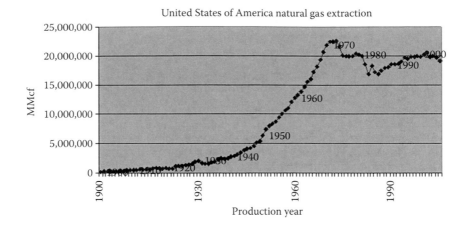

FIGURE 3.10
US NG production, 1900–2005. Note the peak at 1970, the falloff, and now the rebound in the late 1980s due to exploitation of shale gas deposits and fracking. (From US EIA data: www.eia. gov/dnav/ng/hist/n9050us2a.htm.)

has a strong odorant added to it, generally *t-butyl mercaptan* (which smells like sulfury rotting cabbage), to warn people of the presence of NG in the air in parts-per-million concentrations.

There are many domestic uses for compressed NG (CNG) (in tanks): for example, central heating furnaces, boilers, water heaters, clothes dryers, and cooking stoves. CNG is also a cleaner alternative fuel for vehicles than gasoline or diesel. Efficient CNG-specific piston engines require higher compression ratios than found in gasoline engines. The effective octane rating of CNG is 120–130, while regular E10 gasoline is typically 87 octane. Thus, dual-fuel piston engines that are to run on either gasoline or CNG necessarily must have lower compression ratios to accommodate the petrol, which gives them lower efficiency when burning CNG.

NG is a major feedstock for the production of ammonia for use in fertilizer via the Haber process (cf. Glossary). As a jet fuel, liquid methane has more specific energy than the standard kerosene mixes (jet fuel) do. Its low temperature can help cool the air entering the engine for greater volumetric efficiency, in effect replacing an intercooler. Alternatively, it can be used to lower the temperature of the jet exhaust. The Russian aircraft manufacturer, Tupolev, is currently running a research and development program to produce LNG- and H_2-powered jet aircraft.

The dynamics of global NG production are made more complicated because it is produced in various political regions and its export pricing and availability can be manipulated for political reasons.

New NG deposits are constantly being discovered; for example, test wells drilled by Houston-based Escopeta Oil Co. have discovered an estimated NG reserve of 5.5 tft^3 (1.56 × 10^{11} m^3) under Alaska's Cook Inlet (World Briefing 2011).

3.7.2.1 Fracking

Fracking is the process of hydraulic fracturing of deep earth strata (e.g., shale) in order to increase the flow of NG, oil or water from a deep, drilled well (Biello 2010a). Gas wells in deep shale reservoirs have low natural permeability, hence inherently low gas flow rates. In drilling a well, the main borehole is drilled vertically to 5000 to 20,000 ft. (1.52–6.10 km) and then turned horizontally to extend into the gas reservoir. The volume of fracking fluid (FrF) used depends on the well bore diameter and its length and the porosity of the gas containing shale. A typical gas well fracking operation can use approximately 15,000 m^3 (3.96 million gal.) of water per well, and over the life of the well, it may be fracked hundreds of times (Pacinst 2012). Perhaps 90% of the NG wells in the United States use fracking in order to produce gas at economical rates. Water used in fracking is unfit for human or agricultural uses because of the toxic chemicals it contains. It is stored, reprocessed, and used again for fracking.

The fracking process forces the water-based fracking fluid at up to 15,000 psi (100 MPa) into the well, where pressure is maintained for 3–10 days. The fracking fluid itself is generally approximately 98.5% water with dissolved chemicals; the rest is sand or resin-coated sand, intended to prop open the cracks in the shale, permitting better gas flow to the extraction pipe. Besides sand, fracking fluid can contain various proprietary chemicals to expedite the process. According to Chesapeake Energy, the most common fracking fluids contain 0.5% chemicals, including, but not limited to, mixtures of the following:

- *Hydrochloric acid (corrosive):* used to help dissolve minerals and crack the rock
- *Ethylene glycol (poison):* to prevent scale deposits in the pipe that lines the well
- *Isopropanol (toxic):* increases the viscosity of the fracking fluid
- *Glutaraldehyde (poison):* used as a bactericide on bacteria that can cause corrosion
- *Petroleum distillates (flammable):* used to minimize friction
- *Guar gum:* used as a gel to increase viscosity and suspend the sand
- *Ammonium persulfate:* used to delay the breakdown of the guar gum
- *Formamide (toxic):* used to inhibit corrosion of the well casing
- *Borate salts:* used to maintain fluid viscosity under high temperatures
- *Citric acid:* a chelating agent used to prevent precipitation of metal in the casing

- *Potassium chloride (KCl):* used to prevent fluid from interacting with the soil
- *Sodium or potassium carbonate:* used as a pH buffer

Not every drilling company uses the same fracking fluid formula. Some of the 85 fracking chemicals listed by the Pennsylvania Department of Environmental Protection (DEP) include *xylene, toluene,* and *tetramethylammonium chloride:* chemicals with known toxicities to humans and animals (Junkins 2010). *Kerosene* (flammable), *benzene* (a carcinogen), and *formaldehyde* (poison) are also used in the composition of some FrFs. Included in Halliburton's FF mixture are *formaldehyde, ammonium chloride, acetic anhydride, methanol, HCl,* and *proargyl alcohol* (Junkins 2010).

In the tiny rural town of Pavillion, Wyoming, the US Environmental Protection Agency (EPA) tested 19 drinking water wells and found high levels of dissolved *methane, benzene,* and *2-butoxyethanol (2-BE) phosphate* (which causes kidney failure), a solvent used in fracking fluid used in local gas wells. Chris Tucker, a spokesperson for Energy in Depth, a consortium of NG and oil producers, stated that there was "no proof" tying groundwater contamination to fracking (Feldman 2010). If the exact composition of the fracking chemicals were known, analytical chemistry could provide "proof" of the sources of drinking water contamination.

Fracking has the potential to contaminate aquifers and surface freshwater. The geology of most fracking sites puts the gas/shale deposits at over 5000 ft. in depth (1.52 km); aquifers are around 1000 ft. (305 m) or shallower, and normally, there is no leakage of NG or fracking fluid up into the aquifer. Sometimes, the vertical gas well pipe passes through an aquifer on its way down. Leakage around the outside of the well pipe is one way toxic drilling chemicals and NG can enter the aquifer.

Once the considerable volume of fracking fluid from the well is removed, it must be stored on the surface and not be allowed to get into the environment and contaminate surface water and shallow wells. Fluid leaks that have occurred during the fracking process have come from high-pressure blowouts of the seals at the NG wellheads (Litvak 2010). Accidents in handling the withdrawn fracking fluids on the surface can also occur.

Another problem is gas escaping upward through porous ground following fracking of a relatively shallow gas deposit that is too close to a deep aquifer. This improbable event took place on January 1, 2009, in Dimock Township, Pennsylvania, when a water well exploded because of NG in the water.

The exact composition of the fracking fluids used is a closely guarded secret of the well drilling companies. If it were known, and there were water contamination from the fracking process, it would be easy to analyze the chemical "fingerprint" of the water. If it matched the FF composition used by the gas company in its wells, it would be the basis for group lawsuits and government-ordered shutdown of the contaminating wells (Melzer 2010).

Fracking was exempted from the US Safe Drinking Water Act of 2005, something that should be reversed (Lustgarten 2009).

On December 9, 2011, the US Environmental Protection Agency released a 121-page draft report entitled *Investigation of Ground Water Contamination Near Pavillion, Wyoming.* This report was the subject of public and peer review and was finalized in 2012 (EPA 2012). The EPA draft report said that pollution from 33 abandoned oil and gas (surface) waste pits is responsible for some degree of groundwater pollution in 42 shallow wells in the area. The EPA drilled two, 1000-ft. water-monitoring test wells in the Pavillion gas field and found fracking chemicals such as carcinogenic benzene and 2-BE. The benzene was at 50 times the safe level, and there were also acetone, toluene, naphthalene, isopropanol, diethylene glycol, triethylene glycol, tert-butyl alcohol, ethylbenzene, xylenes, and traces of diesel fuel. These chemicals are all used in fracking. The water samples were also saturated with dissolved NG that chemically matched the NG composition from the gas wells (Lustgarten and Kusnetz 2011). In addition, the water pH in the two test wells was found to be very alkaline (11.2 ≤ pH ≤ 12.0). (The pH of domestic well water typically ranges from 6.9 to 10.) The source of this extreme alkalinity was suspected to be the KOH used as a fracking cross-linker or solvent (EPA 2011a).

The presence of these chemicals in the deepwater wells was not the result of the surface water contamination. The Canadian NG driller/fracker company, EnCana, refused to give the EPA a detailed list of the chemicals they used in their fracking operations in Pavillion, found fault with the EPA studies, and "refuted" the findings in the draft report (Vanderklippe 2011). Clearly, deep underground geological conditions around Pavillion were such that rock faults caused by fracking allowed the fracking fluids and NG to migrate upward into deep aquifers. In the Extended Abstract of the EPA Draft Report, it was stated that fracking occurred as shallow as 372 m (1220 ft.) below ground, not around 5000 ft. The NG industry would have us believe that fracking is perfectly safe because aquifers are found at approximately 100-m depths, while the coal-bed methane and shale gas begin at 1000 and 2000 m, respectively (Economist 2012g). Indeed, this is true in most fracking scenarios, which have zero impact on air and water pollution.

In conclusion, the process of fracking is ubiquitous in shale NG production. The fracking fluid is a proprietary mixture of poisonous, toxic, flammable, and carcinogenic chemicals in water. Most fracking has not had bad environmental results. However, in some instances, there has been evidence that the fracking fluid has contaminated aquifers and surface water, and the NG released by fracking has dissolved in well water, causing an explosion hazard. Fracking is not without health and environmental risks, and its users need close governmental regulation and supervision. There should be a minimum gas well depth (dependent on local hydrology) at which fracking is permitted, based on the known, local, deep aquifer depth. Also, the exact composition of a driller's fracking fluid should be made public as part of the permitting process.

Potable water is too precious a commodity to make toxic. America needs energy, but not at the expense of contaminated water and public health.

Incidental methane emissions from all kinds of NG production are hard to estimate. They come from pipe leaks, valve leaks, wellhead leaks, and excess gas escape during incomplete flaring of NG at petroleum wellheads. Also hard to quantify are the natural emissions of methane from MH deposits in warming tundra and from deposits under the bottoms of the arctic seas. As more prospecting, drilling, and production of shale NG occur, there will be more methane released into the atmosphere (Economist 2012g). While methane molecules have a much shorter half-life in the atmosphere, they are at least 21 times more potent than CO_2 as a GHG.

3.7.3 Coal

The FF coal is the largest source of energy for the generation of electric power worldwide, as well as one of the largest sources of anthropogenic CO_2 emission and air pollution. (There is no "clean coal.") Coal has many forms: (1) *Peat,* considered to be a precursor to coal, is mined from near-surface deposits in bogs, mostly in Ireland, Scotland, and Finland. It has importance as a domestic fuel, for power generation, and in making certain single-malt scotch whiskies. (2) *Lignite,* or brown coal, has low energy content. It is used mostly for power generation. Lignite's heat content is approximately 28.47 MJ/kg. (3) *Sub-bituminous* and *bituminous* (soft) coal is used for power generation and as an important source of petrochemicals. It also is used to make coke. Its heat content is approximately 33.9 MJ/kg. (4) *Steam coal* is a grade between anthracite and bituminous coal. It was once widely used as fuel for steam locomotives and ship boilers, as well as domestic heating. (5) *Anthracite* is a hard, glossy, black coal, used primarily for residential and commercial space heating. Anthracite's heat content is approximately 35.3 MJ/kg.

Viewing coal as a nonrenewable resource in the light of sustainability, we are concerned about how much is left on the Earth. Some 930 billion tons (1 ton = 2000 lb. = 907.2 kg) of recoverable coal worldwide should be compared to the 7.08 billion tons burned in 2007, releasing 133.2 quadrillion British thermal units (BTUs). In terms of heat content, this coal burned has the heat equivalent of 57 million barrels (9.1 Mm^3) of oil per day. By comparison, NG provided the energy equivalent of 51 million barrels of oil/day, and oil itself provided 85.8 million barrels/day (Mbbl/day). At the 2010 rate of coal consumption, world reserves could last approximately 118 years (see Table 3.3). As the world's population grows, as oil and gas reserves are depleted, and as developing countries become more industrialized, the 2007 rate of coal consumption should grow, so perhaps 100 years is a better guesstimate. British Petroleum (BP), in a 2007 report, stated that there were approximately 909.1 billion (9.091×10^{11}) tons of *proven* coal reserves worldwide. (*Proven reserves* are those identified by exploratory drilling by mining engineers.)

TABLE 3.3

Estimates of Total Proven Coal Reserves (2008), 2010 Coal Production, and Estimated Reserve Life of Reserves for Top 10 Countries

Country	Total Reserves (Bituminous + Anthracite + Lignite) (Million Tons)	2010 Production (Million Tons)	Life of Proven Reserve (Years)
United States	237,296	984.6	241
Russia	157,010	316.9	495
China	114,500	3240	35
Australia	76,500	423.9	180
India	60,600	569.9	106
Germany	40,699	182.3	223
Kazakhstan	33,600	110.8	303
South Africa	30,156	253.8	119
Poland	5709	133.2	43
Indonesia	5529	305.9	18
EU	*–*	*537.5*	*105*
World Total	*860,938*	*7273.3*	*118*

Source: World Energy, 2010, *Survey of Energy Resources,* World Energy Council. www.worldenergy.org/documents/ser_2010_report_1.pdf.

Note: 1 US ton = 2000 lb. = 907 kg; 1 tonne = 2000 kg.

Of the three FFs, coal has the most widely distributed reserves; it is mined in over 100 countries, on all continents except Antarctica. The largest proven recoverable coal reserves (of all kinds) by country at the end of 2008, in billion tons, are shown in Table 3.3. Proven coal reserves estimates vary, depending on date compiled, who compiled them, and the sources of the numbers. The data are noisy.

Russia has the longest coal reserve time because they can use energy from their abundant gas reserves, decreasing their rate of coal production and consumption.

3.7.4 Oil

The world's *proven crude oil reserves* are approximately 1.332 trillion barrels (2008). The top 14 ranking oil producers, consumers, exporters, importers, and estimated reserve lives of various oil-producing countries are given in Table 3.4.

The dates in parentheses are the data dates from the CIA table. Note that the three major oil-producing countries that are most rapidly depleting their oil reserves are China, Algeria, and Russia. The major oil importers are the United States and China. The estimated reserve life in years was found by

TABLE 3.4

Data on World and Country Oil Reserves, Production, Consumption, Exports, and Imports for the 15 Countries with the Largest Oil Reserves

Country	Proven Oil Reserves	Production/Day (Mbbl/day)	Consumption/Day (Mbbl/day)	Exports (Mbbl/day)	Imports (Mbbl/day)	Est. Reserve Life, Years (from 2011)
Saudi Arabia	266.8 billion bbl (2008)	10.250 (2007)	2.311 (2007)	8.9 (2007)	41,680 bbl (2005)	71
Canada	178.6 billion bbl (2008)	3.425 (2007)	2.371 (2007)	2.225 (2006)	1.229 (2005)	143
Iran	138.4 billion bbl (2008)	4.033 (2007)	1.679 (2006)	2.520 (2006)	167,000 bbl (2005)	94
Iraq	115 billion bbl (2008)	2.094 (2007)	0.295 (2007)	1.670 (2007)	NA	150
Kuwait	104 billion bbl (2008)	2.613 (2007)	0.3347 (2006)	2.356 (2005)	8022 bbl (2005)	109
United Arab Emirates	97.8 billion bbl (2008)	2.948 (2007)	0.381 (2006)	2.703 (2005)	232,300 bbl (2005)	91
Venezuela	87.04 billion bbl (2008)	2.667 (2007)	0.7383 (2007)	2203 (2006)	0 (2005)	89
Russia	60.0 billion bbl (2008)	9.876 (2007)	2.858 (2007)	5.080 (2007)	73,140 bbl (2005)	16.6
Libya	41.46 billion bbl (2008)	1.854 (2007)	0.2787 (2006)	1.445 (2005)	575 bbl (2005)	61
Nigeria	36.22 billion bbl (2008)	2.352 (2007)	0.312 (2006)	2.473 (2005)	154,300 (2005)	42
Kazakhstan	30 billion bbl (2008)	1.445 (2007)	0.2431 (2006)	1.236 (2005)	127,600 bbl (2005)	57
United States	20.97 billion bbl (2008)	8.457 (2007)	20.68 (2007)	1.165 (2005)	13.710 Mbbl (2005)	68
China	16.0 billion bbl (2008)	3.725 (2008)	7.578 (2007)	79,060 bbl (2007)	3.190 Mbbl (2007)	12
Qatar	15.210 billion bbl (2008)	1.125 (2007)	108,900 bbl (2006)	1.026 (2005)	0 (2005)	37
Algeria	12.2 billion bbl (2008)	2.173 (2007)	0.2798 (2006)	1.844 (2005)	13,110 bbl (2005)	15
EU	*6.865 billion bbl (2008)*	*2.674 (2007)*	*14.39 Mbbl (2007)*	*6.979 Mbbl (2001)*	*17.71 Mbbl (2001)*	*70*
World	*1.332 trillion bbl (2008)*	*84.79 (2007)*	*85.27 (2007)*	*66.19 (2005)*	*65.41 (2005)*	*43*

Note: Many more oil-producing countries (200 total) are listed in the *CIA World Factbook*: www.cia.gov/library/publications/the-world-factbook/rankorder/2173rank.html. bbl = US barrel; petroleum = 159 liters.

dividing a country's proven reserve by its yearly production rate. (Note: 1 US bbl petroleum = 42 US gal. = 159 L = 0.159 m³.)

The Santos Basin, a deepwater oil deposit stretching approximately 500 mi. along the southern coast of Brazil, is capped with a salt dome. Geological studies and an exploratory well drilled in 6000 ft. (1.83 km) of ocean and down 15,000 ft. (4.63 km) below the sea bottom led to estimates that the Santos Basin reserve held from 5 to 8 billion barrels of recoverable oil. This deposit is being developed and may produce *close to* 6 million barrels per day (9.54 × 10⁵ m³/day) by 2025 (Yergin 2011, Chapter 12).

Present total world proven oil reserves can also be described by oil types: about 30% is light crude, 30% is in oil sands bitumen; heavy oil is 15%, and extra-heavy oil is 25%. Extraction of oils from tar sands is energy intensive and expensive. Now that oil is on the market at US $108/bbl (4/07/11), $94.89/bbl (8/02/11), $82.26/bbl (8/20/11), $102.59/bbl (11/17/11), $104.87/bbl (5/01/12), $97.08 (5/11/12), $84.10/bbl (6/9/12), $93.35/bbl (8/09/12), and $96.01/bbl (8/17/12), it pays to process tar sands and oil shale. Estimated proven oil reserves (land) are shown on the map in Figure 3.11.

In addition to proven oil reserves, there are many *prospective oil resources:* a 2008 US geological survey estimated that areas north of the Arctic Circle have 90 billion barrels (1.43 × 10¹⁰ m³) of *undiscovered*, technically recoverable liquid oil and 44 billion bbl (7.0 × 10⁹ m³) of NG liquids in 25 geologically defined areas (USGS 2008). This represents 13% the expected undiscovered oil on Earth. (There are estimated to be approximately 1.03 trillion bbl of recoverable, *undiscovered* oil on Earth.) Figure 3.12 illustrates the locations of these estimated, prospective Arctic oil reserves. Note that the densest prospective oil deposits are in the oceans along the northern, northeast, and northwest costs of Greenland, northern Canada and Alaska, and Russia, where the methane clathrate deposits are located.

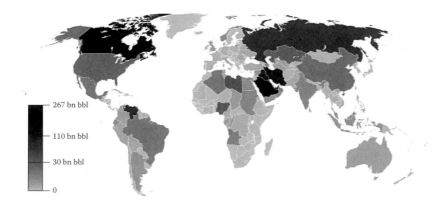

FIGURE 3.11
World oil reserves by country. (Courtesy of the CIA.)

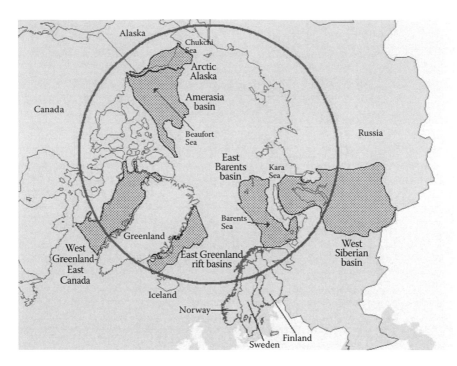

FIGURE 3.12
(See color insert.) Oil reserves above the Arctic Circle. (Courtesy of the USGS.)

The Orinoco tar sands in Venezuela may hold up to 513 billion bbl oil (81.6 Bm³) (in 2009). Cuba also has *undiscovered oil*, estimated to be 20 billion bbl (3.18 Bm³) by the Cuban government. [The current Cuban proven oil reserves are only approximately 124 million bbl (19.7 Mm³); their production rate is 61,300 bbl/day (9.75 km³). Without new wells in new oilfields, Cuban proven reserves may last only approximately 6 years!] Exploratory drilling is currently being conducted in the deep ocean off the coast of Cuba, financed by China and other non-US countries (Yergin 2011).

Oil produced from prospective reserves will be more expensive than current oil because expensive new wells (test wells and production wells) will need to be drilled in challenging locations (e.g., the high Arctic, deep oceans). As oil deposits in proven reserves are depleted, the outputs of individual wells decline with time due to reduced reservoir pressure from reduced oil and gas volumes. The yearly production decline curve for a well can be plotted, and this curve can be used to mathematically extrapolate to the production rate that gives zero profit and thus estimate the effective life of the well. The amount of oil that can be recovered from a deposit ranges widely, from 10% to 80% of the proven reserve. This wide range is due to the geology of individual wells, the quality of their oil, the amount of NG present, and so forth.

The United States imported 4.3 billion barrels of oil in 2008 (51% of the total used). Of these imports, 21.2% came from Canada, 10.4% came from Mexico, 9.1% from Venezuela, 8.6% from Saudi Arabia, 6.9% from Nigeria, 4.8% from Russia, 4.2% from Algeria, 3.9% from Angola, 3.9% from Iraq, 2.6% from Brazil, only 0.7% from Libya, and 24.4% from all other world sources. The market trading prices for oil are volatile (see above); supply uncertainties caused by the political and social unrest in Egypt and certain Arab nondemocracies in February 2011 caused the price of oil to jump 6% in US markets to $91/bbl on February 21, 2011 (Banerjee and White 2011). It rose to $109/bbl on April 7, 2011 and then went down to $84/bbl (September 28, 2011) and back up to $102.59/bbl (November 17, 2011) (USEIA 2011b). Figure 3.13 shows the power-law growth in spot oil prices from 2002 to 2008, with the precipitous drop from approximately $140/bbl to approximately $35/bbl during the 2008 US and global economic recession/panic, and its power-law climb back to its April 7, 2011, value. (The market oil price was hovering around $100/bbl on February 10, 2012, is now at $93.35/bbl as of August 9, 2012, and is rising.)

The bottom line on petroleum production in the world is that it will slowly decrease as present proven reserves become exhausted; this decline will be partially offset by oil production from newly discovered deepwater fields and oil from tar sands and oil shale. The price of oil, and hence *all* petroleum

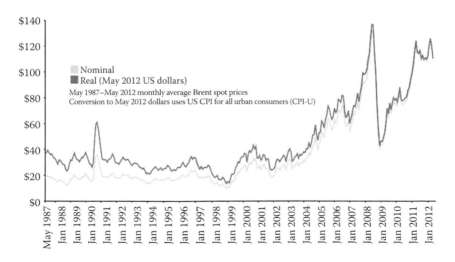

FIGURE 3.13
Brent Spot Monthly oil prices: May 1987 to January 2012. Both nominal and real prices (in November 2011) in US $ are given. (Courtesy of the US EIA and the US Bureau of Labor Statistics.)

products, will steadily (monotonically) increase with many short-term fluctuations. Human population growth, coupled with the industrialization of
countries like Brazil, China, and India, will drive up the demand for oil, further increasing its price. Eventually, the high cost of gasoline and diesel fuel
will trigger widespread changes in our energy usage in the United States.
More electric vehicles powered by storage batteries or hydrogen or methane
fuel cells will be used; there will be more electric public transportation; and
biodiesel, methanol, ethanol, NG, dimethyl ether (DME), and hydrogen will
gain more prominence as alternate fuels for internal combustion engine–
powered transportation. We will rely less on plastics in the food marketing industries and more on recycled and recyclable paper and cardboard
(remember butcher's paper?). Alternate, green sources of electric power
(wind, solar, geothermal, tides) will be developed at a higher rate as will be
new electric power distribution networks. The development of new, expensive, safer, nuclear power plants will occur as well, but at a much slower rate
than green sources. Efficient, low-cost electric energy storage means will also
have to be designed and implemented to support alternate energy sources.

3.7.5 Oil Shale

Oil shale is another worldwide oil reserve that may contain as much as 2.8 to
3.3 trillion barrels (4.45×10^{11} to 5.25×10^{11} m^3) of recoverable oil. This exceeds
the proven conventional, in-ground oil reserves, estimated at 1.317 trillion
bbl (2.094×10^{11} m^3) (Dyni 2010).

Oil shale is an organic-rich, fine-grained sedimentary rock that contains
significant amounts of kerogen (a mixture of organic chemical compounds,
some of which are found in petroleum). There are extensive deposits of oil
shale in the Green River formation in Colorado, Utah, and Wyoming that
are estimated to contain 1.5 to 3 trillion barrels of oil, an amount equal to
the world's proven oil reserves. "Green River oil shale contains between 5%
and 40% kerogen by weight, with the higher end of this range being rare"
(Brandt 2008).

While oil sands originate from the biodegradation of oil, heat and pressure
have yet to transform the kerogen in oil shale into liquid petroleum. There
are oil shale deposits all over the world, including but not limited to the following: the United States, Estonia, Australia, Germany, Russia, the United
Kingdom, Brazil, Israel, Jordan, Egypt, Turkey, and China. Oil shales were
classified as to the origin of their kerogen by A.C. Hutton of the University
of Wollongong. They include terrestrial, lacustrine (deposited at the lake
bottom), and marine (deposited at the ocean bottom). This classification has
proven useful in estimating the yield and composition of the extracted oil
deposits. Total deposits in the United States alone constitute approximately
62% of the world's shale oil resources (Dyni 2010).

Oil shale can be burned directly, or the shale oil can be extracted by
resource-intensive processes. Presently, oil shale is mined by environmentally

destructive open-pit or strip mining and then transported by FF-burning trucks or trains to locations where it can be burned directly, or processed for its oil. Estonia has an installed capacity of 2.97 GW of directly oil shale–fired power plants; other countries also burn oil shale directly (Israel 12.5 MW, China 12 MW, and Germany 9.9 MW). The burned oil shale rock has to be disposed of.

To recover the oil from shale, the surface-mined shale rock is crushed and then subjected to chemical pyrolysis that converts the kerogen to shale oil [a synthetic crude oil (SCO)] and combustible oil shale gas. The shale is heated to between 450°C and 500°C in the absence of oxygen. In situ extraction of oil shale and gas is under development. They offer the advantage of being able to reach deeper deposits of oil shale than can surface mining. The shale must be fractured and then heated, and the oil and gas must be recovered from wells. Shale oil is best suited for refining into middleweight distillates such as diesel fuel, jet fuel, home heating oil, and kerosene. Further refining can produce lighter-weight hydrocarbons suitable for gasoline.

Oil shale processing is a water-intensive activity, which has a potential severe environmental impact on ecosystems, agriculture, domestic consumption, and so forth. Water is used for five activities in shale oil extraction:

(1) *Extraction and retorting:* It is used for cooling equipment and extracted shale oil, producing steam and in situ fracturing of retort zones (fracking), preventing fires, building roads, controlling dust, and so forth.

(2) *Upgrading shale oil:* The quality of the extracted shale oil needs to be improved for transportation to a refinery.

(3) *Reclamation:* Water is used to reclaim mine sites by cooling, compacting, and stabilizing the waste piles of retorted shale and to revegetate disturbed ground surfaces, including the surfaces of the waste piles.

(4) *Power generation:* Oil shale production requires electric power, generally generated by steam turbines turning alternators. The steam production is generally closed cycle, but water is needed to cool condensers. The amount of water used in power generation depends on the fuel used (NG, coal oil) and the power plant design.

(5) *Domestic use:* Potable water is needed to support oil shale workers and their families.

According to the estimates in the US GAO (2010) report on water resources on Green River formation shale oil production, the mining and surface retorting oil extraction processes may consume as much as 49.32 acre-feet of water (60.83 × 10³ m³) per day to produce 100,000 bbl (1.59 × 10⁴ m³) shale oil per day. (This translates to 3.8 m³ water for each 1 m³ of shale oil.) Estimates of water consumed in in situ (underground) shale oil extraction could be as much as 773 acre-feet (9.535×10^5 m³) of water per day to extract 500,000 bbl (7.95×10^4 m³) of oil per day. (The in situ water/oil ratio is approximately 12.0 m³/m³ on a daily basis.) (Note: 1 acre-foot = 1233.5 m³; 1 US petroleum bbl = 44 US gal. = 0.159 m³.)

Three key questions are the following: (1) Where will this water come from? (2) Can this water be recycled? (3) How will global climate change affect these future water supplies? Surface water is available in the Colorado–Utah–Wyoming oil shale regions from the Green, White, Yampa, and Colorado Rivers. Additional water may be withdrawn from wells in aquifers. The bottom line is that the size of the oil shale mining and extraction industry probably will be limited by water availability. Water in the shale oil industry will no doubt have to be heavily recycled for mining applications, mitigating water needs for shale oil production. Water resources in the area are already subject to water rights for the existing demands of agriculture, domestic use, and industry. The GAO report estimates that 297,000 acre-feet/year (3.66×10^8 m³/year) of water would be physically and legally available in 2030 from the White River in Meeker, Colorado. We argue that it will probably be significantly less because of the slow reduction in annual precipitation due to GW, plus direct demands from population growth and agriculture (they will probably still be growing corn for ethanol). When shale oil production takes off, the water shortages produced will be an excellent example of nonsustainability. Which will be more valuable to our society for ecological sustainability, oil or water? Which will be commercially more valuable?

Finally, we summarize the findings of Brandt (2008) on the energy efficiency of in situ shale oil extraction: energy outputs from refined liquid fuel from shale oil are only 1.2 to 1.6 times greater than the total primary energy inputs to the processes. If the CO_2 produced generating electricity used in shale oil production is *not captured*, the well-to-pump GHG emissions are in the range of 30.6–37.1 g of carbon equivalent per megajoule of liquid fuel produced. Brandt stated: "These full-fuel-cycle emissions are 21%–47% larger than those from conventionally produced petroleum-based fuels."

Thus, in spite of the great estimated shale oil reserves in Colorado, Utah, and Wyoming, the extraction of the shale oil is not without sustainability-threatening "side effects." It probably will be limited by available water resources in the area, it will not be particularly energy efficient, and it will be more GHG intensive. As the shale oil industry grows, there should be strong pressure from environmental and farming groups to have the shale oil production processes made more efficient; water must be recycled where possible, and green sources of electric power (wind, solar) must be used to supplement electricity needs. US shale oil has to compete in the market with oil extracted from Canadian oil sands. In the next 10 years, the Canadian oil sands are estimated to produce 4 billion barrels of oil per year, larger than the estimated shale oil production from the Green River oil shale deposit (Oil sands 2008).

3.7.6 Tar Sands

Tar sands are also referred to as bituminous or oil sands. Major deposits are found in the United States in eastern Utah [reserves estimated to be between 1.43×10^9 and 2.27×10^9 m^3 (Tar Sands 2011)], and great quantities are present in east central Alberta in Canada (about 2.09×10^{11} m^3). The Alberta tar sand deposits contain about 85% of the world's known reserves of natural bitumen (Titman 2010; Veazey 2006). There are also large deposits of tar sands in Venezuela (about 3.18×10^{10} m^3 of recoverable reserves in the whole Orinoco belt, assuming 22% recovery) (Cleveland and Roman 2007). A web report by the USGS (2009) reported that their estimate of technically-recoverable heavy oil in the Orinoco Oil Belt Assessment Unit of the East Venezuela Basin Province lies between 380 to 652 billion barrels (4.39×10^{10} to 7.54×10^{10} m^3). Smaller tar sand deposits are also found overseas in the Republic of the Congo, and in Madagascar.

In Alberta, Canada, shallow oil sand deposits are typically 40 to 60 m thick, sitting on top of flat limestone rock. In the Canadian Athabascan sands, there are very large amounts of bitumen sand covered by relatively thin layers of peat bog (1–3 m) on top of clay and barren sand (0–75 m). Trees, topsoil, and peat are bulldozed off the oil sands, creating open pit mines. This light overburden makes surface mining of oil sands feasible. Modern extraction methods permit recovery of over 90% of the bitumen in the sand. Typically, approximately 2 tons of oil sands are required to produce one US petroleum barrel (42 US gal.) of heavy oil.

One common method for extracting bitumen from the sand is the Clark hot water extraction (CHWE) process: the mined oil sands are crushed for size reduction; then hot water from 50°C to 80°C is added to the sands, and the slurry is transported to a primary separation vessel (PSV) where bitumen is recovered by flotation as bitumen froth. The recovered froth is about 60% bitumen, 30% water, and 10% solids by weight. The water and solids are removed, and the bitumen is next processed to decrease its viscosity. It may be heated or mixed with lighter petroleum (liquid or gas), or chemically split (at 482°C and 100 atmospheres), before it can be transported by pipeline to a refinery for upgrading into SCO, thence to make oil products (gasoline, kerosene, diesel fuel, naphtha, etc.).

There are several other means of extracting oil from tar sands; all are water and energy intensive. It is estimated that 2 to 4.5 bbl of water are needed to extract 1 bbl of SCO, or syncrude. Greenpeace estimated that approximately 3.49×10^8 m^3 of water per year is required for *all* Canadian oil sands operations (Greenpeace Canada 2007). Despite recycling, almost all of the water ends up in tailings ponds, which in 2007 covered an area of 50 km^2! This water is highly polluted and toxic; it cannot be returned to its source. Just sitting in the tailing ponds, it is evaporating, leaving behind the less volatile contaminants. What do you think will happen to this polluted water?

Because SCO production from oil sands is energy intensive, it emits more GHG than the production of conventional crude oil. SCO production has been identified as the largest contributor to GHG production in Canada, approximately 40 million tons of CO_2 emissions per year. "Environment Canada claims [that extracting SCO from] the oil sands make up 5% of Canada's greenhouse gas emissions, or 0.1% of global greenhouse gas emissions. It predicts that oil sands will grow to make up 8% of Canada's greenhouse gas emissions by 2015" (CAPP 2008).

About 1.0 to 1.25 GJ of energy is needed to extract a barrel of bitumen and upgrade it to SCO. Most of this energy comes from burning NG. Since a barrel of SCO contains approximately 6.12 GJ of chemical energy, the ratio of energy extracted to energy consumed is approximately 5 to 6. In addition, the surface mining of tar sands is damaging land and riverine (i.e., the Athabasca River) ecosystems (Kunzig 2009). It remains to be seen whether SCO extraction and refining from tar sands turns out to be more environmentally damaging and less sustainable than extracting and refining kerogen from US oil shale. Both processes are environmentally very abusive.

A second means of extracting bitumen from deep oil sand deposits uses drilled wells and a process called steam-assisted gravity drainage (SAGD). This process was initiated around Christina Lake, Canada, in 2002. A set of steam injection wells are drilled down to the top of the deep bitumen sand deposit and then turned horizontally and run along the deposit. A second set of extraction wells are drilled deeper than the steam injection wells and then turned to run horizontally parallel and below the steam injectors. Steam at approximately 250°C causes the bitumen to liquefy and trickle down to the collectors, where a mixture of hot water and melted bitumen is pumped to the surface. Because the SAGD process requires lots of steam, it is energy inefficient; NG must be burned to make the steam with a corresponding emission of CO_2. To improve the efficiency of this deep bitumen extraction process, butane gas ($CH_3CH_2CH_2CH_3$) has been mixed with the injected steam to dissolve the bitumen. With steam alone, the steam-to-extracted oil ratio was approximately 2.38; with the butane solvent gas, extraction efficiency is expected to improve to 1.7, that is, less steam (hence less energy and water) will be required for a given amount of oil. What fraction of the injected butane gas will be recoverable? It is a fuel as well as a bitumen solvent. Further evolution in the solvent-based oil extraction process will eliminate steam entirely and use just heated solvents. A $60 million, solvent-extraction pilot plant is being tested at Suncor Energy's Dover site, northwest of Fort McMurray, Canada (Fairley 2011).

In July 2010, an Enbridge pipeline carrying heavy crude oil (diluted bitumen) in Marshall, Michigan, ruptured and spilled an astonishing 819,000 gal. of toxic sludge into Talmadge Creek and over 30 mi. of the Kalamazoo River (Pearce 2012). The light organics floated, and those that did not evaporate could be skimmed up, but the more dense tar-based hydrocarbons sank to the bottom of the watercourses. It has taken Enbridge, a Canadian Company, about 2 years to "clean up" the mess. US investigators knew about the crack in the pipeline that led to the spill 5 years before the disaster, yet nothing was done. (Not in my box?)

A current (January 2012) controversy exists on the siting of the proposed Keystone Oil Pipeline, to run from the Canadian tar sands, down across the central United States, crossing the northern part (main recharge area) of the important Ogallala Aquifer, thence south to the existing Texas Gulf Coast refineries. This pipeline will also carry the liquefied bitumen from the tar sands, a nasty material to get into groundwater if the pipeline fails for any reason.

Also, Enbridge is currently proposing a diluted bitumen pipeline to run from Alberta to Portland, Maine, across northern New England (Huffington 2012). The pipeline would run through northern Vermont, New Hampshire, and across Maine to the coast (Portland), endangering many rivers, lakes, and ecosystems, including Sebago Lake, which provides water for approximately 200,000 persons in Maine. Since the proposed pipeline would cross the US/Canadian border, and several state borders, both US and state regulatory agencies will be involved in the permitting. Let us hope more attention is paid to pipeline cracks in the future.

3.7.7 Methane Hydrate

MH is a clathrate hydrate in which a large amount of NG (mostly methane [CH_4]) is trapped in ice-like cages of water molecules, forming a milky-appearing, ice-like solid. MH has been found under ocean floor sediments, on the deep ocean floor, and also below the frost line in arctic tundra. Antarctic ice cores, taken in archaeoclimate studies, have found large concentrations of MH dating back approximately 800,000 years (by isotopic dating), indicating high atmospheric methane concentrations around that time. See the excellent description of the physics and properties of methane clathrates in the paper by MacDonald (1990).

MH crystals are viewed as a potential source of FF methane. In an early feasibility study on the extraction of ocean-bottom clathrate methane, Max and Dillon (1998) reviewed the physical properties and the structural integrity of the Blake Ridge MH deposits, about 500 m deep in an area approximately 300 km south of Cape Hatteras in the Atlantic, covering approximately 34°–31° N. latitude and 77°–74° W. longitude. They commented that care must be taken in the extraction process to prevent spontaneous methane escape.

The stoichiometry of "typical" MH is 1 mole of methane for every 5.75 moles of water, but this can vary with formation temperature and pressure. One liter of MH contains, on the average, 168 L of methane gas at standard temperature and pressure (STP). An MH phase diagram is shown in Figure 3.14. The diagonal dashed heavy phase boundary line divides the pressure–temperature regions where methane is a gas or a hydrate; the vertical boundary marks the

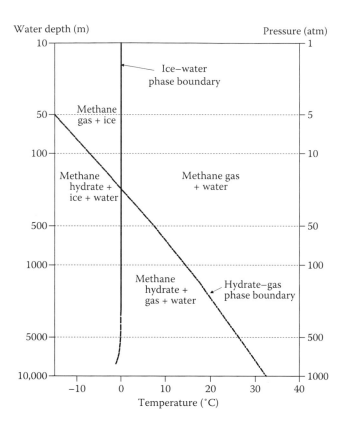

FIGURE 3.14
P–T phase diagram for combined water and methane, showing MH regions.

phase boundary for water–ice. Thus, at water pressures above 5 atmospheres, and temperatures above 0°C, it is possible to find MH along with water and gas. For example, at 10°C, MH forms in water at depths below 700 m, or pressures above 6 atmospheres (73.5 psi); also, MH is stable at 20°C at pressures above 200 atmospheres (2939 psi). Other physical properties of MH are described in an article by Walter (2011).

MH is found all over the Earth, particularly in arctic regions around the permafrost boundary in tundra, and also in the oceans in bottom sediments along continental margins. When methane gas from natural decay or leakage from NG deposits dissolves in water and then is subjected to high pressures at low temperatures given by the MH–water P–T phase diagram, MH forms and is sustained.

Considerable FF energy is trapped in MH deposits. In 2008, the US Minerals Management Service [now the Bureau of Ocean Energy Management, Regulation and Enforcement (BOEMRE)] gave an estimate of the in-place MH deposits in the Gulf of Mexico. This estimate did not consider whether the methane in the MH is technically or economically recoverable. Their estimate was that between 11×10^{12} and 34×10^{12} ft.^{3} (3.12×10^{11} and 9.63×10^{11} m^3) of methane was contained in the in-place MH deposits, with an expected value of 21.444×10^{12} ft.^{3} (6.07×10^{11} m^3). The assessment also stated that approximately 6.7×10^{12} ft.^{3} (1.90×10^{11} m^3) of methane was in MH in relatively high concentrations in sandy sediments, better suited for mining.

In addition to the rich Gulf of Mexico deposits of MH, in 2008, the US Geological Survey (USGS) reported that there is approximately 85×10^{12} ft.^{3} (2.41×10^{12} m^3) of undiscovered, technically recoverable NG resource in MH deposits on the North Slope of Alaska (NETL1 2011). Putting these MH gas reserves into perspective, the total US NG resource, excluding MH amounts, is estimated to be approximately 2.074×10^{12} ft.^{3} (5.87×10^{10} m^3). Thus, even partial recovery of methane from known MH deposits would greatly increase the US NG reserves.

What do these vast MH NG resources mean for sustainability? First, the NG/methane carbon is fossil carbon. Burning NG from MH has the same negative environmental impact as burning NG from wells on land. Yes, methane from MH will supply the world with more energy, and the combustion products from burning it are far cleaner than those from burning coal and oil, the dirty FFs. Methane can also be made into DME, an effective diesel fuel that burns more cleanly than diesel fuel made from fossil petroleum.

Deepwater methane recovery mining technologies are described in the report *Gas Hydrates in New Zealand—A Large Resource for a Small Country?* (NETL2 2011). New Zealand's offshore gas hydrates resources (GHRs) may contain approximately 20×10^{12} ft.^{3} (5.66×10^{11} m^3) of NG, enough to sustain the country's energy needs for several decades, even if partially extracted, and also provide energy export revenues.

MH deposits are found under offshore seabeds around every continent, even Antarctica. MH is also found under arctic tundra in Alaska, Russia, and Canada. Prospective MH deposits on and under the ocean floor are easily located with sonic imaging techniques. There is no doubt that as other FFs become exhausted, more attention will be paid to recovering the energy from this source. Since most MH is undersea, the experience gained in deepwater oil drilling will be put to good use. MH deposits are generally on the seafloor or below it by as much as 700 m.

Magiawala et al. applied for a US Patent on November 22, 2007 (pub. no. US 2007/0267220 A1) for *Methane Extraction Method and Apparatus Using High-Energy Diode Lasers or Diode-Pumped Solid State Lasers*. Heat energy from near infrared (NIR) laser photons delivered at the end of a drill located in an undersea bottom methane clathrate deposit would melt the clathrate and release CH_4 gas (and water and WV), which would be piped to the surface and separated, and the methane would be compressed. We note that other less technical sources of heat energy, such as steam, could also be used, and the water and WV would also have to be separated from the methane.

There is no doubt that future extensive NG production from MH deposits will lead to incidental methane leaks into the atmosphere, even from the sea bottom. Methane is a potent GHG. Taken with the CO_2 from burning MH gas, solar energy absorption and planetary warming will increase more rapidly—unless the CO_2 is sequestered.

Due to silting and heating of water, the extraction of methane from undersea MH deposits can potentially be very harmful to the coastal marine ecosystems that we depend on for food. If all the methane extraction takes place well under the ocean bottom, these effects will be lessened. Energy companies are currently in the exploration stage to locate the richest MH deposits that are the easiest to process in situ. An arctic Alaska project, led by Conoco Phillips, plans experiments in the Prudhoe Bay region to test the effectiveness of a methane extraction method that injects CO_2 gas into a sub–ocean bottom MH reservoir, where the CO_2 molecules are exchanged for methane in situ (NETL1 2011). If successful, this process could lead to a means of underground CO_2 sequestration while methane fuel is produced. If the methane is burned nearby to make electricity, this extraction/sequestration method could be made very efficient while reducing the carbon load from burning the mined gas (ORNL 2002). Still, methane from ocean hydrates is an FF, and burning it releases CO_2. The availability of this new source of FF energy will add to the atmospheric CO_2 burden and may delay our shifting to noncarbon energy sources. Capture of CO_2 from methane-burning power plants, compressing it, liquefying it, and injecting LCO_2 into the methane extraction wells to swap it physically for CH_4 gas is an expensive way to make clathrate methane carbon neutral.

3.7.8 US Oil Addiction

The 2009 US Energy Information Administration figures for daily oil consumption for various countries are given in Table 3.5.

Oil price increases were triggered by the recent unrest in the Middle East (e.g., civil war in Libya; protests in Egypt, Syria, etc.) and global catastrophes such as the earthquake/tsunami/nuclear meltdown in Japan. Americans still drive their low-mpg SUVs to the grocery store and to commute in and put little effort toward asking for development of alternative energy sources. Politicians in Congress respond by taking the short view, clamoring for more deepwater offshore oil exploration and the release of oil from the US Strategic Oil Reserve to drive pump fuel prices down: always the quick fix. For America to be economically sustainable, it must begin a strategic, slow, steady transition from its FF-dependent economy toward alternative energy sources. This requires careful long-range planning and firm government leadership. Our high oil dependence makes our overall economy fragile to any small disruption of our current 18.8 million barrels per day oil supply. Over half of US oil is imported (see Table 3.4), enhancing this fragility. Increases in the price of crude oil per barrel and gasoline at the pump

TABLE 3.5

Oil Consumption of the World's Largest Oil
Consumers

Country	Oil Consumption, Million Barrels/Day (2009 Data)
United States	18.8
China	8.3
Japan	4.4
India	3.1
Russia	2.7
Brazil	2.5
Germany	2.44
Saudi Arabia	2.43
South Korea	2.19
Canada	2.15

Source: White, R.D. and Lee, D., 2011. Quake could roil
energy costs, *Hartford Courant*, March 12, 2011.

Note: China is rapidly catching up to the United States.
Note that 1 US petroleum bbl = 0.159 m^3 = 42 US gal.

quickly follow any interruption in supply, real or perceived, and are soon
reflected in other market sectors such as the cost of foods, heating, and transportation. Natural disasters in the United States and abroad (earthquakes,
tsunamis, asteroid strikes) and wars abroad must be included in US strategic
energy planning. Effective government energy policy change and investment in research on alternative energy sources will result only from loud
public outcry, and this will happen at a tipping point determined by the high
costs of fuels, transportation, and goods.

One measure of the global economic sensitivity to the price of oil is that
a 10% increase in oil price will reduce global economic growth by approximately 0.25%. With the average world economy growing at approximately
4.5% (in 2011), this sensitivity suggests that the price of oil would have to leap
to over US $150/bbl to quash all economic growth. Smaller increases will sap
growth and raise inflation (Economist 2011c).

Whether from foreign or domestic wells, oil sands, or oil shale, as the population grows and crude oil supplies taper off, basic economics will put the
SUVs and pickup trucks in their garages again and force us to use alternative fuels based on ethanol, methanol, and hydrogen, if we can afford them.
There will be a slow but predictable increase in the use of public transportation, carpooling, and telecommuting by the Internet. We as a country in
the future must consume, on the average, less energy per year per person,
or else we may be faced with severe stagflation and a steady drop in the US
standard of living.

One prediction for global oil production (net of processing gains) is that it will reach 96 Mbbl/d in 2035, an increase of 13 Mbbl/d over 2010 levels. Much of this production will be from NG liquids, tar sands, oil shale, and new deep ocean wells (IEA 2011).

3.7.9 Emissions

The law of unintended consequences writ large applies to the combustion of all FFs and all wood and woody (green) fuels. The effects of carbon dioxide are well known; its IR absorption spectrum contributes significantly to the greenhouse effect. The natural sinks of CO_2 are also well known; it is taken up and metabolized in photosynthesis by phytoplankton and higher plants. It is also dissolved in water, and the resulting carbonic acid is taken up by plankton and minerals. The latter process is slow. Other fuel combustion products include the oxides of nitrogen (NO_x) and sulfur (SO_x), as well as a whole array of volatile organic compounds (VOCs), and polyaromatic hydrocarbons (PAHs). Combustion also can produce particulates: ash containing toxic metals such as lead, cesium, and chromium, as well as soot (black carbon) can be emitted. Many of these combustion products are health hazards, causing obstructive lung disease and cancer. NO_x and SO_x gasses dissolve in cloud moisture and cause acid rain, which damages crops, forests, fisheries, buildings, etc. These gasses also lead to the formation of clouds with high albedos that reflect sunlight back into space, counteracting the effect of GHGs such as CO_2 and methane. The complexity of the heating/cooling effects of emissions is underscored by the poorly understood thermodynamic properties of black carbon (Economist 2011b).

Climate forcings by emissions can be expressed in terms of watts per square meter; the more watts per square meter, the greater the warming. Some forcings are as follows: CO_2, 1.7 W/m^2; other GHGs, 1 W/m^2; black carbon, 0.7 W/m^2, with a range of 0.4–1.2 W/m^2. When erupting volcanoes emit SO_x aerosols into the stratosphere, they last longer than those created by industrial pollution lower down in the troposphere, perhaps years, rather than weeks. This persistence of SO_x particulates in the stratosphere multiplies its cooling effect by about 25 times. It was suggested by Dr. Crutzen (at the Scripps Institute of Oceanography) that if man were to inject SO_x into the stratosphere in the "proper amount" while phasing out industrial greenhouse emissions, it could counteract current GW trends and give extensive health benefits (Economist 2011b). Beware the LUC! $SO_2 + H_2O$ yields sulfurous acid, and $SO_3 + H_2O$ yields sulfuric acid, hence acid rain: Chemistry 101.

One modeled (estimated) global temperature increase in 2070 is 2.75°C if nothing is done about emissions. Black carbon and methane reduction will give a 2.25°C rise, as will CO_2 reduction. If both CO_2 and methane as well as black carbon are reduced, the projected global temperature rise is 1.75°C [from a graph in Economist (2011b)].

3.8 Chapter Summary

In this chapter, we have described the more important complex issues that affect human sustainability. The many direct and indirect effects of population growth; GW and its causes and many effects; the importance of freshwater supplies for sustainability and how future climate changes will challenge its supply; and the potential adverse effects on our food supply of pollinator loss by honeybee CCD and bumblebee population loss are discussed. The little-appreciated effects on food sources of species size reduction caused by increasing habitat temperature are also described. Finally, we examine the FF energy sources and their impacts on the climate, economics, our energy supplies, and our overall sustainability.

The impact of the depletion of critical minerals and ores on sustainability should also be considered, in particular, rare earth elements used in making permanent magnets used in electric motors. Most of these lanthanide elements (e.g., neodymium, dysprosium, cerium, terbium, didymium, terbium) are found in China (Bourzac 2011). Shortages will have a dampening effect on the development of efficient electric vehicles.

4

Mitigations of Human Impacts through Technology

4.1 Introduction

Technological societies adept at mathematics, engineering, chemistry, physics, botany, agronomy, and molecular biology have recently developed tools that have the potential to mitigate the impacts on sustainability caused by our population growth and activities. Many of these technological innovations also have had untoward "side effects," requiring that the innovators rethink their approaches to avoid the more dire effects of the *law of unintended consequences*. The promise of technological mitigation of problems based on population growth has led some economists (e.g., Julian Simon 1998) to assert that population growth is not important; technology will save us from our excesses. The fallacy of this assertion was revealed by US anthropologist and historian Joseph A. Tainter in his 1988 book, *The Collapse of Complex Societies*. Tainter considered the collapses of the Western Roman Empire, the Mayan civilization, and the US Chaco Canyon culture. We consider societal collapse as the ultimate manifestation of unsustainability. Tainter outlined four principles about why complex societies collapse. These are the following:

1. "Human societies are problem-solving organizations;
2. Sociopolitical systems require energy for their maintenance;
3. Increased complexity carries with it increased costs per capita; and
4. Investment in sociopolitical complexity as a problem-solving response often reaches a point of declining marginal returns."

A central hypothesis in Tainter's book is the *principle of declining (diminishing) marginal returns* and its effect on sustainability. He makes the point that this is a recurring theme in the growth and decline of most societies, both ancient and modern. Societies become more complex as their populations grow and they attempt to solve growth-limiting problems. This social complexity is characterized by the numerous specialized social and economic roles that develop and by the many socioeconomic pathways through which they are coordinated, as well as by contemporary reliance on symbolic and abstract communication (e.g., radio, TV, the Internet, cell phones, texting, etc.) and the existence of a class of information producers and analysts who are not involved in primary resource production (e.g., bureaucrats). Such burgeoning complexity requires a substantial "energy" subsidy in which resources and other forms of wealth (e.g., taxes) are consumed. When a society is confronted with a challenge, such as a shortage of fossil fuel (FF) energy, and difficulty in gaining access to it, it tends to create new layers of bureaucracy and infrastructure to address the challenge. Such added complexity may not lead to a solution to the challenge and, indeed, can put additional burden on citizens with increased taxes and restrictive legislation. The society's economic growth is damped by the cost of the additional complexity. The globalized modern world is subject to many of the same stressors that caused older societies to wane. Tainter remarked: "When some new input to an economic system is brought on line, whether a technical innovation or an energy subsidy, it will often have the potential at least <u>temporarily</u> to raise marginal productivity." (We have underlined "temporarily.") (See *Diminishing marginal returns* in the Glossary.)

In this chapter, we examine the important factors our technological society has developed and is developing to make human existence sustainable, given human population growth and its many consequences, specifically the exhaustion of natural resources, anthropogenic climate change, and shortages of food and water.

4.2 Biofuels

4.2.1 Introduction

To be a green fuel, a biofuel feedstock must not be derived from any FF. It must come from recently growing and CO_2-consuming plants, including their seeds, grains, stems, trunks, needles, and leaves. Thus, the CO_2 produced by combusting or fermenting this biofuel "short-cycles" the CO_2 back to the growing plants. However, it is still CO_2 put into the atmosphere, and once there, one CO_2 molecule is just like another.

"Green" biofuels include the following:

(1) Ethanol, distilled from fermented corn mash, sugarcane sap, and other sugar-rich crops such as sorghum and sugar beets. Ethanol can also be produced by genetically modified (GM) bacterial biodegradation of cellulosic plant materials to their component sugars, which can then be fermented by GM yeasts and bacteria. Ethanol production from fermentation chemical reactions produces CO_2 gas in a 1:1 ratio.

(2) Methanol can be produced by direct synthesis by microorganisms and also by direct chemical synthesis pathways (e.g., catalytic oxidation of biogenic methane).

(3) Biodiesel, manufactured from the natural oils in plant seeds such as crambe, canola, corn, oil palm, cottonseeds, peanuts, rapeseeds, sunflower seeds, safflower seeds, and so forth.

(4) Methane gas (CH_4), a fuel, is a natural product of recent bacterial decomposition of organic material and is commonly recovered (along with CO_2) from wells driven into in landfills and so forth. Methane also occurs in natural gas (NG), but this is not "green," nor is the methane recovered from methane hydrates (MHs) green.

(5) Char carbon from the hydrothermal carbonization (HTC) process (see Section 4.2.4) is green.

4.2.2 Energy Densities of Fuels and Batteries

The chemical energy stored in fuels is generally released as heat upon stoichiometric oxidation. The energy released can also include light (photons) and sound (e.g., pressure waves in an explosion). The maximum energy inherent in a fuel that can be released by chemical oxidation can be described by its *specific energy* in megajoules per kilogram or its *energy density* in megajoules per liter. (1 MJ $\cong$ 0.28 kWh). Table 4.1 lists some common fuels' energies.

The energy densities of some common batteries and electrical energy storage devices are listed in Table 4.2.

4.2.3 Ethanol Fuel from Plant Starches

Ethanol (EtOH, C_2H_5OH), the drinkable alcohol, has many uses: as a solvent, a reagent in many chemical synthesis pathways, a fuel and fuel additive to replace FFs in transportation and heating applications, and an antiseptic. It is a direct competitor to methanol (see below) in many green, non-FF (NFF) energy applications.

Ethanol-based internal combustion engines have been around since 1826, when Samuel Morey used alcohol in the first American internal combustion engine prototype. Henry Ford designed his ethanol-fuelled "Quadricycle" in 1896. Twelve years later, in 1908, when Ford introduced the Model T automobile, it had an adjustable carburetor that enabled it to run on ethanol *or* gasoline, or a combination of both—it was the first flex-fuel vehicle (Yergin 2011)! Fast-forward to the second decade of the 21st century: An E100-fueled internal combustion engine requires a compression ratio of approximately 13:1 to take advantage of ethanol's high octane rating and uses multiport fuel injection. Direct injection raises the ethanol octane rating to 130. However, the use of high ethanol/gasoline blends (e.g., E85) presents a problem in cold weather. Ethanol tends to increase the enthalpy of vaporization of gasohol.

TABLE 4.1

Specific Energies and Energy Densities of Oxidizable Fuel Substances

Fuel Substance	Specific Energy (MJ/kg)	Energy Density (MJ/L at STP)
Hydrogen, liquid	142	10.1
Hydrogen compressed at 10,150 pound-force per square inch (psi) = 700 bar	142	5.6
Methane at 1 atm at 15°C: (CH_4)	55.6	0.0378
NG	53.6	0.0364
Compressed NG at 3600 psi	53.6	9
Liquid NG at −160°C.	53.6	22.2
Liquefied petroleum gas (LPG), propane	49.6	25.3
LPG, butane	49.1	27.7
Acetylene (C_2H_2)	48.241	—
Gasoline (87 octane)	46.4	34.2
Diesel fuel/residential heating oil	46.2	37.3
Gasohol (E10) (87 octane)	43.54	33.18
Biodiesel oil	42.2	33
Gasohol (E85)	33.1	25.65
Butanol	36.6	29.2
Coal, anthracite	32.5	72.4
Dimethyl ether (DME)[a]	30.75	20.63
Ethanol	30	24
Methanol	19.7	15.6
Glucose (sugar)	15.55	23.9
Wood (dry)	*ca.* 18	—
Peat sod	12.8	—
Compressed air at 4350 psi (300 bar)	0.5	0.2
Potential energy (PE) of water at 100 m dam height	0.001	0.001

Source: Thomas, G. and J. Keller, 2003. *Hydrogen Storage—Overview.* Presentation at Sandia National Laboratories' Hydrogen Delivery and Information Workshop, May 7–8, 2003. Available at: http://www1.eere.energy.gov/hydrogenandfuel cells/pdfs/bulk_hydrogen_stor_pres_sandia.pdf, accessed July 5, 2012.

Note: Note that molecular hydrogen tops the list. STP = standard temperature and pressure for gasses.

[a] Semelsberger, T.A., R.L. Borup and H.L. Greene. 2006. Dimethyl ether (DME) as an alternative fuel. *Journal of Power Sources.* 156: 497–511.

When fuel vapor pressure is below 45 kPa, starting a cold (below 11°C) engine becomes difficult. Hence, E85 is the maximum blend used in flex-fuel vehicles in the US and European markets. In places with severe cold weather, the United States has a seasonal reduction to E70, although it is still sold as E85. In warm Brazil, flex-fuel vehicles (e.g., 2008 Honda Civic). can operate with

TABLE. 4.2

Specific Energies and Energy Densities of Various Batteries

Storage Device	Specific Energy (MJ/kg)	Energy Density (MJ/L)
Battery, lithium ion nanowire	2.54	
Battery, lithium sulfur	1.80	1.80
Battery H$_2$ closed-cycle *fuel cell* (FC)	1.62	
Battery, Zn–air (hearing aids)	1.59	6.02
Battery, lithium ion	0.72–0.46	3.6–0.83
Battery, Zn–Mn (alkaline), long life	0.59–0.4	1.43–1.15
Battery, silver oxide	0.47	1.8
Battery, Ni–metal hydride (consumer)	0.4	1.55
Battery, Ni–metal hydride, used in cars	0.250	0.493
Battery, Pb/H$_2$SO$_4$	0.14	0.36
Battery, Ni/Cd	0.14	1.08
Battery, Zn–carbon	0.13	0.331
Ultracapacitor	0.0199	0.050
Capacitor (energy is stored in an electric field, rather than chemical potential)	0.002	—

Source: Thermoanalytics, 2012. *Battery Types.* Available at: www.thermoanalytics.com/ support/publications/batterytypesdoc.html, accessed July 5, 2012.

E100, which is actually the 95.6% ethanol azeotrope. During cold starts, pure gasoline is injected from a small secondary gasoline tank near the engine. Volkswagen redesigned its 2009 VW Polo E-Flex engine to cold-start without gasoline; this is the first Brazilian flex-fuel model without an auxiliary gasoline tank for cold starts. Brazil also boasts the Honda CG 150 Titan Mix, the first flex-fuel motorcycle sold in the world (in 2009).

In the United States, E10 is a standard motor fuel (10% EtOH + 90% gasoline); the ethanol is added to stretch the fuel volume and to serve as an oxidizer to reduce unwanted emissions (NO_x, SO_x, CO, particulates). In Brazil, most automobiles are "flex models" whose engines can run on blends of ethanol/gasoline, ranging from E20–E25 gasohol to E100. E100 is an azeotrope composed of 95.6% EtOH + 4.4% water; however, E85 is most commonly used. Ethanol/gasoline blends are becoming increasingly more available in the EU countries. Germany, France, and Sweden lead in the consumption of bioethanol (DOE 2012d).

There are several pathways whereby ethanol can be synthesized, including the *Fischer–Tropsch process*, which uses syngas. However, to be a green fuel, ethanol must be distilled from fermented plant products. The substrates now used are corn kernels, sugarcane, sorghum cane, and sugar beets. In the United States, cornmeal mash is cooked, then fermented by yeasts, and distilled; in Brazil, sugarcane is crushed, and the sugary sap is fermented. Brazil makes 6800–8000 L/ha of EtOH from sugarcane; the United States makes

3800–4000 L/ha from ground maize. The total EtOH production in 2009 in Brazil was 24.9 billion L (6.578 billion US gal.), coming from sugarcane grown on 3.6 million ha. In the United States, the total EtOH production in 2009 was 40.693 billion L (10.750 billion US gal.), coming from corn grown on 10 million ha. Brazil had 35,017 ethanol (E100) fueling stations (or 100%) in 2010. The United States had only 2321 (or 1%).

Currently, the first-generation process to produce ethanol in the United States uses only a small part of the corn plant, that is, the corn kernels. The corn stover is generally used for cattle feed (e.g., as silage). The corn kernels contain starches (about 50% of the dry kernel mass), which must be thermally and/or enzymatically broken down to the component sugars (glucose, fructose, and xylose), which can be used by yeasts to make ethanol. The basic fermentation reaction is

$$\overset{\text{(Glucose)}}{C_6H_{12}O_6} \xrightarrow{\text{(Zymase)}} 2\,\overset{\text{(EtOH)}}{C_2H_5OH} + 2\,CO_2 + \text{heat} \tag{4.1}$$

The glycolysis metabolic process in yeast metabolism makes two 3-carbon pyruvate (CH_3COCOO^-) molecules from one D-glucose molecule. In one biofermentation pathway, the enzyme *pyruvate decarboxylase* catalyzes the formation of one 2-carbon acetaldehyde (CH_3CHO) molecule plus one CO_2 gas molecule, given one pyruvate molecule. Next, the enzyme *alcohol dehydrogenase* acts on the acetaldehyde molecule to form one molecule of ethanol (CH_3CH_2OH) (Northrop and Connor 2009). The residue following the fermentation of the corn mash is called *distillers' dried grains with solubles* (DDGS). DDGS are used as a high-protein, low-carbohydrate livestock feed.

The liquid containing the ethanol is separated from the corn mash and yeast, and the alcohol is separated from it by fractional distillation. Fractional distillation of the ethanol-containing water can concentrate ethanol to only approximately 95.6% by volume (89.5 mol%), because of the formation of an ethanol–water *azeotrope* with a boiling point at 78.1°C at atmospheric pressure. Because *both* the water and alcohol in the azeotropic mixture boil at this temperature, no further separation by distillation can be accomplished. A number of processes have been developed to obtain 100% ethanol from the ethanol–water azeotrope. Desiccation of the ethanol azeotrope can involve water absorption by hydrophilic substances such as starch, corn grits, zeolites (clays), or glycerol. Industrial absolute alcohol is made by adding a small quantity of benzene to the azeotrope and redistilling. Absolute ethanol is obtained in the third distillation fraction at 78.3°C. This absolute ethanol contains a small amount of benzene, a carcinogen, and if diluted is not fit for human or animal consumption; it is useful only as a fuel, industrial reagent, or solvent. Do not drink 200-proof lab alcohol!

In an economics mathematical modeling study, Taheripour and Tyner (2007) addressed the problem of how the 51 cents/gal. federal subsidy for ethanol producers will affect the price of E10 gasoline, the prices of corn and other foods, and the price of corn used in livestock and poultry feed. They showed that ethanol producers pass subsidy benefits to farmers when they become a major corn buyer and the supply of corn is limited and how the farmers will pass these benefits to landowners (a "trickle-down" model). Eventually, these subsidies cause a rise in all agricultural land leases and land prices.

About 1% of the annual US corn crop is eaten directly by humans, a larger portion goes into processed foods (cornmeal, corn bread, high-fructose corn syrup, etc.), and about half the crop is used to feed livestock and poultry; in 2009, 41% went into making ethanol (Yergin 2011). We expect that US corn-based ethanol production will saturate in the next decade or so; future droughts will limit corn growth in the US Midwest. There is only so much land to grow corn, and demand is growing with the number of motor vehicles (and the population), as is the demand for affordable food.

4.2.4 Cellulosic Ethanol

Second-generation ethanol production processes remove the need for corn kernels as a feedstock, thus liberating much of the US corn crop for food uses and perhaps reducing the rate of inflation of meat prices (corn-fed beef, chicken, pork, etc.). Cellulosic ethanol (CE) production requires that plant lignocellulose (a structural component of plant cell walls—wood is 40%–50% cellulose) be broken down into its component carbohydrate polymers: lignin, cellulose and xylan. Figure 4.1 illustrates the plant molecules relevant to CE production. The three-component, monolignol molecules that are used in the biosynthesis of lignin—a lignin subunit, a xylan polymer subunit, and a cellulose polymer subunit—are shown. Lignin, xylan, and cellulose have considerable hydrogen bond cross-links between chains. The cellulose and xylan polymers are digested by acid hydrolysis and/or enzymatically into their component sugars, glucose and xylose, respectively. After isolation, a neutral solution of the sugars is then fermented by suitable GM yeasts to ethanol. The ethanol is then separated, distilled, and dried into 99.5% EtOH (NREL 2007). The first steps in the production of CE are to physically macerate the plant feedstock and then treat it with heat and acid or enzymes to break down the lignocellulose into its component sugars, which can then be fermented by yeasts.

One advantage of making EtOH from lignocellulose is that any plants will do, for example, corn stalk stover, straw from harvested grains (wheat, rye, oats, and rice), spent peanut and soybean plants, as well as *Miscanthus* sp., wood chips, sawdust, switchgrass, citrus peels, and waste from lawn and tree trimming. Another advantage is that CE can be produced far more efficiently than corn-based ethanol. Combustion of corn-based ethanol provides only about 26% more energy than is required for its cultivation and production, while CE yields approximately 80% more energy than is required for its growth and production (Ratliff 2007).

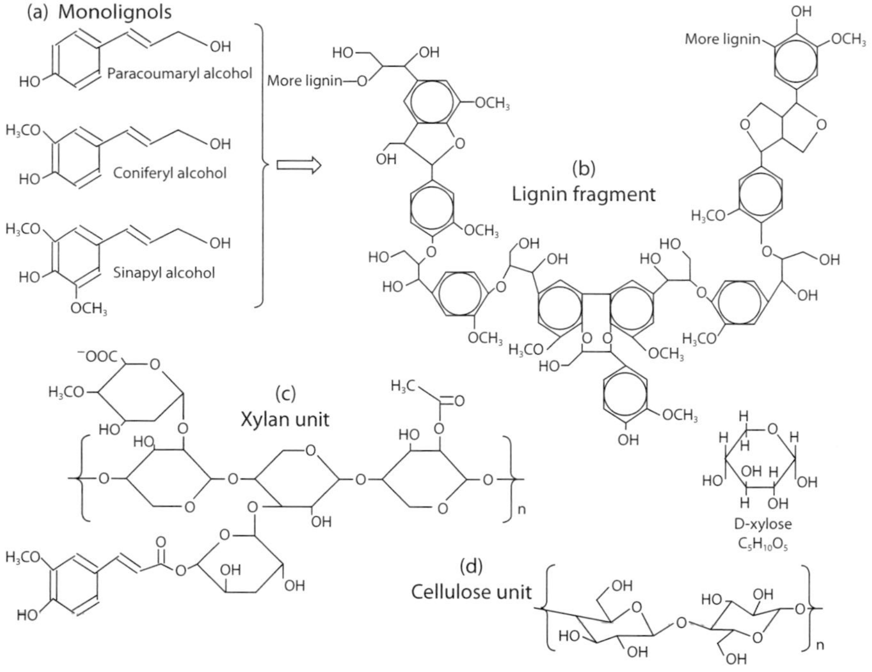

FIGURE 4.1
Plant biopolymer molecules lignin (b), xylan (c), and cellulose (d). Three alcohols that are used in the biosynthesis of lignins are shown in a. The pentose sugar D-xylan is used in the biosynthesis of cellulose.

The production of CE is a complicated, multistep process that is continually being made more efficient. (The original CE process devised in Germany in 1898 could yield 7.6 L EtOH per 100 kg wood waste.)

First, a cellulose feedstock, such as paper pulp, stover, sawdust, wood chips, switchgrass, and so forth, is pretreated to separate the cellulose and xylan polymers from the rest of the plant matter. *Second,* cellulase enzymes or GM fungi that metabolize cellulose and xylan to sugars are mixed with the cellulosic slurry to break them down into their component sugars. *Third,* the sugar solution is separated from the lignin and other residues. *Fourth,* yeasts are added to ferment the sugars to ethanol. *Fifth,* the ethanol–water product is distilled to concentrate the ethanol to the 95.6% azeotrope. *Sixth,* the ethanol can be further dehydrated to an approximately 99.5% concentration.

A second-generation CE production process being developed by Mascoma, a start-up biofuels company in Lebanon, NH, promises to be less expensive. The company is developing three GM organisms that will combine the first two steps of the first-generation process: two bacteria and a yeast. They break down the cellulose and convert the sugars into ethanol at high temperatures. *Clostridium thermocellum* and *Thermophilus saccharolyticum* are thermophilic bacteria that break down cellulose and xylan polymers into glucose and xylose sugars, respectively, at high temperatures. Only the glucose is metabolized by *Saccharomyces* sp. yeast, however, and the ethanol yield is low because of metabolic inhibition of the fermentation pathways by metabolic products such as acetate, lactate, and the end product, ethanol itself. Mascoma's molecular biologists have genetically engineered both bacterial strains able to ferment the pentose wood sugar, D-xylose ($C_5H_{10}O_5$) (see Figure 4.1). They also modified both bacteria's metabolic pathways so they do not produce the unwanted lactate and acetate by-products.

A yeast was also genetically engineered by Mascoma to produce cellulolytic enzymes (cellulases) while growing on cellulose, breaking it down to its component hexose sugar. Also, genes were inserted to enable the yeast to ferment the pentose sugar xylose, further increasing the process' ethanol yield. In experiments with paper sludge, the GM yeast broke down and converted 85% of the cellulose into sugars and produced ethanol without the help of added enzymes. In February 2009, Mascoma announced that its pilot CE plant in Rome, NY, had begun producing ethanol. The demonstration facility has the flexibility to run on various biomass feedstocks, including wood chips, corn stover, and sugarcane bagasse. The pilot plant provided performance engineering data sufficient to support construction of a commercial CE refinery in Kinross, MI (Mascoma 2009). The commercial-scale Kinross CE plant (capacity of 20 million gal./year) was built with funds from the US Department of Energy (DOE), the state of Michigan, Valero Energy corporation, and of course, Mascoma. Kinross Cellulosic Ethanol LLC is expected to be completed by the end of 2013. Its feedstock will primarily be hardwood pulpwood (Mascoma 2011).

An innovative green process to make ethanol from syngas ($CO + CO_2 + 5H_2$) is being developed by a start-up company, Coskata, in Warrenville, IL. Coskata scientists claim that their process can make ethanol out of a variety of organic feedstocks including (but not limited to) wood chips, household garbage, grass, and old tires.

The first step is pyrolysis of the feedstock with an electric plasma torch to produce syngas (CO, CO_2, H_2). Instead of using the Fischer–Tropsch process to make mixed alcohols from syngas, Coskata uses GM, anaerobic bacteria, such as *Clostridium ljungdahli,* found in commercial chicken wastes. The isolated bacteria in an aqueous culture medium are fed directly by the syngas, and their metabolism produces ethanol. The ethanol is separated at the reactor output from the water by vapor permeation. Hydrophilic membranes draw off the water, leaving relatively pure ethanol behind. This process consumes one-half as much energy as distillation, per liter of fuel. Coskata is building a pilot plant intended to make 151,442 L/year (40,000 gal./year) (Bullis 2008). Another start-up company using GM *Clostridium* sp. bacteria to make ethanol from syngas is BRI Energy in Fayetteville, AR. While the feedstock for syngas production may be "free," the electric energy required for plasma torch pyrolysis is a debt that must be charged against the free energy of the ethanol produced. Pyrolysis by focused solar energy is also a possibility for this process.

gas2.org (2009) listed 19 startup companies actively building CE pilot plants and doing research to improve the efficiency of cellulosic ethanol production in the US and Canada. In March 2007, the US government awarded $385 million in grants to help start-up companies get into cellulosic ethanol production. Balboa (2012) stated that in 2011, there were no large-scale, commercial US CE plants on line in spite of all this "priming". There were, however, a few operational pilot plants operated by certain energy and biotechnology companies. Some major EtOH producers have commercial-scale CE plants expected to come online by 2014. For example: INEOS Bio (UK) was expected to commission an 8 million gal./year CE plant in the second quarter of 2012 at Vero Beach, FL. Valero and Mascoma were expected to complete a 20 million g/year CE plant in Kinross, MI by the end of 2013. DuPont is building a 27.5 million g/year [MGY] (104.1 million L/year) CE plant in Nevada, Iowa, expected to be operational in mid-2014. The $200 million DuPont plant will use corn stover as feedstock (Dittrick 2012; Green Car Congress 2012a). GraalBio plans to build a $145 million CE plant in Alagoas, Brazil. The plant will have a nominal production capacity of ca. 22 million US gallons (82 million liters). It will use straw and sugarcane bagasse as feedstock; it may begin production in the last quarter of 2013 (Green Car Congress 2012b). On the down-side of CE plant development, British Petroleum (BP) subsidiary, BP Biofuels Highlands, announced on October 25, 2012 that it is ending its plans to produce CE in a proposed 36 MGY plant in Highlands City, FL. The project was originally scheduled to begin operation in 2013, and was unique in that it included construction of a 20,000 acre dedicated biomass energy grass farm around the CE plant. BP also stated that it will no longer be pursuing CE production in the US, suggesting that it is also abandoning plans for a 72 MGY CE facility in the Gulf Coast (Seekingalpha 2012).

The investment in pilot plants has been largely driven by US Government grants, and the fact that cellulosic ethanol production in North America will be more cost effective than producing ethanol directly from corn or other grains. So why are there no large CE plants on-line in the US in January 2013? The problem is complex and involves lack of investment capital, energy market uncertainties due to the rise in domestic NG production, the lack of government support, etc. (Thomasnet 2012).

In his state-of-the-union address on January 23, 2007, US President G.W. Bush announced a proposed mandate of 35 billion gal. of ethanol per year (132.5 billion L/year) to be produced by 2017, of which 20 billion g/year would be from cellulosic sources. Perhaps by 2017, US ethanol production will meet the 35 billion gal. target, and most of it will be cellulosic.

We also expect that ethanol from corn will gradually be phased out as a fuel source in the United States for economic reasons. Ethanol from corn can be viewed as primarily recycled FFs because of the large amount of FF energy required to plow, plant, cultivate, fertilize, suppress weeds, water, harvest, obtain the corn kernels, convert them to ethanol, and distill the ethanol. Table 4.3 illustrates the estimated *net energy gain ratio* (the "energy balance") for various fuels, that is, the ratio of the fuel energy output to the total energy input in producing that fuel.

Searchinger et al. (2008) concluded that once direct and indirect effects of land use changes were considered in the life cycle assessment of biofuel additives in gasoline, instead of saving, both corn kernel and CE use *increase* net carbon emissions compared to pure gasoline, by 93% and 50%, respectively.

In a second paper published in the same issue of *Science* as Searchinger et al., Fargione et al. (2008) reported on a study that examined six scenarios of wilderness being converted to grow biofuel crops: Brazilian Amazon to grow soybeans for biodiesel feedstock, Brazilian Cerrado to grow soybeans for biodiesel, Brazilian Cerrado to grow sugarcane for ethanol, Indonesian and Malaysian lowland tropical rainforest to grow oil palms for biodiesel, Indonesian and Malaysian peatland tropical rainforest to grow oil palms for biodiesel, and US central grasslands to grow corn for ethanol. Fargione et al. found that a net carbon debt is created when natural lands are cleared and converted to biofuel production. This carbon debt applies to both direct and indirect land use changes.

Other less quantifiable disadvantages of using corn kernels for ethanol production are the loss of arable land that could be used to grow other food crops for humans and animals (e.g., soybeans, potatoes, grains, peanuts, etc.) and also that the extra water used to irrigate the fuel corn crops deplete local aquifers at a higher rate than if the ethanol corn were not grown. The higher costs of food (meats in particular) exist because less corn is available for farm animal feeds, driving up the cost of food production. This increased food production cost is reflected on the shelf prices in supermarkets. When the alternative, green, CE synthesis methods described above are widely implemented, ethanol will continue to be produced and consumed in the United States at high rates, and food prices should rise at a lower rate.

At present, CE production promises to be expensive: although the feedstock is cheap, its processing into ethanol is not, and collection and transportation of feedstock to the CE plant is energy intensive. In our opinion, tuning the process with GM bacteria and yeasts promises to reduce cost of CE to below that of corn-based ethanol.

TABLE 4.3

Estimated Values of Energy Balance for Several Sources of Ethanol

Country	Fuel Type	Energy Balance
United States	Corn ethanol	1.3
Brazil	Sugarcane ethanol	8
United States	CE	2–36[a]

Source: Bourne, J.K. Jr. and R. Clark. Green dreams. *National Geographic*. October 2007. p 41.

[a] Depends on production method.

4.2.5 Methanol as Energy Source

Methanol (methyl alcohol, MtOH, CH_3OH) is the poison one-carbon alcohol (its vapor is also toxic if inhaled in quantity). [The LD_{50} of liquid methanol in mice is about 7 g/kg (Smith and Taylor 1982).] Methanol is emerging as a strong competitor with hydrogen and ethanol as an NFF. Below we discuss the many uses of methanol, the sources of methanol (green and nongreen), and its advantages and disadvantages as a fuel.

Methanol has several uses as a fuel: (1) It can power internal combustion engines as a direct gasoline substitute. It has a high octane rating [research octane number (RON) = 107 and motor octane number (MON) = 92]; it has a higher flame speed than gasoline, leading to higher efficiency, as well as a higher latent heat of vaporization (3.7 times that of gasoline), meaning that the heat generated by the engine can be removed more efficiently, leading to air-cooled engines. Methanol also burns cleaner than gasoline. However, methanol has an *energy density* of only 15.6 MJ/L versus 24 MJ/L for ethanol and 34.2 MJ/L for E10 gasoline. Methanol is blended with 15% gasoline to form M85 fuel; this gives burning methanol a color, contributing to its safer use. (2) It cannot be used directly in diesel engines. However, it can be converted chemically by catalytic dehydration to a gas/liquid, DME (H_3COCH_3), which is a good diesel fuel with a cetane number of 55–60 (compared to 40–55 for regular diesel fuel). DME has a good cold-start ability and reduces engine noise. DME also has much lower emissions of NO_x, SO_x, and CO pollutants. China is a major producer of commercial DME. On May 5, 2012, Alibaba Prices listed Chinese 99.99% DME as costing US $1000–1060/ton Free on Board (FOB) (to the destination port), with a 14-ton minimum order. This translates to a wholesale cost "at the ship's rail" of approximately US $1.10/kg, or $0.50/lb. See the Glossary for properties of DME.

Methanol is also currently used in the production of biodiesel by transesterification of vegetable oils (e.g., rapeseed oil). (3) Methanol can be used in direct methanol FCs (DMFCs) (cf. Section 4.5.6) to generate direct current (DC) electric power to power vehicles and alternating current (AC) inverters for backup power. (4) Methanol and DME can be used to power gas turbines that generate electricity. (5) It has the potential to be used as a domestic heating fuel. (6) Methanol can be used to synthesize a number of industrial chemicals: it can even be transformed into gasoline. In the methanol-to-olefin process, it can be catalytically converted to ethylene and propylene, building blocks in the polymer industry (this is a nonpetrochemical source of these chemicals). Methanol is also a good solvent.

The sources of methanol include direct synthesis by microorganisms and direct chemical synthesis pathways (e.g., catalytic oxidation of methane). Currently, approximately 90% of the worldwide production of methanol is derived from methane from nonrenewable NG. This methanol production proceeds in two stages: methane is converted catalytically into syngas, thence to methanol. The catalyzed reactions are as follows:

$$\underset{\text{(Methane)+(steam)}}{} \quad \underset{\text{(Cat.)}}{} \quad \underset{\text{(Syngas)}}{}$$
$$2\,CH_4 + 3\,H_2O \rightarrow CO + CO_2 + 7H_2 \qquad (4.2)$$

$$\underset{\text{(Syngas)}}{} \qquad \underset{\text{(Cat.)}}{} \underset{\text{(Methanol)}}{}$$
$$CO + CO_2 + 5\,H_2 \rightarrow 2\,CH_3OH + H_2O + Heat \qquad (4.3)$$

The syngas step accounts for up to 70% of the methanol production cost, and current research is seeking to streamline the production and lower its cost (Shekar 2006). Methane can also be created by "cooking" biomass; a wide variety of biomass sources can be used, including, but not limited to, wood chips and sawdust, forest waste, grasses, crop by-products, stover, animal waste, aquatic plants (algae, seaweeds), and municipal waste.

Although conventional NG sources supply the feedstock for nongreen methanol production, other FF sources such as coalbed methane, tight sand gas, and MH deposits in the deep ocean and arctic tundra can be used. (Effective means of collecting this distributed methane gas and bringing it to market will have to be developed.)

Biogenic methane (and thus methanol) is produced by methanogenic bacteria, belonging to the *Euryarchaea* division of *Archaea* bacteria. They generate methane gas from CO_2 and hydrogen gas in a complex biosynthetic pathway. The enzymes of this pathway are extremely oxygen sensitive and use some unique cofactors, one of which is coenzyme F420, which is highly fluorescent under ultraviolet (UV) light. The H_2 gas comes from commensal, H_2-producing anaerobic bacteria. In addition to using CO_2 for a carbon source, some methanogens can use formic acid and acetate (Methane 2011).

A source of "green" methanol is the direct, catalytic hydrogenation of CO_2 with H_2. The CO_2 is a ubiquitous combustion product; the hydrogen can be "green," obtained by electrolysis of water by DC electricity from solar cells (SCs), wind turbines (WTs), nuclear energy, and so forth. The reaction is

$$\underset{\text{(Cat.)}}{}$$
$$CO_2 + 3\,H_2 \rightarrow CH_3OH + H_2O \qquad (4.4)$$

Methanol also can be produced by electrochemical reduction of CO_2, a process that uses DC electrical power. The catalyzed reactions are

$$\underset{\text{(Cat.)}}{}$$
$$CO_2 + 2\,H_2O + electrons \rightarrow CO + 2H_2 + (3/2)O_2 \rightarrow CH_3OH \qquad (4.5)$$

Woo (2001) described an experimental, in vitro, biosynthetic pathway used to make methanol in three steps from CO_2 gas using biogenic enzymes.

$$CO_2 \xleftarrow[\text{(F}_{\text{ate}}\text{DH)}]{\text{NADH}\rightarrow\text{NAD}^+} \xrightarrow{\text{FA}} HCOOH \xleftarrow[\text{(F}_{\text{ald}}\text{DH)}]{\text{NADH}\rightarrow\text{NAD}^+} \xrightarrow{\text{FLD}} H_2CO \xleftarrow[\text{(ADH)}]{\text{NADH}\rightarrow\text{NAD}^+} CH_3OH$$

$$(4.6)$$

where NAD^+ is nicotinamide adenine dinucleotide; $F_{\text{ate}}DH$ is formate dehydrogenase; $F_{\text{ald}}DH$ is a formaldehyde dehydrogenase; ADH is alcohol dehydrogenase; FA is formic acid; and FLD is formaldehyde gas.

The three reactions above are equilibria, that is, they can also proceed in the reverse direction. The reduction of CO_2 was shifted to the right by adding excess CO_2 and nicotinamide adenine dinucleotide (NADH) (reduced form) and entrapping the enzyme in a silica sol gel. The enzyme formaldehyde dehydrogenase ($F_{\text{ald}}DH$) catalyzes the conversion of formic acid (FA) to formaldehyde (FLD). In the third step, the enzyme alcohol dehydrogenase catalyzes the conversion of formaldehyde to methanol. In all three steps, NADH is oxidized to NAD^+. Woo argued that it is feasible to convert CO_2 to methanol, citing a paper by (Obert and Dave 1999). Woo actually synthesized methanol in the laboratory using the above reactions. She concluded that enzymatic production of methanol will not be economical or time saving until a feasible way to regenerate NADH from NAD^+ is developed and a way is found to provide electrons for the reactions, which must be scaled up.

As in CE production, there is promise for the use of GM plant cells to directly synthesize methanol. In a pioneering study, Hasunuma et al. (2003) reported on the use of GM tobacco plant cells to biosynthesize methanol. They transformed the genome of tobacco cells (*Nicotiana tabacum*) by inserting the gene for the enzyme *pectin methyl esterase* (PME) taken from the mold *Aspergillus niger*. Overexpression of PME in tobacco plants was shown to lead to the overproduction of methanol in these cells. The methanol produced is emitted mainly through the leaf stomata in vapor form and can be collected from the atmosphere by condensation. PME in normal cells leads to production of small quantities of methanol, which is metabolized to serine, methionine, phosphatidylcholine, and methyl-β-D-glucopyranoside, all used by the plant.

There are a number of advantages of using methanol for fuel:

(1) Methanol offers efficient volumetric energy storage compared to compressed H_2 gas. The volumetric energy density of methanol is considerably higher than liquid H_2, in part because of the low density of liquid hydrogen of 71 g/L. Hence, there is actually more hydrogen in a liter of methanol at 99 g/L than in a liter of liquid H_2, and methanol needs no expensive cryogenic container held at −253°C, or the energy overhead to compress and liquefy H_2 gas.

(2) A liquid hydrogen infrastructure promises to be very expensive. Methanol can use the existing gasoline infrastructure with limited modifications.

(3) Methanol can be blended with 15% gasoline (M85) to make it a better vehicular fuel.

(4) Methanol (and other alcohols) can be made from any organic material using the Fischer–Tropsch method using syngas (Zhu et al. 2011). Importantly, there is no need to use food crops (e.g., corn) and compete with food production for humans and farm animals. The amount of methanol that can be generated from biomass is much greater than ethanol.

(5) Methanol obtained from FFs has a lower price than ethanol (Olah 2005).

(6) Methanol can be converted directly into DC electric energy in the DMFC (see Section 4.4.6).

There are also some disadvantages to a methanol fuel economy. These include, but are not limited to, the following:

(1) The energy density of methanol as fuel is one-half that of gasoline and 24% less than that of ethanol.

(2) Methanol, like biodiesel, is corrosive to some metals used in vehicular fuel systems, including aluminum, zinc, and manganese.

(3) Methanol increases the permeability of some plastics to fuel vapors [e.g., high-density polyethylene (HDPE)]. This property of methanol could increase emissions of volatile organic compounds (VOCs) from fuel, contributing to increased tropospheric ozone and a possible human health risk.

(4) Methanol is toxic, generally lethal when 30 to 100 mL is ingested by humans, but so are gasoline and diesel fuel. Methanol itself is not carcinogenic, nor does it contain any carcinogens; however, it can be metabolized in the body to formaldehyde, which is both toxic <u>and</u> a carcinogen.

(5) Methanol has low volatility in cold weather, which means that pure methanol (M100)–fueled engines are hard to start and will run inefficiently until warmed up. Two cures for this are to preheat the fuel on its way to the injectors and to mix methanol with 15% gasoline, creating M85, which gives the fuel better low-temperature properties.

(6) Pure methanol fires burn with a nearly invisible flame, making their detection difficult after an accident. However, unlike gasoline, plain water will extinguish a methanol fire rather than spread it. M85 burns with a yellow flame.

(7) A methanol leak from an underground storage tank has the potential to contaminate aquifers and nearby well water. The LD_{50} of methanol is approximately 7 mg/g for mice. The symptoms of methanol poisoning in humans include headache, fatigue, nausea, mydriasis, and chronic visual impairment including blindness, convulsions,

circulatory collapse, respiratory failure, and death (Windholz 1976). Methanol is totally soluble in water, unlike petrochemical fuels, and would be rapidly diluted to an innocuous, low concentration at which microorganisms can biodegrade it (Olah 2005).

Certain nonfuel, chemical uses of methanol will compete with its energy applications. Methanol can be used in processes to make the chemicals below (Cheung and Kung 1994).

- Formaldehyde
  - Urea–formaldehyde resins
  - Phenolic resins
- Acetic acid
  - Vinyl acetate
  - Acetic anhydride
- Chloromethanes
  - Methyl chloride solvent
  - Methylene chloride
  - Chloroform
- Methyl methacrylate
  - Acrylic plastics
- DME diesel fuel
- Dimethyl terephthalate
  - Polyester
- Methylamines
- Glycol methyl ethers
- Miscellaneous chemicals
  - Methyl formate
  - Sodium methylate
  - Nitroanisole
  - Dimethylaniline

Methanol is also used as a solvent and as an antifreeze/solvent in windshield washer fluid. It is used also as a hydrate inhibitor in NG processing and as an inhibitor for formaldehyde polymerization.

4.2.6 Biodiesel from Plant Oils

The petrodiesel fuels currently used in diesel engines, home heating, and jet engines are made by distilling crude oil, a nonrenewable FF. Thus, the CO_2

produced in their combustion is "nongreen," a major source of air pollution, and a greenhouse gas (GHG) contributor to global warming.

Some of the advantages in using biodiesel in compression ignition engines include over a 90% reduction in total unburned hydrocarbons, a 75%–90% reduction in polycyclic aromatic hydrocarbons (PAHs), and a reduction in particulates and carbon monoxide over petrodiesel. Only traces of SO_2 are emitted; however, NO_x emissions can either increase or decrease, depending on the biodiesel composition and the engine (Demirbas 2007).

Biodiesel fuel consists of the monoalkyl esters of plant oils or animal fats. Most biodiesel is a light fuel oil produced from oils extracted from plant seeds; it is green or renewable because the CO_2 produced by its combustion comes from recently grown plants, and growing oilseed plants will uptake this CO_2 for photosynthesis and the production of more oil seeds. The plant oils generally used to make biodiesel are from oilseed rape (in Australia, Canada, Europe, and the Ukraine); canola oil; palm oil (Malaysia, Indonesia); coconut oil; soybeans (United States, Brazil, Argentina); corn oil; peanut oil (United States); used cooking oils; and perhaps someday certain algae, GM bacteria, or fungi such as *Cunninghamella japonica* and *Gliocladium roseum*. Animal fats (tallow) are sometimes used to manufacture biodiesel. Used coffee grounds are approximately 10%–15% oil by weight. In a feasibility study, coffee ground oil has been extracted and made into biodiesel by researchers at the University of Nevada at Reno (DOE 2012; Biodiesel 2012).

The production of biodiesel is relatively simple (Demirbas 2007): *First,* the plant oil, cooking oil, or animal fat feedstock is filtered to remove any particulates. *Second,* the presence of unwanted *free fatty acids* (FFAs; carboxylic acids) is assayed. The FFAs are then esterified into biodiesel, esterified into bound glycerides, or removed through neutralization. *Third,* a strong base [e.g., KOH, NaOH, or sodium methoxide ($NaOCH_3$)] is dissolved in the absolute methanol reactant. (Absolute ethanol can also be used, but it is more expensive than methanol.) The base is the catalyst for the biodiesel esterification process. An excess of six parts alcohol to one part triglyceride may be added to drive the reaction to completion. Heat is used to accelerate the reaction. Reaction products include not only the biodiesel esters but also some glycerol, soap, excess alcohol, and trace amounts of water. These by-products are removed, leaving the fuel, biodiesel methyl (or ethyl) esters. The glycerol can be sold as an industrial chemical.

A basic biodiesel synthesis chemical reaction is shown in Figure 4.2. Note that the alkyl groups (R_x) on the triglyceride feedstock can be variable, depending on the feedstock oil source. When methanol is used as the alcohol, the products are glycerol plus three different methyl esters comprising the biodiesel. *Methyl linoleate* is a common methyl ester produced from soybean or canola oil reacted with methanol. *Ethyl stearate* is another ester produced from stearic acid from soy, rapeseed and canola oils, and ethanol.

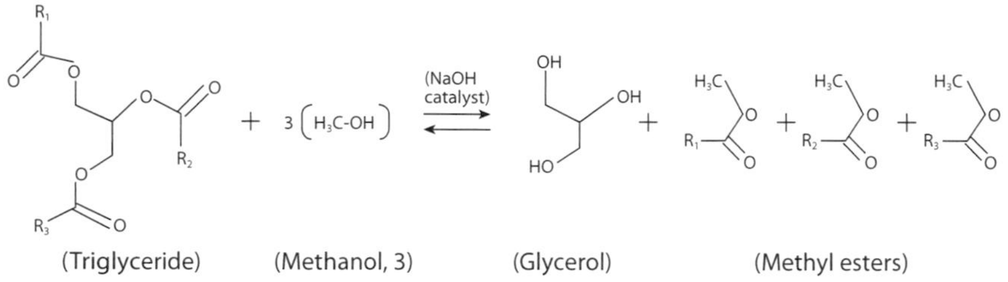

(Triglyceride)   (Methanol, 3)   (Glycerol)   (Methyl esters)

FIGURE 4.2
Basic catalyzed chemical reaction for biodiesel synthesis from methanol and various triglycerides = (RCOO⁻).

Like ethanol in gasoline, biodiesel can be blended with regular petrodiesel fuel: B2, B5, B20, B50, and B100 being 2%, 5%, 20%, 50%, and 100% biodiesel, respectively. B20 can generally be used in a standard diesel engine without modification. B100 requires some engine modifications. The energy value of B100 is approximately 37.3 MJ/L; this is only approximately 9% lower than regular No. 2 petrodiesel.

Because of its alkaline production pathway, biodiesel has some materials compatibility problems. *Plastics:* Biodiesel slowly degrades polyvinyl chloride (PVC) and dissolves polystyrenes. It does not affect HDPE. *Rubbers:* Fluorinated elastomers (FKM) cured with peroxide and base–metal oxides can be degraded when biodiesel loses its stability caused by oxidation. All natural rubbers are degraded by biodiesel. Commonly-used synthetic rubbers FKM-GBL-S and FKM-GF-S used in modern vehicles are biodiesel compatible. *Metals:* Copper-based metals are degraded. Stainless steels (316 and 304) and aluminum are substantially unaffected.

Other problems that accompany the use of biodiesel include gelling at low temperatures and degradation from water contamination. When biodiesel is cooled below a certain point, some of its varied molecules aggregate and form crystals. Once the crystals become larger than a quarter wavelength of visible light, the oil appears cloudy (the *cloud point*). A more quantitative measure of biodiesel low-temperature crystallization is the *cold filter plugging point* (CFPP). This is the temperature where B100 can no longer pass through a 45-μm pore filter. The CFPP temperature varies widely for different composition feedstock oils. For example, the CFPP temperature for canola feedstock biodiesel is approximately −10°C; when tallow is used, the CFPP temperature is approximately +16°C. Clearly, B100 use in winter requires fuel heaters.

Water is a problematic contaminant in biodiesel, even in small quantities. Biodiesel is hygroscopic because of the existence of monoglycerides and diglycerides left over from an incomplete forming reaction. These molecules can act as emulsifiers. Water can come from condensation in the biodiesel tank from moist air. Water in biodiesel can cause the following adverse effects:

(1) The heat of combustion is reduced, and there is harder starting, less power, and smoke in the exhaust.

(2) Water expedites corrosion of fuel system components.

(3) Water supports microbial growth that clogs paper fuel system filters and leads to fuel pump failure.

(4) Water freezes and forms ice crystals that provide sites for gelling of the fuel.

(5) Water can cause pitting in the pistons of a diesel engine.

As petroleum reserves have become exhausted and oil has become more expensive, there has been a rise in the worldwide production of biodiesel. The current yearly demand for petrodiesel in the United States and Europe is approximately 490 million tonnes (= 147 billion gal.). The 2008 production of biodiesel in Europe was 7.8 million tonnes. The total world production of vegetable oil for all purposes in 2006/2007 was approximately 110 million tonnes, including 34 million tonnes each of palm oil and soybean oil. US industrial production of B100 rose exponentially from 0.5 million gal. in 1999 to a peak of 691 million gal. in 2008 and then fell off rapidly to 315 million gal. in 2010 (Soystats 2011). A detailed history of B100 production by month may be found in the US Energy Information Administration (USEIA) report, DOE/EIA0642(2009/12) (USEIA 2010). In June 2008, the largest biodiesel plant in the United States, the GreenHunter Energy Plant, opened in Texas at the Houston Ship Channel. Its capacity is 105 million gal./year, and it uses a variety of feedstocks, including animal fats, vegetable oils, or a blend of the two.

The US Navy Secretary, Ray Mabus, recently ordered the Navy to get ready to use alternative energy for half its power at sea and on shore by 2020. According to Schoof (2011), this will require more than 300 million gal. of biofuels a year for blending with conventional FFs (gasoline, petrodiesel, jet fuel). The US Air Force also plans to fly on a 50:50 blend of biodiesel and conventional jet fuel (e.g., B50) by 2016, translating into an annual require- ment of approximately 400 million gal./year. The obvious motivation for this shift in fuel composition is national security, that is, a reduced depen- dency on politically fragile Middle Eastern and South American petroleum sources. Commercial aviation may follow suit. With the military having a future need for upward of a billion gallons of biodiesel per year for blend- ing, there will be a scramble to develop large-scale plants for mass produc- tion of this important biofuel, just as there has been for ethanol production. The major biodiesel feedstock in the United States is soybean, which contrib- utes approximately 80% to the fuel's cost. The net effect of a food crop (soy, corn) being used for fuel (biodiesel, ethanol) is to raise the cost of the crops, hence the cost of foodstuffs made from them. No one eats oilseed rape. The present on-road average US cost of biodiesel (B100) ranges from $4.29/gal. (with taxes) (Yokayo 2011) to $4.85–4.95/gal. (B100 FOB Midwest) on April 7,

2011. One year ago, ICIS cited the B100 price range of $3.25–3.31/gal. One reason for the price increase may be the current shortfall in soy oil (soy methyl esters) (ICIS pricing 2010).

An Israeli company, *TransBiodiesel*, has developed a more efficient and less costly way to manufacture biodiesel (TransBiodiesel 2009). TransBiodiesel has innovated a new biocatalyst enzyme to convert plant oils to biodiesel. About 13.3 tonnes of their enzyme is required to produce 40,000 tonnes of biodiesel. Their enzyme costs $150–300/kg, so approximately $2 million is required to produce 40,000 tonnes of biodiesel. (This works out to $0.05/kg of biodiesel!) The TransBiodiesel enzymatic process can run at 10°C–35°C; conventional biodiesel production involves heating to 60°C–70°C. The glycerol produced is high quality, transparent, and salt free. Their biocatalysts are water tolerant; up to 5% water in the feedstock can be reacted. Hopefully, the TransBiodiesel bioenzyme approach to making biodiesel will lead to reduced costs per liter over conventional biodiesel manufacturing using $NaOCH_3$ and heat.

The adverse environmental effects of producing biodiesel mirror those from producing bioethanol; they will include the following:

(1) Use of limited freshwater to grow oilseed crops used to make biodiesel.

(2) Use of arable land to grow oilseed crops rather than food crops (this decision probably will be made on an economic basis). Obviously, this crop shift will further increase the cost of food.

(3) Deforestation will occur as woodland is converted to arable land to grow oilseed crops, causing ecological unbalances.

When rapeseed is grown in the United States, the average oil production is 1029 L/ha (110 US gal./acre); high-yield rapeseed fields produce approximately 1356 L/ha (145 US gal./acre) (data from Wikipedia 2012—biodiesel). In addition, rapeseed stover can be made into fuel by HTC, composed into fertilizer (see below), or used to make CE. The National Biodiesel Board has estimated biodiesel production in the United States over 7 years: 2005, 112 million gal.; 2006, 224 million gal.; 2007, 500 million gal.; 2008, 691 million gal.; 2009, 545 million gal.; 2010, 315 million gal. and 2011, 1100 million gal. (Biodiesel 2013). Note that production fell off after 2008, probably due to the recession and the high cost of feedstock.

4.2.7 Biofuel from Microalgae

Private and governmental laboratories [e.g., the *National Renewable Energy Laboratory* (NREL)] are currently engaged in research on maximizing the efficiency of the extraction of neutral lipids [triacylglycerols (TAGs)] from certain eukaryotic, photosynthetic microalgae (Pienkos et al. 2011). TAGs are composed of three molecules of fatty acids (FAs) that are esterified to one molecule of glycerol. Nearly 100% of their mass can be converted to biofuels. FAs are the building blocks of lipids; they are synthesized by enzymes in the chloroplasts of algae, of which the enzyme acetyl-CoA carboxylase is a key in regulating lipid synthesis rates. Some of the photosynthetic microalgae being studied for lipid production include *Chlamydomonas vulgaris, C. reinhardtii, Scenedesmus* sp., *Nannochloropsis* sp., *Botryococcus braunii, Dunaliella tertiolecta, Chlorella salina,* and *Chlorella* sp.

Biofuel algae growth is generally done in a continuous, batch process. It requires algae, water, sunlight, and nutrients. Nutrients consist of phosphorus, nitrogen, CO_2, carbonates, and ions of metals including Ca, Mg, Na, K, Fe, Mn, Zn, Cu, and Co, as well as sulfur (sulfates, sulfites, sulfides, etc.) (Fertguide 2011).

When a sufficient density × volume of algae is reached, the algae are separated from the growth medium, which is recycled, and the algal lipids are extracted. In one procedure, the algal cell walls are ruptured using ultrasound or mechanical or chemical means, and a solvent such as hexane is used to dissolve the lipids. The lipids are separated by distillation, and the solvent is recycled. Other means of breaking down algal cell walls can use osmotic shock (for marine algae) or enzymatic treatment. Origin Oil has patented Quantum Fracturing, a technique using ultrasound to rupture cell walls (Algae Oil 2011). Cavitation Technologies Inc. (CTI) has developed another ultrasound-based technology that ruptures algal cells with the shock waves from collapsing cavitation bubbles. The raw lipids are further treated chemically to yield biodiesel, naphtha, and so forth. Ferrentino et al. (2006) described their research to develop and improve the process economics of producing biodiesel from microalgal oil. They planned to eliminate the extraction step altogether with an in situ biodiesel production method. They used a chloroform/methanol, two-solvent, oil extraction technique. Some 10 different components of extracted algal oil were described by Govindarajan et al. (2009). These included such long-chain olefins as *eicosane* ($C_{20}H_{42}$), *heneicosane* ($C_{21}H_{44}$), and *docosanoic acid* ($C_{22}H_{44}O_2$).

Japanese scientists working in the Central Research Institute of the Electric Power Industry (CRIEPI) have developed a pilot system for extraction of algal lipids that eliminates the need to break down algal cell walls before the oil extraction process. Their pilot process used concentrated, wild-type green algae taken from Hirosawa Pond in Kyoto City. The concentrated algae were treated with a solvent, liquefied DME (H_3COCH_3). DME has a boiling point at −25°C; they used liquefied DME at under −20°C at 0.5 MPa pressure. DME does not form peroxides, is nontoxic, and has no effect on global warming or ozone depletion; it appears to be an environmentally friendly solvent (and fuel). At the temperature and pressure used, DME has a good affinity for oil and lipids and is also partially soluble in water. Thus, it can penetrate algal cell walls, bind with the lipid components inside the algal cells, and more importantly, diffuse out bound to lipids from the cells. This exudate is then heated to vaporize the DME, leaving behind the algal oil. The DME is condensed and reused. The "green crude oil" showed a molecular weight range of 200–400 and a heat value of 10,950 cal/g (Kanda and Mimaki 2010). The green crude is then processed into biodiesel.

In 5 to 10 years, the large-scale processing of lipid-producing microalgae may be tuned to a point where the processes reach a point of diesel parity where the production cost of algal biodiesel equals the cost of petrodiesel. At present, it is estimated that a 10 million gal. algal diesel facility could produce fuel for between \$10/gal. and \$20/gal. There is a long way to go. One obvious step is to create GM algae strains that produce more lipids.

It is estimated that a mature US algal biofuel industry could produce approximately 57 billion gal. (2.158×10^{11} L or 2.158×10^8 m^3) of biodiesel per year (USDOE 2011; Pienkos et al. 2011).

4.2.8 Hydrothermal Carbonization

HTC is a thermochemical process in which biomass is heated in a closed vessel with weakly acidic water in the absence of O_2, at high pressures and moderately high temperatures, to create a *char product* (biochar) and water. HTC was first described in 1544 by Valerius Cordus and in 1592 by Balthasar Klein, and more recently by Friedrich Bergius in 1913 (cited in Titirici et al. 2010). Current research on this important process is being done by workers such as Berge et al. (2011), Sevilla et al. (2011), Funke and Ziegler (2011), Titirici et al. (2010), Heilmann et al. (2010), Sevilla and Fuertes (2009), Cui et al. (2006), Bobleter 1994), and so forth. HTC is a chemical process that works by the dehydration and condensation of a carbohydrate feedstock, which can be as varied as microalgae (Heilmann et al. 2010), various sugars, cellulose (Sevilla and Fuertes 2009), and various crop waste products including crushed sugarcane (bagasse), macerated stover, spent tomato plants, sewage sludge, fermentation waste, leaves, pine needles, and so forth. The crop waste plus water and citric acid was heated in a closed reaction vessel in the absence of atmospheric O_2 to temperatures between 200°C and 250°C for times ranging from 1 to 12 h, depending on the input stock. [In the ease of microalgae, the algae was heated to approximately 200°C for approximately 1 h at a pressure not exceeding 2 MPa (290 psi).] The HTC reactions, once started, are self-sustaining and exothermic, so means must be taken to remove excess heat from the reaction vessels and limit the internal pressures to prevent reaction vessel explosions from overpressure. At the end of the reaction period, the vessels are cooled and opened, and the char product is separated from the liquid and dried.

The char product has many uses: it can be used to make *terra preta*, an artificial char-enriched soil that promotes plant growth. As *terra preta*, the char product is in a carbon sink (Titirici et al. 2007). *Terra preta* can exhibit an almost perfect spongelike cubic mesoporosity with a surface ideal for water sorption and ion binding.

Pelletized char product can also be stored and used as a clean, carbon-neutral fuel. The CO_2 emitted by its combustion is in a sustainable, green "short cycle" with CO_2 uptake by crops and forest growth.

HTC is actually a complicated, multistep series of chemical reactions, as described by Sevilla and Fuertes (2009). The hydrochar particles produced are the result of (1) hydrolysis of cellulose chains, (2) dehydration and fragmentation into soluble products of monomers that come from the hydrolysis of cellulose, (3) polymerization or condensation of the soluble products, (4) aromatization (formation of ring molecules) of the polymers thus formed, (5) appearance of a short burst of nucleation of the hydrochar particles, and (6) growth of the nuclei so formed by diffusion and linkage of molecular species from the solution to the surface of the nuclei. The hydrochar particles have a hydrophobic, aromatic core coated with a hydrophilic outer shell. What is amazing about the HTC process is that no CO_2 is produced! No combustible carbon is lost. Crop waste is converted directly to a solid, dense, high-carbon fuel.

In summary, HTC provides a viable alternative to disposing of human-produced plant (and other organic, carbohydrate-containing) waste. Plant waste can also be composted, a process that releases GHGs (CO_2, CH_4) and that may take weeks to months to reach an end product, depending on temperature and moisture content. The compost produced is generally used for fertilizer and to augment soil. The HTC reactions form a carbon-neutral fuel or soil conditioner, without CO_2 release. They are also exothermic, permitting the harvesting of heat energy, which can be used to "bootstrap" the next HTC reactor. Funke and Ziegler (2011) did a scientific study in which they measured the heat of reaction for the HTC of certain pure, feedstock substances in megajoules per kilogram. These were glucose (1.06), cellulose (1.07), wood (0.76), cellulose at 260°C (1.08), and cellulose with acetic acid at pH 3 (0.86). Depending on the feedstock mixture in the reactor, it is probably safe to say that from 0.75 to 1.07 MJ/kg, heat is released in the biochar process.

The *International Biochar Initiative* (IBI) "is a non-profit organization supporting researchers, commercial entities, policy makers, farmers and gardeners, development agents and others committed to sustainable biochar production and use" (IBI 2011). For those readers interested in pursuing HTC technology and do-it-yourself biochar production, this Web site is an excellent place to start. IBI supports research projects, conferences on biochar, biochar standards, international and domestic policy, biochar project resources, and so forth. Also visit the New England Biochar LLC Web site, http://new englandbiochar.org/Home.html (accessed February 4, 2011).

4.2.9 Solar Thermochemical Reactors

An experimental, solar-powered, thermochemical reactor has been developed in a joint project by scientists from ETH Zurich, the Paul Scherrer Institute (PSI) in Switzerland, and Cal Tech. The prototype reactor uses mirror-collimated solar energy to achieve high reaction temperatures (about 2000°C), which permit the continuous catalytic splitting of gaseous CO_2 plus H_2O, using metal oxide redox reactions. The reactor uses a two-step cycle consisting of a thermal reduction of nonstoichiometric cerium oxide at approximately 2000°C and then reoxidizing it with H_2O and CO_2 at below 900°C to produce hydrogen gas and carbon monoxide, known as *syngas* (CO, CO_2, H_2), a precursor to various liquid hydrocarbon fuels. An experimental 2000 W solar reactor gave a measured efficiency of 0.8%, measured as the ratio of the heating value of the fuel produced divided by the solar radiative power input (Solarbenzin 2011). The prototype solar reactor consisted of a funnel-like reflecting cavity, the aperture of which is covered with a quartz window approximately 75 cm in diameter. "The [internal] dimensions of the cavity ensure multiple internal reflections and efficient capture of the incoming solar energy. A porous, monolithic ceria cylinder inside the cavity is subjected to multiple heating and cooling cycles to induce fuel production. Reacting gasses flow radially across the porous ceria cylinder, while products from the reaction exit the cavity through an axial outlet port." (Reactor 2011).

Ongoing research on the PSI syngas solar reactor is seeking to improve its efficiency (over 15% is theoretically possible) and scale up the system to megawatt-sized systems.

4.3 Desalination

4.3.1 Introduction

Human sustainability is critically dependent on adequate supplies of freshwater. As the Romans found, cities cannot exist without freshwater. It is clear that water is required for human consumption, sanitation, agriculture, power generation, and industry. As climate change causes changes in rainfall patterns, and aquifers and rivers are exhausted by overconsumption linked to population growth, man has turned to the technology of desalination to remove salt and mineral pollution from remaining water sources and augment regional freshwater supplies.

According to the US Geological Survey (USGS 2010a), the salinity of water is defined by the following scale: *freshwater,* <1000 ppm dissolved salts; *slightly saline water (brackish),* from 1000 to 3000 ppm; *moderately saline water,* from 3000 to 10,000 ppm; *highly saline water,* from 10,000 to over 35,000 ppm; *seawater,* approximately 35,000 ppm salt.

"A typical American uses 80 to 100 gallons of fresh water a day, according to the US Geological Survey. The entire country consumes *ca.* 323 billion gallons/day of surface water, and another 84.5 billion gallons of ground (aquifer) water/day. Depending on local energy prices, and the means used, 1000 gallons of desalinated sea water costs around $3 or $4" (Schirber 2007). Schirber also stated that current desalination methods require approximately 14 kWh of energy to produce 1000 gal. of desalinated water. (Note: 1 US gal. = 3.7854 L; 264.2 gal. = 1 m³.)

Desalination is a process whereby dissolved minerals in water above a certain concentration (usually > 2000 ppm) can be removed, providing "fresh" water. The salt water can be from the oceans or a lake, or aquifer water contaminated with fertilizers or salt. There are a number of means for desalination; all require the input of energy. Some means are "green" or sustainable. Desalination means include *distillation,* including flash (vacuum) distillation; multiple-effect evaporator; vacuum freezing; and vapor compression; as well as various ion-exchange processes using semipermeable membranes, including electrodialysis reversal, reverse osmosis (RO), and nanofiltration. Whatever the means used, a desalination plant is expensive to operate efficiently (energy is increasingly expensive), and the cost goes up with the salinity of the input feedstock.

A significant, local ecological disturbance can result from the discharge of the brine residue from desalination back into the ocean. Yes, the high salt content will eventually be mitigated by the "principle of infinite dilution," but local marine ecosystems will be disturbed. The brine discharge is denser than normal seawater, and care must be taken in the design of the discharge system to ensure rapid mixing and dilution. The high-saline discharge zone may be as much as 150 m in diameter around the discharge pipe's opening in a well-designed discharge system before near-normal sea salinity is seen. Another strategy of brine disposal is to pump it into shallow lagoons on land and let the sun finish the job. The salt crystals can be harvested for human use and for use by certain industries (Miller 2003).

4.3.2 Distillation

Distillation is a process whereby thermal energy input to saline water causes the water molecules to go into the vapor phase; the vapor is then cooled and condensed into nearly pure H_2O. (The heat of vaporization of freshwater at 100°C and 1 atm is 5390 cal/kg.) The thermal energy required for vaporization can come from geothermal (GT) energy; combustion of FFs; electric energy from various sources including nuclear, wind, and solar; and directly from heat in solar energy. Distillation of steam to make freshwater has been used on marine naval ships since the end of the 19th century.

Vacuum distillation offers the advantage that lower temperatures are required for boiling, and thus less energy is required to boil saline water in a partial vacuum. The energy saved in vacuum boiling must be balanced against the energy taken by the air pumps required to achieve the partial vacuum in the boiling vessel. Condensing the steam can make use of cold seawater pumped from deep in the ocean, eliminating the need for refrigerated chilling. Multistage flash distillation accounts for approximately 85% of worldwide desalinated water, that is, approximately 10–13 billion gal. of water per day, or only about 0.2% of global water consumption (Schirber 2007).

As the world's freshwater needs rise, technology has been busy devising more energy-efficient means to desalinate water. Better membranes for RO are being designed, and an electroosmosis system was developed by Siemens Water Technologies in 2008 that uses electric fields that can desalinate 1 m^3 of seawater using only 1.5 kWh of energy. Freezing seawater can also yield low-salt ice, and icebergs are a potential source of freshwater.

4.3.3 Reverse Osmosis

RO operates by forcing saline water through special nanopore membranes at high pressure. The smaller water molecules pass easily through the membrane's pores, while larger, ionized dissolved mineral molecules are substantially blocked. Most of the energy in RO desalination goes into driving the pumps that pressurize and circulate saline water. Several RO cycles are required to achieve the desired water purity. A problem seen in some RO plants is the filters being clogged with organic microparticles from the seawater; prefiltering the seawater is essential for efficient RO plant operation.

4.3.4 Humidification/Dehumidification

This process operates on the principle of *mass diffusion*. Dry air is passed over warm saline water and picks up water vapor (WV). The humid air is next chilled to condense out the WV, yielding freshwater and dry air, the latter of which is recirculated (Bourouni et al. 2001). Energy is used to warm the saline, circulate the air, and run the refrigeration to chill the air in the condensers.

4.3.5 Diffusion-Driven Desalination

The diffusion-driven desalination (DDD) process uses low-grade waste heat from industrial processes or power generation, or solar thermal energy. Thus, given the availability of low-grade waste heat, its main cost is the electric power needed to run pumps and blowers, resulting in a lower operating cost than conventional distillation technologies. The main components of a DDD system include a direct-contact evaporator (diffusion tower) and a condenser, both of which utilize countercurrent flow through packed beds. Khan et al. (2010) described a prototype DDD system that desalinates cold seawater feedstock. The salt water flows downward in the diffusion (evaporation) tower while hot flue gasses containing sulfur dioxide (SO_2 and CO_2) are forced upward in the tower. The humidified, exiting flue gasses are chemically scrubbed of CO_2 and SO_2 [flue gas desulfurization (FGD)]. They then pass upward into a direct-contact condenser, where contact with downward-flowing, cold freshwater precipitates the moisture from the scrubbed flue gas. The water is collected and passed through a heat exchanger to cool it before being used in the condenser. There is a net gain of freshwater because of the WV condensed from the scrubbed flue gas (Klausner et al. 2004).

4.4 Carbon-Free Energy Sources

4.4.1 Introduction to Wind Energy

Wind works as an energy source because moving air can exert pressure on surfaces, and pressure times area equals force. The force can move mass plus reactive (viscous) forces, doing work. Up until the end of the 19th century, wind-powered sailing ships were used in commerce and war all over the world. In this section, we focus on the use of wind to do mechanical and electrical work on land.

Windmills have been used by man to grind grain or pump irrigation water as early as 200 BC. The first practical windmills were built in Sistan, a region between Afghanistan and Iran, in the 7th century AD. These windmills, called *panemone*, used vertical-axis turbines with 6 to 12 rectangular sails covered in matting or cloth and were used to pump water or grind corn and sugarcane. Horizontal-axis WTs (HAWTs) first appeared in Europe in the middle ages. The first historical records of their use in England date to the 11th and 12th centuries. By the 14th century, Dutch windmills were used to drain marsh areas of the Rhine delta. In 1887, a Scot, James Blyth, built a vertical-axis WT (VAWT) connected to an electrical generator in Marykirk, Scotland. Blyth used its output to light his home. Also, in 1887, US inventor Charles F. Brush built the first automatically operated WT for electricity production in Cleveland, OH. Brush's WT was a horizontal-axis, multiblade design, 18 m tall, weighing 3.6 tonnes. It drove a 12 kW capacity generator.

A forerunner of modern, HAWTs was in operation in Yalta, USSR, in 1931. It drove a 100 kW alternator and was on a 30 m tower, connected to the local 6.3 kV distribution system. In the fall of 1941, the first megawatt-capacity WT/alternator was synchronized to a utility grid in Vermont. It ran for 1100 h before breaking down; due to World Warr II (WW II) material shortages, it was not repaired. [For a summary of the history, technology, and worldwide development of WT energy, see Robinson (2009).]

4.4.2 Energy in Wind

While wind energy, per se, is free to use, it is not constant. Wind velocity over land increases with altitude, due to friction drag from the oceans, bushes, trees, and geological surface features. Thus, windmill and WT designers put their machines on tall towers located over plains, on mountain ridges, and offshore (over water) to realize the highest wind velocities. The wind velocity at height z is approximately proportional to $\ln(z/z_o)$, where z_o is a roughness factor (about $4°10$ m over still water to 1 m in cities) (Grogg 2005).

Wind is air molecules that move from high-atmospheric-pressure areas to low-pressure areas on the Earth's surface. For example, anyone who has sailed in coastal waters in the summer knows the phenomenon of an onshore breeze. As the sun heats the landmass, it warms and, in turn, warms the air over it, making it less dense. This warm air rises, creating lower pressure over the land. The cooler air over the water flows toward the land and the lower atmospheric pressure. At night, the land cools, and the onshore breeze dies and then reverses into an offshore breeze. Onshore and offshore breezes are generally light winds.

Cyclonic storms are another source of wind. Over North America, the air circulation over a cyclonic low-pressure area is counterclockwise (CCW). Over the United States, the center of a low-pressure cyclone generally moves in a west-to-east direction. The wind directions around a cyclone are due to the high-pressure gradient from outside the low to its center, plus the Coriolis forces, which act on the centripetal (radial) wind vectors in a net CCW manner, causing a CCW spiral of wind flowing toward the center of the low (Cannon 1967, Section 5.3).

North Atlantic hurricanes are intense cyclones that derive their energy from higher water temperature over which they form, usually in the Atlantic near the west coast of tropical Africa. Hurricane trajectories are varied; some head northwest, north, and then northeast and stay at sea until they encounter cooler water, where they dwindle. Others head northwest into the Caribbean Sea thence to the warm Gulf of Mexico, and others may follow the US east coast north where they can damage US coastal cities. Hurricanes pack intensive wind energy, and because of their high wind velocities, they are a threat to Gulf and Atlantic coast WT structures, rather than being a source of energy.

Anticyclones are high-pressure areas around which the winds circulate in a clockwise direction. The winds in the northern United States are generally westerly: from the southwest in warmer months and from the northwest in the colder months. A nor'easter is generally a violent east coast cyclonic storm whose high winds come from the northeast, at least until the center of the low moves to the east, far offshore into the Atlantic Ocean.

4.4.3 Physics of WTs as Energy Sources

WTs provide the physical interface between the fluid dynamics of moving air and the rotating machinery that drives electric generators, hydraulic pumps, and so forth. In this section, certain properties of WT operation are reviewed.

It was shown in 1919 by physicist Albert Betz that the theoretical maximum energy extraction from wind kinetic energy (KE) by any WT is 16/27

(59.3%). Modern WT designs can approach from 70% to 80% of the Betz limit. Grogg (2005) derived an equation for the power P extracted by a HAWT from moving air:

$$P = 2A\rho U^3 a(1 - a)^2 \text{ watts} = \text{J/s}$$

where $A = \pi R^2$ = disk area swept by turbine blades, R is the blade radius; ρ = air density; U = mean air speed over A; a = induction factor or the fractional decrease in U once it has reached the rotor, $a = (U - U_1)/U$; and U_1 is the air velocity at the front of the rotor. The International System of Units (SI) is used.

The important property is that the wind power imparted to the rotor, P, is proportional to the incident wind velocity cubed. For any WT to function effectively, U must exceed approximately 5 m/s. Any site considered for a WT must have a wind speed distribution measured. That is, by sampling U with an anemometer at turbine height over a year, a histogram is constructed where the number of hours per year at speed versus wind speed is plotted, and then by applying the formula for power in a moving mass of air, $P_A =$ ½ $A\rho U^3$, a wind power density histogram is plotted where power from each wind speed is plotted versus wind speed. Examples of these two histograms are shown in Figure 4.3. Note that because the wind power is proportional to its speed cubed, the wind power density histogram is skewed to the right from the wind speed density histogram. A WT to be located on this hypothetical site should be designed for optimum aerodynamic efficiency for wind velocities from 8.5 to 12.5 m/s where maximum power is available for the HAWT.

Doubling the shaft height of a HAWT increases the expected wind speeds by 10% and hence the expected turbine power by 33%. Taller towers must be more robust to withstand horizontal forces on the turbine blades; doubling the tower height generally requires doubling the tower diameter, increasing the amount of steel used by a factor of four, and causing a commensurate increase in cost.

WT aerodynamic design is complicated and will not be treated here. Grogg (2005) gives a good introduction to WT aerodynamics. For an illustrated summary of WT designs, also see Ragheb (2010).

4.4.4 Types of WTs

The two types of WTs are as follows:

(1) The HAWT. Most modern HAWTs use three variable-pitch (featherable) blades and are situated on tall towers.
(2) The VAWT has several subtypes:
   a. The *Savonius,* which uses three scooplike blades.

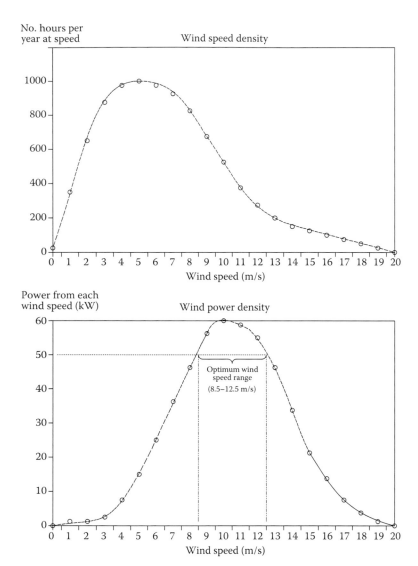

FIGURE 4.3
(Top) Typical wind speed density graph at a WT site. Note the "long tail." (Bottom) Wind power density graph for the wind speed density plotted above. Wind power is proportional to the speed cubed. See texts for discussion.

    b. The *Darrieus,* which uses two or more curved, "eggbeater" blades, each attached at the top and bottom of the vertical shaft.

    c. The *Giromill,* which is like a Darrieus, except the blades are straight instead of curved. Giromill VAWT blades can have variable pitch.

Figure 4.4 illustrates these three VAWT types. A Finnish engineer designed the Savonius VAWT in 1922. The Darrieus "eggbeater" VAWT design was by a French engineer in 1931. The two-blade Darrieus VAWT gives a pulsatile torque output, which means extra wear on blades, bearings, and gears. Curiously, the fixed-blade Darrieus VAWT must be spun by a starter motor to start; this generates symmetrical aerodynamic torque forces that maintain rotation and generate mechanical power. One Giromill/Darrieus design, the Cycloturbine, varies the blade pitch as the vertical blades revolve, giving near-constant (smooth) output torque. Another Dutch Darrieus VAWT design, the Turby, eliminates torque pulses by using three blades, each given a helical twist around the spin axis of 60° (cf. Figure 4.4). VAWTs offer the advantage of being able to extract more wind energy per square meter of land on which the VAWT farm is located than equivalently rated HAWTs. That is, their aerodynamics allows them to be more densely packed without losing efficiency, and *they do not need to be actively pointed into the wind*. VAWTs are more subject to wind damage because it is harder to feather them in high winds than HAWT designs with blade pitch control and axis azimuth control.

A modern, three-blade HAWT is a complicated system. Power P_w from the input drive shaft can be measured in watts (joules per second) and is given by $P_w = \tau_w \dot{\theta}$, where τ_w is the torque in newton meters from the wind forces on the blades and $\dot{\theta}$ = the angular velocity of the shaft in radians per second. This power, minus various mechanical and electrical losses, is converted to electrical power by an alternator designed to run at line frequency or by a DC generator connected to an inverter that makes line-frequency AC power.

WTs are designed to operate over a wide range of wind speeds. The *survival wind speed* of most commercial WTs ranges from 144 km/h (90 mph) to 259 km/h (161 mph). The most common survival wind speed is 216 km/h (134 mph). Neglecting mechanical losses, the angular velocity of the blades is approximately given by $\dot{\theta} = P_o / \tau_w$, where P_e is the electric power output. If an alternator is the driven generator, it must have a constant angular velocity in order to produce line-frequency (e.g., 60 Hz) power in order to supply the local AC power distribution network. This requires that τ_w be adjusted by various means such as regulating blade pitch. Wind turbines are not 100% efficient: $(P_w - P_e) = P_{loss}$, the power lost in converting wind to electrical power.

Modern HAWTs generally have two or three blades. Increasing the number of blades from one to two increases the aerodynamic efficiency by 6%; increasing the blade count from two to three increases efficiency by only 3%. A single blade creates mechanical vibrations in the system, in spite of a counterweight, and two-blade HAWTs also experience cyclic, asymmetric loads at the blades' bases. (The wind velocity is higher on the upper blade, producing a larger horizontal force on the hub and tower.) Three blade turbines are more symmetrically balanced and run more smoothly with minimum vibration. Most modern HAWTs use three blades.

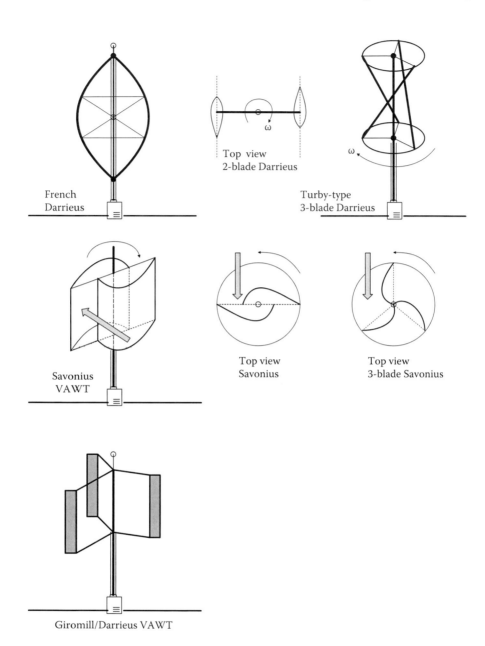

FIGURE 4.4

Some designs for VAWTs. The basic *Darrieus* "eggbeater" uses two blades. To minimize vibrations and torque pulsations, the three-blade *Turby* design twisted the blades 60° from top to bottom. The *Savonius* design uses two or more "scoop" blades. The *Giromill/Darrieus* design uses three flat blades; in some versions, the blade pitch can be continuously varied as the turbine spins to extract maximum wind power.

The turbine drive shaft is generally coupled to the generator through a gear-box. This is required because 60-Hz, two-pole, three-phase alternators generally run at exactly 3600 rpm; four-pole alternators run at 1800 rpm. The HAWT input shaft angular velocity (turbine rotor speed) is designed to run at 5–20 rpm for maximum aerodynamic efficiency, so the gear ratio must be between 1:3600/5 = 1:720 and 1:3600/20 = 1:180. Alternators require DC in the rotor winding or large permanent magnets in the rotor (as found in automobile alternators) to induce an electromotive force (EMF) in the stator windings. A description of the relationships between the torque-angle, electrical load, and shaft torque for an alternator running at constant angular speed is beyond the scope of this section. For details of alternator performance, the interested reader should consult the venerable text on electric machinery by Fitzgerald and Kingsley (1952).

In order to capitalize on letting the WT turn at the speed that is most aerodynamically efficient for the transfer of wind energy (at a given speed) to electrical energy, DC generators can be used that can run over a wide range of speeds. Alternatively, the variable-speed WT can drive an alternator whose AC output is rectified to DC, as in an automobile's DC power system. The DC outputs can then be inverted back to AC power at the appropriate line frequency, phase, and voltage to enter the local power grid.

The speed and torque (power) input to a WT must be regulated in order to do the following:

- *Optimize the aerodynamic efficiency* of the turbine blades in light winds.
- *Keep the generator within its speed and torque limits.* An alternator connected directly to the local 60 Hz AC power grid *must run* at 3600 (or 1800) rpm. Thus, the WT rotor blades must be coupled to the generator by a mechanical speed-reduction transmission.
- *Keep the rotor blades and hub within their centrifugal force limits.* The centrifugal force on the blades increases with the square of the rotation speed. (For a point mass M rotating about a center at radius R with angular velocity $\omega r/s$, the centrifugal force is $F = MR\omega^2$ newtons.)
- *Keep the rotor and tower within their strength limits.* The force of the wind increases as the cube of the wind speed.
- *Enable maintenance.* The turbine must be stopped (taken off-line), blades furled, and the rotor braked. (A furled rotor blade has the edge of the blade facing into the wind.)
- *Reduce noise in populated areas.* The noise from a WT increases approximately with the fifth power of the relative wind speed (seen from the moving tips of the blades). In noise-sensitive environments, the tip speed can be limited to approximately 60 m/s (200 ft./s).

Modern, high-power HAWTs driving 60 Hz alternators maintain constant rotational speed under a variable torque load and variable wind speed by two means: (1) By adjusting the blade pitch. Many HAWTs use hydraulic pitch angle controls working against springs, so if hydraulic power fails, the blades automatically furl. Other designs use electric servomotors to control the pitch of each blade. The servomotors have battery backup. Smaller HAWTs (<50 kW) may have their blade pitch regulated mechanically by a centrifugal governor. The yaw angle ϕ ($0°$ with rotor blades facing into the wind) of the HAWT also controls the turbine power output; it falls off approximately as $\cos^3(\phi)$. (2) If there is a sudden drop in the alternator's load torque on the blades, constant rotor speed can be maintained by temporarily reloading the alternator electrically with an in-tower resistive load. The electrical load provides a dynamic braking torque on the alternator and blades. The resistor dissipates the load energy as heat. The advantage of the resistor load is that it can be rapidly switched in and out electronically, in a graded manner. The heat may be viewed as wasted wind energy.

Typical modern HAWTs have rotor diameters of 40 to 90 m (130–300 ft.) and are rated at between 500 kW and 2 MW. As of 2010, the most powerful HAWT (the Enercon E-126) is rated at 7.58 MW. The Enercon HAWT has an overall height of 198 m (650 ft.) and a rotor diameter of 126 m (413 ft.). The largest VAWT is located in the Le Nordais Wind Farm on Cap-Chat, Quebec, Canada. This Éole turbine is 110 m tall, is rated at 3.8 MW, and has a Darrieus design. The tallest HAWT is the Fuhrländer WT Laasow; its rotor axis is 160 m (525 ft.) above the ground, and its rotor tips extend to a height of 205 m (673 ft.) (Franken 2006).

Figure 4.5 illustrates the yearly increase in worldwide wind generating capacity from 1996 to 2008. Note that the rise has an exponential, power-law shape, and the 2008 wind power capacity was approximately 120.8 GW. Even so, this is a very small fraction of the total world's energy consumption.

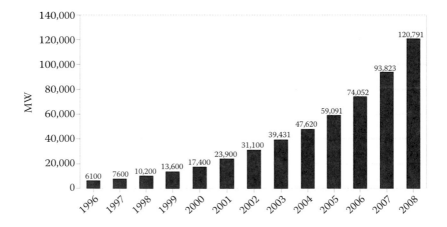

FIGURE 4.5

Bar graph showing the exponential growth in the generation of wind power in North America, Europe, and the rest of the world over 1996–2008. A great deal of this growth (70.4%) has occurred in northern Europe. (Courtesy of Splettstoesser, T., Wikimedia Commons. http://en.wikipedia.org/wiki/File:WorldWindPower.png.)

4.4.5 Solar Energy

4.4.5.1 Sun Flux on Earth

The Earth receives approximately 174 petawatts (PW) (174×10^{15} W) of incoming solar radiation (called insolation) at the upper atmosphere. About 30% of this insolation is reflected back into space, while the rest is absorbed by clouds, WV, landmasses, and the oceans. About 100 useful petawatts of solar flux reach the Earth's surface [about 2.5×10^{24} J/year (Larkum 2010)]. The *maximum* effective solar power concentration per square meter on the Earth's surface is approximately 1 kW, at noon (Miyamoto 2012). Figure 4.6 illustrates the Earth's solar power budget, adapted from a National Aeronautics and Space Administration (NASA) (2012c) educational Web article. Note that the power in (the insolation) to the upper atmosphere equals the total power radiated back into space, which mostly leaves as long-wave infrared (IR). (This diagram, which uses two-significant-figure precision, does not include the solar power taken up by plant photosynthesis, i.e., absorbed into the chemical reactions that fix carbon.)

Figure 4.7 illustrates the emittance spectrum of sunlight measured at the top of the Earth's atmosphere in watts per (square meter × micrometer). This emittance spectrum from the sun approximately follows a blackbody (BB) radiation emission format with a temperature of 5250 K (see Planck's law in the Glossary). At $T = 5250$ K, the peak W_λ occurs at $\lambda = 552$ nm. Also shown in Figure 4.7 is the solar radiation spectrum measured at sea level. Note the overall attenuation by the atmosphere and the notches caused by energy absorption by the major GHGs, CO_2, and WV in the IR wavelengths. Figure 4.8 illustrates a family of ideal BB radiation spectra plotted on a log–log coordinate system. Relative values of W_λ are shown for different BB Kelvin temperatures in the figure.

The sea-level solar radiation spectrum is full of notches in the IR due to energy absorption by atmospheric gas molecules; however, at UV wavelengths (about 250 nm), absorption is from the ozone layer and tropospheric ozone (O_3).

The total solar energy absorbed by the Earth's atmosphere, oceans, and landmasses in a year is approximately 3.85×10^{24} J (an area of ca. 5.1×10^8 km²) (Smil 2006). Photosynthesis captures approximately 3×10^{21} J/year (FAO 2006). Some of the incoming solar energy that does not go into plants acts to raise the temperatures of air, land, and water, and some long-wave IR power is reflected back into the atmosphere, where some is trapped by GHGs (e.g., methane, CO_2), and some radiates into space (see Figure 4.6).

The *useful* solar power on the Earth's surface amounts to approximately 568×10^{18} J/year (568 exajoules/year) or 18 terawatts (TW). The insolation (solar power flux) for most people is approximately 150 to 300 W/m² or 3.5 to 7.0 kWh/day. This is an awesome amount of power. How can this "free" solar power be used? Many ways have been developed to capture and use solar energy. These can be grouped into three categories: (1) direct solar thermal

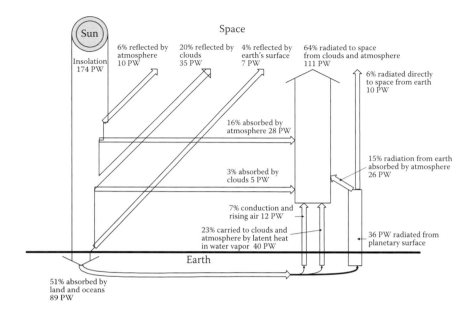

FIGURE 4.6
Earth's solar energy budget. Estimated heat flows are shown between sources and sinks.
(1 PW = 10^{15} W.)

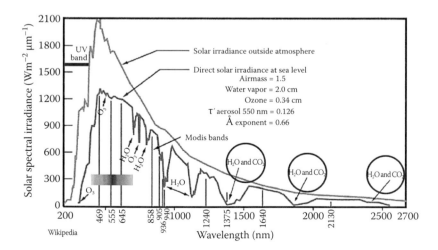

FIGURE 4.7
Graphs of solar spectral irradiance in watts per (m² μm) outside the atmosphere and at the
Earth's surface. The notches in the sea level spectrum are from power absorption by atmo-
spheric gasses such as WV, CO_2, and ozone (O_3). (Courtesy of NASA.)

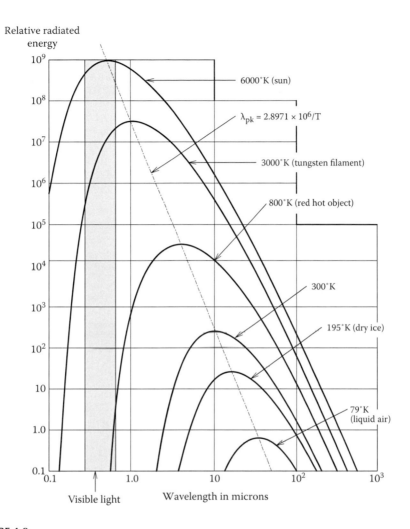

FIGURE 4.8
Family of ideal BB spectral radiation curves as functions of the BB temperature. Note log–log scale. (From Northrop, R.B., *Introduction to Instrumentation and Measurements*, 2nd ed., CRC Press, Boca Raton, FL, 2005. With permission.)

capture, (2) solar thermal electric power generation (STEPOG), and (3) photo-electric power generation.

In category 1, we have greenhouses used in agriculture, passive solar houses, solar heating of water (for domestic and industrial uses), solar cooling (by air convection), solar water desalination (the solar still), solar water disinfection (by UV photons in sunlight), solar cooking, and the catalyzed reaction of $CO_2 + H_2O$ in a solar furnace to make syngas fuel (cf. Section 4.2.7). In category 2, mirror-array (heliostat) focused solar energy is used to heat water in high-powered (tens to hundreds of megawatts) boiler/turbine/

generator AC power systems. In category 3, solar heat is also concentrated to power large thermopile [TP; multiple, series thermocouples (TCs)] arrays to make DC power. Solar photon energy is used to produce DC electric power with photovoltaic (PV) panels (SCs). (The DC output of SCs can be converted to line-frequency AC by inverters and coupled to the power grid.) We will describe some of these applications below.

4.4.5.2 Solar Thermal Electric Power Generation

Solar thermal electric generation (STEG) systems are simple in concept (Bowman et al. 2011). The site for a STEG system is generally chosen to lie in a flat, arid (desert) location, close to the equator, where there are few clouds, little rain, and longer days in winter. Such sites exist in the American southwest, Mexico, North Africa, Spain, China, India, the Australian desert, and so forth. One type of STEG plant consists of a large array of mirrors (thousands, in some cases) surrounding a central tower on the top of which is a water boiler. The mirrors, called heliostats, are on azimuth and elevation-controllable mounts, so that each individual computer-controlled heliostat can track the sun throughout the day and keep its rays exactly focused on the boiler. Steam from the boiler passes in a closed cycle to the turbines driving generators, thence to an air-cooled condenser, and then the hot water is recirculated back to the boiler. Figure 4.9 shows an air view of the National Solar Thermal Test Site (NSTTS) at Sandia National Labs in Albuquerque, NM.

FIGURE 4.9
(See color insert.) Aerial photograph of the STEPOG test facility at the Sandia NSTTS at Kirtland Air Force Base, NM. The heliostat mirrors are steerable. (Courtesy of NASA.)

FIGURE 4.10
Artist's version of the three completed Ivanpah solar STEG sites in the San Bernardino Desert, CA. (Courtesy of California Energy Commission, Preliminary Staff Assessment: Ivanpah Solar Electric Generating System. Application for Certification 07-AFC-5 to the State of California, December 2008. www.energy.ca.gov/2008Publications/CEC-700-2008-013/CEC-700-2008-013-PSA.pdf.)

Note the arrays of directable heliostat mirrors that direct solar energy to the boiler on the tower. Figure 4.10 shows an artist's high-altitude rendering of the to-be-completed Ivanpah 1, 2, and 3 STEG systems in the San Bernardino desert in California; the large arrays of heliostats surround each of the six boiler towers for maximum solar efficiency.

Another STEG design uses arrays of long, parabolic mirrors; each mirror focuses the sun on a long pipe running the length of the mirror. Each pipe contains oil that absorbs the solar energy focused on it by the mirrors. The hot oil is pumped to a heat exchanger/boiler where water is boiled to make steam to run the turbogenerators. Luz Industries built a total of 354 MW STEG power plants using this design in California's Mojave Desert beginning in the 1980s (Ausra 2007). The advantage of the long parabolic mirrors is that they do not need to be moved hourly like heliostats to track the sun; they face south at a fixed-angle elevation (in a given season). Ausra developed a Fresnel-type mirror array that is close to the ground and that focuses solar energy on long, water-filled pipes above the array, giving direct steam generation. Their compact linear Fresnel reflector (CLFR) mirror array was conceived in the early 1990s. It requires fewer foundations and positioning motors per square meter of mirror, and because the flat mirror strips are close to the ground, there are reduced wind loads on the mirrors, thus less steel usage: a clever design.

Still another medium-power (5–25 kW peak) STEG system design uses a single, large, parabolic dish mirror (picture a large radio telescope antenna), 7–10 m in diameter, to focus concentrated sun energy on a boiler

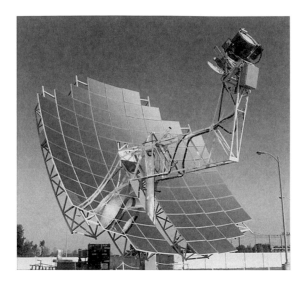

FIGURE 4.11
Photograph of a 25 kW capacity, solar energy–powered Stirling engine/generator system located at the Sandia, NM, NSTTS. (See more on Stirling engines in the Glossary.) (Courtesy of US DOE.)

at its focal point (Bedi and Falk 2000). The mirror is mounted on a motor-powered, azimuth-elevation, sun-tracking mount. A variant on the single large parabolic dish mirror is the use of 16 smaller dish mirrors in a concentric array. These dish/engine systems have a peak efficiency of 29.4%, compared to approximately 23% for a large-scale power tower system like Ivanpah. (Efficiency is defined here as peak solar power input divided by peak electrical power output.)

Figure 4.11 illustrates a prototype, 11.6 m diameter, parabolic dish mirror powering a Stirling heat engine. This 25 kW system at the Sandia US NSTTS operated at approximately 20% efficiency using hydrogen gas as the working fluid. The entire assembly (mirror dish + Stirling heat engine + generator) must track the sun in azimuth and elevation, making it technically complicated and expensive to construct. The Stirling engine is an external heat-source, reciprocating piston engine that uses the thermal energy–caused expansion of a closed-cycle gas such as air, hydrogen, or helium to power its cycle. (Helium is better because the oxygen in superheated, high-pressure air can oxidize lubricants, and hydrogen has a propensity to leak out of containment at high temperature and pressure.) Multiple arrays of solar Stirling/AC generator systems, each producing approximately 25 kW peak power, can have their outputs added together, so 40 units could produce 1 MW of electric power for the grid. The high thermal efficiency of a solar/Stirling system, plus the fact that no FF is required, is offset by the system complexity, the practical limit to system size, and the fact that a

FIGURE 4.12
Aerial view of one Ivanpah STEG system. The mirrors are fixed, forming a giant, multi-element, long-focal-length parabolic mirror focused on the heat collector in the tower. (Courtesy of California Energy Commission, Preliminary Staff Assessment: Ivanpah Solar Electric Generating System. Application for Certification 07-AFC-5 to the State of California, December 2008. www.energy.ca.gov/2008Publications/CEC-700-2008-013/CEC-700-2008-013-PSA.pdf.)

solar/Sterling system cannot go online at once; there is a few minutes' delay while it warms up and comes up to speed (warm-up could be expedited by initially burning NG). (See a further description of Stirling engines in the Glossary.)

An alternative STEG system design philosophy, seen in Figure 4.12, uses a large array of individual, fixed mirrors focused on a single, central heat collector tower as exemplified by the Ivanpah Solar Electric Generating System (ISEGS) under construction in Southern California. The constructor, BrightSource Energy, has been financed by private investors, now a total of $2.2 billion, including $1.6 billion in loans from the US DoE, and an investment from Google. Bright Source Energy stated in August 2012 that the Ivanpah Solar Energy Generating System (SEGS) has reached the halfway point in construction. It is on-track to completion in 2013 (Marketwatch 2012; NREL 2012b). Figure 4.12 illustrates one of the STEG units in the California Ivanpah 1 System, located in the Ivanpah Dry Lake basin in San Bernardino County (California 2010a & b). Three STEG systems are being built in the Ivanpah site to have a total peak power output of 400 MW. They are scheduled to be completed in 2013. Power from the ISEGS will be transmitted on 115 kV lines to a substation. Each of two 100 MW sites would occupy approximately 850 acres; the 200 MW STEG would require approximately 1600 acres. A total area of 5.3 mi.^{2} (13.73 km^2) will be used for the ISEGS, or 0.0343 km^2/ MW.

Each Ivanpah STEG system also includes the following:

- *An NG-fired, start-up boiler* to provide heat for plant start-up in the morning and during temporary cloud cover.
- *An air-cooled steam condenser* to permit closed-cycle operation.
- *A Rankine-cycle reheat steam turbine* that receives live steam from the solar boiler and reheat steam from a solar reheater located on the top of its own tower adjacent to the turbine.
- *A reserve water tank* with 250,000 gal. capacity.
- *Auxiliary equipment* including feed water heaters, a deaerator, an emergency diesel power generator, and a diesel fire pump.
- Some *groundwater from wells* is required for all three plants: less than 100 acre-ft./year will be used for boiler makeup water, as well as washing heliostats and personal use.

All STEG systems require thermal energy storage to compensate for clouds and permit operation late into the day after sunset. This can be accomplished with batteries, compressed air energy storage (CAES), or energy storage in a thermal mass. Sustainability issues with STEG systems are discussed in Section 4.4.5.6.

4.4.5.3 Solar PV Electric Power Generation

SCs capture solar photon energy and convert it to photoelectrons, which can do electrical work. The first SC was developed in 1954 by researchers at Bell Laboratories; it used a diffused, silicon *pn* junction. It could be considered to be a large photodiode. Since then, solar PV cells (PVCs) have been steadily developed to increase their efficiency, power-to-weight ratio, and output, as well as to decrease their cost. Solar panels are arrays of SCs; they are generally composed of modules. In a solar panel, the modules are connected in series to obtain high DC output voltage and then in parallel to boost the available current. Large, commercial solar panels are made from crystalline silicon (c-Si). Monocrystalline Si SCs have higher efficiencies than multicrystalline types. However, multicrystalline SCs have lower costs to compensate for their lower efficiencies. SunPower, in San Jose, CA, makes the highest-efficiency crystalline SCs on a commercial scale (Shah 2011; SunPower 2011).

The first SCs cost approximately $250/W peak power. In order to compete with FF sources of electric energy, SC prices had to be reduced to near grid parity. As SC manufacturing technology was perfected, SC prices per peak watt dropped (about $100/W in 1971). By 1973, Solar Power Corporation (owned by Exxon) was producing solar panels for $10/W and selling them at $20/W. By late 2011, efficient production in China coupled with a drop in

European demand reduced the per-peak watt prices to approximately $1.09 in October 2011. This is close to grid parity.

To increase the light-capturing efficiency of SCs, the glass or plastic covering the front of the cells is generally given an antireflective (AR) coating similar to that used on camera lenses. In the simplest case, when light meets a transmissive (clear) medium (of refractive index n_2) perpendicular or normal to it, a fraction of the incident light intensity I_1 is reflected back toward the source, and a fraction of the light intensity is transmitted into the medium. The *reflectance* or *reflection coefficient*, R_{ag}, gives the incident intensity fraction reflected, and the *transmittance* $T_{ag} = 1 - R_{ag}$ gives the fraction of I passing into the medium:

$$R_{ag} = \frac{(n_a - n_g)^2}{(n_a + n_g)^2} \quad \text{and} \quad T = \frac{4n_a n_g}{(n_a + n_g)^2}. \tag{4.7}$$

where n_a is the refractive index of air ($\cong 1$), and n_g is the refractive index of the transmission medium ($n_2 \cong 1.5$ for common glass).

For example, given $n_a = 1$ and $n_g = 1.50$, $T_{ag} = 0.96$ (96%) and $R_{ag} = 0.04$ (4%). That is, 4% of the light intensity is reflected and lost.

AR coatings can lower R, hence raise T, improving SC efficiency. Lord Rayleigh discovered the principle of the simple AR coating in 1886! See Figure 4.13. (Nonnormal ray incidence is shown for clarity.) An AR material with refractive index n_{ar} is placed on the surface of the glass. Thus, the incident ray is reflected at the air-AR surface with reflectance $R_{a/ar} = (n_a - n_{ar})^2/(n_a + n_{ar})^2$. Its intensity in the AR medium is $I_{ar} = I_1 \times T_{a/ar}$, and at the AR-glass surface, it is reflected back into the AR coating, and the light intensity continuing into the glass has intensity $I_t = I_1 \times T_{a/ar} \times T_{ar/g}$. It can be shown that $(T_{a/ar} \times T_{ar/g})$ has a maximum for $n_{ar} = \sqrt{n_a n_g}$. If $n_a = 1.0$ and $n_g = 1.5$, the optimum AR refractive index is $n_{ar(opt)} = 1.225$. The overall transmission into the glass medium is calculated to be 98%, rather than 96% with no simple AR coating.

Refinements of AR coating technology include the use of a quarter-wave thickness of AR coating and multiple AR coatings to make destructive interference of reflected waves and enhance antireflection over the visible range of wavelengths (40–700 nm) and a wide range of incidence angles. SCs need to maximize solar energy transmission over 450 to 1100 nm.

The spectral response [also known as (aka) responsivity] of a typical silicon SC under glass rises linearly from near zero at approximately 300 nm to a peak of 0.55 A/W at approximately 900 nm and then falls off abruptly to zero at 1200 nm. The spectral response $\mathbf{R}(\lambda)$ is above 50% of peak from approximately $450 < \lambda < 1020$ nm. See Figure 4.14. The short wavelength response of the SC below 400 nm is attenuated by the absorption of UV photons by the glass covering the SC.

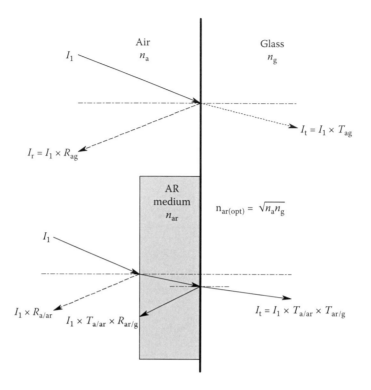

FIGURE 4.13
(Top) Behavior of an oblique ray of light impinging on a flat surface of a medium with a higher refractive index than air. (Bottom) Behavior of an oblique ray impinging on a thin AR coating on a glass surface. The optimal AR coating refractive index can be shown to be the geometric mean of the refractive indices of air and glass. See text for analysis.

There are many materials used to make SCs; the most common is c-Si of various types; cadmium telluride (CdTe) has also been used to make inexpensive, thin-film SCs. A square meter of CdTe thin-film SC contains about the same amount of toxic Cd as a single Ni–Cad C-cell battery in a more stable and less soluble form. The direct-bandgap material, copper–indium–gallium selenide (CIGS), has the highest efficiency (about 20%) of other thin-film, SC materials. SCs have been made from the following: (1) monocrystalline silicon (c-Si); (2) polycrystalline or multicrystalline silicon (poly-Si or mc-Si); (3) ribbon silicon, a type of mc-Si; (4) CdTe; (5) CIGS; and (6) gallium arsenide multijunction.

First Solar, an Arizona-based company, manufactures SC panels using CdTe using mining waste. Their thin-film, CdTe on glass SCs have an average efficiency of 11%–12%, compared to 14%–15% for silicon-based SCs. They are currently building a solar plant with 290 MW peak power in Agua Caliente, AZ. The panels of the Agua Caliente plant will cover approximately 1750 acres. First Solar is also building two PV plants in California that will have

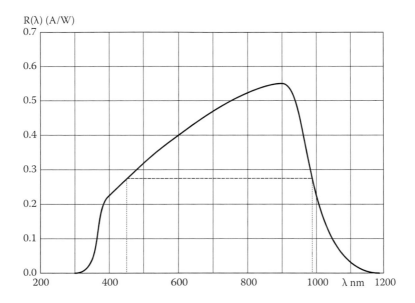

FIGURE 4.14
Approximate spectral responsivity of a typical silicon SC. Note that the responsivity peaks in the near-infrared (NIR) wavelength.

peak capacities of 550 MW. First Solar is also developing SCs using the more efficient CIGS material. The price per watt of solar panels is steadily falling as the technology improves and demand increases. In early 2011, the price was approximately $1.75/W. In 2012, the price per watt of solar panels was as low as $0.91/W [250 W panels: Astroenergy 250, Wholesale solar (2012)]. The cheapest solar energy now costs approximately $120–140/MWh. Compare that with approximately $70/MWh for the latest US offshore wind energy and approximately $70–90/MWh for NG-fired power generation (Economist 2011f).

One of the newest SC technologies with great promise for low cost makes use of light-absorbing organic dyes [dye-sensitized SCs (DSSCs)] (Hoppe and Sariciftci 2004). Interestingly, in organic SCs, C_{60} (carbon) buckminsterfullerene molecules are used for electron transfer because of their higher electron affinity. Organic SCs do not have the efficiencies of c-Si SCs, however. Organic polymers are also being developed for SCs. Konarka Power Plastic polymer SCs have reached efficiencies of 8.3% (Konarka 2011). These "plastic" SCs are unique in that they are flexible and can be made to follow curved contours. Unlike junction devices, organic polymer SCs (OPSCs) do not rely on the inherent electric field of a *pn* junction to separate the photon-generated electron-hole pairs. The active region of an OPSC consists of two materials, one that acts as an electron donor and the other as an acceptor. When a captured photon generates an electron-hole pair in the donor material (e.g., polyphenylene vinylene or copper phthalocyanine), the charges tend to remain bound in the form of an exciton and are separated when the exciton diffuses to the donor/acceptor interface. The widely used acceptor molecules are electronegative, C_{60}, buckminsterfullerene "soccer ball" molecules.

The major impact of large, solar panel "farms" on sustainability appears to be the large areas of ground they shade, altering plant and animal ecology. Since the major US installations are in the desert southwest, it appears that this impact will be low (this is not arable land or forest). In order to interface solar panels with the grid, their low-voltage DC power outputs must be inverted (i.e., changed to 60 Hz AC power). Inversion can be accomplished by having a DC motor drive an alternator or by a solid-state system. The inversion process has power losses, decreasing the overall efficiency of the conversion process from solar energy to AC power.

Maximum energy capture by flat solar panels requires that they be positioned with their surfaces perpendicular to the incident sunlight. This requires that the panel arrays be continuously adjustable in azimuth and elevation or that some optical means be employed to increase the energy capture from sunbeams obliquely incident on fixed panels. Clearly, steerable solar panel arrays are expensive, and require maintenance, while south-facing solar panels mounted on a roof at some optimum, fixed angle are far more economical.

4.4.5.4 Solar Thermoelectric Power Generation

This makes use of the Seebeck effect with TPs. A TP is an array of TCs connected in series, as shown in Figure 4.15b and c. A basic TC is shown in Figure 4.15a. It is made from two dissimilar metal wires, such as iron and constantan (an alloy), or platinum and nichrome (another alloy), or *p*-silicon and *n*-silicon, and so forth, that are pressure- or spot-welded together (never soldered) in two junctions. Its open-circuit output EMF can be shown to be a nonlinear function of the temperature difference between the two junctions, ΔT:

$$V_{oc(TC)} = A(\Delta T) + B(\Delta T)^2/2 + C(\Delta T)^3/3 + \cdots, \ \Delta T \equiv T_2 - T_1 \qquad (4.8)$$

For example, an iron–constantan TC has an $A = 57 \ \mu V/°C$ at $(T_2 + T_1)/2 = 750°C$.

Figure 4.15b shows a simple, 10-junction TP in which the output EMF is five times the EMF of the single TC circuit. In a power-generating TP, many hundreds of TC junctions are sandwiched between two thin, heat-conducting but electrically insulating plates, generally a black ceramic material. The "cold" plate touches the cold TC junctions; it also makes contact with a heat sink kept cool by circulating water or forced air. Incoming solar power heats the top plate to temperature $T_2 \gg T_1$. Thus, the open-circuit EMF of the TP is $V_{oc(TP)} \cong KA(\Delta T)/2$, where K is the total number of TC junctions in the TP. ΔT is proportional to the solar power absorbed by the TP's upper plate. Metal wires have resistances, which increase with temperature, and the wires

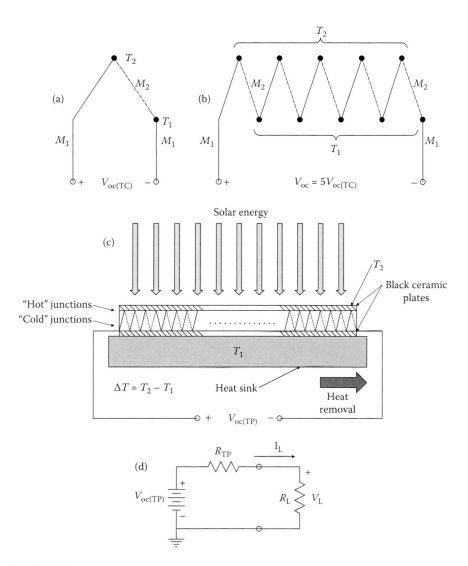

FIGURE 4.15
Thermoelectric power generation basics: (a) TC. Two metals are used; $T_2 > T_1$. T_1 is reference temperature. (b) Series array of five TCs makes a simple TP. (c) Solar thermoelectric power generation using a multi-TC TP. A heat sink is used to maximize $(T_2 - T_1)$. (d) Thévenin equivalent circuit of a STEPOG system. DC output power is maximized when the load resistance equals the Thévenin equivalent resistance of the TP, R_{TP} (i.e., the total series resistance of the N TCs making up the TP). R_{TP} is desired to be small.

in the TP, however short, also do, giving the TP a temperature-dependent Thévenin equivalent (internal) resistance R_{TP}, which increases as $(T_2 + T_1)/2$ increases. The equivalent circuit of the TP is shown in Figure 4.15d. It is easy to show that the power (P_L) in the load is

$$P_L = \left[\frac{V^2_{oc(TP)}}{(R_{TP} + R_L)^2} \right] R_L \text{ watts} \qquad (4.9)$$

Maximum power can be shown to be transferred from the TP to R_L when R_L is made equal to R_{TP} (the matched load condition). This is

$$P_{Lmax} = \frac{V^2_{oc(TP)}}{4R_{TP}} \cong \frac{(A_{TP}\Delta T)^2}{4R_{TP}} \text{ watts} \qquad (4.10)$$

where

$$A_{TP} = KA(\Delta T)/2$$

Commercial, off-the-shelf thermoelectric generators (TEGs) come in square, flat packages a few millimeters thick. For example, the TEC Company offers a Bi-Te–based TEG, model TEG1-12611-8.0, that is 56×56 mm^2. Its specifications are based on $\Delta T = 300 - 30 = 270$°C. The open-circuit voltage (OCV) at this $\Delta T = 9.5$ V. This gives an A_{TP} parameter of 38 mV/°C. The internal resistance and matched load resistance is 1.8 Ω, the output voltage across the 1.8 Ω load is $V_o = 4.75$ V (half the OCV), the matched load output current is 2.7 A, and the output power dissipated in the 1.8 Ω load resistor is 12.5 W. The heat flow across the TEG is approximately 290 W, and the heat flow density is approximately 9.2W/cm^2. Crystal Ltd. offers a TEG model G-127-14-16-L-S that can supply 2.42 W to a 2 Ω matched load resistor with $\Delta T = 150 - 50 = 100$°C. The output voltage is 2.2 V across the 2 Ω load resistor at rated power output. $A_{TP} = 44$ mV/°C for this TEG. Solar power can be focused on the top TEG plate of a TP by a parabolic mirror or a lens.

It is clear that solar energy is not the only possible source of heat for thermoelectric power generation. Heat can be derived from thermal energy storage masses, including solar ponds, and GT sources. Heat from combustion of NG can be used directly, but more importantly, "waste heat" from electric power generation using FFs, internal combustion engines' exhausts and cooling water, nuclear plants' steam condensers, gas turbines' exhausts, jet engines' exhausts, and so forth can be used to generate ancillary DC power up to hundreds of watts. TEGs can even be used in wireless patient monitoring (WPM) systems to power sensors and short-range radio modules in lieu of batteries. There, the modest temperature difference, $\Delta T \cong 37 - 25 = 12$°C, is derived from human body temperature. The heat source is human metabolism. A Linear Technology LTC3108-1 energy harvester integrated circuit (IC) can be used to convert the millivolt-level output of a TEG to a higher, regulated, DC voltage suitable to run micropower WPM devices.

4.4.5.5 Thermophotovoltaic Power Systems

We have seen that improving the efficiency of solar PV power generation is an active quest. SC surface etching, AR coatings, and new semiconductor compositions have been used to increase power conversion efficiency. Recently, thermophotovoltaic (TPV) systems have been developed that use engineered BB radiation absorber–emitters ("sun-traps") that rise to a high equilibrium temperature (about 800°C–1300°C) and reradiate IR energy in a relatively narrow band of wavelengths that corresponds to the most sensitive band of the underlying PVC.

The spectral efficiency of silicon PVs lies between 450 and 1020 nm, spanning the visible and near-infrared (NIR) wavelength bands (see Section 4.4.5.3). The solar energy spectrum at the Earth's surface basically has a solar BB radiation spectrum with notches in it in the IR from the atmospheric photon energy–absorbing gasses (e.g., CO_2, WV). This spectrum peaks at approximately 469 nm and then falls off slowly with increasing wavelength. The "long tail" of the Earth surface solar spectrum contains significant photon energy out to approximately 2500 nm and is ineffective in generating PV power.

Recently, scientists at the Massachusetts Institute of Technology (MIT) have devised an ingenious means to make use of the longer wavelengths in the solar spectrum to improve SC efficiency. Peter Bermel and colleagues (Bermel et al. 2011; Chandler 2011) developed a tungsten "sun-trap" absorber–emitter that absorbs long-wave infrared (LIR) energy and increases its temperature (to as high as 1227°C) and then reradiates IR mostly in the NIR band where the underlying indium–gallium–arsenide PVCs are most efficient. Bermel's experimental tungsten sun-trap can use sunlight or thermal energy from combustion. The sunside (input side) of the tungsten absorber–emitter has a dense array of holes in it: they are 0.75 μm in diameter and 3.0 μm deep; they are spaced 0.8 μm on centers in a rectangular grid. The side of the emitter facing the PVC has geometric patterns etched in it to cause it to reradiate IR only in the desired band effective for the PVC material used (Economist 2012a). The optimal output-side patterns are an active research area. Micron-dimension crosses have been used to give radiation bandwidth selectivity, and quite possibly, we will see simple fractal, NIR "antenna" patterns used in the future (Northrop 2012). The front surface holes give the tungsten absorber–emitter directional sensitivity to the IR source; hence, the TPV system must track the sun. Bermel et al.'s sun-trap TPV system, operating at temperatures over 1000°C, was capable of overall energy conversion efficiencies of over 37% in unconcentrated sunlight.

We note that the improved thermal trapping efficiency of Bermel's sun-trap also gives it application in STEPOG, where instead of PVCs, thermal-to-electric power conversion is done directly by *TPs* (see Section 4.4.5.4 above) attached to a high-temperature heat source.

To improve absorber–emitter efficiency in metallic sun-traps, it may be possible to microetch repeated fractal, optical, NIR antennas on the emitter side tuned to the wavelength bands that optimally excite the underlying PVCs.

Andreev et al. (2005) used parabolic mirrors or quartz lens solar concentrators to increase the efficiency of their experimental TPV power systems. Plain tungsten or tantalum absorber–emitters were used that reached temperatures between 1500 and 2000 K. Peak overall TPV cell efficiencies of 12%–13% were obtained for emitter temperatures between 1500 and 1900 K. These authors showed overall conversion efficiencies in the range of 30% when a tungsten 3-D photonic crystal was used for the absorber–emitter.

4.4.5.6 Solar Energy Storage

Because solar energy is available only during the day, there must be an efficient means to store solar energy from large STEG systems. One means is to directly store solar heat captured by the system. Heat storage materials must have high specific heat capacities. Commonly used heat storage materials are water, stone, packed earth, and phase-change materials that store energy in their heat capacity as well as in their heat of fusion; these include paraffin wax, Glauber's salt (hydrated sodium sulfate $Na_2SO_4 \cdot 10H_2O$), and certain low-melting-point metal alloys. When the molten material cools, it fuses (turns solid) and gives off its heat of fusion. For example, the heat of fusion of water is 80 cal/g (334,560 J/kg).

Other means of temporary energy storage make use of the electric output of the STEG system: these include flow batteries (FBs) and CAES (cf. Section 4.6 in this text); pumped hydro storage (PHS) is unlikely because there is little water in the desert regions where large-scale solar energy systems are generally operated.

Figure 4.16 illustrates the typical energy from sunlight, energy in storage, and STEG plant power output versus time over a day. Peak power is delivered over 80% of a day (19.2 h) because the heat stored in molten salt is used to make steam for approximately 4 h after the sun has set. Solar power is only available for approximately 10 h at that season and latitude of the STEG plant.

Solar thermal energy can also be stored in shallow ponds (about 1–2 m deep) of salt water. Just the surface area of the pond traps solar energy. Such solar ponds have been constructed on the shore of the Dead Sea in Israel. Ormat Systems installed a 20 ha pond in Israel capable of generating 5 MW. Bottom temperatures can vary between 70°C and 80°C. Bottom temperatures can approach 100°C under optimum conditions. A demonstration solar salt

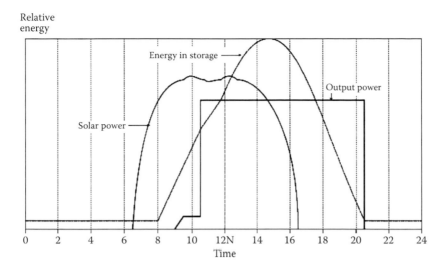

FIGURE 4.16

Energy storage in a typical STEG plant. The solar power input to the system starts at approximately 6:30 AM (sunrise), rises to a maximum about 9 AM, and then begins to fall as the sun begins to go down around 3 PM and falls to zero at sundown at 4:30 PM. Thermal energy in storage begins to rise at 8 AM, peaks at 3 PM, and is exhausted approximately 8:30 PM. Electric power output from the STEG plant with thermal storage reaches maximum at approximately 10:30 AM and runs at maximum capacity after the sun sets, powered by the stored thermal energy until no more useful stored heat is available, at approximately 8:30 PM. With a higher thermal storage capacity, the STEG electric power output could run longer. (Redrawn from Sandia National Laboratory, Desirable features of power towers for utilities, 2012. http://energy.sandia.gov/?page_id=1437.)

pond at Beit Ha'aravah has 250,000 m² area and powers a 5 MW turbine. Such ponds are often operated at high power at night and allowed to "rest" and "charge" in the sunlight (Faiman 2012).

The largest solar pond in the United States is in El Paso, TX. This 0.3 ha pond runs a 70 kW Rankine-cycle turbogenerator that runs a 20,000 L/day desalinating plant and supplies process heat to an adjacent food processing company (Bedi and Falk 2000). Strangely, there are no convection currents in a solar pond because both the salinity and water temperature increase with depth. The brine temperature can reach about 90°C at the bottom. By tapping the heat in this bottom layer, a solar-to-electric efficiency of as much as 31% can be reached (Sciencedaily 2008). Solar ponds are perhaps the least expensive solar energy harvesting systems to build and run, but these economies must be balanced against problems with siting and their low efficiencies.

4.4.5.7 Solar Energy and Sustainability

Fortunately, solar energy capture has little negative impact on human sustainability. Using the ISEGS figures above, approximately 0.034 km²/MW of land surface is used for STEG system solar energy capture. This is land that is generally not used for agriculture because of extreme aridity and high temperatures, for example, desert land in North Africa, middle eastern countries, the US southwest, Mexico, Australia, and so forth. One problem with the desert siting of STEG facilities is the transmission of the electric power to centers of population. In most cases, new transmission lines must be built to existing main grids. An effective use of STEG power generated near the ocean is to keep it on site to desalinate seawater for agriculture (make deserts bloom). The capital costs of STEG plant construction and transmission line installations will be met when a *grid parity* tipping point is met. Grid parity is the point where the cost of PV- or STEG-generated electric power is equal to or less than the cost of AC grid power, which mostly comes from burning FFs. In sunny parts of the US southwest, grid parity may be reached around 2015. It may be sooner if FF prices continue to escalate. Grid parity has already been reached in Hawaii and other islands that otherwise use diesel fuel to produce their electricity (Farrell 2012).

Recently, two MIT scientists, Alexander Mitsos and Corey Noone, reported on their research on solar mirror (heliostat) placement to improve the efficiencies of STEG systems. The results of their modeling study showed that instead of placing mirrors in concentric semicircles around the solar collector tower, better use could be made of the land if the mirrors were made smaller and arranged in a Fermat spiral array in which each element is set at a constant angle of 137.5° relative to its neighbor. This results in a pattern of mirrors around the southern axis of the collector tower similar to the spiral arrangements of the florets in a sunflower around the flower's center. Noone et al. (2011) were able to show that almost 16% of ground space was saved (for the same mirror area), and there was a small, approximately 0.4%, efficiency improvement in solar energy trapping (Economist 2012b). The 16% space saving is important when siting STEG systems in areas where there is a land use conflict.

California is the most active state in developing large-scale STEG systems. As of 2010, California had approved the construction of 10 new STEG plants with a total peak capacity of 4.1925 GW. It also had 2 STEG systems under review with a total peak capacity of 300 MW. California already has 5 Luz-built STEG plants now operational with a total peak capacity of 503.8 MW (California 2010a & b). Spain now (2012) has 26 STEG plants totaling 1.102 GW peak capacity (CSCC 2012). In 2009, Hawaii generated 36.41 GWh of energy by PVs, more than double its PV power in 2008 (Hawaii Data Book 2010).

STEG systems that use flat heliostat mirrors are sensitive to high wind velocities; above some critical value, the mirrors must be put horizontally to minimize wind loading and possible mechanical damage. Of course, this effectively shuts down solar power generation.

Cost of PV SCs: Many are now made at sites with cheap technical labor (e.g., China, Taiwan). PV electric power cost on a commercial scale is now approximately $1.20/W. Efficiencies run between 16% and 24% for normal crystalline Si technology (Shah 2011).

4.4.6 Hydropower, Including Tides and Waves

4.4.6.1 Introduction

A cubic meter of water weighs approximately 1000 kg (1 tonne). Thus, a cubic meter of water moving at 1 m/s (2.24 mph or 3.6 km/h) has approximately 1 kJ of KE, some of which can be harvested and turned into electrical energy. Water can move due to tidal forces, or fall, converting its PE in a pond or reservoir to KE, thence to electrical energy.

Man has harnessed continuously moving water to drive gristmills since before the first century BC. Gristmill technology was simple and effective; moving water drove a waterwheel connected with wooden shafts and gears to a horizontal, moving, round millstone (the runner stone) resting on a stationary stone (the bed stone). Grain was poured in the center of the runner stone. The flour was collected on the outer margin of the stones, directed radially by spiral grooves cut into the bottom face of the runner stone, which typically rotated at approximately 120 rpm. Water-powered gristmills were found throughout the civilized world by 1000 AD, and the *Domesday Book* lists 5624 in England in 1086. There were approximately 17,000 in England by 1300 (Gristmill 2004).

At first, streams and rivers were used to power waterwheels used to drive grain-grinding mills. Simple dams were built to give the water PE to drive overshot and undershot waterwheels. Strong, periodic tidal flows out of saltwater millponds were also harnessed to drive waterwheels connected to gristmills. In one such tide mill located in Westport, CT, several tidal gates were opened to facilitate filling the Compo (aka the Sherwood Island) Millpond on a rising tide, the gates automatically shut when the tide began to ebb, and the impounded water was released through a mill gate to drive the waterwheel. On spring tides, the starting hydraulic head could be over 2 m. Figure 4.17 illustrates the Compo Tide Mill in Westport, CT, originally built in 1705 by John Cable (Jennings 1933). In the middle of the 19th century, the mill was acquired by the Sherwood family, who rebuilt it, as well as constructed a substantial breakwater, wharves, and new tide and sluice gates. The mill was run to grind corn (maize) in the mid-19th century by Henry Burr Sherwood and, after the Civil War, was also used to grind an imported mineral, barite (barium sulfate), used to make such late 19th-century goods as oilcloth, photographic emulsions, wallpaper, lithographic inks, and so forth. The mill used a unique, *horizontal* waterwheel, about 14 ft. in diameter. The original Compo mill burned in 1895 and was rebuilt. The Sherwood Island Millpond continues to be used for recreational boating and was used to raise oysters and clams until the 1970s.

In the 20th and 21st centuries, falling water has been used to drive large hydroelectric turbines. The water is given PE by impounding it behind dams and then releasing it to flow down penstocks to gain high KE to drive turbines.

Most recently, in the 21st century, engineering development has focused on extracting ocean wave energy (OWE) (Vining 2005; Rhinefrank et al. 2006). A variety of innovative designs have been developed to harness OWE,

FIGURE 4.17
Photograph of the old Sherwood Island (Compo) Mill Pond gristmill taken before it burned in 1895. KE from tidal outflow from the millpond drove an unusual, horizontal, 14 ft. diameter waterwheel at the bottom of the millrace. (From a Northrop family photograph archive.)

as described below. The electric power generated by OWE is necessarily periodic in magnitude because wave power itself is periodic. Thus, a means of smoothing out and storing the pulses of wave-generated electric power must be used for it to be usefully interfaced with a terrestrial power grid.

Like wind and solar energy, OWE is "free." Its only cost is in building the OWE capture (OWEC) systems, their physical interface to the power grid, and their maintenance. The largest ocean waves are to be found in regions where the wind blows onshore the strongest, with the longest fetch. Hazardous to OWE systems are hurricanes, typhoons, and extreme storms that generate supersized waves and storm surges that test the systems' robustness.

4.4.6.2 Ocean Wave Energy

The total annual average wave energy off US coastlines (including Hawaii and Alaska) calculated at a water depth of 60 m has been estimated to be 2.1 PWh/year (2.1×10^{15} Wh/year) (White 2006). Waves are caused by wind.

Wave power in deep water where the water depth is $\gg \frac{1}{2}\lambda$ is given by the relation (Vining 2005)

$$P_w = \frac{\rho g^2 H^2}{32\pi} T \, \text{W/m} \tag{4.11}$$

where P_w = power per meter of wave-front length (W/m); ρ = seawater density (1000 kg/m³); T = wave period (s); g = acceleration of gravity (9.81 m/s²); H = peak-to-trough wave height (m); and f = wave frequency = $1/T$.

For example, moderate, deepwater ocean swells with a period $T = 8$ s and $H = 3$ m have $P_w = 68.8$ kW/m of wave front. Just how much of this power can be harvested depends on the design of the OWE converter. In a major ocean storm, $H \cong 15$ m and $T \cong 15$ s, so $P_w \cong 3.22$ MW/m. At a 20 m water depth (d), an ocean wave's energy drops to approximately 1/3 of the level it had in deep water where $d \gg \lambda$.

Wave energy is really due to solar radiation, which warms the oceans and land unevenly, heating the overlying atmosphere unevenly, causing winds to flow from high-pressure areas to low-pressure zones, which, when blowing across water, causes waves. This transfer of solar energy to the atmosphere creates the strongest winds; hence, the largest waves are generally between 30° and 60° latitude and on the western shores of the continents. While the total annual average wave energy off the US coastlines (including Alaska and Hawaii) at a water depth of 60 m has been estimated to be 2.1 PWh (2.1×10^{15} Wh), "[e]stimates of the worldwide economically recoverable wave energy resource are in the range of 140–759 TWh/yr for existing wave-capturing technologies that have become fully mature" (Wave 2006).

According to Rhinefrank et al. (2006), there were (in 2006) more than 1500 wave energy device patents. Natural (economic) selection will determine which robust designs are least costly to build and operate and which will attract investor money and be implemented as practical OWE-harvesting devices.

OWE systems may be first classified by the location of the system as shoreline, near-shore, and deepwater designs. Shoreline devices are fixed on the shoreline and take advantage of breaker waves over the tidal zone. Advantages of shoreline OWE systems include ease of maintenance and shorter lengths of copper power cable needed to connect them to the grid. One disadvantage is that in approaching the shore, waves lose much of their energy "feeling the bottom." Another disadvantage is that shoreline designs are generally highly visible from the shore. Near-shore OWE device sitings enjoy more wave energy than onshore installations. However, they require underwater power cables to shore, maintenance is more difficult, they may be a hazard to shipping, and if visible from shore, siting protests may occur by not-in-my-front-yard (NIMFY) esthetes. Deepwater OWE device designs enjoy full wave energy, and NIMFY protesters ordinarily cannot see them, but they do suffer from the disadvantages of maintenance difficulties and the requirement for long, expensive, underwater power cables to shore, with their losses. They also may pose navigation hazards.

Wave energy extraction technologies are highly innovative and various. We describe below some of the more promising designs; these include oscillating water column–compressed air turbine (OWCCAT), overtopping devices, float (buoy) systems, and hinged contour devices. For a pictorial overview of these technologies, visit the Robinson (2006) Web site.

1. A shoreline OWCCAT system design uses an air column trapped in a chamber over the surging water. The wave action periodically compresses and decompresses the air, driving a bidirectional air turbine. Figure 4.18 illustrates the schematic of an OWCCAT OWE

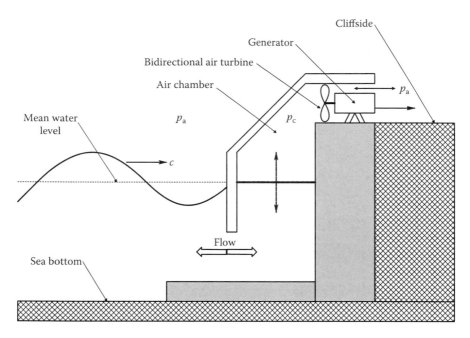

FIGURE 4.18
Elevation schematic of an OWCCAT generator. Trapped air in the air chamber is periodically
compressed and exhausted by wave action, driving a bidirectional, low-pressure air turbine/
generator system. OWCCAT systems have been built with approximately 500 kW generating
capacity. They are practical only on leeward coasts with strong wave action and a low tidal
range.

system. One such OWCCAT system, built on Islay Island, Scotland,
by Wavegen generates a peak power of 500 kW and provides power
for approximately 400 island homes; this LIMPET 500 OWC system
has been operational since November 2000. A more efficient, large
OWCCAT OWE system similar to the Scottish design was built by
Energetech in Port Kembla, Australia. A small prototype OWCCAT
system was operated by Energetech at Point Judith, RI (Robinson
2006). The best siting for such OWC systems is on a coastline having high waves and a low tidal range, so the OWC system intake is
always under water and the water is deep right up to the shore.

A Japanese OWE system design uses onshore wave energy to cause
a steel pendulum flap to swing back and forth, powering a hydraulic
pump that pressurizes a hydraulic tank that runs a turbogenerator.

2. As an example of an overtopping OWE system, a tapered channel
system (TAPCHAN) has been built in Norway. The tapered channel
funnels wave water up into a reservoir on a cliff, converting wave KE
to pond PE. The water then flows back to the sea through a low-head

turbogenerator at a constant rate. The reservoir acts as a smoothing capacitor for the pulsatile wave energy (and volume) (Rhinefrank et al. 2006) and allows for relatively constant electric power generation. Another overtopping wave energy conversion (WEC) system design is moored near shore where wave height is greater. Waves impinging on the seaward edge of the device surge water into a reservoir on top of the device, where the water flows downward through a low-head, water turbogenerator, giving nearly continuous power output as long as the turbine outlet volume is less than the overtopping (input) volume (Vining 2005).

3. There are many types of buoy or point-source OWE systems. They are generally moored in deep water or in near-shore waters to take advantage of deepwater OWE. In one design, wave action powers a hydraulic pump-pressure accumulator system, which in turn drives a hydraulic motor connected to a generator. In another design, wave action provides linear force to a moving permanent-magnet, fixed-coil DC generator, or a linear moving-coil, fixed-permanent-magnet generator (Rhinefrank et al. 2006).

4. The Pelamis OWE system is a good example of the floating, hinged contour device (*Pelamis* is a genus of tropical sea snake). Working and prototype Pelamis OWE systems are found in Portugal, Scotland, Hawaii, Oregon, Maine, Massachusetts, and San Francisco. The Pelamis OWE system, developed by Ocean Power Delivery Ltd., is classified as a *floating attenuator,* that is, it decreases the heights of the waves acting on it by absorbing some of their energy. Physically, each Pelamis device consists of four concatenated steel cylinders, each approximately 130 m in length and 4.0 m in diameter. The 130-m length axis is oriented perpendicular to the wave crests to intercept maximum OWE. Pelamis OWE systems are generally sited in water over 50 m in depth and about 6 km from a leeward shore where the waves are strong, so it is still practical to connect the Pelamis systems to the shore by underwater cables. An artist's rendition of a Pelamis wave power "farm" is shown in Figure 4.19. The four cylinders of each Pelamis unit are joined by three 2-D hinged couplings that drive hydraulic pumps as each cylinder pair is forced by wave pressure to assume a different angle with respect to each other's length axis. The couplings allow pairs of connected tubes to move in both pitch and yaw (heave and sway) in relation to each other in order to extract maximum energy from waves that are not propagating exactly in the direction of the Pelamis units' axes. The high-pressure hydraulic fluid from the pumps drives hydraulic motors coupled to electric generators. Electric energy derived from the three joints of each Pelamis unit is sent to a central, undersea collection unit, thence to shore by a common power cable. The power output of

FIGURE 4.19
Artist's rendition of a Pelamis wave generator power "farm." See text for description. (Courtesy of Robinson, M.C., Renewable Energy Technologies for Use on the Outer Continental Shelf, PPT slide presentation. Ocean Energy Technology Conf., National Renewable Energy Lab. June 6, 2006. http://ocsenergy.anl.gov/documents/docs/NREL_Scoping_6_06_2006_web.pdf.)

a single Pelamis system is necessarily pulsed DC (full-wave-rectified AC), which must be smoothed before being inverted to grid AC. The smoothing can be accomplished by summing many asynchronous unit outputs in the "farm" and using capacitors, a flywheel, or other energy storage means (Rhinefrank et al. 2005; Wave 2006). Pelamis units must be moored fore and aft to resist not only wave forces but also forces from tides and winds.

A single demonstration Pelamis unit installed off the coast of Oregon had the estimated electrical specifications shown in Table 4.4. The Oregon commercial-scale system uses four clusters of 45 Pelamis units (180 total). Each cluster consists of 3 rows with 15 devices per row. This commercial Pelamis system has a calculated yearly energy production of approximately 180 GWh/year (about 493 MWh/day). The electrical interconnections between the devices are by midwater jumper cables; four independent subsea cables carry the four cluster power outputs ashore for interfacing and connection to the AC grid.

The McCabe Wave Pump (by Haydam Ltd.) works similarly to the Pelamis system except the McCabe system has only three pontoons and two hydraulic pumps actuated by the relative motion of the end floats to the central pontoon. The hydraulic pumps charge a hydraulic accumulator to store the wave energy; the pressurized hydraulic fluid drives a hydraulic motor–generator unit.

TABLE 4.4

Demonstration Pelamis Electrical Specifications off the Oregon Coast

Device Rated Capacity (kW)	750[a]
Annual Energy Absorbed (MWh/year)	1472
Annual Energy Produced (MWh/year)	1001
Average Electrical Power Generated (kW)	114
Number of Homes Powered by Device	114

[a] The single-device capacity was derated to 500 kW in the commercial plant (Hammons 2009).

Pelamis and McCabe "farms" constitute distinct navigation hazards. They are marked with warning buoys, radar reflectors, and strobe lights on each individual Pelamis machine. Other moored OWE-harvesting systems also must be similarly marked. They are best used in regions that are not subject to frequent hurricanes or typhoons.

5. Scientific Applications & Research Associates (SARA) has developed a prototype OWE-capturing magnetohydrodynamic (MHD) system. SARA has built and tested a 100 kW MHD laboratory prototype OWEC system. The SARA OWEC system is designed to operate near shore where wave amplitudes are higher.

The MHD generator is moored on the ocean bottom. A shaft connected to a surface buoy is driven up and down by wave action. This pistonlike action forces an electrically conducting fluid at high velocity past powerful, fixed, permanent magnets and electrodes. Figure 4.20 illustrates schematically the vector geometry of MHD electric power generation. The fluid contains a high concentration of ions having high mobilities, giving the generating fluid high conductivity. Faraday's induction law states that the open-circuit EMF between the electrodes, E_F, is given by the vector integral

$$E_F = \int_0^d (\mathbf{v} \times \mathbf{B}) \cdot d\mathbf{l} \text{ volts} \tag{4.12}$$

where the mean flow velocity $\mathbf{v}$ is perpendicular to the mean magnetic field intensity $\mathbf{B}$, dl is the differential distance in the diameter, and d is the insulating conduit diameter.

For orthogonal vectors, Equation 4.12 reduces to $E_F = BLv$ volts. Expressed in terms of the average volume flow, $\bar{Q}$, of the conducting fluid, E_F can also be written as

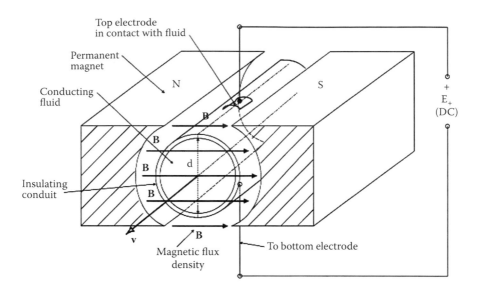

FIGURE 4.20
Diagram illustrating the Faraday MHD effect. See text for description.

$$E_F = \frac{\bar{Q}B(4\times10^{-8})}{d\pi}\text{volts} \qquad (4.13)$$

where the pipe diameter d is in centimeters, B is in gauss, $\bar{Q}$ is in cubic centimeters per second, and E_F is in volts (Northrop 2005).

The open-circuit EMF, E_F, for each electrode pair in a multiple-conduit MHD system appears in series with a low Thévenin series resistance, R_T, whose magnitude depends on the electrode material and area; the fluid's conductivity, ρ; and the electrode separation, d. It is well known that maximum power output from a Thévenin source occurs when the E_F, R_T source is terminated in a load resistance, R_L, equal to R_T. This maximum power in R_L can easily be shown to be $P_{L\max} = E_F^2/(4R_T)$ watts for a single electrode pair. The total induced MHD power is converted from low-voltage DC to commercial-grade 60 Hz AC for export ashore by an inverter/transformer (SARA 2011).

Waves are generally larger in winter months, having approximately six times their summer energy levels. The electric power initially generated by an OWE system of any design is necessarily periodic in time at the wave frequency. Consequently, a means must be used to smooth the wave power pulses. One means of power smoothing is by alternately charging and then discharging an array of supercapacitors (cf. Section 4.6.6) or keeping a massive flywheel spinning.

4.4.6.3 Tidal Energy

The ocean's tides are the result of the vector sum of gravitational forces between the Earth, the moon, and the sun. Because of orbital constraints, some tides are biperiodic; one period (solar induced) is 24 h, and the other (lunar induced) is semidiurnal at approximately 12 h, 25 min. The so-called spring tides, also called perigee tides, occur when the gravitational forces of the sun and moon act together, giving a maximum high-to-low range. On some coastlines, particularly in certain estuaries and bays, the perigee tidal difference can reach over 17 m. Neap tides occur when the tide-generating forces of the sun and the moon are acting at right angles to each other. Tidally caused water movements contain both KE and PE. The latter is in the difference in water height between full high tide and full low tide. Tides are also affected locally by strong winds, which blow with or against their flow, as well as atmospheric pressure differences. In northern latitudes, where tidal flows are channeled between straights or islands, high water flow velocities are seen that can reach 2–3 m/s or more. The global tidal peak energy potential is estimated to be approximately 3 TW, with approximately 1 TW available in comparatively shallow waters. In the EU countries, France and the United Kingdom have tidal ranges over 10 m (32.8 ft.). High tidal ranges are also seen in northern Russia, as well as Alaska, Canada, Argentina, Korea, and Western Australia (Lim and Koh 2009; Soerensen and Weinstein 2008).

Locations with particularly strong tidal currents (water KE) are found around the United Kingdom and Ireland, between the Channel Islands and France, in the Straits of Messina between Sicily and Italy, and in various channels between the Greek Islands in the Aegean. Still other strong marine current assets are to be found in Southeast Asia, the east and west coasts of Canada, and so forth (Soerensen and Weinstein 2008).

UK-based *Tidal Electric* is building a 300 MW tidal power system in China in the mouth of the Yalu River (on the border of northeast China and North Korea) (US DOE 2004) to be larger than the 240 MW French tidal power plant at La Rance, in Brittany, France. The Rance tidal power station (TPS) is the world's first (and largest, to date) TPS. It became operational in November 1966. It consists of twenty-four 10 MW bidirectional, bulb-type turbogenerators, giving it a peak power-generating capacity of 240 MW. The 5.35 m diameter bulb turbines are located along a 700 m barrage dam. The La Rance TPS has a capacity factor of approximately 40% and supplies an average power of 96 MW, giving it an annual energy output of approximately 600 GWh. The average range between high and low tides is 8 m, with a maximum spring tide range of 13.5 m (44.3 ft.). The Rance TPS storage basin has an area of 22 km². The ongoing cost of electricity production by the Rance TPS is below 0.02 euro/kWh (REUK 2008).

The Annapolis Royal TPS is located on the Annapolis River, immediately upstream from the town of Annapolis Royal, Nova Scotia, Canada. It consists of a dam and a powerhouse. This TPS harnesses the large tidal differences of the large tides in the Annapolis Basin, off the Bay of Fundy. It was put online in 1984 and has a peak power capacity of 20 MW and an average annual generation of 50 GWh. The Bay of Fundy has a mean tidal range of 17 m (55.8 ft.) at Burntcoat Head. During the 12.4 h tidal period, approximately 115 billion tonnes of water flow in and out of the bay. There are approximately 6 h, 13 min between each high and low tide (Parks Canada 2012).

Scheduled for completion in 2015, an underwater, 200 MW tidal energy plant is being developed in Cobscook Bay near Eastport, ME (in the southeast corner of the state). The mean tide range (high to low) is 20 ft. (6.1 m) in Cobscook Bay. This creates strong tidal currents in and out of the bay. Ocean Renewable Power Co. of Portland, ME, and Stillwater Metalworks in Bangor, ME, are jointly developing the system, which will have a 46 ft. wide, underwater support structure holding a 10,000 lb. (4536 kg) generator flanked by several hydrokinetic turbines (Farwell 2010). Like the Rance TPS, the Cobscook tidal system will generate power on both rising and ebbing tides and obviously must go offline around high and low tide, when the current is slack.

The millpond strategy, described above for tidal gristmills, is adaptable for electric power generation in coastal regions with high tidal ranges. First, an impoundment pond that can be naturally filled by the incoming tide must be constructed with appropriate tide gates, and then a bidirectional, low-head, high-flow turbine connected to an alternator can be run on the outgoing tide until the pond level is approximately 1 m above the current low tidal level. Then the system must go offline until the tide turns, and the incoming tide rises more than 1 m above the pond level, and then water again turns the turbine and refills the pond.

OWE and TPS energy is gaining acceptance in the world. In 2011, the UK government made $4 million (£2.5 million) grants (total) to three companies to do pilot plant research on wave and tidal energy–generating systems in the United Kingdom. Seven British companies and three universities will be involved in the work, including Bauer Renewables Ltd., Pelamis Wave Power Ltd., and Marine Current Turbines Ltd. (Smith 2011).

In all tidal power generation systems, the possibility of harm to local or migrating marine life exists. Turbines are not "fish friendly." Two humpback whales were trapped upstream of the Annapolis Royal TPS in the Annapolis River in Nova Scotia; one mature animal escaped in 2007, and another immature whale died. Some freshwater hydropower plants have solved the problem of fish versus turbines by the use of underwater 60 Hz AC electric fields to scare fish away from turbine intakes, and also pulsed-DC electric fields used in electrofishing systems can be used to attract fish to the positive electrode (anodal electrotaxis) and temporarily stun them, allowing them to be washed down a penstock to safety in the river below (Northrop 1967, 1981). Gratings used to exclude larger fish from turbine intakes tend to trap fish against them and kill some of them by suffocation. Certain OWE systems may also be deadly to fish.

Any OWE or tidal power-generating system may be subject to storm and tsunami damage and must be built robustly, especially considering that global warming can affect the intensity and frequency of severe storms. If a deepwater OWE system breaks loose from its mooring and is dashed on the shore, or is damaged and sinks, recovery and repair or replacement will be very expensive.

4.4.6.4 Hydroelectric Power

In this section, we address the power produced by the KE of water falling from reservoirs behind dams (as opposed to tidal hydropower). The *available hydropower, P_a,* of a dam of height h and a flow Q in a penstock is given by the relation

$$P_a = \rho Qgh\eta \text{ watts} \tag{4.14}$$

where ρ is the density of water (e.g., about 10^3 kg/m^3), Q = volume flow rate (e.g., m^3/s), g = acceleration of gravity (9.8 m/s^2), h = hydraulic head in meters, and η = efficiency of system (0–1).

The PE of the hydraulic head behind the dam is converted to KE of the water exiting the penstock into the turbine. The *hydraulic input power to the turbine* is given by

$$P_i = \frac{1}{2} \rho Av^3 \text{ watts} \tag{4.15}$$

where A = the penstock area in square meters and v = velocity of the water exiting the penstock in meters per second.

Penstocks are giant pipes that carry the moving water from behind and below impoundment dams to turbines that drive the alternators. As long as rain and snowmelt refill the rivers leading to the reservoirs, the power generated is totally green and low cost. The initial cost of building the dam, powerhouse, transformers, and transmission lines is the major investment; operation of a falling-water hydroelectric plant is relatively inexpensive.

The creation of the reservoir generally has severe environmental impacts on both valley and river ecosystems. The population of any village that may lie in the valley that is flooded must move. There is generally loss of farmland and wildlife. Another negative impact of hydroelectric power generation is silt accumulation behind the dam. Silt from rain and snowmelt runoff, carried by fast-moving rivers and streams above the dam, settles in the still water behind the dam, gradually reducing the effective reservoir volume. It can be removed by dredging or creating a temporary, massive outflow of impounded water. Some of the silt then is carried downstream from the dam and then becomes a downstream environmental problem.

Dams have been known to fail infrequently but catastrophically. In 1963, the Vajont Dam in Italy failed, killing approximately 2000 people. The Banqiao Dam failure in southern China killed approximately 26,000 people directly, and another estimated 145,000 perished from the following epidemics.

Millions were left homeless. Dams can fail because of earthquakes or poor construction and/or design and inadequate maintenance, coupled with heavy rainfall.

Major hydroelectric dams also create problems for migratory fish such as salmon and shad. Many dams in the United States have been licensed contingent on that they provide fish ladders to permit at least some fish to reach upstream water to spawn. The low dam in Holyoke, MA, on the Connecticut River uses a fish elevator (literally) to lift upstream-swimming shad above the dam; it has been highly effective.

The world hydroelectric energy production and consumption has risen nearly linearly: 1.7×10^9 kWh in 1980 to 3×10^9 kWh in 2006. The top 10 hydroelectric power–consuming countries are the following: (1) China (431.43 billion kWh); (2) Canada (351.85 billion kWh); (3) Brazil (345.32 billion kWh); (4) United States (289.25 billion kWh); (5) Russia (173.65 billion kWh); (6) Norway (118.21 billion kWh); (7) India (112.46 billion kWh); (8) Japan (84.90 billion kWh); (9) Venezuela (81.29 billion kWh); and (10) Sweden (61.11 billion kWh) (IndexMundi 2011a, 2011b). Still, hydroelectric power constitutes only approximately 8% of the total electric power consumed in the United States (Yergin 2011) The largest US hydropower facility is at the Grand Coulee Dam on the Columbia River in Washington State. Completed in 1942, the Grand Coulee can produce a *peak power* of 6.80 GW. The Three Gorges Dam in China is the world's largest operating hydroelectric plant, with a peak capacity of 22.5 GW.

4.4.7 GT Energy and Heat Pumps

The primary form of geothermal (GT) energy is in the form of heat from the core of the Earth that has risen through geological faults to a depth where it can be accessed by drilling wells. This core heat originates from energy left from the formation of the planet and from radioactive decay; it is estimated to be on the order of 10^{31} J or 2.778×10^{24} kWh, energy that is virtually inexhaustible by the activities of man. The GT gradient, which is the difference in temperature between the molten core of the planet and the surface, drives a continuous flux of thermal energy (heat) from the core to the surface. The total planetary core-to-surface heat flux is approximately 44.2 TW. Radioactive decay of isotopes in the core is estimated to replenish this lost core heat at a rate of 30 TW.

Three common manifestations of GT energy are geysers, hot springs, and hot lakes. In certain GT locations, surface water can percolate down until it reaches temperatures high enough to flash it into steam, which then expands and rises to the surface. The steam may erupt periodically, driving hot water with it in the form of a geyser, or percolate more gently up into a spring or lake, heating it.

GT power uses steam formed when water is pumped down injection wells into a GT hot zone. The steam is extracted from nearby steam production wells and is used to drive turbines, which turn generators; the condensed steam can be recycled (down the injection wells), or the hot water condensate can be used for domestic or greenhouse heating. Other hot, GT water at the surface can also be used directly for heating. Approximately 70 countries made direct use of 270 petajoules (PJ) of GT energy in 2004 for GT heating. Over half this energy went into space heating, another third into heated pools, and the balance went for agricultural and industrial applications. Worldwide, 88 PJ of space heating was extracted from 1.3 million GT systems with a total capacity of 15 GW (NREL 2012). Heat pumps (HPs) for home heating are the fastest means of exploiting GT energy; their usage growth rate is approximately 30%.

Worldwide, approximately 10.960 GW of GT electrical power was online in 25 countries in 2010. The top GT electricity–producing countries are the following: the United States (California, Arkansas, Nevada, and so forth), 3.086 GW from 77 power plants; the Philippines, 1.904 GW; Indonesia, 1.197 GW; Mexico, 958 MW; Italy, 843 MW; New Zealand, 628 MW; Iceland, 575 MW; Japan, 536 MW; plus many others. In 2010, GT power made up approximately 18% of the electricity generated in the Philippines (Holm 2010).

There are adverse environmental effects associated with producing GT electricity. GT steam and condensate may carry CO_2, toxic hydrogen sulfide (H_2S), CH_4, and ammonia (NH_3). Existing GT power plants emit an estimated average of 122 kg of CO_2 per megawatt-hour of electric energy generated, a small fraction of the CO_2 emission per megawatt-hour of electricity from an FF-generating plant.

In addition to the dissolved gasses, hot water from GT sources may contain trace amounts of dissolved elements such as calcium, iron, mercury, arsenic, antimony, and boron. These elements will precipitate as the water cools, so to avoid environmental contamination, the best practice is to pump this water back into the injection well.

The electric energy required to drive the pumps and compressors in a GT power system may come from nongreen, CO_2-producing sources. However, this parasitic load is always a small fraction of the heat energy output. Ideally, the electric power to drive the pumps and compressors should come from the GT power plant itself.

Other environmental effects that have affected GT power production are Earth subsidences (occurring in the Wairakei Field in New Zealand and Staufen im Breisgau, Germany). Hydraulic fracturing (fracking) associated with high-pressure water injection has produced small earthquakes measuring up to 3.4 on the Richter scale. Operation of a GT power plant in Basel, Switzerland, was suspended because over 10,000 small seismic events occurred on the first 6 days of water injection.

A GT power plant uses 20 L of freshwater per megawatt-hour of electric energy produced. Contrast this with the over 1000 L used per megawatt-hour in a nuclear, coal-generating, or oil-generating plant.

GT power generation requires no fuel, except to drive pumps, and is relatively free from fuel cost rises. In building a GT plant, drilling wells accounts for over half the costs, and exploration of deep heat resources involves financial risk. A typical well doublet (injection and extraction) in Nevada can provide 4.5 MW and costs approximately $10 million to drill, with a 20% failure rate (Geothermal 2009).

HPs are electrically powered systems that can move heat from one volume to another. They can be used to heat or cool a volume. In the cooling mode, an HP works the same as an ordinary air conditioner. It makes use of a closed system containing a refrigerant gas that absorbs heat as it expands and vaporizes, cooling one volume, and then releases heat to another volume when it condenses after being compressed (Cowan 2008).

A ground-source HP (GSHP) (aka GT HP) absorbs heat from a depth of approximately 9–10 m in the earth at temperature T_u and transfers it indoors to temperature T_i for heating; normally, $T_i > T_u$. GSHP operation can be reversed for domestic air conditioning; in this case, heat is absorbed from indoors and released underground at the lower temperature. Figure 4.21

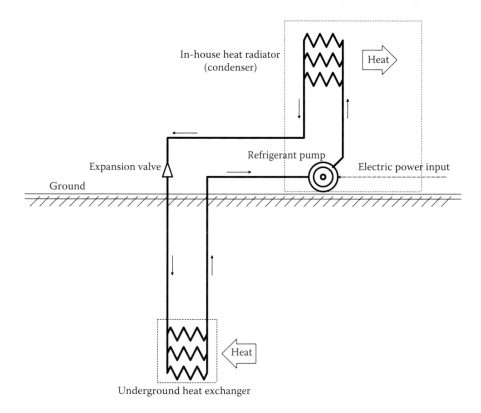

FIGURE 4.21
Schematic of a GSHP. See text for description.

illustrates schematically the components of a GSHP. Various refrigerant gasses have been used in HPs. These include but are not limited to R-134a (1,1,1,2-tetrafluoroethane), R-717 (ammonia), propane, butane, CO_2, R-744, R-22, R-410A, R-600A (isobutane), and DME.

The energy performance of HPs can be compared by calculating their coefficient of performance (COP), which is the *ratio of useful heat movement to electrical work input*. (Ideally, the COP of an HP should be as large as possible.) Working on a mild day (10°C outside), a typical air-source HP (ASHP) has a COP of 3–4, while an (indoor) electrical resistance heater has a COP = 1.0. A GSHP working with an underground heat reservoir temperature of approximately 10°C will have a COP number of 7.2 when heating an indoor heat exchanger coil to 35°C. Its COP drops to 5.0 when the indoor heat exchanger is at 45°C. The electric power required to run the HP compressor generally comes from an FF source in the United States; ideally it should be green power. Subfloor, indoor heat exchangers are generally used for improved heat extraction with HPs, or air forced over the heat exchanger makes warm air for efficient domestic heating.

Besides COP rating, HPs are also rated by a *seasonal energy efficiency rating* (SEER) and their *heating seasonal performance factor* (HSFP). The SEER is the ratio of how much energy (in Btu) is pumped outside in cooling mode divided by the electricity used (in watts) for cooling; it should range between 14 and 18. The HFSP is basically the ratio of energy pumped indoors for heating to energy used for heating, but it also takes into consideration supplemental heating needs and the energy needed to defrost the unit, if required. The HFSP rating should range from 8 to 10 in a well-designed unit (Cowan 2008).

Instead of mechanically compressing the refrigerant gas, then expanding it, an *absorption HP* uses heat from combusting natural gas or LPG to raise the refrigerant gas pressure before expansion. The COPs of this type of HP are lower than compression HPs, typically 1.5. On the other hand, a gas furnace for space heating has a COP that only approaches unity.

HPs work best in moderate climates. Compressor-type HPs use no FFs directly; they use only electric power, the cost of which is rising with demand and with the increasing cost of FF used to generate the power. The choice of an HP to install must take into consideration the local climate, the heating demands, and cost of electric power.

4.4.8 Hydrogen Economy

4.4.8.1 Sources of H_2

Most of the H_2 produced in the United States on an industrial scale is made by *steam reforming* methane (or methanol) over a catalytic surface at high temperature. The first step of the reforming generates $H_2O + CO$. Then a "shift reaction" changes the $CO + H_2O$ to $CO_2 + H_2$. The H_2 must be separated by a

CO_2 gas absorption or liquefaction means. The feedstock methane is generally obtained from FF sources (NG, coal).

The electrolytic decomposition of water to $H_2(g)$ and $\frac{1}{2} O_2(g)$ is an efficient means of producing H_2 from what is basically an inexpensive and generally ubiquitous feedstock (freshwater). The electrolysis redox reactions require a net anode–cathode potential difference of 1.24 V to break down pure water at 25°C and 1 atm. The pure H_2 gas is collected at atmospheric pressure at the cathode (−); the anode (+) reaction produces pure O_2. The smallest amount of electric energy to totally electrolyze 1 mol (18.016 g) of water is 65.3 Wh at 25°C. To produce 1 m^3 of H_2 at 1 atm and 25°C requires 4.8 kWh of energy. Electrolytic H_2 is "green" only if the electricity used comes from a renewable source.

Another "green" source of H_2 gas is from anaerobic, bacterial decomposition of organic waste in landfills, livestock manure, and human sewage treatment plants. (Methane is also produced.) There are many other pilot processes being investigated for H_2 fuel production, too many to list here. See the US DOE Web articles on *hydrogen fuel* for more details (DOE 2012b).

Hydrogen gas can also be produced photochemically from green algae, such as *Chlamydomonas reinhardtii*. A metabolic condition known as "anaerobic oxygenic photosynthesis" is where the algae's photosynthetically generated oxygen is consumed internally by the cell's own respiratory metabolism, causing anaerobiosis of the culture in the light. The anaerobic photosynthesis triggers the induction of a cellular hydrogen synthesis pathway and the release of H_2 gas. This unusual metabolic pathway has been described by several authors (Hemschemeier et al. 2009; Mellis et al. 2000). Ongoing research seeks to improve the efficiency of H_2 production by anaerobic photosynthesis in *C. reinhardtii* to the point where large cultures of the algae might serve as practical, low-cost sources of H_2 gas.

Another means of making H_2 gas "as needed" uses Powerballs, pellets of metallic sodium covered with a waterproof polyethylene plastic "skin." The pellets can be safely transported and then opened in a patented H_2 generator to produce the gas, presumably by adding water to exposed Na metal. The products are NaOH + $\frac{1}{2} H_2$ + heat. The caustic NaOH is a biohazard and must be recycled.

In 2001, Chrysler Corporation developed a prototype FC vehicle, the *Natrium*. The H_2 for the *Natrium* was produced onboard by reacting sodium borohydride (BH_4Na) with borax ($Na_2B_4O_7 \cdot 10H_2O$). The vehicle had a range of approximately 300 mi. before refueling (Rapal 2002). The reactants have a considerable energy cost of production, are not green, and are not environmentally friendly.

4.4.8.2 Storing Hydrogen

Although hydrogen possesses the highest energy by weight of any fuel (143 MJ/kg), it is the lightest and most easily diffusible gas. As a gas, H_2 compressed to 700 bar (10,150 psi) has a low energy per unit volume, only 5.6

MJ/L. When H_2 is stored as liquid hydrogen, it must be kept below $-253°C$. Refrigerating H_2 to this temperature consumes 25%–30% of its PE content (Parliament 2010; afdc 2012). To liquefy 0.45 kg of H_2 gas requires approximately 5 kWh of electrical energy. To avoid the energy loss in the liquefaction process, H_2 can be stored under high pressure in its gas phase. Compressing the gas is also not without energy cost, however.

Other means of storing H_2 gas that do not require such high pressures make use of the affinity of molecular hydrogen to the surfaces of certain metals, forming metallic hydrides. Certain alloys such as magnesium–nickel, magnesium–copper, and iron–titanium adsorb H_2 and release it when heated. Current research is seeking compounds that will be light, adsorb H_2 with a high density, and release it quickly.

In development (Wool et al. 2010) is an organic equivalent to metal hydrides to enable the storage of more H_2 at lower pressures in the fuel tanks of hydrogen-powered vehicles. This material is a unique, nanoporous char product, pyrolyzed chicken feather fibers (PCFFs), made from what is normally waste in the poultry industry (about 2.72 billion kg/year) (Senöz and Wool 2009). When keratin-based chicken feathers are processed by controlled pyrolysis, the specific surface area of the PCFF increases up to 450 m^2/g by the formation of fractals and nanopores, thus enabling increased H_2 adsorption over raw (untreated) feather fibers. The feather fibers were first heated at 215°C to strengthen their structure and then pyrolyzed at 400°C–450°C for approximately 1 h.

The PCFF product is capable of holding 1.5% of its weight in H_2. Since approximately 4.5 kg of H_2 gas is needed to travel 480 km (300 mi.), a large, 284 L (75 gal.) tank stuffed with some 300 kg (661.4 lb.) of PCFFs would be required! This falls short of the 6% H_2 storage target that has been set by the US DOE. The ongoing research on PCFF seeks to improve the H_2 adsorption by PCFFs, perhaps by the use of a catalyst.

4.4.8.3 Distribution of H_2

The obvious ways of distributing hydrogen gas are the same as methane: that is, pipelines and high-pressure, railroad (RR) tank cars. Hydrogen pipelines present special problems: H_2 is approximately three times bulkier in volume than NG having the same enthalpy. H_2, because of its small molecular size, leaks more easily, and it causes hydrogen embrittlement of pipeline steel, increasing maintenance and materials costs. Pipelines cannot be used to transport liquid H_2 any distance because of the extensive thermal insulation required. Insulating RR tank cars for LH_2 transportation is more practical.

4.4.8.4 H_2 Uses

Burning H_2 in air is totally green; one gets H_2O + heat + some NO_x gasses. The WV can be condensed and can reenter the natural water cycle or be collected as a source of pure water. The heat can be used to make steam to drive turbines or be used in other industrial and domestic heating applications. Hydrogen gas can also be used as a fuel in modified internal combustion engines.

Perhaps the most interesting green use of H_2 fuel is in an FC power source for all-electric transportation and for backup electric power. The outputs of an FC are low-voltage DC electric power and heat; its inputs are a fuel (e.g., H_2 or CH_4) and oxygen (in air). The exhaust products of an H_2 FC are generally WV and heat. To deliver the desired amount of power ($P_o = V_oI_L$), FCs can be assembled in series or parallel stacks or series-parallel arrays. The DC output of an FC array can be changed into AC using a solid-state, or rotary, inverter (DC motor + alternator set).

Some of the first H_2 FC applications were to deliver reliable power to US space vehicles. As the cost of diesel fuel and gasoline continues to climb, more attention is being paid to the design of efficient all-electric vehicles powered with FCs. "The tank [H_2]-to-wheel efficiency (TTWE) for a fuel cell vehicle is greater than 45% at low loads and shows average values of about 36% when driving a cycle like the NEDC (New European Driving Cycle) is used as test procedure. The comparable NEDC TTWE value for a Diesel vehicle is 22%. In 2008, Honda released a [prototype] fuel cell electric vehicle (the Honda FCX Clarity) with fuel cell stack claiming a 60% tank-to-wheel efficiency" (Mick 2008). These efficiencies do not consider the energy spent in producing, compressing, or liquefying H_2. Active FC research and development (R&D) areas include the development of new, more efficient catalysts and more efficient means of generating and carrying H_2 fuel.

Figure 4.22 illustrates estimates of the overall energy efficiency of a battery-powered electric vehicle (EV) versus the overall efficiency of the same EV powered by a hydrogen FC (DOE 2012c). The estimated 33% efficiency of the AC electrical generation is common to both vehicles. However, we assume that the H_2 generation is done at the power plant, so the 93% transmission efficiency does not apply to the H_2 FC system. It is also assumed that the H_2 gas is generated by the electrolytic decomposition of freshwater and is not liquefied. The overall efficiency for the battery EV is found by multiplying all the efficiencies in the pathway and is approximately 21%. Losses in both systems are primarily in the form of heat and friction in the motor/drive train. The hydrogen FC–powered vehicle has lower overall efficiency, only approximately 6%. This is because of the relatively low efficiencies of electrolysis and the FC itself (Ridley 2006). Note that the overall energy efficiency of gasoline-powered automobiles varies widely depending on design, from 9% to 36%. Again the losses are from low-grade heat and mechanical friction and drag (Saxton 2009).

A prototype, FC-powered light aircraft was developed by Boeing and European partners in 2008. This *FC demonstrator airplane* used a proton exchange membrane FC (PEMFC) (plus a lithium ion hybrid battery backup).

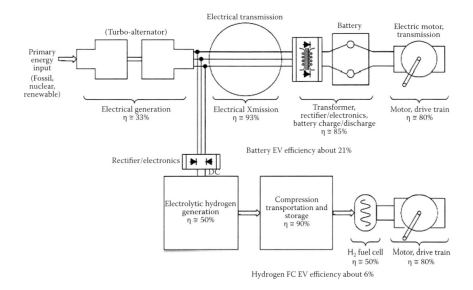

FIGURE 4.22
Block diagram illustrating the overall efficiencies (primary power input to EV electromechanical power output) of a typical battery-powered EV versus a hydrogen FC–powered EV.

The FC was a unique Flat Stack design, which allowed the FC to be integrated with the aerodynamic surfaces of the plane (Spenser 2004). Obviously, the propeller was driven by an electric motor.

4.4.8.5 Cost of Hydrogen

The cost of compressed H_2 gas (at 2000 psi or 14 MPa) in steel cylinders is approximately $100/kg, plus cylinder rental. A large gas cylinder holds approximately 0.6 kg of compressed H_2. *This price is over 50 times the cost of gasoline per unit energy.* Another quotation on H_2 gas in cylinders is $0.15/ft.³. The cost of liquid H_2 in a tanker truck (4300 kg or 15,000 gal.) is approximately $25,800; this is $6/kg (Lipman 2011).

The US DOE has targeted an "at the pump" selling price of $2.60/kg as a goal for making renewable H_2 economically viable. One kilogram of H_2 has approximately the energy equivalent of a gallon of gasoline. Mauro (2003) estimated that H_2 produced from wind-powered electrolysis of water could cost $24/million Btu, or $24 per 292.9 kWh, or $24 per 1.054 GJ, or approximately $3/kg H_2 at the source. Add the cost of liquefaction and distribution, and the cost becomes closer to approximately $32/million Btu, and so forth. If gasoline costs $4/gal., hydrogen's break-even cost would be approximately $8/kg. The factors impeding the development of a working hydrogen energy economy in the United States are as follows:

(1) The cost of H_2 production and liquefaction.
(2) The technology and cost of distribution of H_2 and LH_2.
(3) The lack of standardized "gas station" designs.
(4) If LH_2 is used as an automotive fuel for H_2 FC–electric drive vehicles, appropriate cryogenic storage for the −253°C fuel will have to be developed. This may be expensive and take time.
(5) The "chicken or egg" principle: Why invest in H_2 fuel production and distribution if there is no market (few H_2-powered cars)? Why build H_2-powered cars if H_2 fuel is not easily available?

Mauro pointed out that with an FC-powered car, you need half as much energy in the H_2 stored onboard as is needed in a vehicle powered by a gasoline internal combustion engine. If we consider 1 kg H_2 to have the same energy content as 1 gal. (3.785 L) of gasoline, perhaps one can afford to pay twice as much for a kilogram of H_2 as for a gallon of gasoline.

For a comprehensive treatment of the future of the hydrogen economy, see the online paper by F.D. Doty (2004). Doty considered the economics of H_2 production, storage, and distribution, as well as its use to power FC-powered vehicles (of which, 8 years later, there are very few). Gasoline and diesel fuel prices in 2004 were nostalgically low compared to mid-2012, so their current prices only raise the costs of H_2 production, storage, and distribution. The cost of running H_2 FC vehicles will have to be factored against the increasing availability of green fuels: for example, biodiesel from plant seeds and algae, DME from biogenic methane (an effective diesel fuel), and direct combustion of methane in modified IC engines (why convert CH_4 to H_2 when CH_4 can be used directly?).

4.4.8.6 Competition

In a future, nearly FF-free world, H_2 as an energy source will have to compete with methane, methanol, DME, biodiesel, and ethanol. There are many pros and cons to be considered, some described above. In automotive applications, liquid H_2 will always have a disadvantage because of the high energy overhead required to liquefy it and store it. Methanol *is* a liquid fuel. H_2 FCs are green, generally exhausting water and heat, but the direct methanol FC exhausts $CO_2 + H_2O$ + heat. The CO_2 would be green, however, only if the methanol were derived from recently grown plants.

4.4.9 Interfacing Intermittent Renewable Sources to the Grid

Solar and tidal electric power sources are periodic (diurnal or lunar), and wind energy is stochastic, dependent not only on prevailing winds but on passing weather systems as well. Storms can affect solar, wave, and tidal sources as well. Consequently, some means must be used to store and buffer the nonconstant, excess energy from these sources (see Section 4.6) in order to deliver power at appropriate levels to the AC distribution grid without surges and dropouts. Further complicating the interfacing of renewable sources to the grid is the fact that many sources such as PVs, FCs, wave-MHD, and some wind and tide turbines deliver DC as their primary output. These relatively small-scale DC power sources must be inverted, that is, converted to AC, and then connected robustly to the power distribution grid. These renewable sources are considered to be distributed generators (DGs) on the grid. Smart electronic power inverters make it possible to put an active DC power source, such as a PV array, onto the grid without generating destructive transient power surges. Rajapakse et al. (2009) gave an example of a simulated 250 kW PV array connected through a three-phase inverter to a 125 kV AC grid. The simulated solar radiation goes from a steady-state 500 W/m² to 900 W/m² for about 7 s. There is no apparent change in the 125 kV grid voltage; however, there are corresponding transients in the inverter's real and reactive power outputs and also small transients on its 3-ϕ output currents. In other words, smart inverter and interface electronics were able to compensate for the input power surge.

The case for compensation for grid transients for a large wind farm DG source was also considered by Rajapakse et al. A fast transient short somewhere on the grid causes a drop in the grid voltage at the wind farm connection. The wind farm is initially not disconnected from the grid to prevent instability problems on the grid. This fault ride-through strategy presents design problems for the wind farm's AC–AC interface. Underload or overload of the WTs must be compensated for mechanically by rapid blade pitch adjustments, and sudden underloads can be handled electrically by the use of dummy loads for each WT alternator to increase its load torque; this action prevents the WT blades and alternators from speeding up excessively and doing damage when the WTs are suddenly unloaded. Rajapakse et al. also examined the use of static volt-amperes reactive (VAR) compensators in stabilizing WT outputs in response to grid load transients and variations in wind power.

In summary, the many types of DGs present grid interface design challenges due to the nonstationary fluctuations in their primary power inputs and transients on the grid. The engineering challenge is that both the stability of the grid and the integrity of the DGs must be preserved. It is clear that a central coordinating computer must integrate incoming load, weather, generation capability, and incident data in order to optimally regulate many, diverse, integrated power sources.

Germany has decided to phase out its nuclear reactors, which prior to 2010 produced 23% of the country's power, and build a "smart grid" to take up the slack. This *Energiewende* or energy revolution will make extensive, expanded use of WTs, SCs, FCs, and tidal and wave energy: not just giant SC farms but also the surplus power from thousands of privately owned solar arrays. They are trying to achieve a 35% grid contribution from renewable sources by 2020. (Renewables supplied 17% in 2010.) The article by David Talbot (2012) discusses the problems this transition will encounter, including the expected consequence of more air pollution from burning more coal for extra power generation to take up the slack caused by closing the nuclear plants before the renewable sources are fully integrated online. *Energiewende* also requires the future development of more energy storage systems on the grid (e.g., pumped hydro, compressed air) to compensate for the nonstationary nature of wind and solar energy inputs.

4.5 Carbon-Neutral Energy Sources

4.5.1 Wood and Biomass

In carbon-neutral energy sources (CNESs), the fuels are combusted to extract their heat of oxidation; the CO_2 emitted enters the atmosphere in a "short cycle." That is, it is taken up stoichiometrically by the growing plants that will comprise the fuel. While theoretically, in the steady state, there is no net atmospheric CO_2 gain from use of a CNES, there will be a transient rise in atmospheric CO_2 concentration $[CO_2]$ from its use because the rate of CO_2 release from combustion is orders of magnitude higher than its uptake rate by the growing plants of the same mass or the uptake rate by land minerals and the oceans. Also, the production of wood and biomass fuels itself has a carbon footprint: chain saws, skidders, sawmills, trucks, and so forth all burn FFs, as do wood chippers and pellet-producing machines.

Wood has been burned for home heating and cooking for tens of thousands of years (Mithen 2003). More recently, efficient, airtight, cast-iron wood-burning stoves were developed in the early 19th century, some having air-convection heat extraction means. Steel was used for stoves in the later 19th century. In Europe, stoves were constructed from decorative ceramic tiles laid over cement and/or brick bodies. They had high heat storage capacities. In the later part of the 20th and in the 21st centuries, wood by-products (sawdust, pellets, chips, leaves, branches) have been used as feedstock to make methanol and ethanol, as well as feedstock for the HTC process, the output of which is biochar (a form of charcoal) (see Section 4.2.6).

The combustion of oven-dry, red oak firewood (common in the Eastern United States) releases approximately 21.8 MJ/kg of heat energy. Moisture in the wood seriously degrades its available thermal energy. One estimate is that a 30% moisture content by weight reduces the heat output by approximately 50% (Ince 1979).

All wood smoke contains heat, plus carbon dioxide, and a number of noxious hydrocarbons (CO, HCl, SO_x), as well as NO_x, heavy metals, and VOCs (including dioxins, furans, benzenes, phenols, resins), plus fine particles (soot, ash). The amounts of these noxious gasses depend on the exact wood used and its combustion temperature. Another potential hazard in wood smoke and ashes is radioisotopes. Cesium 137 (Cs^{137}) and Strontium 90 (Sr^{90}) were found in wood ash in Sweden and northern Vermont. The sources of these radioisotopes are evidently fallout from aboveground testing of nuclear weapons and the Chernobyl disaster. Farber (2000) tested wood ash from northern Vermont and found approximately 15,000 pCi/kg of Cs^{137} in 1990. It is not known what the present level of isotopic contamination is. [^{137}Cs is a β particle (electron) emitter with a half-life of approximately 37 years; its electrons have an energy of 0.550 MeV. ^{90}Sr is a β emitter with a half-life of 28.1 years; its electrons have an energy of 0.546 MeV.]

The CO_2 from wood combustion is carbon neutral; it will be recycled slowly into new tree growth. The other gasses and particulates cause air pollution, leading to human health problems such as asthma, emphysema, cancer, and heart disease. The cure for most of the polluting emissions is to burn the wood at over 1200°C where CO and other noxious gasses are largely converted to CO_2. This temperature may not be practical to reach in some home wood stoves. The sale and installation of new wood stoves in some US jurisdictions require the incorporation of catalytic "afterburners" to combust the CO, ash, and noxious gasses in the wood smoke. Besides cleaning the smoke, afterburners radiate additional heat.

The composition of wood ashes from burning wood is variable, depending on the wood used and the temperature of combustion. About 1% of the dry mass of wood becomes ash. The nutrient composition of unleached wood ash is generally 0% nitrogen, 1%–2% phosphate (P_2O_5), and 4%–10% potash (K_2O). Calcium carbonate is the principal component (25% to 50%) of unleached wood ash mass. The ash of hardwoods such as maple, elm, beech, and oak contains one-third more calcium than the ash of softwoods. There can also be traces of iron, manganese, zinc, copper, and some heavy metals. Unleached wood ash has some value as a soil sweetener (deacidifier) and fertilizer, although it is devoid of nitrates. It adds calcium carbonate and a little phosphate. Weathered wood ash has practically no fertilizing or liming value (Utzinger 2011). It is better to spread unleached wood ash as fertilizer than to waste it in a landfill. Soil pH should be monitored yearly, however.

Globally, wood fuels represent about 9% of the world's total energy consumption (FAO 2012). The European Union gets approximately 6.7% of its electricity from wood. However, some EU nations are far ahead of that curve. As of 2007, Sweden obtained 26.5% of its electric power from wood, Latvia 25.5%, Finland 24.9%, and Austria 17.7% (Gomez and Belda 2010). Austria's use of wood to generate electricity is growing rapidly. In 2006, it produced 176 PJ of energy from wood, a 34% increase over 2000's production. Austria produced 23% of its electricity from wood (particularly wood) in 2009, and this is expected to reach 34% by 2020 (Burgermeister 2009). Canada also produces a significant amount of electricity from wood (primarily pulp and other wood wastes). Currently, power plants converting biomass to electricity produce approximately 378 MWh/year. However, this is less than 5% of Canada's total yearly electricity consumption (Centre 2012).

A 2000 list of biomass power plants with start-up dates from 1979 to 1998 was given in a report by Wiltsee (2000). Some plants use sawmill waste (Mill); others logging waste (tops, branches, stumps) (Forest); agricultural waste (stover, bagasse, etc.) (Ag); urban refuse (Urban); manufacturing waste (sawdust) (Mfg); old tires (Tires); tire-derived fuel (TDF); landfill gas (LFG) (mostly CH_4); and recycled material fuels (plastics, paper, cardboard, wood) (RDF). Biomass electricity is generally inexpensive to generate because the fuel is largely free waste, and the costs incurred have to do with transportation and processing. Below is a table of some biomass-burning power stations in 2000. They are mostly in the United States except for one in Canada and one in Finland (see Table 4.5).

As of 2011, the top nine biomass power plants in the world included the following (Green World 2011):

1. Port Talbots, Wales, UK: 350 MW; to use imported fuel from Canada and the United States

2. Teeside, UK: 295 MW; to be completed in late 2012

3. Alholmens, Finland: 240 MW; to use both coal and biomass (pulp and paper) feedstock

4. Port of Bristol, UK: 150 MW; to be completed in 2014

5. Wisapower, Finland: 125 MW

6. Kua Vo, Finland: 125 MW

7. Lieth, Scotland, UK: 100 MW

8. Simmering, Vienna: 66 MW

9. Pecs, Hungary: 65 MW

Another use of wood in the United States is in domestic, exterior, free-standing boilers for home hot water–derived heating. They are generally sited some distance (downwind) from the house and are connected to a heat exchanger in the house using underground piping. The mess of wood, bark, smoke, and ashes is kept outside, and the risk of fire is reduced. The firebox is large enough to hold larger pieces of wood, so less cutting and splitting is required, and fires can burn all night without stoking. In some municipalities in New England, ordinances have been passed banning outdoor boilers because of their rank wood smoke. The water-filled jacket surrounding the firebox keeps combustion temperatures low, leading to incomplete combustion and adding to the pollution emitted. We anticipate a second generation of outdoor, wood-fired boilers that use a forced draft to raise combustion temperature, as well as mandatory catalytic converters in their smokestacks.

A modern, processed wood fuel that offers several advantages over logs for fuel is pelletized sawdust. The sawdust is a natural by-product of lumber mills, modular housing, and wooden furniture manufacturing. *Wood pellets*

TABLE 4.5

List of US (and Two Foreign) Electrical Power Plants That Use Renewable Fuels (Biomass)

Plant	Location	Fuel	MWe	GWh/Year
Williams Lake	British Columbia	Mill	60.0	558
Okeelanta (cogen)	Florida	Bagasse, Urban	74.0	454
Shasta	California	Mill, Forest, Ag	49.9	418
Colmac	California	Urban, Ag, Coke	49.0	393
Stratton	Maine	Mill, Forest	45.0	353
Kettle Falls	Washington	Mill	46.0	327
Snohomish	Washington	Mill, Urban	39.0	205
Ridge	Florida	Urban, Tires, Methane	40.0	200
Grayling	Michigan	Mill, Forest	36.0	200
Bay Front	Wisconsin	Mill, TDF, coal	30.0	164
McNeil[a]	Vermont	Forest, Mill, Urban	50.0	155
Lahti (cogen)	Finland	Gas from Forest, Urban waste	25.0	153
Multitrade	Virginia	Mill	79.5	133
Madera (closed)	California	Ag, Forest, Mill	25.0	131
Tracy	California	Ag, Urban	18.5	130
Camas (cogen)	Washington	Mill	17.0	97
Tacoma	Washington	Wood, RDF, Coal	40.0	94
Greenidge	New York	Manufacturing	10.8	76
Chowchilla II (closed)	California	Ag, Forest, Mill	10.0	53
El Nido (closed)	California	Ag, Forest, Mill	10.0	53
US Total				**4.374 TWh/ year**

Source: Wiltsee, G., Lessons learned from existing biomass power plants. NREL/SR-570-26946. National Renewable Energy Laboratory, Golden, Colorado, 2000. www.doe.gov/ bridge.

Note: The right-hand column gives the yearly average electrical energy produced per year over 1996–1998. Cogen means co-generation; Urban means urban waste; Ag is agricultural waste biomass (corn stalks, dead plants, etc.)

[a] Data from Burlington Electric, Joseph C. McNeil Generating Station, June 2012. www. burlingtonelectric.com/page.php?pid=75&name=mcneil.

are made by compressing the sawdust into a doughlike mass, which is then fed to an extruder die with holes approximately 6 mm in diameter. The high pressure of the extrusion process causes the temperature of the "dough" to rise, which in turn causes the lignin in the wood to plastify, binding the structure of the pellets together as they cool (Spelter and Toth 2009; Pellets 2011).

The energy content of wood pellets is approximately 4.7–4.9 MWh/tonne = 16.9–17.6 MJ/kg. Wood pellets used in Europe are required to have less than

10% water content and have a density of over 1 ton/m^3, so they do not float in water. European standards for manufacturing wood pellets require that they be made from virgin wood sawdust; not allowed are particle board, painted or chemically treated wood, melamine resin-coated panels, and so forth; US standards for wood pellet composition are generally less strict than those in Europe, potentially leading to more air pollution problems. Because pellets can be fed to a stove or boiler automatically with an augur, they make an attractive alternative to oil burners. The air space between them means that a forced draft can cause very complete combustion, releasing maximum heat and producing relatively low amounts of NO_x, SO_x, and VOCs. Wood pellets, properly combusted, are one of the less-polluting heating options available. However, wood pellets appear to be carbon neutral only if they are made from wood harvested from sustainable forests, not clear-cut tracts.

A 2012 study reported on pellet use in Europe: Sweden, 1.85 million tonnes; Italy, 850 kilotonnes (kt); Germany, 900 kt; Austria, 509 kt; Denmark, 1.06 million tonnes; and Finland, 149 kt (Verhoest and Rickman 2012). In the United States, the 2008 cost of heating with pellets was approximately $20 per million Btu (1 Btu = 1054.35 J). This translates into $5.14 per 40 lb. bag or $275/tonne (Buyer's Guide 2009). In Austria, pellet prices for domestic use were approximately € 243/tonne in 2005. In Europe in 2009, the average cost of pellet energy was 4 euro cents/kWh (1 kWh = 3.6 × 10^6 J). The ultimate cost of pellets must include the cost of the labor used in their production and distribution and also the cost of the FF used in the heavy logging machinery and trucks that transport the wood to the sawmills, the energy required to saw the wood, plus the fuel needed to transport the sawdust to the pellet factory, which also uses energy, plus the fuel required to transport the pellets to the end users.

A central problem with the sustainability of wood fuel is the demand rate versus the growth rate of trees. Clearcutting, as practiced in the northwestern United States, is an efficient yet highly ecologically disruptive way to harvest wood. It leads to a glut of wood-derived fuel from the sawmills and cleanup on the tract. But once the logging is finished, that source is turned off until new growth has matured (in tens of years). The demand for wood for domestic construction, furniture, and fuel is continuously growing with the population, as is the demand overseas for US wood and wood products.

4.5.2 Fuel Cells

FCs are electrochemical cells that continuously convert a fuel to a source of DC electrical power. There are many designs and several fuel sources for FCs, for example, hydrogen gas, methane, and methanol. FC oxidizer molecules (electron acceptors) include, but are not limited to, oxygen, chlorine, and chlorine dioxide. FCs have application in augmenting and backing up grid power (their DC output must be inverted to AC). They also are an alternative to storage batteries in powering EVs. FC design is an active research field in electrochemistry and chemical engineering.

Like many technological "jewels," FCs were developed long before their need by our modern society. In 1839, Sir William R. Grove, a Welsh judge, inventor, and natural philosopher, made an FC. He mixed pure hydrogen and oxygen in the presence of an electrolyte and produced an electric current and water. It was a curiosity that produced no appreciable electric power. In 1889, the name *fuel cell* was first coined by Ludwig Mond and Charles Langner, who attempted to build a working FC using air and industrial coal gas. Many others worked on early FC designs. One successful FC design was introduced in 1932 by Cambridge (UK) electrochemist Francis T. Bacon. Finally, in 1959, Bacon perfected the design of a 5 kW, H_2/O_2 FC using a molten KOH electrolyte. (See the detailed summary of FC development and applications in Fuel Cell Today 2011.)

Table 4.6, taken from a US DOE *Comparison of Fuel Cell Technologies* table, summarizes some of the properties of developed FCs to date.

One advantage of FCs is that they have no moving parts. The hydrogen FCs exhaust only water and heat. The MCFC discharges CO_2 + water. However, the CO_2 can be separated and used as an oxidizer along with O_2 in air in this FC. Figure 4.23 illustrates the schematic of an MCFC. The electrolyte charge carrier is molten carbonate ions in an inert, ceramic matrix [lithium aluminum oxide ($LiAlO_2$)]. Other light, energy-dense fuels can also be used in the MCFC; the high running temperature converts these fuels directly into H_2 by a process called internal reforming.

At the anode of all FCs, a *catalyst* aids the oxidation of the fuel and the release of free electrons (as an electric current) to an external circuit where work is done. Probably the most expensive anode catalyst is platinum, but nickel is often used. A catalyst is also used at the FC cathode. Other catalysts, such as cerium, and carbon nanotubes are being investigated, the objects being to reduce FC cost and at the same time to increase their efficiencies. FCs that operate at high temperatures (e.g., MCFCs, solid oxide FCs) can generally use nonprecious metal catalysts.

The typical output voltage of an FC at rated current load is 0.6–0.7 V. Consequently, a number of FCs can be stacked (connected in series) to raise the net output voltage to a value that is most effective in driving output loads such as solid-state inverters.

Below we describe the redox reactions in some common types of FC.

- The *zinc-air fuel cell* (ZAFC) uses an oxidizable electrolyte fuel cassette made from metallic Zn microparticles suspended in a paste of KOH. The cathode is separated from the air (O_2) supply by a gas diffusion electrode, a permeable membrane that allows atmospheric O_2 to pass through and makes electrical contact with the cathode material (Sapkota and Kim 2009).

TABLE 4.6

Nonexhaustive List of Some FCs and Their Characteristics

FC Type	Electrolyte	Fuel	Operating Temp.	Electrical Efficiency/ Output Power	Applications	Comments
Polymer-electrolytic membrane FC	Solid polymer polyperfluorosulfonic acid	H_2 (primary)	< 100°C	25%–58% 1–250 kW	Backup power, portable power, certain vehicles	Low temperature, quick start-up
Alkaline FC	Aqueous KOH soaked in a matrix	H_2	90°C–100°C	>80% 10–100 kW	Military, space vehicles	Can use a variety of catalysts
Phosphoric acid FC	Liquid phosphoric acid soaked in a matrix	H_2	150°C–200°C	>40% 50–1000 kW 250 kW (typical)	Distributed generation	Increased tolerance to impurities in H_2
Molten carbonate FC (MCFC)	Liquid solution of Li, Na, and/or K carbonates	H_2 also uses CO_2 as oxidizer	600°C–700°C	Approximately 46% 1–1000 kW	Large distributed generation, utilities	High efficiency, fuel flexibility, short life due to corrosion
Solid oxide FC	Yttria-stabilized zirconia	H_2, CH_4 or CH_3OH	600°C–1000°C	35%–43% 1–3000 kW	Auxiliary power, electric utility	Fuel flexibility, can use variety of catalysts
Direct methanol FC	Polymer conducting H^+	CH_3OH mixed with steam	50°C–120°C	20%–40% kW range	Transportation vehicles	Methanol can have green source and is easy to export; the DMFC is expensive

Source: US DOE, Comparison of fuel cell technologies, 2008. www.hydrogen.energy.gov.

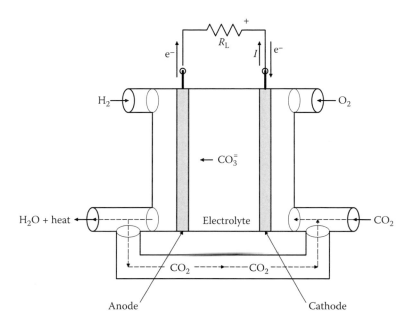

FIGURE 4.23
Schematic of an MCFC. See text for details of operation.

At the *cathodes*, the O_2 is reduced with water to form hydroxyl ions.

$$\underset{reduced}{\overset{\downarrow}{O_2}} + 2H_2O + \overset{\downarrow}{4e^-} \rightarrow 4OH^- \, (E^0 = +0.40V) \tag{4.16}$$

In the zinc paste anode, the following occur:

$$\underset{oxidized}{Zn} + 4OH^- \rightarrow ZnOH_4^= + \overset{\uparrow}{2e^-} \, (E^0 = -0.625V) \tag{4.17}$$

$$ZnOH_4^= \rightarrow +ZnO + 2OH^- + H_2O \tag{4.18}$$

The overall ZAFC reaction is

$$Zn + \tfrac{1}{2} O_2 \rightarrow ZnO \tag{4.19}$$

Typical specifications for a ZAFC (one of several) used to power busses include the following:

- Number of cells per battery module: 47
- Battery open-circuit voltage: 67 V
- Single cell O.C voltage: 1.426 V
- Operating voltage: 57.4 V
- Capacity: 325 Ah
- Energy capacity: 17.4 kWh
- Peak power 8 kW
- Weight: 88 kg
- Volume: 79 L
- Energy density: 200 Wh/kg
- Dimensions: 726 × 350 × 310 mm
- Manufacturer: *Electric Fuel Corporation* (www.electric-fuel.com/EV)

The *DMFC* uses a polymer electrolyte in which the internal charge carrier is the H$^+$ ion (proton). Liquid methanol is oxidized in the presence of water at the anode, generating carbon dioxide, and electrons, which travel through the external circuit, doing electrical work. At the cathode, returning electrons reduce oxygen atoms to form water. The reactions are as follows.

Methanol is oxidized at the anode:

$$CH_3OH + H_2O \rightarrow CO_2 + 6H^+ + 6e^- \tag{4.20}$$

At the cathode:

$$(3/2)O_2 + 6H^+ + 6e^- \rightarrow 3H_2O \tag{4.21}$$

The overall DMFC reaction is as follows:

$$CH_3OH + (3/2)O_2 \rightarrow CO_2(g) + 2H_2O \tag{4.22}$$

DMFCs operate in the range of 50°C to 120°C. They are expensive because of the platinum and/or ruthenium, which must be used as catalysts.

PEMFCs are an efficient type of FC for powering EVs (FCBasics 2011). Figure 4.24 illustrates a PEMFC. Using hydrogen fuel, PEMFCs and electric motors may eventually replace internal combustion engines in many low-power applications. First used in the 1960s for the NASA Gemini Program, PEMFCs are now being developed that produce 1–2 kW. Their electrolyte is a thin, solid polymer (plastic) membrane, which is permeable to protons when saturated with water but does not conduct electrons. The PEMFC's fuel is hydrogen, and its charge carrier is the hydrogen ion (proton). At the anode, the hydrogen molecule is split (oxidized) into protons and electrons. The protons travel through the membrane to the cathode, while the electrons travel in the external circuit and do electrical work. Oxygen from air is supplied to the cathode and combines with the electrons and protons to form water. The electrode reactions are simple; at the anode we have

$$\overset{\downarrow}{2H_2} \rightarrow 4H^+ + \overset{\uparrow}{4e^-} \tag{4.23}$$

At the cathode

$$\overset{\downarrow}{O_2} + 4H^+ + \overset{\downarrow}{4e^-} \rightarrow 2H_2O \tag{4.24}$$

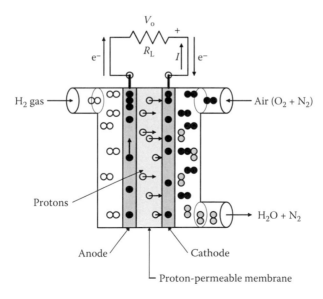

FIGURE 4.24
Schematic of a proton exchange FC. See text for details of operation.

The overall FC reaction is

$$2H_2 + O_2 \rightarrow 2H_2O + heat \qquad (4.25)$$

Compared to other types of FCs, PEMFCs can generate more power for a given volume or weight of FC. That is, they are compact and lightweight. They also operate at less than 100°C, which permits rapid start-up.

The *Bacon fuel cell* was developed in the 1950s by Francis T. Bacon, an electrochemist from Cambridge, UK. The electrolyte is molten KOH; the electrodes are porous sintered nickel powder, permitting gasses to diffuse and contact the electrolyte. The nickel also acts catalytically. At the anode, H_2 is the fuel (the electrons do the work):

$$\overset{\downarrow}{2H_2} + 4OH^- \rightarrow 4H_2O + \overset{\uparrow}{4e^-} \qquad (4.26)$$

At the cathode, the reaction is

$$\overset{\downarrow}{O_2} + \overset{\downarrow}{4e^-} + 2H_2O \rightarrow 4OH^- \qquad (4.27)$$

The overall reaction is

$$2H_2 + O_2 \rightarrow 2H_2O + heat \qquad (4.28)$$

Motivations that drive research on FCs include the following: (1) Hydrogen FCs are green energy sources; they produce DC electric power, water, and heat. (2) FCs oxidize simple fuels such as H_2 gas, methane gas, and methanol that can have green sources. (3) They are useful for auxiliary sources of electric power, and perhaps most importantly, they can power totally green EVs, given green sources of fuel. Many of the large motor vehicle companies have seen the handwriting on the wall and have designed prototype H_2/FC vehicles.

The nation (and the world) may eventually have to switch to a hydrogen fuel economy (see below) in order that FC automotive power sources become ubiquitous as the fossil-fueled, internal combustion engine is today. A switch to all-electric vehicles will mean a heavy sink for the world's copper and aluminum reserves and also intense recycling of these metals (into electric motor windings and electric cables). A plethora of hydrogen–electric vehicles will also strain the rare earth resources used in manufacturing high-flux-density permanent magnets used in certain kinds of electric motors.

The rise of hydrogen gas distribution networks and H_2 fueling stations must parallel the rise in H_2/FC vehicles. Many critics are not enthusiastic about using H_2/FCs to power vehicles. Perhaps the price of gasoline and diesel fuel has not yet risen high enough to activate the social action rate sensitivity law (SARSL) responses required in investors, industry, and government. In May 2009, US Secretary of Energy Stephen Chu announced that since H_2/FC-powered vehicles "will not be practical over the next 10 to 20 years," the government would cut funding for H_2/FC vehicle development. However, the US DOE would continue to fund R&D on stationary FCs. He cited difficulties in the development of the required infrastructure to distribute H_2 vehicle fuel as a justification for cutting research funds. What may drive this infrastructure development in the future will be the inevitable rising costs of FFs, public demand, and the fact that H_2 fuel must compete with methane and methanol-derived fuels. H_2/FC EVs suffer another problem: most FCs are difficult to start at below-freezing temperatures. Some means is necessary to warm the FC to its start-up temperature. This could be a battery-powered electric heater or an external hydrogen flame. Once started, the FC generates its own heat, keeping its exhaust water in vapor phase [see Vine (2011), Fuel Cells (2010), and NREL (2007b) for a summary of H_2 vehicle design issues].

In spite of logistic difficulties in distributing H_2 to "gas stations," automobile manufacturers worldwide have developed over 64 prototype FC EVs since 1991. About nine FC EV bus designs have also been put on the road over the past two decades [see Yvkoff (2011) for a current listing of FC vehicles].

4.5.3 Biogenic Methane

The fuel gas, methane (CH_4), is lighter than air, colorless, odorless, noncarcinogenic, and explosive in high concentrations [over 53,000 ppm by volume (ppmv) in air with 21% O_2]. [The triple point (TP) of methane is at $T_t = 90.68$ K and $P_t = 11.72$ kPa = 1.70 psia. The normal boiling point for methane (at 1 atm) is 117.7 K. Methane's critical point (CP) is at $T_c = 190.4$ K and $P_c = 45.4$ atm.]

Biogenic methane is formed at relatively shallow depths underground by the bacterial decomposition of organic matter. Methanogenic bacteria are anaerobic and belong to the domain of *Archaea*. There are over 36 described species of methanogens. The organic matter can be anthropogenic, that is, from human activities (e.g., from landfills, pit silos, manure pits, etc.), or from natural sources, (e.g., dead vegetation in marshes, swamps, lake bottoms, etc.) (EPA 2010). A first step in the production of methane gas is the action of bacteria on the organic substrate to produce CO_2 and H_2. The dissolved CO_2 forms bicarbonates, which are combined by methanogenic bacteria with hydrogen ions to form methane and water (MacDonald 1990). Thus, under natural conditions, methane is generated along with CO_2. This decomposition gas moves to the surface (air) by mechanisms of molecular diffusion and convection. Diffusion is the result of a concentration difference between two different locations. Diffusive flow is in the direction of decreasing concentration. Convective flow occurs when a pressure or temperature gradient exists between two locations (Hariri and Chou 2008).

The biogenic methane from methanogenic bacteria–derived energy from garbage in landfills, manure from animal feedlots, and sewage from cities is a valuable source of energy. LFG tapped from wells driven into the center of landfills is 45%–60% CH_4; 40%–60% CO_2; 2%–5% N_2; 0.1%–1% O_2; 0.1%–1% NH_4; and 0.01%–0.6% nonmethane organic compounds [NMOCs; 0%–1% sulfides (SO_2, NO_x, H_2S, H_3CSCH_3, mercaptans, etc.), 0%–0.2% H_2, and 0%–0.2% CO]. The composition and generation rate of LFG depends on temperature, moisture, and, of course, the organic composition of the landfill (e.g., food and garden wastes, street sweepings, textiles, and wood and paper products), as well as the age of the landfill. The composition of LFG is seen to vary with the phase of bacterial decomposition of the landfill waste. In phase III, acid-producing bacteria create compounds for the methanogenic bacteria to consume. The methanogens consume the CO_2 and the acetate, too much of which would be toxic to the acid-producing bacteria (these anaerobes make acetic and formic acids, as well as alcohols and CO_2). During phase III, CH_4 production steadily rises to a steady state of 45%–60%. In phase IV (the final phase), LFG emission is in a steady state: 45%–60% CH_4; 40%–60% CO_2, and 2%–5% N_2. Phase IV can last over 20 years and begins approximately 5–7 years after the waste is buried (EPA 2008).

The methane from capped garbage landfills can flow as much as 500 ft.^{3}/min (850 m^3/h) from a single "well," and a large landfill might support as many as 10 methane wells (EPA 2011a). Thus, the total methane yield from a large, 10-well, phase IV landfill might be as much as 204,000 m^3/day. Another source estimates total landfill emissions of CH_4 to be 45 × 10^9 kg/year or 1.1 × 10^8 kg/day (IPCC 2001). By contrast, the total natural methane emissions are approximately 213 × 10^9 kg/year, and methane from cattle totals approximately 92 × 10^9 kg/year. The total anthropogenic methane emission, including landfills, is approximately 337 × 10^9 kg/year. The grand total yearly biogenic methane emission is approximately 550 × 10^9 kg (IPCC 2001) or approximately 503–610 × 10^9 kg (EPA 2010).

Peer et al. (1992) developed a simple, empirical, linear regression model for methane emission from landfills. They considered data from 21 landfills with gas recovery systems. Their model gave a regression line predicting methane recovery rate as a function of landfill mass. This was $Q_{CH4} = 4.52M$, where Q_{CH4} = methane flow rate (m^3/min) and M = mass of refuse (10^6 tonnes). They found no functional statistical model linking CH_4 production to climate variables. They compared their regression model with the Environmental Protection Agency's (EPA's) Landfill Air Emissions Estimation Model, which was developed for regulatory purposes. The EPA model is

$$Q_{CH4} = L_oR[\exp(-kt_c) - \exp(-kt)] \tag{4.29}$$

where Q_{CH4} is the CH_4 generation rate at time t in cubic feet per year, L_o is the potential CH_4 generation capacity of the refuse in cubic feet per million grams (Mg) of refuse, R is the average annual refuse acceptance rate during active landfill life in Mg per year, k = CH_4 generation rate constant in units per year, t_c = time since landfill closure in years (t_c = 0 for an active landfill), and t = time since initial refuse placement in years.

US Power Partners (2009) described the EPA's 1996 *Landfill Rule* requiring that wells be used to collect methane from licensed landfills that meet the following criteria: (1) does or did accept municipal solid waste; (2) was active on or after November 8, 1987; (3) has a total permitted capacity of at least 2.5 million metric tons of waste; and (4) has NMOC (e.g., CO_2, dimethyl sulfide, etc.) emissions of at least 50 metric tons per year. Power Partners (2009) also listed a number of operational power plants run from landfill methane; for example, in 2006, Exelon Power completed a 60 MW LFG-powered generation station in Fairless Hills, PA (Fairless 2012). In many cases, we see energy wasted by burning off the LFG as it is collected. Burning methane converts it to CO_2 (plus wasted heat energy). The CO_2 has approximately 1/20 the GHG effect as methane, so given the lack of initiative to collect the gas for fuel, burning it at the landfill is the next best thing, as it will eventually leak into the atmosphere.

Table 4.7 summarizes biogenic emission estimates. It uses data from several sources (see notes).

If untapped, most landfill methane will find its way up into the atmosphere or dissolve in an underlying aquifer's water. Methane is about 21–23 times as effective as CO_2 at trapping solar energy, so burning biogenic methane as a fuel mitigates global warming by converting the methane otherwise leaked into the atmosphere to CO_2. Because it contains significant amounts of CO_2, LFG has approximately half the heating value (16,785–20,495 kJ/m³) as that of NG (35,406 kJ/m³) (Bade Shrestha et al. 2008).

The pure methane combustion reaction is well known:

$$CH_4 + 2\,O_2 \rightarrow CO_2 + 2\,H_2O + heat\ (1\ kg\ of\ CH_4\ at\ STP\ yields\ 55\ MJ) \tag{4.30}$$

The same argument can be made for methane in MH deposits. As the seas and tundra gradually warm over the next several decades, methane (and other gasses) will be released from the near-surface deposits of MH into the atmosphere, where it will exacerbate global warming, creating a positive-feedback cycle that will be difficult to quantify. If the methane from the

TABLE 4.7

Estimated Sources of Atmospheric Methane

Atmospheric CH$_4$ Origin	CH$_4$ Emission (Tg/year)
Natural Emissions	
Wetlands	170
Oceans, estuaries, and river bottoms	9
Termites	20
Lakes	30
Hydrates	5.5
Living plants (trees, crops, etc.)[a]	Negligible
Wild animals	8
Permafrost	0.5
Uptake by upland soils and riparian areas[b]	−30
Natural total	**213**
Anthropogenic CH$_4$ Emissions	
Cattle	92[c]
Energy (coal mines, gas wells)	96[c]
Landfills	45[c]
Biomass burning	39[c]
Waste treatment	25
Rice agriculture (paddies, not plants)	40
Anthropogenic total	**337**
Grand total	**550**

Note: NB: 1 Tg = 10^{12} g = 10^9 kg = 10^6 tonnes.

[a] Data from Nisbet, R.E.R. et al., *Proc. Royal. Soc. B.* 276, 1347–1354, 2009.

[b] Data from EPA, Methane and nitrous oxide emissions from natural sources, *EPA 430-R-10-001*, April 2010.

[c] Average figure from IPCC, Trace gasses: Current observations, trends and budgets; 4.2.1 Non-CO$_2$ Kyoto gasses; 4.2.1.1 Methane, *Climate Change 2001* (Ch. 4). www.grida.no/climate/ipcc_tar/wg1/134.htm.

near-surface MH deposits (the easiest to tap) is collected and burned for its thermal energy, the resulting CO$_2$ will have only approximately 5% of the GHG effect (on a mole-per-mole basis) than the leaked CH$_4$ (see Section 3.5.7).

Methane release into the atmosphere from warming in the Arctic tundra and in the Arctic Ocean is accelerating with global warming. Russian scientists have observed huge plumes of methane bubbling up from thermally decomposing clathrate deposits in the Laptev Sea off the northern coast of Siberia, above the Arctic Circle. In the late summer of 2011, the Russian oceanographic research vessel, *Academician Lavrentiev,* conducted an extensive survey of approximately 10,000 mi.^{2} off the East Siberian Coast, collaborating with researchers from the University of Georgia (Athens, United

States). They counted more than 100 fountains, or torchlike structures, bubbling through the water column and putting huge amounts of methane directly into the atmosphere. A Russian research team estimated the new Arctic methane release to be well over the 8 million tonnes/year estimated in 2010. Indeed, the Arctic atmospheric [CH_4] was approximately 1.9 ppm (December 2011). This rise will certainly accelerate Arctic warming (Club of Rome 2012; Vast 2012).

4.6 Energy Storage Means

4.6.1 Introduction

Energy storage is a means of distributing and smoothing the electric energy generation level and the demands for electric power over time and area on an electric power distribution grid. In this section, we will consider energy storage by mechanical means: in particular, PHS and CAES, electrochemical means (batteries), purely electrical means (supercapacitors), and thermal means (such as heat stored in molten salt and salt ponds). Flywheels can also store rotational KE. However, a mechanoelectric means must be used to transform this KE to useful electrical energy. Energy can also be stored by electrolysis of water to make hydrogen gas, which can power FCs or be combusted directly for its heat. Energy storage is particularly needed for such time-variable generation systems as SCs, solar furnaces, WTs, and tidal power generation, where the primary energy input is intermittent, diurnal, or periodic (Barnes and Levine 2011). A comprehensive review of a broad spectrum of energy storage technologies can be found in the Naish et al. (2007) report to the European Parliament's committee on Industry, Research and Energy (ITRE).

4.6.2 Pumped Hydro Storage

The physics of PHS is simple; 1 kg of water pumped up 100 m (328.1 ft.) gains a PE of 980 J = 0.7624 Wh, so 10^7 kg or 10^4 tonnes (ca. 10^4 m^3 or 8.11 acre-ft.) pumped up 100 m gains a PE of 7.62 MWh, and so forth. PHS accounts for approximately 99% of bulk storage capacity worldwide, about 127 GW total. PHS has 70%–87% efficiency (Economist 2012c).

In PHS, excess (cheaper) electrical power from the grid is used to pump water (from a river, lake, or ocean) up several hundred meters to an artificial lake where it is impounded. Pumping is generally done late at night, when the grid is most lightly loaded. PHS systems can typically supply approximately 2 GW generating capacity. A PHS system is run cyclically; that is, it is charged at night, and its stored water is released to run turbines/generators at peak load times in the morning and evening. Subject to evaporation and leakage into aquifers, the impound lake can hold its "charge" for months. There are approximately 295 PHS systems worldwide. The Ffestiniog PHS system in Llyn Stwlan, Wales, UK, commissioned in 1963, has four turbogenerators with 360 MW total capacity that come online within 60 s (Ffestiniog 2012). (The delay in coming online in a PHS plant is the time it takes for the water in the penstocks to reach full velocity, spinning the turbogenerators at synchronous speed (428 rpm).) Another PHS system in northern Wales is the Dinorwig power station, completed in 1984 (Williams 1991). It has six 300 MW turbines, giving an installed peak capacity of 1.65 GW. The Dinorwig PHS plant runs at 74%–75% efficiency, that is, it uses approximately 33% more electric power when pumping water back up to Machlyn Mawr (storage lake) than it actually produces. Of note is the fact that the Dinorwig PHS is connected to the UK national grid at Pentir by approximately 6 mi. of buried 400 kV cables. The cables were buried for aesthetic reasons. Imagine that in the United States.

In one PHS system design, called the Green Power Island, being developed by Gottlieb Paludan, a Danish architectural firm working with researchers from the Technical University of Denmark, an artificial island is built having a deep central reservoir (well below sea level). When the wind blows, some of the WT power is used to pump water *out of the reservoir* into the ocean. When extra grid power is needed, seawater is allowed to flow back into the reservoir, driving low-head turboalternators (Economist 2012c).

An excellent example of a functioning, US PHS system is found in Northfield, MA. Northeast Utilities Service Corp. maintains a pumped storage facility, the Northfield Mountain Station (NMS); its 300-acre upper storage reservoir is 244 m (800 ft.) above the Connecticut River and holds 21.2×10^6 m^3 (5.6×10^9 gal.) of water. The powerhouse is built into a cavern in the mountain; it houses four turbines driving four 270 MW alternators, with a total peak capacity of 1.08 GW. When excess power is available on the grid, the turbogenerators become motor pumps and pump water from a lower reservoir (or the Connecticut River in emergencies) up into the upper storage reservoir. Normally, the NMS does not affect the river flow or temperature, and its operation is nonpolluting. The NMS went into service in 1972 (Northfield 2011).

4.6.3 Batteries

A rechargeable battery or storage battery (also known as a secondary cell) belongs to the group of electrochemical cells (pairs of half-cells) in which the electrochemical reactions at their electrodes are reversible. Storage batteries come in all sizes, shapes, and energy storage capabilities. They are used in many consumer products such as cell phones, portable phones, iPads, Kindle readers, laptop computers, automobile ignition, motorized wheelchairs, golf carts, forklifts, power tools, hybrid and all-electric vehicles, and so forth. Our interest here is primarily in grid energy storage applications where the variable DC outputs of solar (PV) arrays and some WTs can be stored and inverted to line-frequency AC as needed. A table of rechargeable battery specifications is given in Table 4.8.

Because of their importance in sustainable transportation, research on rechargeable storage batteries is a very active area. Objectives include minimizing weight while maximizing the energy density (megajoules per kilogram, watt-hours per kilogram, or watt-hours per liter), minimizing cost per watt-hour, minimizing the self-discharge rate in percentage per month, maximizing the number of C/D cycles possible, and so forth. Figure 4.25 compares various secondary cell types for watt-hour per kilogram versus watt-hour per liter. Note that the venerable lead–acid (automobile) battery is near the origin. The lithium-ion battery (LIB) group has the highest energy density performance. LIBs are ubiquitous; they are used for portable electronics, laptop computers, power tools, and now in electric automobiles such as the Nissan Leaf, the Tesla Volt, and so forth.

Lithium ion batteries are a family of rechargeable batteries in which lithium ions (Li^+) move internally from the negative electrode (anode) to the positive electrode during discharge and in the other direction when charging. In the external (load) circuit, charge is carried by electrons, by convention moving in the opposite direction to the load current, I_L. Unlike lithium (primary) batteries that are disposable, rechargeable lithium ion (secondary) batteries use an intercalated lithium compound as an electrode material instead of metallic lithium. The cathode of an LIB cell is typically carbon (graphite); the electrolyte is a lithium salt in a nonaqueous organic solvent; and the anode is a metal oxide, such as a layered oxide (lithium cobalt oxide), a polyanion (lithium iron phosphate), or a spinel (lithium manganese oxide). The electrolyte is generally a mixture of nonaqueous organic carbonates (e.g., ethylene carbonate or diethyl carbonate) containing complexes of lithium ions. Lithium ion batteries must be sealed from water, which reacts with lithium to form LiOH and H_2 gas, which can cause the battery to overheat and explode.

One example of a Li ion cell is the lithium thionyl chloride (LTC) cell (PowerStream 2003). Its half-cell reactions when discharging (supplying electric power) are given as follows:

(anode)
$$4Li \rightarrow 4Li^+ + 4e^- \quad \uparrow \tag{4.31}$$

The cathode half-cell reaction is

(cathode) ↓
$$4Li^+ + 4e^- + 2SOCL_2 \rightarrow 4LiCl + SO_2(g) + S(s) \tag{4.32}$$

TABLE 4.8

Summary Compilation of the Properties of Some Common Rechargeable Batteries (Secondary Cells)

Battery Type	Electrolyte	Operating Temp. (°C)	Cell OCV	Energy Density Achievable (Wh/kg)	Power Density (Sustained) (W/kg)	Charge/Discharge (C/D) Efficiency, %	C/D Cycles, Life, #
Lead–acid	H_2SO_4	−20–60	2.1–2.2	30–40	25	70–90	200–2000
Nickel–iron	KOH	20–30	1.2	60	—	65	2000
Zinc–iron	KOH	50–60	1.65	90	—	45	600
Na–sulfur	β–Al_2O_3	300–400	1.76–2.08	120	120	70	2000
Lithium–iron sulfide	LiCl–KCl	400–450	1.6	150	—	75	1000
Nickel–cadmium	KOH	−40–60	1.2	40–60	140	70–90	500–2000
Nickel–metal hydride	KOH	10–50	1.2	60–80	220	50	<3000
Nickel–zinc	—	—	1.7	60	900	—	100–500
Lithium ion	$LiPF_6$	−20–60	3.6	100–200	360	70	500–2000
Lithium sulfide	AlN	430–500	2.0	130	140	75	200
Zinc–chlorine	$ZnCl_2$	—	—	<120	<100	75	—
Li ion polymer	—	—	3.7	130–200	>3000	—	500–1000

Source: Engineering ToolBox. Rechargeable batteries, 2012. www.engineeringtoolbox.com/rechargeable-batteries-d_1219.html.

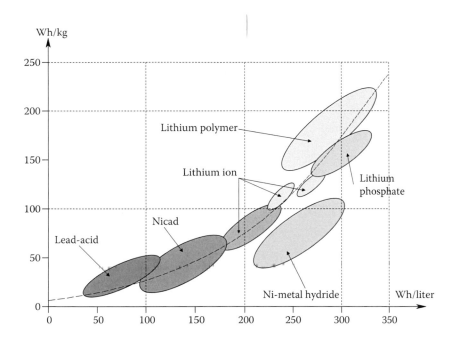

FIGURE 4.25
Plot of energy density (watt-hours per kilogram) versus energy compactness (watt-hours per liter) for some common types of secondary cells. (Based on ICCNexergy, Battery energy density comparison, www.iccnexenergy.com. Accessed January 25, 2013.)

The overall LTC cell reaction when supplying power is

$$4Li + 2SOCl_2 \rightarrow 4LiCl + SO_2 \text{ (g)} + S \text{ (s)} \qquad (4.33)$$

The electrolyte in the LTC cell is liquid thionyl chloride (Cl_2OS) in which the electrolyte salt, lithium aluminum chloride ($LiAlCl_4$), is dissolved. This cell has a very high energy density (about 500 Wh/kg) and a cell voltage of 3.3–3.5 V. The interior of an LTC cell is not environmentally friendly.

When a lithium-ion cell is discharging (supplying power), the Li moves from the anode to the cathode; the reverse occurs when the cell is charging. In a Li–Co cell, the transition metal, cobalt (Co) in $LiCoO_2$ is oxidized from Co^{3+} to Co^{4+} during charging and reduced from Co^{4+} to Co^{3+} during discharge. The liquid, nonaqueous electrolytes in lithium ion batteries contain lithium salts such as $LiPF_6$, $LiBF_4$, or $LiClO_4$ and conduct lithium ions. The open-circuit voltage of a lithium-ion cell varies with the cell type and can range from approximately 1.5 to 3.5 V. LIB development is an active area. There are now over 16 variations on the LIB theme.

LIBs can rupture, ignite, or explode when exposed to high temperatures. If a cell is short-circuited, it will overheat and possibly catch fire. A domino effect may occur where adjacent cells heat up and fail, possibly causing the entire battery to rupture or ignite. LIBs used for automotive applications and other large-capacity backup power supply applications have the lithium–cobalt oxide cathode replaced with lithium–iron phosphate to improve cycle counts, shelf life, and safety.

Zinc–air batteries are another popular, nonrechargeable, small battery. They are used for hearing aids, cameras, small portable electronic devices, and so forth. A porous anode consists of an electrolyte paste containing zinc metal particles, water, and KOH. O_2 from the air reacts at the cathode and forms hydroxyl ions (OH-), which drift into the zinc paste and form zincate $\left(Zn(OH)_4^{-}\right)$ ions, releasing electrons to travel to the cathode. The zincate forms ZnO and water in the electrolyte. The water and the hydroxyls from the anode are recycled at the cathode, so there is no net water loss/gain. The cell open-circuit voltage is theoretically 1.65 V. The half-cell reactions are as follows.

At the anode:

$$Zn + 4OH^- \rightarrow Zn(OH)_4^{=} + 2\,e^- (E_0 = -1.25V) \qquad (4.34)$$

In the electrolyte:

$$Zn(OH)_4^{=} \rightarrow ZnO + 2\,OH^- + H_2O(pH = 11) \qquad (4.35)$$

At the cathode:

$$\tfrac{1}{2}\,O_2 + H_2O + 2e^- \rightarrow 2\,OH^- \; (E_0 = +0.34\ V) \qquad (4.36)$$

Overall:

$$2\,Zn + O_2 \rightarrow 2\,ZnO\ (E_0 = 1.59\ V) \qquad (4.37)$$

The electrochemistry of the small, zinc–air, primary cell has been scaled up into the ZA/FC. A single ZA/FC module (containing 47 cells) can discharge approximately 17.4 kWh before it has to be refueled with fresh zinc metal/electrolyte cells (Electric Fuel 2012). See Section 4.5.2 for description of the ZA/FC.

See the PowerStream (2003) online paper detailing the half-cell reactions, electrolytes, and properties of many types of primary and secondary batteries.

4.6.4 Flow Batteries

An FB is a type of rechargeable battery in which components in the electrolyte, rather than the anode and/or cathode, are consumed by the production of electrical energy. An FB can be rapidly "recharged" by replacing the spent electrolyte with new, while saving the spent fluid for reprocessing (reenergization). An FB is not an FC. FBs fall into two broad categories: redox FBs (RFBs) and hybrid FBs (HFBs). In RFBs, all electroactive materials are dissolved in the electrolyte. The energy of an RFB can be determined independently of the battery's power capability because the energy is dependent on the electrolyte volume, and the power capability depends on the reactor size. In practical terms, this means that the discharge time of an RFB at full power can be varied from minutes to days, while that of an HFB may typically vary from several minutes to several hours (Savinell 2011).

The vanadium RFB (VRFB) has been widely developed and applied and will be described here as an example of an RFB. Advantages of the VRFB are the following: The same ion (vanadium) is used for both oxidation and reduction reactions. Reactions are aqueous, with no phase change. There is a large open-circuit, cell EMF; there is fast response to loading (350 μs); and VRFBs have high overload capacity. Electrolytes are recyclable even if positive and negative electrolytes are mixed. VRFBs operate over a wide ambient temperature range.

Figure 4.26 shows the schematic of one VRFB cell. When in use, the liquid energy carriers (electrolytes) are continuously pumped in a circuit between the battery and storage tanks. The liquid energy carrier contains sulfuric acid with dissolved vanadium salts in a range of oxidation states. The anode and cathode electrodes in a VRFB are separated by a proton (H^+) exchange membrane. The electrodes are an inert carbon–polymer composite felt that supports the redox reactions and supply/collect the electrons from the external circuit. The arrows inside the cell in the figure indicate the direction of reactions when the VRFB cell is supplying DC power (discharging). Note that the protons move from right to left during discharge. During charging, all the arrow directions are reversed. VRFBs are said to be a socially responsible, "green" energy storage means; no lead, zinc or cadmium, bromine, and so forth is used. The VRFB was pioneered at the Australian National University of New South Wales (UNSW) in the early 1980s. Since then, they have been manufactured in Australia, Canada, Japan, and Thailand, to name a few locations. They have been used for voltage sag protection, load leveling, and stabilization of WT outputs.

The following reactions take place at the electrodes when the battery is supplying power:

$$\overset{(V^{5+})}{VO_2^+} + 2H^+ + e^- \overset{\downarrow}{\to} \overset{(V^{4+})}{VO^{++}} + H_2O$$

(reduction of V^{5+} at the *positive electrode*)

(4.38)

$$V^{++} \overset{\uparrow}{\to} V^{+++} + e^- \text{(oxidation of } V^{++} \text{ at the } \textit{negative electrode,}$$
$$\text{a source of electrons in the external circuit)}$$

(4.39)

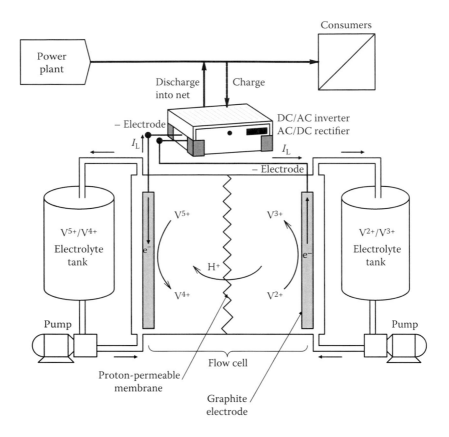

FIGURE 4.26
Schematic of a VRFB cell. See text for description.

$$VO_2^+ + 2H^+ + V^{++} \rightarrow VO^{++} + H_2O + V^{+++} \text{ (overall reaction)} \quad (4.40)$$

The cell voltage ranges from 1.0 V discharged to 1.6 V when fully charged. The VRFB cell specific energy is 15 to 25 Wh/kg depending on the vanadium concentration in the electrolytes (Cellstrom 2010). Tokuda et al. (2000) described the development of a 450 kW VRFB battery system. Their VRFB battery specifications are given in Table 4.9.

Coulombic efficiency remained at 95% over 12,000 C/D cycles, voltage efficiency converged on 85%, and energy efficiency also fell from approximately 82% to 80% over the 12,000 cycles. An ideal application for a 450 kW VRFB would be backup power in a nuclear power station to run cooling pumps, lighting, and motors used to move fuel rods, and so forth.

The *zinc–bromine FB* (ZBFB) is an example of an HFB. This is an environmentally hazardous FB because it uses zinc ions and liquid bromine. The characteristics of the ZBFB are listed in Table 4.10.

TABLE 4.9

Characteristics of a 450 kW VRFB

Output Power	450 kW
Battery Capacity	900 kWh (450 kW × 2 h)
DC Voltage	450 V (average)
DC Current	1000 A (average)
Cell Module	1 module: cell stack 4P × 2S Series of 3 modules
Electrolyte	Vanadium 1 mol/L Sulfuric acid solution
AC/DC	PWM control

Note: P = parallel, S = series, PWM = pulse width modulation.

The redox reactions in the ZBFB *supplying power* are as follows: Zinc metal is used at the negative electrode. The reaction is the reversible oxidation, in which electrons are supplied to the external circuit.

$$Zn(s) \rightarrow Zn^{++}(aq) + 2e^{-} \uparrow \tag{4.41}$$

At the positive electrode, bromine is reversibly reduced to bromide:

$$Br_2(aq) + 2e^{-} \downarrow \rightarrow 2Br^{-}(aq) \tag{4.42}$$

The overall cell reaction is thus

$$Zn(s) + Br_2(aq) \rightarrow 2Br^{-}(aq) + Zn^{++}(aq) \tag{4.43}$$

The measured open-circuit, cell potential difference (EMF) is approximately 1.67 V. The cell is separated by a microporous or ion-exchange membrane to prevent Br_2 from reaching the negative electrode where it would react with the zinc metal, causing the battery to self-discharge. To further reduce self-discharging, and also reduce the vapor pressure of toxic Br_2 vapor, complexing agents are added to the positive electrolyte to reduce the Br_2

TABLE 4.10

ZBFB's Specifications

Specific Energy	34.4–54 Wh/kg
Energy Density	124–190 J/g
C/D Efficiency	70%
Energy/Consumer Price	US $400/kWh (US $0.11/kJ)
Cycle Durability	>2000 cycles
Nominal Cell Voltage	1.8 V

concentration (Savinell 2011). During charging, Zn(s) is electroplated onto conductive electrodes by an electrochemical reduction reaction, while at the same time, Br⁻ ions are oxidized, and liquid bromine is formed. These reactions are reversed when the battery is supplying power (discharging).

FBs are generally used for large energy storage applications (1 kWh to several megawatt-hours) such as storing (and buffering) the outputs of intermittent (renewable) power sources, for instance, SC "farms," tidal hydroturbines, and WTs. They can also be used for load balancing on a power grid where they store cheaper nighttime power and, using an inverter, provide AC power to the grid to meet peak loads. They are also used as backup power sources (uninterruptible power sources = UPSs) for critical applications such as hospitals, refineries, and nuclear plants. They are also suitable for high-powered EVs. They are not environmentally friendly.

4.6.5 Compressed Air Energy Storage

Figure 4.27 illustrates a typical CAES system. During times of off-peak (low-cost) electricity availability on the grid, electricity is used to pump air under pressure into a large, leakproof cavern. During times of peak electricity

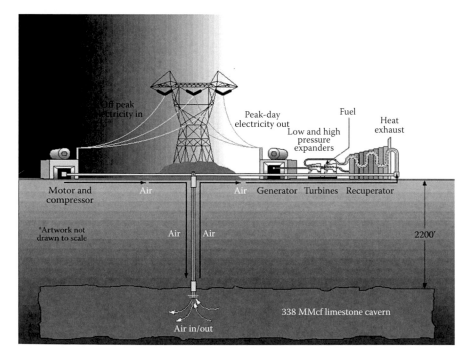

FIGURE 4.27
Schematic of a CAES system. See text for details. (Courtesy of Burroughs, C., Sandia assists with mine assessment, Sandia National Laboratories, Albuquerque, NM, April 24, 2001. http://www.sandia.gov/media/NewsRel/NR2001/norton.htm.)

consumption, the air is released to turn a low-pressure air turbine/generator or feed a gas-fired turbogenerator. In adiabatic storage, a heat exchanger captures most of the heat resulting from air compression into an underground reservoir and then returns this heat to the released, expanding air that is doing mechanical work. The theoretical efficiency of an adiabatic air storage/ release cycle can approach 100%, but the round-trip efficiency is expected to be more like 70% because of heat losses. Isothermal cycling of a CAES requires heat extraction during compression and insertion during expansion in order to maintain constant gas temperature. An isothermal cycle is practical only for low-power-level systems; again their theoretical efficiency can approach 100%, but in practice, it is lower because of heat losses.

Using the ideal gas assumption, it is possible to calculate the work done by isothermally expanding compressed air stored in a reservoir from high pressure, P_A, to a lower pressure, P_B. If we assume the approximation for ideal gas behavior, $PV = nRT$, and then assume that T is constant, we can write $P_A V_A = P_B V_B$, or $V_B/V_A = P_A/P_B$. From the definition of *work*,

$$\mathbf{W_{A \to B}} = \int_{V_A}^{V_B} P \, dV = nRT \int_{V_A}^{V_B} dV/V = nRT(\ln V_B - \ln V_A) = nRT \ln(V_B/V_A)$$

$$= nRT \ln(P_A/P_B) = PV \ln(P_A/P_B) \qquad (4.44)$$

This amounts to approximately $2.271 \ln(P_A/P_B)$ kJ/mol at 0°C (273 K), $2.478 \ln(P_A/P_B)$ kJ/mol at 25°C, or $100 \ln(P_A/P_B)$ kJ/m^3 of gas at 100 kPa $\cong$ atmospheric pressure. Note that *ideal* isothermal processes are thermodynamically reversible. To be isothermal, heat must be extracted or added to the process. In other words, compressing the same volume of gas isothermally from P_B to P_A takes the same amount of energy, that is, $\mathbf{W_{A \to B}} = \mathbf{W_{B \to A}}$, at constant T.

CAES facilities built to date take advantage of natural and man-made underground caverns. Table 4.11 lists some existing and proposed CAES plants. The McIntosh, AL, CAES system was built by Dresser-Rand, a company with extensive engineering experience in "smart" CAES system designs (Dresser-Rand 2010).

As the number of WT "farms" and PV panel facilities grows in the United States and abroad, there will be a high incentive to develop new CAES because of their economics; they are less expensive than PHS and large arrays of batteries. There is a problem in siting a CAES plant. It requires a leakproof, air storage cavern, which may not be found within a *practical radius* of a wind farm or solar array. (A *practical radius* must take into consideration the cost of transmitting wind-generated or solar electric power to the CAES site, and thence to a suitable node in the local grid.) The wind-generated or solar panel electric power, like all electric power, suffers power losses proportional to the power line length when it is transmitted from source to sink (CAES air

TABLE 4.11

Partial List of CAES Systems in Operation or under Construction

Location	Status	Cavern	Capacity
Huntdorf, Bremen, Germany	Operational 1978	2 salt caverns, 3.1×10^5 m^3 total volume	290 MW
McIntosh, southwest Alabama	Operational 1991	Salt cavern, 5.6×10^5 m^3	110 MW
Norton, OH	Under construction 2008	Limestone mine	$800 \rightarrow 2700$ MW
Iowa Stored Energy Park (ISEPA) Des Moines, IA	Under construction 2009	3000 ft.–deep anticline in porous sandstone	270 MW
Matagorda County, TX	Proposed 2008	Salt dome	540 MW

Source: Data from Das, T. and J.D. McCalley, Compressed air energy storage, Iowa State Univ., 2012. http://homeengineering.iastate.edu/~jdm/wind/Compressed%20Air%20 Energy%20Storage_Chapter_TRISHNA%20DAS.pdf.

pumps). It is the consideration of this power loss that governs approximate maximum power line distance from the source to sink. The maximum practical distance may be on the order of a hundred miles or so (161 km).

To sidestep the requirement for a salt cavern for CAES, large-diameter, capped steel pipes can be used to store the compressed air. The company Sustainx, associated with Dartmouth College's Engineering School, has built a 40 kW demonstration CAES using pipe storage and plans to build a 1–2 MW system (Economist 2012c).

A CAES facility does not necessarily need to drive turbogenerators; it can directly power compressed air motors used for light industry and commercial purposes. In 1896, a Paris CAES system had approximately 2.2 MW of compressed air energy distributed at 550 kPa (80 psi) in 50 km of air pipes delivering pneumatic power to air motors used in industry. Usage was measured by air flow meters (Chambers 1896).

A CAES system can be operated at nearly constant pressure by allowing a surface water lake or impoundment to fill the underground air storage hardrock cavern as air is withdrawn to do work. Simple physics tells us the air pressure in the cavern varies between $P_{max} = \rho g y_{max}$ and $P_{min} = \rho g y_{min}$, where ρ is the density of water, g is the acceleration of gravity, y_{max} is the distance from the pond surface to the bottom of the cavern, and y_{min} is the distance from the pond's surface to the surface of the water in the filled cavern. Nearly constant pressure operation can also be obtained by throttling the outflowing air to supply the turbine at constant pressure. "The Huntdorf CAES plant is designed to throttle the cavern air to 46 bar at the hp turbine inlet (with caverns operating between 48 to 66 bar) and the McIntosh system similarly throttles the incoming air [to the turbine] to 45 bar (operating between 45 and 74 bar" (Succar and Williams 2008). (NB: 1 bar = 100 kPa =

TABLE 4.12

Comparison of Capital Costs for Certain Energy Storage Means

Technology (Capacity)	Capital Cost ($/kW)	Capital Cost ($/kWh)	Hours of Storage
CAES (300 MW)	580	1.75	40
Pumped hydro (1 GW)	600	37.5	≈10
Sodium sulfur battery (10 MW)	1720–1860	180–210	6–9
Vanadium redox battery (10 MW)	2410–2550	240–340	5–8

Source: Adapted from Succar, S. and R.H. Williams, Compressed air energy storage: Theory, resources, and applications for wind power, Princeton Environmental Institute, April 8, 2008. www.princeton.edu/pei/energy/publications/texts/SuccarWilliams_PEI_CAES_2008April8.pdf.

14.50 psi = 1 atm.) Operation at constant turbine inlet pressure means the generator can supply peak power over an extended interval.

It is possible to derive formulas relating the electrical energy output of the turboalternator (E_{GEN}) to the storage volume of the CAES system cavern (V_S). Succar and Williams do this for three cases: (1) *constant cavern pressure–constant turbine inlet pressure* (water-pressurized cavern), (2) *variable cavern pressure–variable turbine inlet pressure*, and (3) *variable cavern pressure–constant turbine inlet pressure* (throttling). For all three cases, the electric energy storage density E_{GEN}/V_S increases approximately linearly with increasing reservoir pressure, or equivalently, mass per unit volume. One calculation showed that a constant-pressure cavern could deliver the same output with only 23% of the storage volume required for a case 2 scenario. Detailed algebraic calculations can be found in Succar and Williams (2008, Section 2.3).

The economics of CAES is generally more favorable than other short-term, high-power means of energy storage. Table 4.12, adapted from Succar and Williams, clearly illustrates this advantage.

Finally, a word of caution about CAES and compressed air–powered automobiles: A study by Doty (2009) showed that compressed air automobiles are ultimately inefficient. They state: "The total energy produced by the four, huge, 3800 psi, air tanks could be up to 24 MJ (~6.7 kWhr), or about the same as in 0.8 quarts of gasoline." Compared to electric cars, compressed air cars have only approximately 7% fossil source-to-wheels efficiency, compared to approximately 24% for electric cars and a little more for advanced hybrids. Doty stated: "Unless the energy on the electric grid is over 80% clean (wind, nuclear, hydro...), [operation of] a mid-sized air car results in the release of more pollution than the Prius burning conventional gasoline."

4.6.6 Electric Double-Layer Capacitors

The electric double-layer capacitor (EDLC), also known as a supercapacitor or ultracapacitor, has a much higher power density than batteries. Also, the energy density of EDLCs is typically on the order of hundreds of times greater than a conventional electrolytic capacitor. A typical D-cell (battery)–sized, low-voltage electrolytic capacitor may have a capacitance of several thousand microfarads; an EDLC of the same size and voltage rating may have a capacitance of several farads, an increase of two orders of magnitude. Larger EDLCs can have capacities of up to 5000 F. Certain EDLCs have reached specific energies (aka energy densities) of approximately 30 Wh/kg (1.8 kJ/kg) (Jeol.com. 2007; Zheng 2002). As is well known, a discharged capacitor's voltage rises exponentially when charged from a constant-voltage DC source in series with an equivalent resistance, R_s. The charging time constant is simply R_sC seconds. Likewise, a capacitor discharges its voltage exponentially into a resistive load, R_L, with time constant R_LC seconds. A supercapacitor can also be charged quickly by an electronic current source. The total electric energy stored in a capacitor is given by the well-known relations

$$W = \tfrac{1}{2}CV_c^2 = \tfrac{1}{2}Q^2/C = \tfrac{1}{2}QV_c \text{ joules or watt-seconds} \qquad (4.45)$$

where C is the capacitance in farads, V_c is the capacitor's terminal voltage in volts, and Q is the charge stored in the capacitor in coulombs ($Q = CV_c$).

For example, a 5000 F EDLC charged to 3 V holds 22.5 kJ of energy, enough to run a power tool or emergency lantern or buffer a single solar panel.

EDLCs have a number of applications in "energy smoothing" of alternate energy sources such as smaller WTs and PV panels in homes, vehicles, and boats. EDLCs are low-voltage devices, a property that makes them useful for use with solar panels. Other applications of EDLCs include the following: (1) the power source for diesel engine starting in tanks, locomotives, submarines, and trucks; (2) use as energy accumulators in dynamic braking systems in EVs; (3) use in powering electric busses (capabus), which fast recharge makes possible; (4) use in portable power tools; (5) use in portable communications devices; and (6) backup power for emergencies.

The cost of EDLCs in 2006 was approximately $0.01/F (which translates to about $2.85/kJ) and is expected to fall further as manufacturing technology improves and more producers enter the market. A 3 kF EDLC was priced at approximately $50 in 2011. That is about $0.037 per stored kilojoule assuming a 3 V peak charge (renewableenergyworld 2011). Worldwide EDLC sales were only approximately US $275 million in 2009 and are expected to grow at 21.4% through 2014 as prices fall and applications grow.

The *advantages* of using EDLCs as energy storage buffers include the following: (1) long life, with little degradation over hundreds of thousands of C/D cycles; (2) low cost per C/D cycle; (3) very high rates of C/D (no chemical reactions are involved); (4) very low internal resistance, hence low heating levels during deep C/D; (5) high specific output power, which can exceed 6000 (J/s)/kg at 95% efficiency; (6) no corrosive electrolytes and low material toxicity (e.g., activated carbon from coconut shells); (7) simple charging (no full charge detection needed as with batteries and no danger of overcharging).

Disadvantages of EDLCs include the following: (1) The maximum voltage that EDLCs can operate at is from 2 to 3 V. (2) Complicated electronic interfaces are required to step discharge voltage up and maintain a constant output voltage as the EDLC voltage falls. (3) EDLCs can be operated in series to increase maximum stored voltage, but individual capacitors require voltage-balancing electronics to prevent overvoltages. (4) EDLCs have the highest dielectric absorption of any type of capacitor. (5) Charged EDLCs lose their charge (self-discharge) faster than a typical electrolytic capacitor and faster than a rechargeable battery (secondary cell). (6) The amount of energy stored per unit weight is generally less than that of an electrochemical battery (about 500 J/kg for a standard EDLC vs. 9.6 kJ/kg for a typical LIB).

4.6.7 Flywheels

KE can be stored in a rotating flywheel and then used to do mechanical and electrical work. The stored mechanical energy is given by the well-known relation

$$E = \tfrac{1}{2} J \dot\psi^2 \text{ joules} \tag{4.46}$$

where J is the flywheel's moment of inertia in kilograms square meter, and $\dot\psi$ is its angular velocity of rotation around its spin axis in radians per second.

A typical flywheel geometry has a thick, dense, ring with inner radius R_1, outer radius R_2, thickness d, density ρ, and total mass M. High-tensile steel spokes or a thin metal web connects the outer mass ring to the hub and rotating shaft. It can be shown (neglecting the spokes) that a flywheel with this geometry has a moment of inertia of (Sears 1950)

$$J = \tfrac{1}{2} \pi \rho d \left(R_2^4 - R_1^4 \right) = \tfrac{1}{2} M \left(R_2^2 + R_1^2 \right) \text{kg m}^2 \tag{4.47}$$

Hence, an $M = 1$ tonne $= 1000$ kg flywheel with $R_1 = 1$ m, $R_2 = 2$ m, has a $J = 0.5 \times 10^3 \times 5 = 2.5 \times 10^3$ kg m^2. Now assume that the flywheel has been caused to spin at 24,000 rpm. $\dot\psi$ is found by dividing rpm by 60 and multiplying by 2π (radians in a revolution). Thus, $\dot\psi = 2513$ rad/s, and the flywheel stores $E = \frac{1}{2} \times 2.5 \times 10^3 \times (2.513 \times 10^3)^2 = 7.894 \times 10^9$ J $= 7.894$ GJ, or 7.894 MJ/kg, an appreciable amount of KE. How to release it?

The shaft must be coupled, either physically or magnetically, to a KE recovery system (KERS), which can consist of a continuously variable, speed-reduction transmission, which drives an electrical generator's torque load at constant velocity. The electrical generator, in turn, powers electric motors for propulsion and battery charging. As the flywheel loses its KE by transforming it to electrical kilowatt-hour, it slows down, and the transmission must continuously change its effective gear ratio to keep the loaded generator spinning at constant velocity at that load torque. Similarly, an electric motor coupled through the same transmission can spin up the flywheel to its rated top rpm when it is finished delivering energy. Another design of KERS avoids the problem of mechanically coupling the shafts of the flywheel and generator and, instead, imbeds strong permanent magnets symmetrically around the flywheel. As the flywheel spins in its vacuum housing, the moving magnetic fields are coupled to stationary coils arranged around the exterior of the housing, inducing a high-frequency, polyphase, AC voltage, which can then be transformed, rectified, smoothed, and regulated. This type of no-contact flywheel can be accelerated to its energy storage speed (ω_o) by computer-controlled switching of currents in the external coils, similar to the operation of a brushless DC motor. As the flywheel slows down due to supplying energy to the coils, the frequency and amplitude of the induced EMFs drop in the coils.

It has been suggested that flywheel energy storage may be an effective alternative to batteries for electric drive vehicles (Economist 2011i). Several considerations exist when using large flywheels to store energy in lieu of batteries in automobiles:

1. To be effective, the automotive energy storage flywheel needs to have considerable mass and moment of inertia, J. This added vehicle mass has to be accelerated, braked, and turned when driving.

2. To run at high rotational speeds (up to perhaps ≤ 60,000 rpm), the flywheel itself needs to withstand high centrifugal forces and hence must be constructed from high-tensile-strength materials, including carbon fiber/epoxy laminates. It will be expensive.

3. The high-velocity flywheel should run in a vacuum to minimize frictional losses from air. This requires a hermetically sealed housing and a vacuum pump to exhaust air around the wheel. Use of a vacuum housing adds the technical problem of how to effectively couple flywheel KE out to an external generator/motor with minimum losses.

4. A continuously variable transmission is required to mechanically couple the flywheel's energy to the external generator or motor, which must run at constant speed.

5. A large, heavy, spinning flywheel is in effect a gyroscope and has unique mechanical behavior when there is an attempt to move the axis of spin of the rotor, that is, the phenomenon of induced gyro moments. Gyro moments can act on the vehicle's chassis, seriously affecting the driving behavior of the vehicle in response to its roll, pitch, and yaw (turning).

For example: Assume that a flywheel spinning with its spin axis horizontal to the Earth and directed to the front of the vehicle is spinning CCW viewed from the front of the vehicle. The flywheel is fixed to the chassis of the car. If the vehicle (and flywheel) turns to the left at a rate of $\dot{\varphi}$ radians per second, the flywheel acts like a constrained mechanical-rate gyroscope (Northrop 2005), and a pitch torque couple, $\mathbf{M}_\theta$ newton meters, acting on the gyro's shaft, is generated that tries to tip the front of the flywheel (and the vehicle) up, putting a down force on the vehicle's hind wheels and unloading the front wheels (see Figure 4.28a). This couple is proportional to the angular rate of turning to the left, $\dot{\varphi}$. If the vehicle turns to the right, the gyro couple is reversed; the front wheels are loaded and the rear wheels unloaded. We note that the couple $\mathbf{M}_\theta$ is given by the relation

$$\mathbf{M}_\theta = J\dot{\psi}\dot{\varphi} \text{ newton-meters} \tag{4.48}$$

where J is the moment of inertia of the flywheel, $\dot{\psi}$ is its angular spin velocity in radians per second around its spin axis, and $\dot{\varphi}$ is the angular velocity of the turning vehicle perpendicular to the spin axis in radians per second (i.e., its yaw angular velocity (see the figure).

If an automotive energy-storing flywheel weighs 150 kg and has $R_1 = 0.5$ m and $R_2 = 1$ m, then its rotational moment of inertia is $J = 0.5 \times 150 \, (1.25) = 93.75$ kg m². Assume it is spinning at 20,000 rpm, or $\dot{\psi} = 2.094 \times 10^3 \, \text{rad/s}$. Its *total* KE is $E = 0.5 \times 93.75 \times (2.094 \times 10^3)^2 = 2.056 \times 10^8$ J, or 1.371 MJ/kg of flywheel. (E10 gasoline contains 43.5 MJ/kg, by comparison.)

We now calculate the magnitude of the couple lifting the car's front wheels: $|\mathbf{M}_\theta| = 93.75 \times 2.094 \times 10^3 \times \dot{\varphi}$. Take the turning angular velocity $\dot{\varphi}$ to be a modest 0.1 rad/s. Thus, the pitch moment, $\mathbf{M}_\theta$, is an enormous, very significant, 1.963×10^4 N m, or 1.448×10^4 ft. lb., around the center of the flywheel. Clearly, a vehicle with its flywheel mounted with its spin axis directed horizontally, that is, fore and aft, would be undrivable!

Now examine what happens when the same flywheel is rigidly mounted in a car with its spin axis vertical. Refer to Figure 4.28b. Now roll velocity is converted to a pitch moment and vice versa. In addition, there is negligible gyro response to yaw velocity (turning right or left). Fortunately, unlike single-engine, propeller-driven aircraft, cars do not experience appreciable roll or pitch, but only yaw (turning) motions. There is a possibility that driving fast over a pot-holed road could excite some sort of mechanical resonance in the suspension, so the suspension would have to be well damped.

The answer to the pitch moment problem is to use two identical, coaxial flywheels spinning in opposite directions at the same angular speed. The pitch moments will cancel out. A single transmission would be connected to the matched pair of counter-rotating flywheels with two, coaxial shafts.

The future of flywheels as energy storage devices (on land) looks promising. As we have seen, appreciable KE can be stored in a 1 tonne flywheel spinning at 24,000 rpm, in vacuo. This energy can be used to buffer the outputs of intermittent green sources such as solar panels or WTs. A major technical

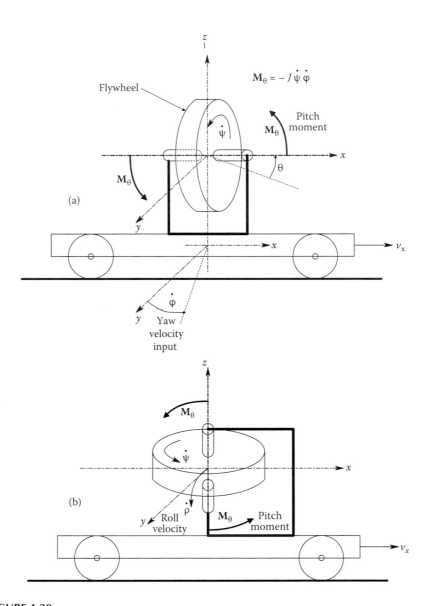

FIGURE 4.28
(a) Flywheel KE storage system on a vehicle moving in the *x* direction. Note how yaw velocity ($\dot{\varphi}$) (in turning) of the vehicle and flywheel generates a gyro pitch moment, $\mathbf{M}_\theta$. (b) Now the flywheel's rotational axis is vertical (z direction), so that roll velocity ($\dot{\rho}$) generates a gyro pitch moment, $\mathbf{M}_\theta$. Yaw velocity now has negligible effect on driving dynamics.

problem to be solved is how to couple the rotational KE from the rapidly spinning wheel to an electrical generator/motor through a computer-controlled, continuously variable transmission, so the generator runs at constant speed, regardless of its electrical load torque, τ. As with PHS, when electrical power on the grid is cheap late at night, it can be used to drive a motor to spin up the flywheel, storing electrical energy as rotational KE.

A simplified scenario for flywheel energy storage uses the flywheel to turn a permanent magnet-alternator/rectifier with a fixed gear reduction (the resultant output DC voltage is proportional to generator speed). The DC generator output is then chopped using silicon-controlled rectifiers (SCRs) (making a variable duty cycle square wave), and this is used to drive a DC motor. The chopping duty cycle determines the motor's mechanical power output (speed and torque).

4.6.8 Energy Storage by Mass PE

Advanced Rail Energy Storage (ARES) of Santa Monica, CA, is developing a gravity-based energy storage system that uses modified railway cars on a specially built track. Off-peak grid power is used to pull the cars to the top of a hill. When energy is needed, one or more cars is released and allowed to roll down the hill. The wheels are coupled to generators, which, as the result of delivering electric power, provide dynamic braking for the cars. The company claims that the round-trip efficiency (the ratio of electric power in to power out) is approximately 85%, compared to 70%–75% for PHS (Economist 2012c). A third rail, or an overhead power line with a pantograph electrode, can be used to couple the output power to power conditioners and the grid or to drive the motors powering the car to the hilltop.

4.7 Fusion Power

4.7.1 Introduction

Nuclear fusion offers, in theory, an almost limitless source of power derived from superabundant elements found on the Earth's surface. Fusion occurs in the sun through a proton–proton chain reaction, converting hydrogen to helium and releasing a huge amount of electromagnetic energy in the form of light and heat (Stern 2004; Bahcall 2000).

Unfortunately, as research has shown, fusion is far more elusive on Earth. A continuous fusion process is difficult to start, contain, and sustain in order to harvest the energy. Nuclear fusion is a process whereby two or more atomic nuclei are forced together to "fuse" and form the heavier nucleus of a new atom. Energy is released in the fusion process in the form of high-energy neutrons, protons, gamma rays, and heat. Most commonly, hydrogen atoms ($_1$H), deuterium atoms (heavy hydrogen) ($_1$H^2), and tritium (the radioactive hydrogen isotope) ($_1$H^3) are used in fusion processes. (Tritium emits 18.6 keV electrons with a half-life of 12.6 years.) Four of these fusion reactions are (1) the deuterium–tritium (D-T) fuel cycle (FC), (2) the deuterium–deuterium (D-D) FC, (3) the deuterium–light helium (D-$_2$H^3) FC, and (4) the proton–boron ($_1$H$^+$-B^{11}) FC. Below we describe the D-D reactions because they not only are used in "hot" fusion but also were assumed to take place in "cold fusion," discussed below.

The actual fusion process requires forcing two positively charged nuclei together with enough energy (kinetic, thermal) to overcome the atomic electrostatic repulsive force and cause their protons and neutrons to feel the attractive nuclear force and fuse into a new atomic nucleus. The fusion of lighter nuclei generally releases more energy than it takes to force the nuclei together. The goal is to create such an excess of released fusion energy so the process can be made self-sustaining. The released energy is kinetic and thermal.

Many complex designs for fusion reactors (FRs) have been developed since work was begun on trying to make fusion power a reality. Reactors have used electrostatic and magnetic fields to contain the plasmas and moving charged particles involved in the reactions. One of the first experimental FRs, called the tokamak, was designed by Russian physicists in 1950. The basic tokamak design uses a toroidal containment chamber surrounded by a helical magnetic field (traveling around the torus in circles) and a poloidal magnetic field (traveling in circles orthogonal to the toroidal field). The effect of the poloidal field is to concentrate plasma containment in the toroidal reaction chamber. The poloidal components of the tokamak fusion containment are generated mainly by ion and electron currents in the plasma. Most FR designs to date have been based on the poloidal tokamak. Figure 4.29 illustrates the geometry of the poloidal and toroidal field coils (only one each is shown for clarity). The poloidal field coils lie in the plane of the tokamak chamber and circle its central axis. The toroidal field coils circle the tokamak chamber and are located around its circumference. The net result of the vector sum of the poloidal and toroidal magnetic fields is to cause the moving charged particles in the plasma to circle inside the containment chamber torus in a spiral manner, minimizing particle and energy loss from the plasma.

For fusion reactions to occur, the plasma must be heated to over 10^{8}°C (equivalent to 10 keV). Plasma heating is accomplished by several mechanisms, including but not limited to the following:

(1) Ohmic heating—by passing an induced electric current through the plasma.
(2) Neutral-beam injection—fast, high-KE atoms are shot into the plasma, become ionized, and are trapped by the magnetic fields.

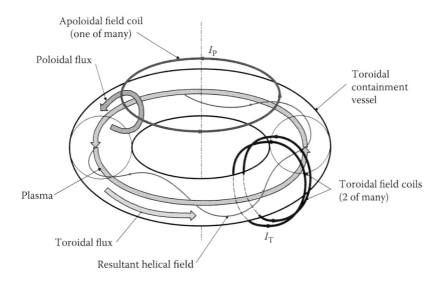

FIGURE 4.29
Schematic of a tokamak reactor torus showing the orientation of the toroidal and poloidal magnetic field coils. Flux vectors are also shown.

Their KE is transferred as heat to the plasma, raising its temperature to 10–30 million °C.

(3) Magnetic compression of the plasma by pulsing up the magnetic fields, forcing the plasma ions closer together (radially inward) in the toroidal reactor chamber.

(4) Radio-frequency (RF) heating—ultrahigh frequency (UHF) (84 and 118 GHz) RF wave energy can be transferred to the plasma, heating it. Photon compression by high-energy laser beams is used to heat deuterium pellets to fusion temperatures in one type of a reactor.

The high neutron flux produced by many fusion reactions is absorbed by the inside wall of the tokamak. This energy causes heating, which is extracted with a cooling fluid, which, in turn, goes to a heat exchanger and boils water to steam, which can drive a turbogenerator.

Note that all FRs to date (2011) have produced lower energy outputs than the energy required to start and sustain fusion, regardless of fusion reaction type and confinement designs.

Experimental FR designs *currently in operation* are listed in Table 4.13.

D-D fusion involves forcing two deuterium (heavy hydrogen) atoms together so their nuclei combine to form one helium atom, plus neutrons, tritium, heat, and a little gamma radiation. D-D fusion is a two-step process, in which an unstable high-energy helium intermediary ($^4He^*$) is formed:

TABLE 4.13

Tokamaks: List of Operational FRs in 2012

FR Name	Date Started	Current Location
TM1-MH: Castor, then Golem	1977	Czech Republic
T-10	1975	Russia (Moscow)
TEXTOR	1978	Jülich, Germany
JET: Joint European Torus	1983	Culham, United Kingdom
Novillo Tokamak	1983	Mexico City
JT-60	1985	Naka, Japan
STOR-M	1987	Canada
Tore Supra	1988	Cadarache, France
Aditya	1980	Gujarat, India
DIII-D	Late 1980s	San Diego, CA, United States
COMPASS	2008	Czech Republic
FTU	1990	Frascati, Italy
Tokamak ISTTOK	1991	Lisbon, Portugal
ASDEX Upgrade	1991	Garching, Germany
Alcator-C-Mod	1992	MIT, United States
TCV	1992	Switzerland
TCABR	1994	São Paulo, Brazil
HT-7	1995	Hefei, China
MAST	1999	Culham, United Kingdom
NSTX	1999	Princeton, NJ, United States
EAST (HT-7U)	2006	Hefei, China
KSTAR	2008	Daejon, South Korea
SST-1		Gandhinagar, India
IR-T1	—	Tehran, Iran

Source: Wikipedia, Tokamaks, 2012.

$$^2H^+ + {}^2H^+ \rightarrow {}^4He^* + 24 \text{ MeV} \text{ (Two deuterons are fused to form an unstable high-energy He intermediary.)} \quad (4.49)$$

High-energy fusion experiments have observed only three decay pathways for the excited-state helium nucleus, with branching ratios showing the probability that any given intermediate $^4He^*$ will follow a particular pathway. The products formed by these decay pathways are known to be as follows (Wikipedia 2011):

$$^4He^* \rightarrow n + {}^3He + 3.27 \text{ MeV} \text{ (probability = 0.5) stable "light"}$$
$$\text{helium product + neutron KE} \quad (4.50)$$

$$^4He^* \rightarrow p^+ + {}^3H + 4.03 \text{ MeV} \text{ (probability = 0.5) tritium product}$$
$$\text{+ proton KE} \quad (4.51)$$

$$^4He^* \rightarrow \gamma + {}^4He + 24 \text{ MeV (probability} \cong 10^{-6}) \text{ helium product}$$
$$+ \text{ gamma ray (high-energy photon)} \tag{4.52}$$

When 1 W of nuclear power is produced from deuteron fusion consistent with the known branching ratios, the resulting neutrons (*n*) and tritium (^3H) production are easily measured. The neutron flux for a 1 W reaction is approximately 10^{12} neutrons/s.

In T-T fusion, the reaction is

$$^3H + {}^3H \rightarrow {}^4He + 2n + 11.3 \text{ MeV} \tag{4.53}$$

The tritium can come from the first D-D fusion reaction. High-energy neutrons plus helium are produced.

Yet another fusion reaction uses two light helium atoms:

$$^3He + {}^3He \rightarrow {}^4He + 2p^+ + 12.9 \text{ MeV} \tag{4.54}$$

Helium plus two positrons at 12.9 MeV are produced. The ^3He is produced in both D-D fusion reactions.

If the technology for producing controlled, D-D fusion reactions on a large scale were available, the "fuel" would be deuterium obtained from heavy water (D_2O). About 0.016% of the water on Earth is D_2O. Pure deuterium gas can be obtained by electrolysis. Then energy is required to separate it from regular H_2. Some of the energy relations concerning controlled fusion were summarized by Rothwell (1997): "Worldwide annual production of all fuels, converted to an equivalent mass of oil, equals approximately 6.8×10^{13} kilograms of oil. This produces 2.7×10^{15} megajoules (at 40 megajoules per kilogram). A kilogram of heavy water [D_2O] contains 200 grams of deuterium. Converted to helium in a d-d fusion reaction, this produces 1.2×10^8 megajoules, with 1.3 grams of matter annihilated. Thus present [1997] world energy needs could be met with 2.3×10^7 kg of heavy water, or ~24,000 metric tons.... To put it another way, a kilogram of heavy water has as much potential energy as 2.9 million kilograms of oil. The Earth has ~2×10^{13} metric tons of heavy water, enough to last 851 million years at this rate,..."

The costs of building and operating large-scale FRs (very expensive), plus the energy cost of extracting D_2O and deuterium ore, would have to be subtracted from the profits from the "free energy" produced by the power fusion reaction.

4.7.2 "Hot" Fusion

To reiterate, to date (March 2012), no experimental FR design has produced a net power gain.

In 2005, the United States, the United Kingdom, the European Union, Russia, Japan, China, and South Korea joined forces to build a very large tokamak-type, experimental FR, called International Thermonuclear Experimental Reactor (ITER), in southern France (Arnoux 2011). The ITER was made large scale because size counts—energy is lost through the reactor surface (which scales as L^2), and energy is produced proportional to reactor volume (which scales as L^3), so the ratio of energy produced to energy lost is proportional to $L^{3/2}$. Bigger *is* better. ITER should be able to produce 500 MW of output power given 50 MW input power, a projected 10:1 power gain.

The ITER reactor will probably be ready for operation by 2026. Initially, the cost of ITER was estimated to be $6.5 billion, and it was to be finished in 2018. Rapidly rising materials and labor costs and problems with supplies from the joint partners have delayed completion of construction and raised the projected cost to approximately $19.5 billion (Grotelüschen 2010).

If the large ITER tokamak is successful in generating continuous power with a power gain of approximately 10, there may be a rush to build more such units around the world. Perhaps by 2050, the world may see some commercial fusion power plants online, feeding the power grids. Their operation will require isolating D_2O from freshwater or seawater and then breaking it down by electrolysis to isolate the deuterium (D_2). Alternately, the water can undergo electrolysis to produce O_2 plus the 1/6000 ratio of H_2/D_2, from which the D_2 can be separated from the H_2, which can be used in FCs to generate DC power and so forth. Ideally, 30 kg of water will yield 1 g of D_2.

Not all FRs use the toroidal tokamak design. For example, a proposed, "CrossFire" FR design was described by M.L. Ferreira, Jr. (2008). This system uses both magnetic and electrostatic confinement of the reaction plasma and uses fusion reactions that generate *no dangerous neutron fluxes*; thus, heavy neutron shielding is not required. The four aneutronic fusion reactions Ferreira suggested for his FR include the following:

$$H^1 + 2\,Li^6 \rightarrow He^4 + (He^3 + Li^6) \rightarrow 3\,He^4 + H^1$$
$$+\ 20.9\ MeV\ (153\ TJ/kg \approx 42\ GWh/kg) \tag{4.55}$$

$$H^1 + Li^7 \rightarrow 2\,He^4 + 17.2\ MeV\ (204\ TJ/kg \approx 56\ GWh/kg) \tag{4.56}$$

$$He^3 + He^3 \rightarrow He^4 + 2\,p^+ + 12.9\ MeV\ (205\ TJ/kg \approx 57\ GWh/kg) \tag{4.57}$$

$$H^1 + B^{11} \rightarrow 3\,He^4 + 8.7\ MeV\ (66\ TJ/kg \approx 18\ GWh/kg) \tag{4.58}$$

Ferreira described how his FR design makes electricity: "A conversion to electricity is relatively simple. The conversion is done during the neutralization by a positive electric voltage to slow down and an electron gun to neutralize. A positive electric field forces the positively charged products to exchange their kinetic energy into potential energy. The positively charged products easily attract electrons from an electron gun, and the electron gun extracts electrons from a positive terminal of a capacitor increasing its positive voltage, which increase its stored energy ($E = \frac{1}{2}CV^2$). A switching-mode power supply sends this energy to a battery bank. The current of electrons and the electric voltage is equal to electric power ($P = VI$). This method of electricity conversion can exceed 95% of efficiency."

The interested reader is encouraged to visit Ferreira's CrossFire Web site.

4.7.3 FRs and Sustainability

There is no possibility of a *catastrophic* accident in an FR; unlike nuclear fission reactors, FRs cannot run away thermally and radioactively. A fusion reaction requires a precise balance between fuel injection, temperature, pressure, and magnetic (and electrostatic) field parameters to maintain a sustained fusion plasma. If any one of these parameters is disrupted, plasma temperature drops, and the fusion reaction would rapidly cease. Although the plasma in a large FR may have a volume in excess of 10^3 m^3, its density is extremely low, and the total amount of fusion fuel in the FR may typically be a few grams. Stop the fuel flow and the reaction stops within seconds.

One disadvantage of running a large FR lies in the disposal of the radioactive materials it produces. The structural materials in a power FR are exposed to high fluxes of high-energy neutrons, which make them radioactive by generating radionuclides. This creates a biohazard and disposal problem if any material has to be replaced. Choice of materials in building an FR dictates that the materials be "low activation," that is, do not easily become radioactive. For example, vanadium becomes less radioactive than stainless steel for a given neutron dose, and carbon fiber composites also have low activation, as well as being light and strong. Of course, ferrous alloys are required for magnet cores and poles. The bottom line is that FRs would create far less high-level radioactive waste materials than do fission reactors, and in general, they will have shorter half-lives. Also, there is no need to store radioactive, spent fuel rods on the premises.

Another disadvantage of using fusion power is the high economic cost of building and operating power FRs. The large ITER power FR in southern France will probably cost over $20 billion to construct. Many scientists, engineers, and skilled workers will be needed to staff it.

4.7.4 Cold Fusion

In 1989, electrochemists Martin Fleischmann and Stanley Pons set off a scientific "chain reaction" of what is now considered by most scientists to be "pathological science." They claimed to have created a heavy water–based, electrochemical cell in which "cold fusion" occurred. Fleischmann and Pons hypothesized that the high compression ratio and mobility of deuterium could be achieved within porous palladium metal using electrolysis to decompose D_2O to deuterium at the electrolysis cell's Pd cathode. Their initial apparatus was an electrolysis cell with a palladium cathode (–) placed inside a calorimeter. (Calorimeters are used in physical chemistry to measure the heat emitted or absorbed by a chemical reaction by measuring the temperature change in a known volume of water surrounding the reaction vessel.) DC current was applied continuously to the cell for many weeks, with the heavy water being replaced at intervals. Some deuterium was thought to be accumulating within the cathode, but most left the cell, as did the O_2 gas from the anode (+). For most of the time, the electrical power ($P_{in} = V_{cell} \times I_{cell}$) into the cell was equal to the calculated power leaving the cell, and the cell temperature was stable at approximately 30°C. But then, at some point in some of the experiments, the temperature rose to approximately 50°C without changes in input power. These high-temperature anomalies would last for 2 days or more and would repeat several times in any given experimental run once they had occurred. The high-temperature anomalies meant that the calculated power leaving the cell exceeded the DC input power. Eventually, the high-temperature phases would no longer occur in a particular cell. Fleischmann and Pons interpreted these high-temperature events to be evidence that "cold fusion" was occurring at the palladium cathode. (Could they have been due to unknown, exothermic chemical reactions?) They published their results in the *Journal of Electroanalytical Chemistry* in April 1989. Before their paper came out, they disclosed their results at a press conference on March 23, 1989.

Their press release and paper created a firestorm of publicity in the media and interest by other research groups around the world who attempted to duplicate (and improve on) their results. Here was, at least in theory, a cheap, clean source of energy that did not require expensive poloidal tokamak containment of a fusion plasma, so people thought. Other workers obtained mixed and null results indicating that cold fusion did not occur. Finally, at a conference of the American Physical Society in May 1989, the consensus of other scientists, based on scientific evidence, was that the Pons and Fleischmann research results were an example of "pathological science." Several scientific journals including *Nature, Science, Physical Review Letters*, and *Physical Review C* published papers critical of cold fusion claims. Such was the faith in Pons and Fleischmann's spin on their research that the state of Utah invested $4.5 million to create the *National Cold Fusion Institute*. The scientific community was split on the issue, with most scientists refuting cold fusion, while another group of faithful continued research on the topic and published in their own group of journals (one of which is *Infinite Energy Magazine*; peer review blocked their papers from appearing in mainstream scientific journals) (Browne 1989; Krivit 2007).

According to Rothwell and Storms (2011), researchers at Mitsubishi have been able to perform the cold fusion experiments several times per year, "with complete success each time." "They observed excess heat, transmutations and gamma rays" but presumably no neutrons.

The bottom line is that Pons and Fleischmann evidently did not use proper control experiments, failed to address errors, and incorrectly used significant figures. Properly conducted experiments by other scientists have not supported their claims (Tulloch 2011). In addition, Pons and Fleischmann (and all the other workers on cold fusion) never detected the energetic neutrons that would be emitted by a D-D fusion reaction (see above).

The existence of practical cold fusion has yet to be conclusively demonstrated.

4.8 Nuclear Energy

4.8.1 Nuclear Reactors

Nuclear power plants work by using the heat generated by nuclear fission to make pressurized steam to drive turbogenerators. Nuclear fission happens when a neutron with sufficient energy hits a fissionable atom's nucleus (such as $_{92}U^{235}$ or $_{94}Pu^{239}$) and causes it to split into various lighter elements and more neutrons and high-energy electrons (β^-).

Two types of fuel are used in nuclear reactors: In one design, hollow zirconium alloy fuel rods are filled with pellets of ceramic, enriched uranium oxide (UO_2). (Enriched uranium is 3.5%–5.0% U^{235} and 96.5%–95% U^{238}. UO_2 has a melting point of 2800°C.) The fuel rods are arranged vertically in an array in the reactor vessel. (Up to 264 rods may form an array.) The second type of fuel is plutonium-239 ($_{94}Pu^{239}$). Both Pu^{239} and U^{235} decay naturally by emitting high-energy α particles (helium nuclei). Pu^{239} is created by bombarding U^{238} with neutrons; this is done in a breeder reactor.

When a neutron with sufficient KE strikes a $_{92}U^{235}$ atom, it splits into an unstable intermediate atom, $_{92}U^{236}$, thence into two lighter atoms ($_{56}Ba^{141}$ and $_{36}Kr^{92}$), and emits two or three new neutrons, gamma radiation (high-energy photons), and heat. Ba^{141} emits β^- (electrons) with energies from 2 to 3 MeV and has a half-life of 18 min; Kr^{92} also emits β^- and has a half-life of approximately 3 s. Xenon 135 ($_{54}Xe^{135}$) is also produced in the fission process and acts as a neutron-absorbing "neutron poison," slowing reactor fission. Xe^{135} is also a β^- emitter at 1.16 MeV and has a half-life of 9.2 h. Normal fission also produces iodine-135 ($_{53}I^{135}$), which emits electrons with a total energy of 2.8 MeV and has a half-life of 6.7 h; $_{53}I^{135}$ decays to Xe^{135}.

The decay of a single $_{92}U^{235}$ atom thus releases approximately 200 MeV of energy. The neutrons produced by the fission of a U^{235} atom can go on to cause fission in other nearby U^{235} atoms, forming a *nuclear chain reaction*. To produce a controlled level of chain reaction in the fuel, neutron absorbers (e.g., boron in boric acid) and neutron moderators are used in reactors around the fuel rods. Moderators reduce the velocity (KE) of fast neutrons and convert them to thermal neutrons that more easily interact with fissionable ^{235}U. Commonly used moderators include regular water (H_2O, in 75% of the world's reactors); solid graphite (C, 20% of reactors); and heavy water (D_2O, 5% of reactors). Beryllium has also been used experimentally as a moderator.

A nuclear reactor core generates heat from several sources:

- The KE of fission products is converted to thermal energy when these nuclei collide with adjacent atoms.
- When the gamma ray photons are absorbed by the reactor, their energy appears as heat.
- Heat is produced by the radioactive decay of fission products and materials that have been activated by neutron absorption. This decay heat source remains for some time after the reactor is shut down.

To prevent overheating or to shut down the reactor, an array of control rods is inserted in the interstices between the reactor's fuel rods. Control rods are made from a strong, thermal neutron-absorbing material such as graphite, borax, or cadmium. In most reactor designs, the control rods are mounted, so if there is an emergency (such as a loss of power in the reactor's control room), they will drop automatically between the fuel rods and stop the chain reaction. The radioactive materials in the reactor will still continue to decay naturally, releasing radioactivity, which is converted to heat by interaction with materials in the reactor.

Heat extraction from a nuclear reactor for generating electric power is generally accomplished by circulating a coolant past the reactor core. The coolant can be water, a gas, liquid metal, or a molten salt. The coolant passes through a closed-cycle, external heat exchanger, where water is turned into high-pressure steam that turns the turbines that drive the alternators. In some reactors, the water for the steam turbines is boiled directly in the reactor core—this is the boiling water reactor. Figure 4.30 illustrates the basic components of the closed-cycle nuclear power system at Three Mile Island. Note that the cooling tower/heat exchanger is used to decrease the temperature of the condenser-cooling water before circulating it back into the condenser. If cooling water is pumped in from the ocean or a river, a cooling tower must extract heat from the hot water leaving the condenser to give it an environmentally safe temperature before discharging back into the water source.

Commercial nuclear power plants have existed since 1954. Designs have evolved for efficiency and safety since then. Some of the current power reactor technologies include the following: pressurized water reactors, boiling water reactors, pressurized heavy water reactors, high-power channel reactors, gas-cooled reactors, liquid metal fast breeder reactors, molten salt reactors, aqueous homogeneous reactors, and pebble-bed reactors (PBRs). [See

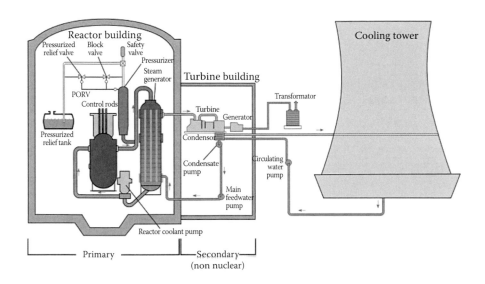

FIGURE 4.30
Schematic of a nuclear power plant similar to one at Three Mile Island. (Courtesy of US NRC, Backgrounder on the accident at Three Mile Island. www.nrc.gov/reading-rm/doc-collections/fact-sheets/3mile-isle.html#tmiview.)

a summary of the detailed descriptions of these reactor designs in Nuclear (2012).]

Note that U^{235} contains 72×10^{12} J/kg of energy; compare this with coal, which, when oxidized, gives 2.4×10^7 J/kg. No wonder nuclear reactor power is an economically viable alternative to burning FFs.

4.8.1.1 Pebble-Bed Reactors

One of the safer nuclear reactor designs is the PBR. Compared to other reactor designs, PBRs have been largely neglected in the second half of the 20th century, and only recently has their design found new interest as a practical, safer, nuclear fission power source. The PBR is a graphite-moderated, gas-cooled, high-temperature nuclear reactor. No water is used in the primary reactor heat extraction system, and no meltdown will occur if the helium cooling gas flow is stopped. In normal operation, heat is transferred from the reactor core by helium gas, which does not form radionuclides when bombarded by high-energy neutrons and other nuclear fission products. Figure 4.31 illustrates the schematic of a basic PBR design. In the core of the PBR are the unique, spherical, nuclear fuel elements called pebbles. Each pebble is 6 cm in diameter and has an approximately 5 mm thick spherical shell of pyrolytic graphite (PG) surrounding a core in which thousands of tristructural-isotropic (TRISO) fuel kernels are imbedded in a graphite matrix. Each 6 cm

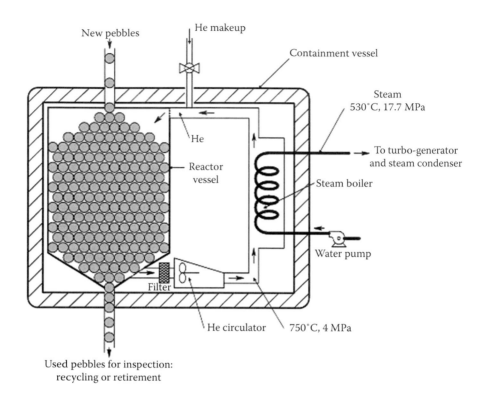

FIGURE 4.31
Schematic of a helium-cooled, pebble-bed nuclear reactor. See text for description of operation.

pebble weighs approximately 210 g and has a total of approximately 9 g of fuel (generally uranium oxide isotopes) in approximately 15,000 fuel kernels.

Each TRISO fuel kernel is a 920 μm sphere consisting of a 40 μm PG outer shell over a 35 μm thick shell of silicon carbide, which in turn coats a 40 μm thick shell of PG. A 95 μm thick porous carbon buffer layer lies under the inner pyrolytic carbon coating. Inside the porous carbon buffer shell is an approximately 500 μm diameter fuel sphere of fissionable uranium oxides (Williams 2009).

PG is used for pebble and fuel kernel coatings because it is a neutron damper, and it will not burn in air (O_2) unless WV (-OH⁻) catalyzes combustion. In a helium atmosphere, PG is stable to a temperature of 4000°C, above which it sublimates. In an emergency loss of coolant gas, the reaction can be cooled by inserting damper rods into the reactor core (not shown in the figure).

In PBR operation, the pebbles continuously slowly flow into the top of the cylindrical reactor chamber and, slowly, under the influence of gravity, work their way down in the reaction chamber into a funnel-like collection chamber where they are collected and are inspected for physical defects and their remaining nuclear fuel content. If good, a pebble can be recycled approximately 10 times over about 3 years. A damaged and/or fuel-exhausted pebble is removed to a safe nuclear waste storage area, and a new pebble is inserted. Pebble flow through a PBR chamber is a complicated mechanical phenomenon involving dense granular flow, which has been studied by Rycroft et al. (2006). Rycroft et al. used a to-scale model modular PBR chamber that was 3.5 m in diameter and approximately 10 m high. In the design used, there was a central column of moderating, pure PG moderator (reflector) pebbles, surrounded by an annulus of dummy fuel pebbles. They also did finite element computer simulations to predict mechanical pebble behavior in the reactor. A solid graphite moderator pebble may last approximately 12 years. Moderator pebbles can be used instead of control rods.

We note that some 370,000 fuel pebbles are used to fuel a 120 MW PBR. If the desired PBR's electrical power output is designed to be 116.3 MW, and assuming a conversion efficiency of 44%, a reactor thermal output of 264.3 MW would be required (Gee 2002). To create 1 GW of total PBR power for the grid, 10 "modular" PBR reactors of approximately 116 MW would be needed.

In operation, inert helium gas is forced through the interstices between the hot pebble spheres with an inlet pressure of approximately 8.4 MPa (1218 psi) and an inlet temperature of approximately 500°C to extract the heat released by nuclear fission. The He leaves the PBR at approximately 900°C. The hot He gas can directly run turbines, be passed through a heat exchanger to boil water to make steam to drive turbines, or be used in industrial processes, such as making hydrogen gas from water. The cooled He gas is then recycled; it is cooled and recompressed to approximately 8.4 MPa.

The first operational PBR was a 10 MW pilot system designed and built by the German *Arbeitsgemeinschaft Versuchreaktor* (AVR) in Jülich, West Germany, in 1966. The AVR PBR was cooled with helium gas, which has a low neutron cross section. This means that few neutrons are absorbed, making the He coolant have low radioactivity. The AVR PBR was operated until 1988, when it was disassembled and inspected. During removal of the pebbles, it was found that a neutron reflector under the pebble bed core had cracked sometime during operation. About 100 pebbles were found stuck in the crack. It turned out that the AVR PBR was heavily beta-contaminated with Sr^{90}, which has a half-life of 28.1 years, and this contamination was present in the form of dust.

The operating temperature of a PBR can range between 750°C and a stable maximum of approximately 1600°C if the coolant gas is stopped for any reason. The 1600°C safe high temperature inside the PBR cannot be exceeded for a given PBR geometry because of an effect called Doppler broadening (DB). In DB, at high temperatures, the fissionable fuel "sees" a wider range of relative neutron speeds; U^{238}, which forms the bulk of the uranium in the kernels, is much more likely to absorb fast or epithermal neutrons at higher temperatures. This reduces the number of neutrons available to cause fission in the U^{235} and causes the reactor's high-temperature power to plateau due to an effective negative-feedback phenomenon. This is the main passive safety feature of the PBR, and it makes the PBR design unique relative to conventional light water reactors, which require active safety controls (that do not always work—e.g., the 2011 Fukushima Daiichi reactor incident).

As described above, the first PBR operated from 1966 to 1988 in Jülich, Germany. The Union of South Africa began development of a pebble-bed modular reactor (PBMR) in 2004. (The modular design runs several smaller PBRs with their power outputs connected together.) Because of the world economic slowdown and cost overruns, in early 2010, the South African government decided to stop funding the PBMR, followed by withdrawal of other nongovernment investors (Nordling 2010).

At present, the only active pilot PBR plant is being designed by China, the HTR-10. This largely follows the German AVR design. It is a 10 MW prototype plant that uses He cooling and a He-driven turbine. The HTR-10 is being designed by Tsinghua University in Beijing. China's first 250 MW PBR is scheduled for commissioning in 2013. China has plans for 30 such plants by 2020 (a total of 6 GW) if the HTR-10 meets test criteria (Bradsher 2011). The Nuclear Science and Engineering Department at MIT has been designing a PBR with 250 MW thermal and 120 MW electrical outputs. The reactor He outlet temperature is to be 900°C with a full-power He mass flow rate of approximately 128 kg/s. He pressure in the reactor is approximately 80 bar (1160 psi = 8 MPa). The MIT PBR design will use 360,000 pebbles and has six control rods. The hot He flow drives three turbocompressors and one turbogenerator. Details of the MIT PBR can be found in the paper by Kadak (2007).

4.8.1.2 Importance of Helium, a Nonrenewable Resource

While He is the second-most abundant element in the universe (after hydrogen), it is found at only 5.2 ppmv in the Earth's upper atmosphere. The good news is that approximately 1.8% of NG found in Texas and Oklahoma is He, and there is plenty of NG in worldwide reserves. He has also been found in NG from wells in Algeria and Qatar and will no doubt be found in other NG wells yet to be drilled.

There are seven He isotopes, helium 2–8. He isotopes 2, 5, 6, 7, and 8 are radioactive; helium-7 and -8 are created in certain nuclear reactions. The most common nonradioactive helium isotope is $_2He^4$, with atomic mass 4.00260. $_2He^3$ (atomic mass 3.01603) is only about one millionth as common in the atmosphere as $_2He^4$.

Helium is a nonrenewable resource, formed over the eons by the slow radioactive decay of elements such as uranium and polonium in the Earth's core. [For example, 1 kg of uranium decaying into its ultimate form, lead, creates 756 L of He (at STP).] He was first discovered in NG from a Dexter, KS, oil well in 1903 (about 2% He). Because of its application to military lighter-than-air vehicles (zeppelins, dirigibles, blimps), the US government set up a monopoly on helium extraction and storage in the National Helium Reserve in 1925 in Amarillo, TX. (Only crude, unrefined He was stored.) By 1995, a billion cubic meters of unrefined He had been stored, and the reserve was

US $1.4 billion in debt, prompting the US congress to phase out the reserve. The resulting *Helium Privatization Act of 1996* directed the US Department of the Interior to begin emptying the reserve by 2005. The United States lost its dominance in the world helium market in the mid-1990s. NG in Arzew, Algeria, was found to contain extractable amounts of He; 17 million m³/year was extracted at first. Other NG wells in Algeria and Qatar were also found to be helium rich. Algeria is now the second-leading world producer of He (Smith et al. 2003).

Three reasons He gas is used to extract heat from pebble-bed nuclear reactors are the following: (1) He has the highest thermal conductivity of any gas (0.1513 W m^{-1} K^{-1}). (2) He has a high molar heat capacity (20.768 J mol^{-1} K^{-1}). (3) He is "transparent" to neutrons. Other scientific, medical, and industrial applications of He include the cryogenic cooling with liquid He of superconducting magnets used in particle accelerators, tokamaks, mass spectrometers, and magnetic resonance imaging (MRI) machines; also, it is used in certain welding applications, and so forth. Medical-grade He is used in deep-sea diving (heliox mixtures). One source on the Internet quoted the price of a "T" cylinder (337 ft.³) of medical grade helium as approximately $69 (January 1, 2011). The US Bureau of Land Management (BLM) was charging $75 per thousand cubic feet for unrefined He from the US reserve in 2011. This is the minimum price established by law and not the open market value. As of October 2010, He sales from the Bush Dome He reserve had paid $804 million, or 58.6% of the original He storage program debt (McBride 2010).

Because the world's He supply depends on its extraction from NG, and the world's NG reserves are finite, we will eventually run out of He (it cannot be made chemically or recycled like carbon or N$_2$). At present, the price of He is kept artificially low by the US government. Eventually, it should be deregulated to dampen demand and extend reserves. We expect that the deregulated price of refined He will rise in inverse relation to our remaining helium-containing NG reserves. This will mean added expenses in the operation of future PBRs and the big ITER FR.

Cai et al. (2010) devised a detailed systems (node-and-branch) model of helium resources that was used to predict that there will be no serious depletion of economic helium resources before 2060. This assumes that the He lost into the atmosphere from venting the NG that accompanies oil well production, or burning the NG emission from oil wells, will be sequestered in the near future. They predicted that the overall future price of He will rise due to the increase in production costs and, later, the increasing scarcity of He resources (as NG resources are exhausted). He demand will continue to increase through to 2030, when the annual global consumption may be around 10^{10} ft.³/year (2.83 × 10^8 m³/year). By 2100, they predict that He production will have run down until the only He resources are low-grade NG fields and the upper atmosphere. (One way we can compensate for the shortage of cryogenic He is to develop high-temperature superconductors for electromagnets.)

4.8.1.3 Hazards of PBRs

What are the environmental hazards of operation of PBRs? Perhaps the most common criticism of PBRs is that the fissionable fuel is encased in combustible graphite, operated at high temperatures ($\leq 1600°C$). Any incident (earthquake, terrorist bomb, etc.) that breaches the inner containment and causes loss of the inert gas coolant (He) and allows air (O_2) to contact the pebbles could precipitate combustion of the outer graphite layer of the pebbles if sufficient WV ($-OH^-$) is present. The TRISO fuel kernels are imbedded inside the 5 mm outer PG shell. Each fuel kernel has a thin, 35 µm, outer shell of silicon carbide for thermal and structural protection. If it were to happen, a PBR fire would release radioactive gasses into the atmosphere from the combusting pebble cores. Fortunately, in the limited history of PBR operation, a fire has never occurred. To minimize the threat from terrorism, all new PBRs should have robust containment vessels just like conventional nuclear reactors. Such designs will make them more expensive to build but will afford protection from crashing aircraft, earthquakes, tsunamis, and most bombs.

Another hazard that was observed in the Jülich AVR is the production of graphite dust that absorbs fission products if they escape from damaged pebbles and TRISO kernels. This radioactive graphite dust can be filtered from the circulating He gas. However, if the gas containment is damaged, the He plus radionuclides will escape into the environment.

A third consideration is the amount of helium, and its cost, required to operate a PBR. He is extracted from NG; it cannot be made chemically like H_2 or methane. Only 0.00052% of the Earth's atmosphere is He. Needless to say, extracting He from the atmosphere is not practical. Both H_2 and He gasses are prone to leakage into the atmosphere from pipelines and containment vessels, which means that He-cooled PBRs will periodically need "makeup" He.

If designed with redundant safeguards, PBRs may contribute to sustainable, NFF energy in the coming five decades. Per megawatt output, they appear less expensive to construct and offer a lower environmental threat than water-based reactors. At present, we are waiting to see how successful the Chinese are with their HTR-10 PBR operation.

4.8.2 Hazards of Nuclear Power Generation

4.8.2.1 Radioactivity and Ionizing Radiation

Many isotopes of the elements are spontaneously radioactive. That is, they randomly emit atomic components such as electrons (β^-) or alpha particles (α) (which are helium atom nuclei), neutrons (n), or protons (p^+), often with sufficient KE to interact with other atoms and molecules, causing ionization (loss of electrons), or they emit photons (visible, UV, γ-radiation). Some isotopes capture electrons.

Natural or induced radioactivity can be damaging to animal and plant life. Critical biological molecules (e.g., RNA, DNA, proteins, enzymes) can be damaged by high-energy β^-, $p+$, α, n, and γ radiation. Mutations caused by irreversibly damaged DNA can cause cancers and birth defects. For example, $_{53}I^{135}$ emits electrons (β rays) with a total energy of 2.8 MeV and has a half-life of 6.7 h. Iodine is used by the mammalian thyroid gland to make two critical thyroid hormones that regulate body metabolism and other important body functions, that is, thyroid hormone (thyroxine) (TH) and 3,5,3'-triiodothyronine (3-IT) (Guyton 1991). Radioactive iodine is taken up by cells of the thyroid gland where the high-energy electrons are in intimate contact with cell DNA. The I^{135} radioactivity can cause cell damage and thyroid cancer, even though it has a relatively short half-life. The strategy for treating I^{135} "poisoning" is to immediately ingest an excess of normal, nonradioactive I^{126}, which, by mass action, will displace most of the high concentration of $_{53}I^{135}$ in the thyroid, reducing radioactivity in the gland. The $_{53}I^{135}$ is diluted to a relatively harmless level in the rest of the body and is eliminated in the urine and feces, as is the excess normal iodine atoms. Another isotope, strontium-90 ($_{38}Sr^{90}$), is also a β^- emitter. However, Sr^{90} has a longer half-life (28.1 years). Sr^{90} electrons are emitted with an energy of 0.546 MeV. Sr^{90} can metabolically end up in bone, where its radioactivity can damage stem cells in the marrow, often causing leukemia. It is much more difficult to eliminate from the body.

The March 11, 2011, earthquake and tsunami that damaged the four Fukushima nuclear power plants in Japan underscore the dangers of radioactivity, even from well-designed reactors. These reactors did not explode violently, as did the Chernobyl reactor in the Ukraine in 1986. The Chernobyl reactor blew approximately 50 tons of highly radioactive material into the surrounding area, contaminating millions of acres of forest and farmland. The Chernobyl disaster forced the evacuation of at least 30,000 people from a number of villages and farms and eventually caused thousands to die from cancer and other illnesses (Brain and Lamb 2011). Some of the radioisotopes emitted from the Chernobyl plant explosion were Te^{132}/I^{132}, I^{131}, Ba^{140}/La^{140}, Zr^{95}/Nb^{95}, Cs^{134}, and radioactive xenon gas (Xe, many radioisotopes), all high-energy β emitters.

4.8.2.2 Radioactivity Measurement

When a small ball of a radioactive element or compound emits its characteristic particles or rays, the times of emission are random, characterized by a Poisson distribution. The directions of emission are also random, spanning a solid angle of 4π steradians (a sphere around the radioactive ball). Thus, a radiation sensor distant from the radioactive ball will respond only to the fraction of the total radioactivity that lies within its directional sensitivity function (DSF) (Northrop 2001). The DSF will vary with the orientation of the radiation sensor relative to the source. A cylindrical Geiger tube will count γ photons at a higher rate if it is held with its axis perpendicular to a radial line from the source than if it is held end on. A beta counter with an end window will count higher when directed toward the source. Particles such as α, β^-, p^+, a γ photon, or a neutron (n) striking a radiation sensor with sufficient energy generally produce a narrow pulse of electric current, the peak and area of which are generally proportional to the particle's energy. The simplest form of radiation counter merely counts the pulses per unit time and ignores any variation in their height. More sophisticated radiation meters pass the pulses through pulse–height windows (in a multichannel analyzer) that actually scale the pulses according to height and count the number of pulses in a given height (energy range) over time. By examining the energy spectrum of the particles counted, it is generally possible to identify the elemental isotopic sources.

There are four principle types of sensor used to measure radioactivity: (1) the ionization tube–type Geiger–Mueller tube (GMTs); (2) the scintillation crystal/photodetector–type detector; (3) the reverse-biased, *pin* diode radiation sensor; and (4) the film badge dosimeter.

Figure 4.32 illustrates two types of GMTs. In Figure 4.32a, a thin aluminum case cathode surrounds a central tungsten anode. The tube is filled with an inert gas such as He, Ar, or N_2, at pressures ranging from a few mm Hg to atmospheric, plus a quenching gas such as CH_4 or alcohol vapor, which prevents energetic photons emitted from the activated inert gas molecules from accelerating to the cathode and stimulating the release of secondary electrons by photoelectric emission. In Figure 4.32b, some GMTs have thin mica or mylar end windows to count lower-energy, low-penetration particles (α and β^-). GMTs are run at DC potentials ranging from 100 to 1300 V, depending on the tube design, the particles to be counted, and the counting characteristics desired. A more detailed description of GMTs and Geiger counter circuits and how they work can be found in Northrop (2005).

The second type of radiation sensor counter makes use of scintillation crystals optically coupled to a photosensor such as an avalanche photodiode (APD) or a photomultiplier tube (PMT). With liquid nitrogen cooling to reduce noise, a PMT can resolve scintillation events as small as 2×10^{-16} lm. The pulse output of the APD or PMT is scaled and counted (Scionix 2011). Figure 4.33 illustrates a block diagram of a scintillator radiation detector. The delay line is used to generate a very narrow pulse to enable high-speed counting.

Scintillators are made from many materials: certain organic crystals, organic liquids, plastics, a variety of inorganic crystals, and certain glasses. An incoming, high-energy particle can excite either an electron or vibrational level in a scintillator molecule. The singlet excitations decay in <10 ps to the S* atomic state without the emission or radiation. The S* state then decays to the ground state S_0 by emitting a scintillation photon. Most scintillation crystals and materials are very effective at sensing β^- and γ photons and less effective with heavy ions and neutrons (Northrop 2005, Section 6.9.3). Scionix (2011) gave several graphs and tables listing the properties and compositions of scintillators.

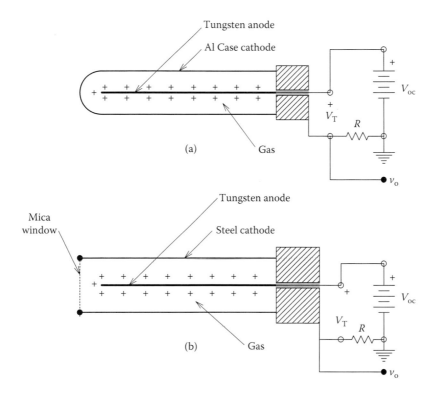

FIGURE 4.32
Schematics of two types of Geiger tubes used for counting radioactive decay events. (a) This tube counts high-energy neutrons and gamma rays. (b) This tube can count β and α particles that can more easily penetrate the mica end window. (From Northrop, R.B., *Introduction to Instrumentation and Measurements*, 2nd ed., CRC Press, Boca Raton, FL, 2005. With permission.)

There are several units of radiation dose measurement. The most basic is the number of radioactive emission events per minute [counts per minute (CPM)]. *Curie* (Ci) is defined as that quantity of any radioisotope undergoing 3.70×10^{10} events per second. Also, 1 becquerel = 27 picocurie (pCi) = 1 cps. *Roentgen* (R) measures the energy produced by gamma radiation (photons) in 1 cm^3 of air. The *rad*, or radiation absorbed dose, recognizes that different materials that receive the same exposure may absorb different amounts of the radiation's energy. A rad measures the amount of radiation energy transferred to some mass of material, typically humans. One roentgen of gamma radiation exposure results in about 1 rad of absorbed dose in humans. The *rem* stands for roentgen equivalent man. This unit relates the dose of any radiation to the biological effects of that dose. The dose in rem = Q × dose in rad, where Q is a "quality factor" dependent on the animal species and type of radiation. For man, given γ radiation or β particles, 1 rad → 1 rem, so Q =

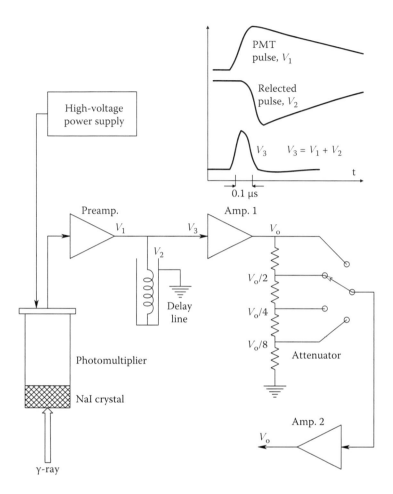

FIGURE 4.33
Block diagram of a scintillation-type radiation counter. What is actually counted are bursts of photons released inside the crystal in response to absorbing the energy of a γ- or x-ray photon, or other-high energy particles. An APD can also be used to count scintillations. (From Northrop, R.B., *Introduction to Instrumentation and Measurements*, 2nd ed., CRC Press, Boca Raton, FL, 2005. With permission.)

1. In the SI system of measurements, 1 sievert (Sv) ≡ 100 rem, and 1 gray (Gy) ≡ 100 rad (Quayle 2011; CDC 2011).

4.8.2.3 Sustainability and Nuclear Power

Beginning in 1955, worldwide development of nuclear power reactors has grown exponentially until the 1986 Chernobyl accident, at which time the world total number of active reactors had reached approximately 400. From

1986 to 2007, the number has held a plateau at about 435 active reactors with an installed power capacity of approximately 375 GW (Nuclear power history 2007). The growth of the number of new nuclear plants slowed because of several factors: fear of nuclear accidents and radioactive contamination, concern about where and how to dispose of spent fuel rods (or "pebbles"), and the expense of constructing new, safer plants.

One of the largest concerns about having a nuclear power plant "in your backyard" is what happens if there is an accident. Serious nuclear reactor accidents have occurred in the Chernobyl plant in the Ukraine in 1986, at the Three Mile Island plant in Pennsylvania in 1979, and most recently in Japan on March 11, 2011 at the four Fukushima Daiichi nuclear power reactors. A nuclear reactor meltdown releases many toxic radioisotopes, including I^{135}, Sr^{90}, Cs^{134}, and so forth, which can enter the food chain by farm animals eating food contaminated with radioisotopes or fish swimming in a region of sea contaminated by fallout. They can also directly enter the body by a person touching contaminated surfaces, breathing radioactive dust, eating plants contaminated with radioactive dust, or drinking contaminated water. These are singular events we have little control over. People and animals must evacuate the contaminated area around the plant accident to a safe distance, as determined by radiological measurements, and be careful about the food they eat and water they drink.

In day-to-day, normal operation, nuclear power plants do not emit any radiation of consequence. They do present two environmental challenges, however. One is from the use (intake and discharge) of large amounts of cooling water used in normal operations. The basic nuclear power plant uses a closed-cycle, fluid loop to carry heat from the reactor to the steam generator/heat exchanger and a closed-cycle steam line to the turbine. The steam exiting the turbine must be cooled and condensed back to water through another heat exchanger that carries cold water from a river or the sea. This condenser water must also be cooled before discharging it back to its source. The environmental problems that must be addressed include keeping fish and mollusks out of the cooling water intakes and making the coolant discharge water cool enough so fish and other marine or aquatic life are not harmed. In modern nuclear power plants, these concerns are generally well met. If a closed-cycle cooling water cycle is used, the hot discharge from the steam condenser is cooled by forced air or air convection in a tower. There is evaporative water loss in air cooling, so loss of cooling water must be made up, so the process is not strictly closed cycle like turbine steam generation. The newer Westinghouse AP300 nuclear reactor design places an emergency water supply tank above the reactor, so that if all cooling pumps fail, there is 3 days' supply of cooling water by gravity feed. In 2011, there were 23 old-style nuclear power plants in the United States that relied only on electric pumps for cooling.

A second, great challenge is what to do with spent fuel rods that have nearly exhausted their useful U^{235} content. The spent rods are still naturally highly radioactive. At present, there is no operational, safe, underground, long-term storage facility for spent nuclear plant fuel rods in the United States. Creating one is a "can that keeps being kicked down the road" by politicians and environmentalists. Yucca Mountain, in South-Central Nevada, was studied as a storage site for spent nuclear rods by the DOE in 1978. In 2002, President G.W. Bush signed a joint congressional resolution allowing the DOE to go further in establishing a safe repository in which to store the country's nuclear waste. On July 18, 2006, the DOE proposed March 31, 2017, as the target date to open the facility and begin accepting waste, based on full congressional funding. Due to subsequent congressional wrangling and a facility budget cut in 2008 to $390 million, progress toward opening the facility was again delayed. In March 2006, the US Senate Committee on Environment and Public Works Majority Staff issued a 25-page white paper entitled *Yucca Mountain: The Most Studied Real Estate on the Planet*. They concluded that

- Extensive studies consistently show Yucca Mountain to be a sound site for nuclear waste disposal.
- The cost of not moving forward is extremely high.
- Nuclear waste disposal capability is an environmental imperative.
- Nuclear waste disposal capability supports national security.
- Demand for new nuclear plants also demands disposal capability.

(A comprehensive history of the controversies, costs, progress, and obstacles in the development of the Yucca Mountain nuclear waste storage facility can be found in Mineral County 2012.)

In May 2009, US Energy Secretary Steven Chu stated: "Yucca Mountain as a repository is off the table. What we're going to be doing is saying, let's step back. We realize that we know a lot more today than we did 25 or 30 years ago. The NRC is saying that the dry cask storage at current sites would be safe for many decades, so that gives us time to figure out what we should do for a long-term strategy... We're looking at reactors that have a high-energy neutron spectrum that can actually allow you to burn down the long-lived actinide waste. These are fast-neutron reactors."

(Conversion of long-lived radioactive materials to others with shorter half-lives by neutron bombardment promises to relieve some of the long-term storage burden.) Secretary Chu concluded: "Yucca was supposed to be everything to everybody, and I think, knowing what we know today, there's going have to be several regional areas" (Secretary 2009).

Because of a lack of developed, geologically stable, secure underground storage sites, US nuclear power plants now store their spent fuel rods in arrays, on premises, in secure, "leakproof" tanks of water. In the four Japanese Fukushima Daiichi reactors, we saw that the spent fuel rods were stored in water on the reactor's second floor. Consequently, when the 9.0-Richter earthquake occurred, the tanks cracked and lost their water, and the rods heated up from undamped radiation.

The fuel rods, if they melt or burn, release toxic radioactive particles with half-lives of hundreds to thousands of years. Perhaps the best way to store radioactive, spent fuel rods is in modules encased in a boron-containing ceramic matrix, that is, dry, and then store them underground in a geologically stable region, well away from any groundwater. Locations for nuclear waste storage facilities were successfully found in Scandinavia by letting local communities with veto powers become involved in the decision-making processes for siting.

4.9 Carbon Capture and Storage

4.9.1 Introduction

The Earth's surface atmosphere contains 78.08% N_2, 20.95% O_2, 0% to approximately 4% WV, 0.93% argon, 0.390% CO_2, 0.0018% neon, and so forth. Other trace gasses are also present in parts per trillion (ppt) concentrations (He, CH_4, H_2, N_2O, O_3) (Pidwirny 2010; NOAA 2010).

Why is there a perceived need to sequester point-source, anthropogenic CO_2 emissions? They are a small fraction of the total CO_2 flux into the atmosphere. One argument is that by not putting this CO_2 into the air, we will slow down and attenuate the global warming process and thus mitigate the threats to human sustainability that global warming is producing. This assumes that CO_2 acting as a GHG is a major causative factor in driving the complex global warming process. However, it neglects the role of other anthropogenic and natural GHG emissions (e.g., WV and methane) and the fact that the planet's climate is an immense, complex, nonlinear system that has tipping points in its behavior. Certainly, the Earth's climate is affected by many factors, in particular, the retention of long-wave IR (heat, originally from the sun's radiation) trapped by GHGs. Solar radiation flux on the Earth's surface varies with sunspot conditions, the Earth's axial inclination relative to its orbital plane, and also the Earth's orbital distance from the sun. Solar heat is stored in the masses of land, oceans, plants, and atmospheric gasses. The Earth has photosynthetic plants, algae, and plankton that use solar radiation to sequester atmospheric CO_2 and release O_2, landmasses that absorb thermal energy and reradiate LIR, and oceans that absorb thermal energy and transport it around by currents. The ice caps and glaciers melt and absorb the heat of fusion of water. The cold freshwater from melting ice caps and glaciers affects the ocean's salinity and circulating currents, hence influencing the land temperature distributions. The atmosphere transports energy in the form of WV; when the temperature of WV decreases, it condenses (undergoes a phase change), releasing its heat of vaporization (2.255 MJ/kg) that initially produced it. WV is an effective and dominant GHG; unlike CO_2 and methane, WV can easily store, transport, and release heat energy through its phase changes, as well as absorbing LIR from the Earth's surface that would otherwise radiate back into space, cooling the planet at night.

By one estimate, in 2009, the combustion of *all* FFs worldwide [e.g., oil and oil products (including gasoline, diesel, jet fuel), coal, and NG] generated approximately 8.4 billion tons (7.62×10^{12} kg) of carbon into the atmosphere (Riebeeck 2011). Some point sources of CO_2 emission include the fermentation of sugarcane, sugar beets, or corn mash to make ethanol for use as a fuel, industrial chemical, or drink. Jet aircraft are also a significant contributor to anthropogenic, FF-derived CO_2 in the atmosphere. They also contribute WV at high altitudes from their exhausts. It was estimated that the approximately 16,000 yearly worldwide aircraft flights contribute approximately 600 million tonnes of CO_2 (6×10^{11} kg) (Mulchandani 2006). Chèze et al. (2011) projected the world jet fuel demands from present to 2025. In their study, they gave historical data from 1981 to 2007, and their extrapolated demand from 2007 gave values for several scenarios. World jet fuel consumption in 2007 was approximately 230 million tonnes/year. Considering air traffic efficiency improvements of 1.9%/year, the consumption in 2025 would be approximately 300 million tonnes/year. If no air traffic efficiency improvements occur, the 2025 consumption of jet fuel could be approximately 500 million tonnes/year.

In an interesting web article, Seat61 (2010) compared the CO_2 emissions per passenger for round-trip air travel for a number of European destinations; for example, London to Paris and return by air takes 3.5 h and injects 244 kg CO_2 per passenger into the upper atmosphere, while a Eurostar train trip takes 2.75 h and releases 22 kg CO_2 per passenger into the lower troposphere. Significantly, 91% less CO_2 is emitted by train travel to and from Paris. The average reduction of CO_2 emission by train versus plane for seven European round-trips was 83.7%!

In an innovative modeling study, Burkhardt and Kärcher (2011), investigated the global radiative forcing from aircraft contrail cirrus clouds in the stratosphere. The young (<5 h old), line-shaped contrails slowly morph into irregularly shaped contrails and result in changes in contrail-induced cirrus (CIC) cloudiness. Contrail cirrus clouds are composed of tiny ice crystals that reflect incoming short-wavelength solar radiation and trap outgoing long-wave IR radiation. They reported that the contrail cirrus radiative forcing offset by the natural cloud feedback results in a net radiative forcing of approximately 31 mW/m². This number is affected by several factors including natural high-altitude air currents, stratospheric air temperature, humidity, and natural clouds. It also varies with time. The authors stated that contrail cirrus clouds cause a significant *decrease* in natural cloudiness, partially offsetting their warming effect. They concluded that net radiative forcing due to contrail cirrus remains the largest single radiative forcing component associated with aviation.

Other, harder-to-quantify, diffuse sources of atmospheric CO_2 include the natural decay of vegetation, burning of rainforests to make agricultural land, animal respiration, and volcanoes. Some CO_2 sources are considered "green," that is, their combustion releases CO_2 that has recently been incorporated into their biomass (plants, trees, wood, leaves, etc.), or, in the case of animals, their breathing releases CO_2 that came from them metabolizing eaten plant food or from eaten meat from animals that ate plant food. Plants, of course, incorporate atmospheric CO_2 into their biomass by photosynthesis.

One estimate of the oxidized carbon in the CO_2 flux into the atmosphere averaged over the period from 1997 to 2006 gave the figure of 1.2 petagrams of carbon (PgC) per year = 1.2×10^{12} kgC/year = 1.2 GtC/year. Since CO_2 is 27.27% carbon by mass, we can calculate that 4.4 $GtCO_2$/year was emitted by worldwide forest clearing using fires to destroy woody mass (van der Werf et al. 2009).

Animal respiration is another significant distributed, biogenic source of atmospheric CO_2 that is often overlooked. Considering only human beings, each of the 7 billion humans on Earth today exhales an average of 0.7 kg CO_2/day, or 256 kg CO_2/year; this gives a total green, anthropogenic (in the strictest sense) yearly emission of 1.79×10^{12} kg CO_2 or 1.79 $GtCO_2$/year, or $(12/44) \times 1.79 = 0.488$ GtC/year (Jana et al. 2010). Thus, human breathing alone contributes a significant amount of CO_2 into the atmosphere: it put approximately 6.4% of the carbon into the atmosphere as did our total FF combustion in 2009 (7.62 GtC/year). Even more atmospheric GHGs come from all breathing animals on Earth (other than man), which include cattle (about 1.5 billion), swine, goats, horses, chickens (about 19 billion), and so forth plus all wild animals, deer, elk, moose, birds, insects, and so forth. One revised estimate for worldwide total farmed livestock GHG respiration [in CO_2 equivalent mass (CO_2eq)] by Goodland and Anhang (2009) was 8.769 $GtCO_2$eq/year. This included methane, which is 23 times more effective than CO_2 as a GHG. ("Livestock" in the Goodland and Anhang article included cattle, buffalo, sheep, goats, camels, horses, pigs, and poultry—i.e., all domestic warm-blooded vertebrates.) Compare their livestock GHG emission number to the 2006 United Nations (UN) Food and Agriculture Organization (FAO) estimate of 7.516 $GtCO_2$eq/year. Clearly, the total yearly worldwide livestock respiration is significant and probably lies somewhere around the FAO and Goodland and Anhang estimates. For the case of the livestock respiration, we must add the CO_2 from all wild animals. The paltry, approximately 1.79 $GtCO_2$/year from human respiration alone may be considered to be green, but it is still a significant fraction of the major anthropogenic CO_2 flux into the atmosphere. It appears that total animal respiration (TAR) is a significant GHG input to Earth's atmosphere.

One reference (CDIAC 2011b) discounted the importance of CO_2 emission from humans because it is "green" (i.e., the CO_2 from humans comes from food that recently integrated CO_2 metabolically into its mass). This may be seen as a zero sum in terms of short-term carbon fluxes, but a CO_2 molecule put into the atmosphere has no "color," green or otherwise—it may be immediately recycled by photosynthesis, be dissolved in the oceans, or stay in the troposphere and act as a GHG for many years. Atmospheric CO_2 from animal respiration is as important as CO_2 from any other source! It may be "green" in origin, but its fate is not necessarily "green," and its mass is not negligible.

From many reliable sources, the average atmospheric $[CO_2]$ has risen from preindustrial levels of approximately 280 ppm (in 1744) to a present-day (April 2012) peak value of 396.18 ppm (dry air mole fraction) at the Mauna Loa observatory. This is an approximately 44% increase over 268 years and a 0.72% increase over the April 2011 $[CO_2]$ at Mauna Loa (NOAA 2012). $[CO_2]$ increased from 327.1 ppmv in 1970 to 384.6 ppmv in 2007, a 17.6% increase (at the La Jolla, CA, pier) (Keeling and Whorf 2008; Cook 2011; Blasing 2010). The University Corporation for Atmospheric Research (UCAR) projected that if anthropogenic CO_2 emissions continue unabated in their present trajectory, the average atmospheric $[CO_2]$ may reach 750 ppmv by 2100 if unchecked (UCAR 2009). This will be 2.68 times the preindustrial level.

The global total mass of oxidized carbon put into the atmosphere from FF burning alone was estimated to be 8.75 billion metric tons in 2008 or approximately 1.30 tonnes per capita (CDIAC 2011c). CDIAC stated that in 2010, total global oxidized carbon emissions from FF combustion and cement manufacture had risen to 9.14 billion metric tons. Global CO_2 emissions reached about 33.4 Gt in 2010, and if land use changes and deforestation are included, 36.8 Gt CO_2 was released in 2010 (Global Carbon Emissions 2010). (To convert CO_2 mass to carbon mass, multiply by 0.2727.) All available graphs and tables of global, yearly, oxidized carbon and CO_2 emissions show a positive first derivative. If nothing is done to reduce anthropogenic CO_2 atmospheric emissions worldwide, we may indeed have "the tragedy of the commons" (Hardin 1968).

Besides its putative role as a causative agent for global warming, an elevated atmospheric CO_2 level also causes another insidious, very serious, slow-acting problem for human sustainability, that is, acidification of the oceans, lakes, rivers, and aquifers. The slow drop in pH due to carbonic acid formation from dissolved CO_2 has been shown to slow down the growth rate of zooplanktons and phytoplanktons, reducing available food needed to maintain normal populations of fish, cetaceans, crustaceans, and other marine and aquatic organisms. The more acid water plus increased sea surface temperatures are blamed for contributing to coral reef die-offs (coral bleaching) that destroy reef ecosystems, in particular, the loss of *Symbiodinium*, a group of important endosymbiotic dinoflagellates that reside in the endoderms of corals, sea anemones, and jellyfish, where they provide their hosts with the metabolic products of their photosynthesis and, in return, receive organic nutrients (e.g., CO_2, NH_4^+) from their hosts (Doak 2012).

The oceans can buffer the dissolved CO_2 in the form of relatively insoluble metallo-bicarbonates (calcium, magnesium, silicon, iron); however, these chemical processes occur too slowly to keep up with the rise in atmospheric $[CO_2]$ over the past 20 years. More acid oceans are a significant threat to human (and marine ecosystem) long-term sustainability.

Finally, to put our preoccupation with carbon dioxide in perspective, it is useful to consider the global carbon cycle (GCC) that approximates the stored carbon mass in various graph nodes in gigatonnes of carbon and describes the carbon mass fluxes between certain nodes in gigatonnes of carbon per year. The carbon mass in the atmosphere is largely in the form of CO_2 gas (398 ppmv). The GHG methane (CH_4) accounts for only approximately 0.46% of atmospheric carbon (at about 1.8 ppmv concentration) (Sussmann et al. 2012; Webster et al. 2012; CDIAC 2011b; Blasing 2012). The total (natural plus anthropogenic) methane flux into the atmosphere is approximately 550 million tonnes/year (cf. Table 4.7). Note that each gigatonne of methane has $(12/16) \times 1 = 0.75$ GtC. Methane has a relatively short atmospheric time constant of approximately 12 years and a global warming potential of 23 (compared to 1 for CO_2 and 22,800 for sulfur hexafluoride; SF_6 has an atmospheric time constant of 3200 years) (CDIAC 2011c).

There are several good Internet references that describe GCC systems: Riebeek (2011), Globe (2011), and Global 2011. In addition, see the GCC diagrams at http://cdiac.ornl.gov/pns/graphics/c_cycle.htm and http://www.learner.org/courses/envisci/visual/img_lrg/global_carbon_cycle.jpg (both sites accessed June 30, 2011). One thing that emerges from inspection of these GCC diagrams is that the numbers for stored carbon (in whatever form) and the carbon fluxes (mass transfer rates) vary between diagrams yet are fairly consistent in scale. In Figure 4.34, we have blended the data from the four sources cited above in a GCC diagram in nodal form. The graph nodes represent compartments holding carbon; the branches describe the directional fluxes of carbon in gigatonnes of carbon per year between connected node pairs. There are five significant sources of atmospheric carbon: anthropomorphic FF combustion and forest clearing, animal respiration, volcanoes, erosion and weathering of carbonate rocks, and decomposition of dead vegetation by microorganisms.

Atmospheric carbon is largely in the form of CO_2; however, there are also some organic gasses such as methane, and other organic pollutant gasses are found at the parts-per-trillion-by-volume (pptv) level (CDIAC 2011a). In our GCC diagram, we have added a new node for terrestrial animals (man, cattle, sheep, pigs, horses, goats, dogs, cats, rodents, cervids, insects, etc.). While the CO_2 they exhale can be considered to be green (short-cycle) CO_2, it is a significant input to the atmosphere. The anthropogenic input from burning FFs and cement production is seen to be a fraction of the atmospheric input from other diffuse sources, including decay of dead terrestrial vegetation and exchange with the oceans. Note that 1 ppmv of CO_2 in the atmosphere translates into 2.13 Gt of carbon, and one source claims that approximately

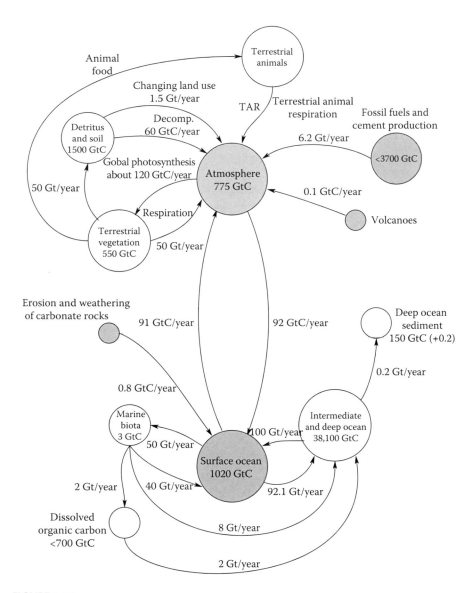

FIGURE 4.34
Planetary carbon cycle. Carbon in the atmosphere is mostly in CO_2 gas and some methane. On land, it is locked in plants, animals, detritus, soil, carbonate rocks, and FFs. In the oceans, it is dissolved, locked in marine biota, and trapped in sediments and in carbonate rocks. We do not have a good estimate of the contribution of total (living) animal respiration (TAR) (terrestrial and oceanic animal CO_2) to the atmosphere. The UN FAO in 2006 estimated the total *equivalent* (includes CH_4) CO_2 released globally by raising livestock (cattle, buffalo, sheep, goats, camels, horses, pigs, and poultry) was 7.516 Gt/year, or approximately 18% of annual worldwide GHG. emissions. However, in a recent *World Watch* paper, Goodland and Anhang (2009) arrived at the robust figure of 32.564 $GtCO_2$eq/year, or over 51% of the worldwide total! Obviously, a very significant amount of GHGs is released through TAR; the preceding figures were based only on farm animals. Wild animals must also be included.

14% of the atmospheric carbon can be attributed to all (cumulative) anthropogenic FF combustion (CDIAC 2011a).

The *Economist* Intelligence Unit (Briefing 2011) estimated that *worldwide,* there were 62.3 million passenger car registrations in 2011, rising to approximately 86.0 million registrations in 2015. Also, there were 20.6 million commercial vehicle registrations in 2011, rising to 30.6 million in 2015. Their calculated world petrol consumption in 2011 is 922.7 million tonnes (9.227×10^{11} kg), rising to approximately 1007.6 million tonnes (1.0076×10^{12} kg) in 2015. Using 0.75 as the specific gravity of petrol (it ranges from 0.72 to 0.76), 1.23×10^{12} L will be burned in 2011, and 1.34×10^{12} L will be burned in 2015. It can be shown that the combustion of 1 L of petrol produces approximately 2.32 kg of CO_2; the combustion of 1 L of diesel fuel puts 2.69 kg of CO_2 into the atmosphere (EPA 2005). Thus, worldwide petrol combustion in 2011 can put approximately 2.85×10^{12} kg CO_2 = 2.85 Gt CO_2 into the atmosphere, and petrol combustion in 2015 will put approximately 3.11×10^{12} kg CO_2 into the air. Since carbon fluxes are used in describing the GCC, we can convert kilograms of CO_2 into kilograms of carbon by multiplying by (12/44) = 0.2727, the mass fraction of elemental carbon in carbon dioxide. Thus global total petrol combustion is predicted to put approximately 0.78 Gt oxidized carbon into the atmosphere in 2011 and 0.85 Gt carbon into the atmosphere in 2015 (Briefing 2011). This is about 1/10 the carbon from *all* terrestrial animals' respiration, however. The predicted motor gasoline energy consumption in the US in 2015 is ca. 16.46 quadrillion Btu (1.74E+19 joules) (EIA 2012).

The Economist (2007) illustrated daily petrol consumption by country in 2003. The United States beat all other countries combined with approximately 1.35×10^9 L/day, or 4.93×10^{11} L/year. *All other countries* together consumed approximately 1.23×10^9 L/day, or 4.47×10^{11} L/year. Unlike CO_2 from point sources, petrol and diesel carbon emissions cannot be directly sequestered, nor can CO_2 from animal respiration.

There are two major natural sinks for atmospheric CO_2. These are direct uptake by plants and phytoplankton for photosynthesis and the physical/chemical uptake by the hydrosphere. In the latter case, there are four relevant reactions:

1. CO_2 gas into aqueous solution:

$$CO_2 \text{ (g)(atmospheric)} \leftrightarrow CO_2 \text{ (dissolved in water)} \qquad (4.59)$$

2. Conversion to carbonic acid:

$$CO_2 \text{ (dissolved)} + H_2O \leftrightarrow H_2CO_3 \qquad (4.60)$$

3. First ionization:

$$H_2CO_3 \leftrightarrow H^+ + HCO_3^- \text{ (bicarbonate ion)} \qquad (4.61)$$

4. Second ionization:

$$HCO_3^- \leftrightarrow H^+ + CO_3^{2-} \text{ (carbonate ion)} \qquad (4.62)$$

Note that the H^+ ions from reactions 3 and 4 above lower the oceans' pH (acidify it). The carbon dioxide storage sensitivity of the ocean is 0.1%/%; that is, a 10% increase in atmospheric CO_2 causes an approximately 1% increase in total ocean storage of carbon as H_2CO_3, HCO_3^-, plus CO_3^{2-}. The bicarbonate and carbonate ions are taken up by plankton, mollusks, and crustaceans to form shells, and they slowly react chemically with dissolved minerals.

At present, human sequestration of atmospheric CO_2 from diffuse sources is under consideration. The worldwide anthropogenic CO_2 emitted by motor vehicles is estimated to be approximately 2.85×10^{12} kg/year (see above). This vehicle CO_2, and that from aircraft and forest burning, cannot be captured at the sources; it must be scrubbed directly from the air if the CO_2 from these sources is to be sequestered. Some problems to ponder: If this is done, who will do it? How much atmospheric CO_2 should be sequestered per year? Do we stop atmospheric CO_2 "scrubbing" when the tropospheric $[CO_2]$ reaches a certain value, or when approximately 2.85×10^{12} kg CO_2/year is sequestered? More importantly, who pays for the atmospheric carbon capture and storage (CCS)? (There are multiple sources in multiple countries.) Scrubbing atmospheric CO_2 appears to be a Sisyphean task. Taxing citizens to pay for atmospheric decarbonization will not be popular; there may be only a slow mitigation of global warming, if any, to reward the process, its expenses, and its unexpected consequences. The companies developing and operating the scrubbing systems will make money, however.

One hears about how "green" EVs are. Yes, they remove the CO_2 and other air pollutants from areas of dense use (cities), but the electricity they consume must come from renewable sources (wind, PV, solar thermal, tidal, etc.) for them to be truly green. The majority of US electric power comes from the combustion of coal today (2012), and it has been calculated that the production of 1 kWh of electric energy releases about 0.454 kg CO_2 gas (CDIAC 2011). Thus, if an EV charges at a rate of 60 W for 24 h, approximately 1.5 kg CO_2 will be released into the air at the coal-burning power plant per day. Multiply this by the number of "green" EVs on the road now and in the future. There is no "free" energy.

Atmospheric methane is approximately 21–23 times as effective a GHG as CO_2; the tropospheric CH_4 concentration is approximately 1.8 ppm in the northern hemisphere. Nitrous oxide (N_2O), also a GHG, is 298 times as bad as CO_2; however, N_2O is found at only approximately 0.32 ppm in the troposphere (Blasing 2010).

4.9.2 Carbon Dioxide Capture from Point Sources

Clearly, reducing all anthropogenic CO_2 emissions to zero has the potential to significantly reduce the rate of atmospheric $[CO_2]$ growth, mitigating the effects of CO_2 on driving global warming and ocean acidification. It will also be expensive and add to the cost of energy (Spath and Mann 2004).

Benson and Cole (2008) commented that if CO_2 sequestration and storage (CCS) is implemented on the scale needed to make noticeable reductions in atmospheric $[CO_2]$, a billion tonnes (10^{12} kg) or more must be sequestered annually. This is a 250-fold increase over the amount of CO_2 actually sequestered in 2008. CO_2 sequestration requires significant energy, the cost of which will be added to the commodity whose production gave rise to the CO_2 (electricity from FFs, cement, steel, electricity from burning wood, pellets, etc.). A large coal-fired power plant emits approximately 8 million tonnes of CO_2 annually. Benson and Cole (2008) calculated that if the yearly CO_2 output of the power plant is stored as a dense, supercritical fluid, a volume of about 10^7 m^3 would be required. Thus, over a 50-year lifetime of the plant, a total volume of 5×10^8 m^3 is needed (0.5 km^3), just for one power plant! And what about the CO_2 emitted by motor vehicles burning gasoline, diesel, ethanol, DME, and so forth? It is not practical or cost effective to capture individual vehicle CO_2 emissions, yet as a class, they are very significant carbon sources. Catalytic mufflers only oxidize carbon monoxide and the remaining fuel vapor in the exhaust to CO_2.

CCS is best done at major point sources of CO_2 emission, that is, the stacks of power plants, steel mills, cement plants, and fermentation plants. Once collected, there are several means of carbon storage, described in Section 4.10.4. They can be categorized as physical, chemical, and biological.

Two possible alternative geological storage options considered for CO_2 are the use of high pressure to pump it underground to enhance oil and gas recovery from depleted fields and in enhanced coal bed methane recovery. In the former case, this is called enhanced oil recovery (EOR). EOR is planned for the Eagle Ford oil-rich shale deposit located southeast of San Antonio, TX. EOR using high-pressure CO_2 injection is expected to produce 800,000 barrels per day from the played-out Eagle Ford shale oil field. Preliminary EOR results were so encouraging that Denbury Resources is building a 320 mi., $825 million pipeline to carry CO_2 from combustion sites in Louisiana to a site south of Houston, where it will be used in EOR. In the Texas Permian Basin oil deposit, EOR improved yields for Occidental Petroleum, KinderMorgan, and ExxonMobile (Muska 2011).

In these EOR means, it is assumed that a significant fraction of the CO_2 remains trapped underground and little comes out, contaminating the oil and/or methane or deep aquifers. Benson and Cole (2008) asked four key questions about underground CO_2 storage efficacy: (1) "Will geological storage reservoirs leak?" (2) "If leakage occurs, what are the health, safety, and environmental risks?" (3) "Can leakage be predicted, detected, and quantified?" (4) "What can be done to stop or slow a leak, should it occur, and how much would it cost?" Note that CO_2 is a toxic gas to mammals in atmospheric concentrations of over approximately 10%. (In 1986, a massive natural CO_2 gas release from the bottom of Lake Nyos in Cameroon, Africa, asphyxiated approximately 1700 people and killed many cattle.)

Chemical CO_2 sequestration is another technology under development. One example is the use of the biologically derived enzyme carbonic anhydrase (CA) to catalyze the conversion of dissolved CO_2 into bicarbonate ions (Ramanan et al. 2009). The well-known reaction is

$$CO_2 (g) + H_2O (xs) \xrightarrow{(CA)} HCO_3^- + H^+ \qquad (4.63)$$

A next step might be to add magnesium hydroxide solution. The reaction is

$$HCO_3^- + Mg^{++} + 2\,OH^- + H^+ \rightarrow MgCO_3(s) + 2\,H_2O \qquad (4.64)$$

The magnesium carbonate precipitate can be collected and used for a number of industrial applications. The CA enzyme (catalyst) can be harvested from GM bacteria; energy is required to produce the magnesium hydroxide, as well.

Another approach to the chemical sequestration of CO_2 gas was described by Huijgen (2007) and Lackner (2003). CO_2 gas is reacted with the crushed minerals wollastonite ($CaSiO_3$) and olivine (Mg_2SiO_4); no enzymes are required. The exothermic reactions are done in aqueous media. The heat can be harvested:

$$CaSiO_3\ (s) + CO_2\ (g) \rightarrow CaCO_3\ (s) + SiO_2\ (s)\ (\Delta Hr = -87\ kJ/mol) \quad (4.65)$$

$$Mg_2SiO_4\ (s) + 2\ CO_2\ (g) \rightarrow 2\ MgCO_3\ (s) + SiO_2\ (s)\ (\Delta Hr = -90\ kJ/mol)$$
$$(4.66)$$

All three reaction products have industrial uses. The reactions are ex situ, that is, done above ground. The product carbonates are very insoluble and easy to isolate. The energy costs of the processes were considered in the Huijgen thesis; the main energy costs were for excavating, transporting, and grinding the mineral feedstock to particle sizes <100 μm and compressing the CO_2 gas. The magnesium and calcium carbonates can be used for plasterboard manufacturing, road construction, grading, and so forth. Certain industrial wastes can also be used for carbonization: slags from coal, steel, and blast furnaces; fly ash; cement and concrete construction and demolition waste; and even mine tailings can be effective sequesters (Huijgen 2007).

Another approach to point-source CO_2 sequestration that has been developed by ATMI Inc. of Danbury, CT, uses a proprietary BrightBlack carbon microbead technology to physically adsorb CO_2 gas. The ATMI process demonstrates capture efficiencies greater than 90%, with CO_2 purities as high as 99%. The ATMI process can be used at coal- or NG-fueled power plants (ATMI 2012). Presumably, the adsorbed CO_2 gas can be released from the carbon microbeads, and the microbeads can be recycled and the CO_2 used industrially or be put into "permanent" geological storage.

Ideally, CO_2 should be catalytically converted to methanol, so the carbon can be recycled as a fuel, instead of the expensive, irreversible solution of hiding it away. The two reactions given in Section 4.2.4 illustrate the catalytic production of methanol:

$$CO_2 + 3H_2 \xrightarrow{\text{(Cat)}} CH_3OH + H_2O \qquad (4.67)$$

and

$$CO_2 + 2H_2O + \text{electrons} \xrightarrow{\text{(Cat)}} CO + 2H_2 + (3/2)O_2 \xrightarrow{\text{(Cat)}} CH_3OH \qquad (4.68)$$

The hydrogen in the first reaction must come from a green source such as the electrolytic decomposition of water using electricity from wind, solar, tidal, or other sources. The electrons in the electrochemical reduction of CO_2 in the second reaction must also come from a noncarbon source. It makes far more sense to burn the carbon in the captured CO_2 in the form of methanol, rather than hide it away. Both CO_2 sequestration and the production of methanol by the means described above cost money. Once the supercritical CO_2 is pumped underground, the chemical energy of its oxidized carbon is, for all practical purposes, lost.

Even if our worldwide CCS efforts are 100% effective in sequestering anthropogenic CO_2, will it have a significant effect in mitigating global warming? Using the data from Figure 4.34, there is a total carbon (CO_2) flux into the atmosphere, exclusive of anthropogenic sources, of approximately (0.1 + 91 + 50 + 60 + 1.5 + TAR) = 206.6 + TAR GtC/year. The CO_2 flux out of the atmosphere is approximately 92 + 120 = 210 GtC/year. The total anthropogenic carbon flux into the atmosphere is estimated to be approximately 6.2 GtC/year, and humans are estimated to add 0.69 GtC/year in the form of CO_2 from their breathing. If we add these numbers to the input flux, the net atmospheric carbon gain is approximately +3.49 GtC/year, which is 0.45%/year of the estimated atmospheric carbon mass (775 GtC). (We have ignored the carbon input flux from the other terrestrial animals in doing these calculations, e.g., the approximately 1.3 billion cattle in the world.)

Finally, consider the current status of CCS in China. In 2008, China became the world's largest GHG emitter. The emissions from its rapid industrialization and urbanization (including automobiles) will continue to grow with the possibility of doubling the current level of GHG emissions by 2020 in the absence of any intervention (Heinz Center 2010). According to an International Energy Agency (IEA) report cited by the Heinz Center report, if CCS were implemented globally, it could account for up to one-fifth of the 50% reduction in worldwide GHG emissions needed from the baseline emissions scenario by 2050. This carbon capture may not be enough to slow global warming enough to justify the large investment in carbon sequestration technology worldwide in the minds of people.

In December 2011, Canada, a signer of the 1997 Kyoto Protocol Agreement on reducing GHG emissions, announced it was pulling out of the Kyoto treaty. When Canada signed, it agreed to cut emissions to 6% below its 1990 emission levels by 2012. By 2009, its emissions were 17% above its 1990 levels. The then major GHG emitter, the United States, never signed the Kyoto treaty (Ljungren and Palmer 2011; Victor 2011). Evidently, for Canadians, not spending money now on point-source carbon reduction trumps what will have to be spent later (e.g., in 2050) to counteract the slow effects of global climate change, including sea level rise. This is another example of the "tragedy of the commons."

Another concern about CCS is the possibility of leakage of the sequestered CO_2 back into the atmosphere, possibly as the result of tectonic activity acting on underground CO_2 reservoirs, or CO_2 sourced from new volcanic activity. From Shenhua (2010), we see that in mid-2010, China had one operational pilot CCS plant. This plant was designed to capture and store 10^5 tonnes of CO_2 annually. It was installed at the Shenhua Coal Liquefaction and Chemical Co. in Ordos, Inner Mongolia, and opened in June 2009. The Shenhua coal liquefaction projects emit approximately 3 million tonnes of CO_2/year. The Lustig (2010) map of China's future CCS plants shows 3 CCS plants under construction, 10 plants preapproved, and 3 in the early planning stage.

So what will be the long-term effects on human sustainability of widespread sequestration of CO_2 from point-source CO_2 emitters? (1) We predict that the cost of electric power from FFs will rise significantly, as will the cost of cement and alcohol fuel. This should encourage energy conservation. The net economic effect will be inflationary. Gross domestic products (GDPs) will decrease. (2) There will be a decrement in the rate of increase in the atmospheric [CO_2]. Recall that anthropogenic point sources are a small fraction of the yearly CO_2 flux into the atmosphere. (The CO_2 flux from animal respiration, plant decay, and the millions of FF burning vehicles worldwide cannot easily be sequestered.) (3) Also, there will be little or no change in the trajectory of global warming—the process appears to be autocatalytic, once past its tipping point. No one can say exactly what the atmospheric [CO_2] tipping point for global warming is. Has it already been reached? (For example, melting tundra permafrost releases methane trapped in hydrates, and methane is about 21–23 times more effective a GHG than CO_2. Warmer oceans release more CO_2 gas and dissolve less atmospheric CO_2.) We can encourage the development of biological CO_2 sequestration means on land and sea by proliferating vegetation and marine algae. However, we know that marine phytoplankton blooms have untoward side effects: when an algae bloom dies, the bacteria feeding on the decomposing phytoplankton cells rob the water of O_2, depleting zooplankton populations needed for healthy ocean ecosystems; they also release the toxic gas H_2S, which poisons zooplanktons and fish. (4) If the United States and Western Europe do CCS, and developing, CO_2-emitting countries (e.g., Brazil, Russia, India, and China) consider CCS systems too expensive, we will have "plowed the waters." Note that the Heinz Center (2010) report stated: "Currently, however, CCS development is not a priority item on the country's [China's] technology advancement agenda. One reason is technological: there remains uncertainty about the safety and permanence of CO_2 storage. Another is economic, given that a full-scale CCS demonstration plant would require a significant initial investment of about $1 billion USD. And finally, pursuing CCS could even have negative environmental effects, not only because the technology itself involves some impacts, but also because it might slow or replace the development of renewable and clean energy sources." These concerns are valid universally, not just for China.

The care of the Earth's atmospheric "commons" *requires worldwide cooperation* if a CCS strategy is to work (even partially) and mitigate the "tragedy." The use of alternative energy sources and the increasing cost of FFs may have a greater effect in reducing the anthropogenic carbon flux into the atmosphere than a partial, expensive attempt at CCS in developed and developing politically diverse countries.

4.9.3 CO_2 Storage and Recycling

In one means of physical CO_2 storage, the concentrated gas is pumped into underground storage volumes under high pressure, generally as a supercritical fluid. Quoting Benson and Cole (2008), "Large sedimentary basins are best suited, because they have tremendous pore volume and connectivity and they are widely distributed. Vast formations of sedimentary rocks with various textures and compositions provide both the volume to sequester the CO_2 and the seals to trap it underground." Other repositories include depleted gas and oil reservoirs (see above) and saline aquifers. In fact, in 1996, the Norwegian company StatoilHydro began the first pilot CCS project in which a million metric tons per year of CO_2 was injected beneath the North Sea into a saline aquifer sand layer called the Utsira formation (Torp and Gale 2003). To monitor the injected CO_2, a separate project called the Saline Aquifer CO_2 Storage (SACS) project was established in 1998. The SACS project used sonic seismic imaging to investigate the geological effects of injection of supercritical CO_2 into the Utsira sand at the rate of 1 million tonnes/year. The Utsira formation is covered with gas impermeable caprock layers; it measures approximately 400 km (North–South) and between 50 and 100 km (East–West) and is approximately 30 km deep at its center. When

CO_2 is pumped under very high pressure down into saline or brackish aquifers, some of it dissolves, and some remains in the supercritical liquid state. The dissolved CO_2 gas reacts very slowly with noncarbonate-, calcium-, magnesium-, and iron-rich minerals, forming solid carbonates. In underground and aquifer CO_2 storage, the high-pressure pumping is not without energy cost, however.

Norway has recently completed and opened an experimental CCS plant at Mongstadt—a billion-dollar development jointly owned by the Norwegian government, Statoil, Shell, and Sasoil of South Africa (Economist 2012e). The Mongstadt CCS facility consists of two plants, with a total capacity of 80,000 tonnes of carbon per year. The CCS plants are connected to the exhaust stacks of a refinery and a nearby gas-fired power station. The CCS plants are designed so the operators can experimentally adjust the CO_2 input concentrations and also the composition of the carbon capture solutions of amines or ammonium carbonate. These chemicals react with the dissolved CO_2 gas to form soluble carbamates and bicarbonates. The remainder of the stack gas, now mainly N_2 and O_2, is exhausted into the atmosphere. The carbon-rich solution is now piped to a reactor where it is treated to release its carbon as pure CO_2 gas, which is then piped away for undersea disposal. The amines and ammonium carbonate are regenerated and recycled. The whole process is energy intensive and expensive to operate. Consider the overall cost/benefit ratio (operating cost vs. mitigation of global warming and its effects).

Quoting Benson and Cole (2008) again: "In a recent assessment of North American [CO_2 CCS] capacity, oil and gas reservoirs are estimated to be able to contain ~80 GtC, saline aquifers between 900 and 3300 GtC, and coal beds about 150 GtC, for a total of about 1160 to 3500 GtC. If these estimates are correct, there is sufficient capacity to sequester several hundreds of years of [point-source CO_2] emissions. Only time and experience will tell whether these estimates are correct."

Another proposed form of CO_2 disposal is in the oceans. By all indications, this method could be fraught with bad environmental consequences, challenging the sustainability of oceanic ecosystems. In the "dissolution method," high-pressure CO_2 gas is pumped from carrier ships or pipelines into seawater at depths between 1000 and 3000 m, forming an upward plume from which the CO_2 dissolves into seawater, forming carbonic acid, bicarbonate ions, carbonate ions, and hydrogen ions. Thus, the acidity of the local seawater increases (its pH drops), harming plankton and marine vertebrates that feed on them. The dissolved CO_2 is also toxic to fish and invertebrates on which larger fish feed. It is unknown how much of the unreacted gas reaches the surface and reenters the atmosphere. If the CO_2 gas is pumped deeper than 3000 m, "lake deposits" are formed, where the high-pressure CO_2 liquefies (denser than water) and forms a downward plume that may settle on the ocean bottom as a liquid CO_2 "lake." In the liquid phase, the CO_2 is expected to react very slowly with the water. In another "dump it in the ocean" scenario, calcium bicarbonate [$Ca(HCO_3)_2$] is made on land and then dumped in the ocean. With calcium bicarbonate, the pH would not drop as low, and it would enhance the retention of CO_2 in the ocean, but this method is more expensive than just pumping in the gas, and there will still be untoward environmental effects.

Dumping any chemical in large quantities into ocean ecosystems automatically invokes the law of unintended consequences; the ecological effects of oceanic CO_2 disposal on planktons, fish, invertebrates, and so forth must be studied before any active ocean disposal of CO_2 is implemented. It makes far more sense to trap CO_2 gas in geologically sound, underground, saline aquifers or make carbonates on land.

Recycling the CO_2 from combustion and fermentation is most likely to provide the most environmentally friendly and financially sustainable means of handling our globally growing $[CO_2]$. Recycling can turn the CO_2 into food, chemicals, and fuel (such as methanol). Combusting fuel, of course, produces more CO_2. Syngas, a mixture of hydrogen, carbon dioxide, and carbon monoxide ($CO + CO_2 + 7H_2$), can be catalytically reacted to form $2CH_3OH$ (two methanol molecules). As a fuel, methanol offers a number of advantages (cf. Section 4.2.5). Another way of recycling CO_2 is through enhanced photosynthesis. CO_2 and other GHGs from power plant smokestacks are fed into membranes in a bioreactor containing wastewater and GM algae, which, when exposed to sunlight, grow rapidly and make a lipid-rich biomass that can be harvested to make biodiesel, plastics, and animal feed. This Bio CCS Algal Synthesis is being investigated in three Australian coal-fired power plants: Tarong in Queensland, Eraring in New South Wales, and Loy Yang in Victoria. (A detailed description of algae-based biofuel applications and products may be found in the FAO 2010 paper.)

Solar thermal energy can be used to convert $CO_2 \rightarrow CO + \frac{1}{2} O_2$ at 2400°C. The well-known Fischer–Tropsch (F-T) process can then be used to catalytically convert the carbon monoxide into the various paraffin hydrocarbons ($C_n H_{(2n+2)}$), which can be used as fuels or industrial chemicals (Torres Galvis et al. 2012). Hydrogen gas is also a reactant used in the F-T process; it must come from electrolytic decomposition of water driven by solar PV current or current from WTs to be carbon-free H_2. (See the Glossary regarding the F-T process.)

4.9.4 Cost of CCS

It is generally true that the processes of CCS require energy and thus will incur expenses that will be passed on to the cost of the commodity being produced (e.g., electric power, cement, steel, ethanol, etc.). Just how much prices will rise depends, of course, on CCS method used, the commodity being produced, the cost of energy used in the CCS process, and the local rate of inflation.

The total cost of CCS (in US \$/tonne CO_2) has several components: The first component is the capital cost of the CC plant. The second component is the cost of operating the CC plant and collecting the harvested CO_2 gas. The third cost component is from compressing and liquefying the CO_2. The fourth cost component is from transportation of the CO_2 to the site where it will be stored. This may be by a gas pipeline or a truck, railroad tanker, or ship. The fifth component is the cost of pumping the gas or liquid CO_2 into its storage site. The storage site must be explored geologically for its potential not to leak gas and its estimated storage capacity, and appropriate high-pressure injection wells must be installed. A sixth cost is associated with continuously monitoring the storage site's integrity. The research report by Christie (2009) gives cost estimates considering the components of CCS. Christie's report gives a detailed overview of the status of CCS technologies and CCS projects in the United States, Europe, and Australia.

There have been many other economic studies to estimate the capital and operating costs of CCS methods on point-source CO_2 emitters. For example, Kolstad and Young (2010) published a cost analysis of CCS for brown coal–fired power plants in the Latrobe Valley, Victoria, in southeastern Australia. Evaluating a number of methods of CCS, they concluded that the cheapest option for retrofitting an existing brown coal–fired plant for CCS was the approach designed by the Calera Corp., Los Gatos, CA, in which the CO_2 is converted to calcium and magnesium carbonates, which have industrial use, or a sodium bicarbonate slurry that can be pumped underground. Their estimate for total real capital cost for transportation and storage using the Calera CCS system was approximately US $142 million, and the estimated real operating cost was US $1.8 million/year. The actual CCS retrofit capture plant was estimated to cost US $800 million on a 500 MW power plant. (2009 US $ was used in their estimates.)

Lindner et al. (2009) examined the economic and environmental factors associated with CCS in the city of Kiel, Germany, and attempted to answer the question of whether new FF power plants should include integrated CCS systems. Since Germany has gone nonnuclear, this is now a very important question because CCS would permit continued use of coal as fuel. They considered an 800 MW pulverized coal (PC)–fired power plant with postcombustion CCS, an 800 MW PC-fired power plant with oxyfuel technology, and an 800 MW integrated gasification combined cycle (IGCC) plant for their analyses. The conventional, 800 MW, PC-fired plant was derated to 709 MW to account for the energy taken by the CCS system. They found that the capital cost of a new CCS plant had a mean investment cost of 1040 million euro, transport costs of CO_2 were 1.2 million euro/year, and storage costs were 11 million euro/year. The other plants were similarly analyzed.

What we conclude is that CCS is not cheap; in fact, it can be an appreciable fraction of the value of the electricity produced from burning the FFs (about 12% in the Lindner study).

Finally, if essentially no action is taken on CCS, the anthropogenic global CO_2 emissions could reach 58 $GtCO_2$/year by 2050. Long-term global temperature rise in the do-nothing scenario may reach 6°C. If countries make good their pledges on point-source CCS, the global emissions could be 40 $GtCO_2$/year. Efficiency gains in FF uses and the increased use of NG could further reduce emissions to approximately 20 $GtCO_2$/year (Orcutt 2012).

4.9.5 CO_2 Capture from the Atmosphere

A more ambitious strategy for CCS is to capture CO_2 directly from the atmosphere. This strategy has its pros and cons, and no doubt some eventual unintended consequences. It has been proposed to mitigate the anthropogenic oxidized carbon emitted by burning non–point-source FFs, particularly the CO_2 from motor vehicles, trains, ships, and airplanes and also CO_2 from forest clearing. Presumably, the various means of sequestering atmospheric CO_2 would ideally collect yearly a CO_2 mass only equal to that calculated for the mass of petrol, diesel fuel, and wood consumed the previous year. Let us consider the numbers: The Earth's atmosphere has a total gas mass of approximately 5 petatonnes (5×10^{18} kg); only approximately 0.039% of that (390 ppm) is CO_2 gas, or 1.95×10^{15} kg or 1.95 teratonnes is CO_2. To achieve a global total of 100 ppm reduction in atmospheric [CO_2] to 290 ppm, one would have to capture a total of 50×10^{12} kg (50 gigatonnes) CO_2 from the air in a year.

In an important early paper, Herzog (2003) reviewed the technologies and did cost estimates of certain CO_2 "air scrubber" technologies. Herzog performed a detailed estimate of the complete CCS process cost (in 2003 US $) using a $CaCO_3$-based reaction set; this was $480/tonne of atmospheric carbon sequestered. Clearly, removing CO_2 gas from the atmosphere has potential merit; it will be expensive, however, and new technologies for economical atmospheric CO_2 sequestration will have to be developed. Do the math: $50 \times 10^9 \times \$480 = \$24 \times 10^{12} = \$24$ trillion. Who will pay to directly clean up the "atmospheric common"? The fuel producers? The consumers? The general public (taxpayers)? What countries? Be concerned; we predict it will never happen, given the world's present economic problems. In his book *Global Warming Gridlock,* David Victor (2011) discussed the international stalemate on CCS and recommended some bottom-up initiatives that should be implemented on national, regional, and global levels.

The most significant challenge to direct atmospheric CCS is the extremely low concentration of CO_2 in the air. One cost estimate for atmospheric CCS by the American Physical Society, cited by Rudolf (2011), is approximately $600/tonne of CO_2 sequestered, versus approximately $80/tonne for scrubbing CO_2 directly from the flue gas of a coal-fired power plant. Power plant flue gas is approximately 10% CO_2. K.S. Lackner, director of the Lenfest Center for Sustainable Energy at Columbia University's Earth Institute, has developed a prototype atmospheric CO_2 absorbing system that uses a plastic that absorbs CO_2 when dry and releases it when wet. Lackner predicts that his system can sequester atmospheric CO_2 for far less than $600/tonne. Another prototype system for capturing atmospheric CO_2 was developed by Stolaroff et al. (2008); it used a sodium hydroxide spray to form sodium carbonate. The Na_2CO_3 solution can be mixed with slaked lime [$Ca(OH)_2$] to precipitate out flakes of limestone ($CaCO_3$). The limestone flakes can be heated (~900°C) to release CO_2 gas for the syngas production of methanol, if desired. Lackner set up his own company, Global Research Technologies (GRT), LLC, which is trying to develop inexpensive CCS for atmospheric CO_2. GRT is developing atmospheric CCS using $NaHCO_3$ as the end product, a technology that uses significantly less energy (Van Noorden 2007). See more about GRT LLC and its ACCESS air-capture technology product at PRWeb (2010).

Recently, scientists at the ETH Zurich and the PSI in Switzerland described a prototype solar energy–powered means of atmospheric CO_2 CCS. They used trapped solar energy to drive the following fluidized reactions: $CO_2 + CaO \rightarrow Ca(OH)_2 \rightarrow CaCO_3$. Five consecutive 1.3 s cycles were used to completely remove 500 ppmv CO_2 from a synthetic airstream also containing 17% H_2O. No estimate of cost was given (Nikulshina et al. 2009).

Perhaps we should view the popular concern about CCS as a temporary measure to mitigate anthropogenic CO_2 while we learn to stop burning FFs and rely on renewable energy from wind, sun (solar thermal, solar PV), water (hydro, tides, waves), hydrogen FCs, and so forth. As a bridging technology, CCS can carry the global energy producers into the mid-21st century until they morph into providing low-carbon-emission systems. The use of methane, methanol, DME, biodiesel, and ethanol as vehicle fuels will produce CO_2 but less than that from burning coal and FF oil derivatives. Fortunately, some of these fuels can be synthesized from CO_2. Most of this CO_2 will, by then, have a green history—that is, be from carbon in recently grown plants. We Earthlings must learn to live using less energy per capita.

4.10 Water Vapor

4.10.1 Introduction

Over 70% of the Earth's surface is covered by water. The total water volume on Earth [in the atmosphere (as vapor, clouds), rivers, lakes, oceans, groundwater, glaciers, ice caps, etc.] is approximately 1.37×10^9 km^3. Over 97% of this total is stored in the oceans as salt water, or approximately 1.34×10^9 km^3 (Elert 2008). Expect the ocean volume to rise in the future due to melting glaciers and ice caps.

The water cycle (cf. Figure 3.7) describes the "compartments" (aka nodes) and fluxes of H_2O between them. WV enters the atmosphere from the solar-induced evaporation of oceans, lakes, rivers, and wetlands, and a small amount by sublimation of ice at high altitudes. Another source of atmospheric WV is evapotranspiration from all growing plants. Water moves through the atmosphere in vapor, liquid, or solid states (advection) propelled by air currents and atmospheric pressure gradients. WV precipitates as rain, snow, sleet, or graupel, on land and seas. Liquid water can enter groundwater storage (aquifers), and surface runoff from streams and rivers can return to the oceans or lakes. Snowmelt is another significant source of freshwater for aquifers, streams, rivers, and lakes. Figure 4.35 illustrates the

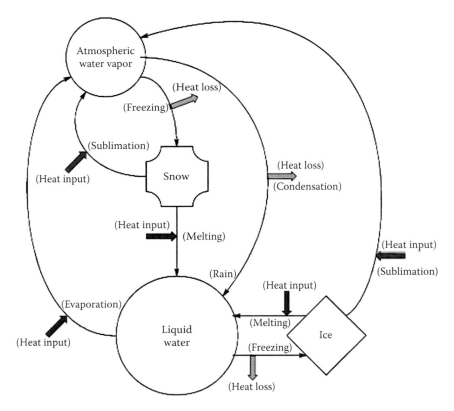

FIGURE 4.35
Diagram illustrating heat flows when water changes state. These heat flows are important in calculations of planetary warming.

heat flows associated with water phase changes, that is, atmospheric water vapor (AWV), liquid water, ice, and snow, at atmospheric pressure. Note that both snow and ice can sublime, that is, go directly to AWV if the temperature is cold enough. The figure does not show sources and sinks, but only phase-change pathways. The heat of fusion of ice melting to liquid water is 80 cal/g at 0°C and vice versa; the heat of vaporization to change water to vapor (steam) is 539 cal/g at 100°C.

The atmosphere above the 48 contiguous US states stores about 152 km³/day of WV. A little over 10% of it, or 16.25 km³, falls as precipitation each day. About 5.96 × 10³ km³ of precipitation falls on the continental United States each year (Corps of Engineers 2011).

The Earth's atmospheric WV ranges from traces to approximately 4% by volume; it generally cannot exceed the 4% level because at environmental temperatures, it condenses into rain, snow, fog, or clouds (the latter two are characterized by micron-sized water droplets or ice crystals). Clouds and fog form when WV-containing air is cooled to a temperature below its dew point. Generally, condensation of WV begins on condensation nuclei such as dust, salt, and ice crystals. WV is different from CO_2 in that it is constantly being created naturally by solar vaporization of water surfaces (oceans, lakes) and continually falling as precipitation on land and sea. The higher the surface air temperature becomes, the more water evaporates naturally and enters the atmosphere, and the more precipitation can occur from the more WV-saturated air. Growing green plants also contribute to the atmospheric WV by the process of transpiration. This transpired WV is from excess water the plant sucks up via its roots that is not used in photosynthesis. For example, an acre of corn transpires WV equivalent to 11,400–15,100 L/day, and a large oak tree can transpire approximately 151,000 L/day (USGS 2011).

There are also anthropogenic sources of atmospheric WV. We have all seen jet contrails, frozen WV from jet exhaust in the stratosphere. All fuel combustion, for that matter, produces WV, as well as CO_2 and other exhaust gasses. Spray irrigation in agriculture on hot, sunny days may be good for the plants, but it also puts WV into the air, as well as water into the soil.

WV is just that: the gas phase of water. The physical behavior of pure water can be best explained by consulting the simplified pressure–temperature phase diagram for water in Figure 4.36. (The diagram is not to scale.) To the left of lines OTH in Figure 4.36, water is a solid (ice); to the right of lines OTC, water is in the vapor phase; between lines CTH, water is liquid. Point **T** is

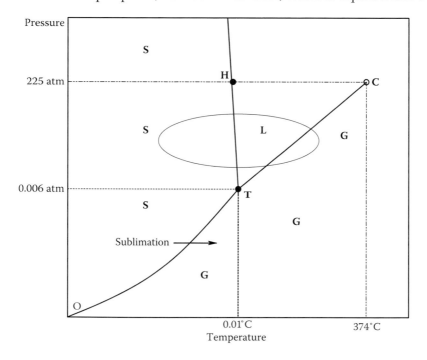

FIGURE 4.36
T-P phase diagram for water (not to scale). **T** is the TP; **C** is the CP. See text for discussion.

the primary TP of water where ice, water, and WV are all in equilibrium (the TP of H_2O is at 0.0098°C and 62.27 Pa = 6.03 × 10⁻³ atm pressure; NB: 1 Pa = 1 kg/m² = 9.68 × 10⁻⁵ atm). (A molecule's TP on a P-T phase diagram is the value of pressure and temperature at which three phases are in thermodynamic equilibrium together.) Interestingly, besides its primary TP, water has 8 more TPs that occur at very high pressures. They demarcate the equilibria between liquid water and many different forms of ice (Ih, Ic, III, II, V, VI, VII, VIII, etc.). (We have not shown these other TPs in the figure because they occur at pressures above 200 MPa and are not relevant in climate studies.)

Point **C** is the CP of water (at 374.1°C and 2.26 MPa = 218.3 atm pressure). At the CP of water, it is impossible to condense vapor into liquid just by increasing pressure; one gets only a highly compressed gas; the molecules have too much energy for the intermolecular attractions to hold them together as a liquid (Clark 2004). Below the TP pressure, P_t, ice Ih sublimates, that is, passes directly from solid to vapor phase if the temperature crosses above the OT line segment boundary in Figure 4.36. Conversely, at pressures below 6.03 × 10⁻³ atm, ice can form directly from WV if the temperature falls to the left of line OT. The heat of vaporization/condensation of pure water is 2.27 MJ/kg.

The mean residence time of a WV molecule in the troposphere is only approximately 9 to 10 days before it gives up its heat of vaporization and changes to liquid water (clouds, fog, rain) or freezes to ice clouds or snow (Pidwirny 2011). Note that these are phase changes. CO_2 gas molecules have much longer lifetimes and can leave the atmosphere by dissolving in water, combining chemically with certain minerals, or being taken up by growing green plants. Unlike WV, CO_2 undergoes no phase changes in the Earth's atmosphere.

4.10.2 WV as GHG

WV absorbs LIR radiation in several IR bands, shown in the bottom graph of "major components" in Figure 4.37. (The absorption spectra of other IR energy–absorbing gasses are shown below.) The atomic bonds of the constituent atoms of the molecules of a substance (in air) absorb photon energy in select wavelength bands, producing a unique absorption spectrum, showing where the incident photon energy is absorbed by the molecular structure of WV.

If a beam of monochromatic light (photons) of wavelength λ having power $P_{in}(\lambda)$ is passed through a sample substance, the emerging ray will have, in general, power $P_{out}(\lambda)$. The *transmittance* of the substance at wavelength λ is defined as $T(\lambda) \equiv P_{out}/P_{in}$, $0 \leq T \leq 1.0$. Also used to describe the selective molecular absorption of photons is the *absorbance*, $A(\lambda) \equiv -\log_{10}[T(\lambda)]$. The absorbance is also known as the *optical density* (OD). Graphs of $T(\lambda)$ and $A(\lambda)$ are also often plotted versus *wavenumber*, ν (the number of waves per centimeter) in units per centimeter; $\nu = k/\lambda$ in units per centimeter, *where* λ is in nanometers, and $k = 10^7$ (the ratio of centimeters to nanometers). The optical

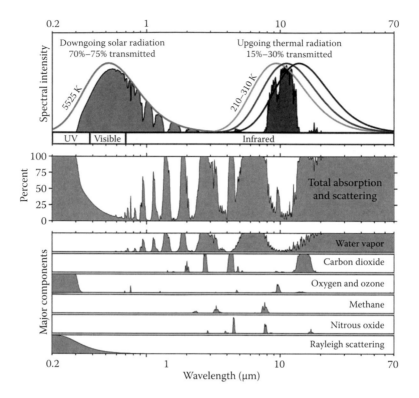

FIGURE 4.37
Radiation transmitted by the atmosphere. (Top graph) Left: incoming solar radiation at the Earth's surface, less absorbed and reflected spectral intensity by atmospheric gasses. Right: reradiated LIR spectral intensity. BB curves for 210–310 K are also shown for reference. (Middle graph) The total atmospheric percent absorption and scattering from atmospheric GHGs is shown. (Bottom graphs) IR absorption spectra of WV, CO_2, O_2 and O_3, CH_4, N_2O, and Rayleigh scattering at short wavelengths. (Courtesy of Rohde, R.A., Wikimedia Commons. http://en.wikipedia.org/wiki/File:Atmospheric_Transmission.png.)

path length, concentration, pressure, and temperature of the substance are noted in taking spectrophotometric measurements.

Note that WV absorbs photon energy strongly in eight major spectral regions, approximately as follows: 0.85, 1.1–1.15, 1.3–1.5, 1.77–2.0, 2.45–2.9, 4.0–4.6, 4.8–8.0, 9.3–10.1, and beyond $\lambda = 13$ μm (Geballe 2011). This leaves "windows" in between some of the WV absorption bands through which the Earth's surface LIR heat radiation can radiate into space, if not blocked by absorption spectra of CO_2 and methane.

The distribution of WV in the Earth's atmosphere is not homogeneous; it is lower over hot deserts and cold polar regions and higher over the equatorial oceans and warm regions of high rainfall. It is carried by moving air currents. More than 50% of all the atmospheric WV is found at altitudes less than 2 km. Some measures of atmospheric WV are as follows: The absolute humidity is the mass of WV in the air per unit volume, and the (%) relative humidity = 100 × (partial pressure of WV)/(WV pressure at same temperature). The temperature at which WV in a given sample of air becomes saturated (condenses) is called the *dew point*. The dew point can be used to find the relative humidity. For example: a 20°C sample of air is in contact with a shiny metal surface that is slowly cooled. When the metal reaches 10°C, dew (water droplets) forms on it. The condensation is detected optically. This means that the WV in the air is saturated at 10°C, and from tables, we find that its partial pressure is 8.94 mm Hg, equal to the WV vapor pressure at 10°C. The pressure required for saturation at 20°C is 17.5 mm Hg. The relative humidity is thus 100 × 8.94/17.5 = 51%.

WV in the atmosphere can absorb and reradiate LIR heat energy. In the form of clouds, condensed WV can reflect solar radiation back into space. About 25% of the solar energy incident on the Earth is reflected back into space by clouds. This has a net cooling effect on the Earth. Clouds also absorb approximately 25% of solar radiation. This energy is reradiated into space as LIR and also back to the Earth's surface as LIR. WV can be considered to form a positive-feedback loop with CO_2-caused warming: as the average surface temperature rises from CO_2-induced warming, more water evaporation occurs, putting more WV into the atmosphere; the WV acts as a GHG, further warming the air and causing more evaporation. This positive-feedback loop amplifies the warming effect of CO_2 alone; the WV and CO_2 act synergistically. But as we have seen, the amount of WV in the atmosphere is self-limiting because of precipitation (phase-change) triggers. When WV condenses as rain, it releases the heat of vaporization that produced it, 2.27 MJ/kg. As this rain evaporates, it again takes up the 2.27 MJ/kg, cooling its surroundings. Again, we stress that atmospheric CO_2 does not enjoy these rapid phase changes and heat fluxes.

It is clear that as global warming occurs, it will cause the evaporation of more planetary surface water and, in particular, will increase the atmospheric WV concentration in equatorial regions. Because of the WV saturation phenomenon, WV concentration in air will be limited to around 4% by the vapor saturation phenomenon, regardless of global warming. Lack of rainfall can occur if the upper air temperature remains above that required to condense the WV concentration in it. That is, the upper atmosphere cannot absorb the WV's heat of vaporization.

One view of the global warming scenario is that small anthropogenic increases in atmospheric CO_2 have a measurable effect on the ability of Earth to radiate heat (LIR) out into space. This is because the natural [CO_2] is low compared to that of WV, and the CO_2 *LIR absorption is not yet saturated* in its four major wavelength bands. Because of the very much larger concentrations of WV in the atmosphere, its seven major absorption bands are generally saturated; little LIR energy can get through them back into space. Thus, any additional CO_2 we release directly contributes to the "darkening" of the CO_2 windows in the total atmospheric WV absorption spectrum, decreasing heat loss from the planet,

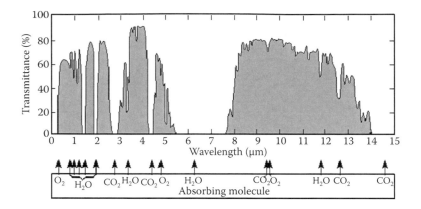

FIGURE 4.38
Percent atmospheric transmittance of the troposphere. Note the large role of the ubiquitous WV in blocking photon transmission between 5.5 and 7.5 μm. (Courtesy of US Navy, *EW and Radar Systems Engineering Handbook*, Naval Air Warfare Center Weapons Division, Point Mugu, CA, 2000. http://www.nawcwpns.navy.mil/r1/ElecWar.htm.)

in particular, the 2.3–3.3 and 4–4.6 μm bands. Thus, the WV absorption "windows" effectively being "closed" effectively amplifies the effect of atmospheric CO_2 in restricting outward LIR radiation, increasing CO_2's effect as a GHG.

Figure 4.38 illustrates an LIR percent *transmittance spectrum* of "typical" Earth's atmosphere, showing LIR radiation "windows" that can be "closed" by atmospheric CO_2 (from any source), particularly at 2.3–3.3 and 4–4.4 μm. WV's big, "closed windows" are from approximately 5–8 and >25 μm. The actual amount of IR spectral absorption in general depends on the partial pressures of the WV and CO_2 gasses, their temperatures, and the IR path length.

4.11 Engineering Energy Efficiency

A significant way to reduce anthropogenic carbon emissions is to burn less FF. This, in turn, requires the greater use of noncarbon energy sources as well as less total consumption of energy per capita. There are many ways we can accomplish this reduction; many appear costly. These include but are not limited to the following:

1. Design new power generation and distribution systems to "be smart" and have lower losses. Design and install higher-efficiency turbines and alternators in new power stations. Use more large energy storage systems (CAES, pumped hydro) to buffer grid loads.

2. Superinsulate our homes and business buildings. New dwellings should also use passive solar designs where possible and strive for compactness (lower volumes to heat or cool).

3. Drive more fuel-efficient vehicles. (The rising prices of FFs will provide the motivation here.) There may be 2 billion automobiles on the planet by 2050. It was estimated that approximately $1 billion is lost per year at present to traffic congestion in the United States (time and fuel). The use of hybrid electric/gasoline automobiles that use dynamic braking to partially charge their batteries is one option. Avoid using all-electric vehicles (EVs) unless their electric grid recharging source is from renewable energy, not FFs. If you alone commute, do not do it in an SUV or a pickup truck.

4. Alleviate traffic congestion by designing "smart" throughway vehicles that can travel safely at high speeds with closer spacing. An antiplatooning algorithm will monitor the speeds, spacing, and absolute positions of nearby vehicles and radio these data to your vehicle's onboard traffic control computer, enabling it to make decisions to adjust its speed and spacing in the traffic queue. GPS data would be used in this process as well. When it is time to leave the throughway, your vehicle's "autopilot" would signal surrounding vehicles, allowing appropriate speed and spacing adjustments to be made. Once off the throughway, you would have manual control of your vehicle. Switching to manual control while on the computer-regulated throughway would be right out except in emergencies.

5. Invest in more public transportation (high-speed trains, buses). It may be cheaper to upgrade than to build new systems.

6. Use greener fuels in our internal combustion vehicles (E85, Dme, Biodiesel, biogenic methane from landfills, etc.).

7. Use more green and low-carbon fuels in electric power generation. These include wood pellets, other biomass (e.g., *Miscanthus*, switchgrass), biogenic methane, biochar, hydrogen from electrolysis of water using electricity generated from renewable sources, and so forth. Power from GT, solar, wind, and water sources is preferred.

8. We should set our thermostats lower in the winter ($\leq 68°F$) and higher in the summer ($\geq 75°F$).

9. Use superinsulated refrigerators, freezers, hot water heaters, and hot water and steam piping.

10. Design more efficient electric motors and their controllers.

11. Replace tungsten-filament electric light bulbs with compact LED lamps and fluorescent bulbs. (The latter contain mercury, an environmental hazard.)

12. Turn off home lights and electronic systems when not in use.

13. Manufacturers should make use of the energy from low-grade heat lost from furnace stacks and steam condensers.

One use of low-grade heat is to power TEGs (TPs). (The TEG outputs are useful for running plant lighting and ventilation fans.) Another use of low-grade heat is to warm winter greenhouses to grow food.

A good example of how implementing energy efficiency literally saved billions for one company was told by Yergin (2011) in his Chapter 31: *The Fifth Fuel—Efficiency*. In 1995, Dow Chemical Corp, the largest US-based chemical company, paid approximately $30 billion for energy and feedstocks. Between 1995 and 2005, "... Dow reduced its energy use on a worldwide basis, per pound of product, by 25 percent. From Dow's point of view, it was more than worth the effort—$9 billion of savings from an investment of $1 billion."

We estimate that engineering 21st-century energy efficiency measures into new factories and FF-burning power plants could realize energy savings from about 15% to over 25%. Retrofitting older factories and power plants for energy efficiency will be more expensive, but significant savings can be realized here, too. Think about how much gasoline would be saved per year if all motor vehicles worldwide were mandated to average 50 mpg, instead of what they get now. The average new car sold in the United Kingdom gets 52.5 mpg (this would be 43.7 m/US gal., if British gallons were meant in the reference) (Economist 2011j). People with large families who justified owning an SUV for its passenger capacity may have to switch to a lighter, more fuel-efficient station wagon, such as we did before Detroit automakers sold the United States on the SUV image.

Many energy reductions in the private sector require personal reeducation about the basics of energy, power, and heat flow and also will need significant lifestyle changes; Items 3, 6, 8, 11, and 12 in the list above are zero- or low-cost; they require only personal behavior changes. Personal energy austerity will happen slowly, largely motivated by education and the energy consumer's pocketbook.

4.12 Chapter Summary

There are many areas where technology can mitigate our impacts on our sustainability. Such remediations all cost money. Money must come from private investors for activities by nongovernmental organizations (NGOs) and from tax dollars for government efforts. The will to spend this money must lie in an informed population who realizes that doing nothing will only make things worse: more inflation, fewer jobs, a higher rate of environmental degradation. However, because most of these challenges to our sustainability occur slowly, it will be hard to marshal public support for expensive remediations.

In 2009, 88% of our civilization's modern energies were derived from oil, coal, and NG, whose global market shares are now 35%, 29%, and 24%, respectively. Annual combustion of these fuels releases approximately 420 exajoules (420×10^{18} J) of energy. "In 2010, ethanol and biodiesel supplied only about 0.5 percent of the world's primary energy, wind generated about 2 percent of global electricity and PVs produced less than 0.05 percent" (Smil 2011a). These ratios necessarily must change. The changes will be driven by public opinion in reaction to the rising costs of all FFs, not an altruistic, save-the-planet motivation.

Energy conservation is part of the solution to our high FF consumption rate. Yes, the government can mandate corporate automotive fuel efficiency (CAFE) miles-per-gallon standards for motor vehicles, but the consumers will force this issue with their pocketbooks as fuel prices continue to grow. Again, our beloved SUVs and drive-around-town pickup trucks will be "put on blocks."

We think that electric utility companies and/or citizens should put "smart" repeater electric energy meters inside houses where the residents can see them every day (rather than in cellars or outside houses). In this digital age, it would be easy to have them display your watt-hours consumed since the last reading, as well as a running power consumption figure *and your cumulative (estimated) electric bill*. Another smart meter should also display your smoothed rate of heating oil or gas consumption, your total consumed, *and the bill estimate*. This information would encourage residents to turn off lights and to use low energy lighting, to air-dry laundry when possible, and to set their thermostats lower in the winter. Locate these energy meters under the home thermostat where people will see them every day. The empty pocketbook is a very strong motivator.

In summary, "As soon as increased demand and improved technology make renewables cheaper than fossil alternatives, the desire to generate energy through the burning of coal and oil will seem perverse, and the transition to a future beyond fossil fuels will become irreversible" (Morton 2011). That is, the tipping point for the green energy versus FF energy systems will be determined largely by parity economics, not by environmental concerns of an educated populace.

Smil (2011) clearly makes the point that implementing a comprehensive carbon capture and sequestration (CCS) system in the United States and the rest of the world would be prohibitively expensive and take tens of years to realize. The tax burden required to implement comprehensive CCS would severely slow economic growth that is slow already. While CCS looks good on paper, it is extremely impractical. First, the CO_2 must be isolated, compressed, liquefied, and stored; then shipped (tank cars or pipelines) to the injection well sites; and then pumped underground. All these steps are energy intensive and expensive. They would add to the already-high costs of extracting energy from FFs. Also, any gas-tight caverns and mines where CO_2 might be put might be better used for CAES to buffer wind and solar energy system outputs. Also, as we showed above, CO_2 gas can also be combined chemically with certain minerals to form stable carbonates and carbides. These processes, too, require energy and need to be developed to scale for world CCS. The best solution is to vigorously develop green energy systems, plus implement lots of conservation, and let Gaia do the work using the oceans, plants, phytoplankton, algae, and minerals.

In conclusion, there is universal agreement that the Earth's climate has been warming over the past 50 years, but not all scientists and atmospheric modelers agree on the cause being primarily anthropogenic CO_2 emission— generally, on the basis that it has been a small fraction of the total CO_2 flux into the atmosphere from natural sources. CCS will allow our continued, profligate consumption of FFs to continue (while they last) with a clearer conscience, assuming that anthropogenic CO_2 emissions are significant in driving global warming. CCS has many disadvantages: (1) It will increase the cost of energy, fueling inflation (pun intended). (2) It may cause delay in the development of alternate energy sources and energy conservation means. (3) It wastes the carbon: CO_2 can be a feedstock in catalyzed chemical processes that transform it into syngas, thence to methane, methanol, other alkanes, and DME (a good diesel fuel). Yes, burning these fuels will release CO_2 into the atmosphere, whence they came, but there is a net energy gain, especially if the H_2 in the syngas comes from electrolysis of water using green electricity and if solar thermal energy is used in the CO_2-to-fuel reactions. (4) Direct sequestration of CO_2 from the air will be expensive; no country will want to sequester someone else's CO_2 emissions (the global atmospheric commons again), so the cost of running atmospheric decarbonization plants, if mandated, will come from grudging taxpayers who will protest when they see no rapid improvement in the Earth's climate. The short-term profit for CCS will go to the chemical engineering firms that design, build, and operate the CCS plants and facilities, paid for by our tax dollars. (5) Even if all anthropogenic GHG emissions were to be stabilized tomorrow, global warming and sea-level rise would continue for centuries due to the time scales associated with climate dynamic processes and system feedbacks. Strong positive feedbacks are inherent in the global warming system: as polar ice and mountain glaciers melt, the Earth's surface albedo falls, and more LIR energy is absorbed, rather than reflected, accelerating melting. Also, as the arctic tundra thaws, clathrate methane will be released into the atmosphere; CH_4 is 21–23 times as effective a GHG as is CO_2, and thus, 1 tonne of CH_4 is equivalent to 21–23 tonnes CO_2 equivalent (CO_2eq) (IPCC 2007a).

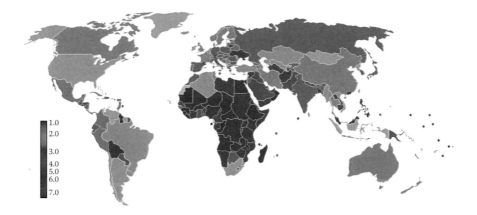

FIGURE 3.2
Map of fertility rates of countries of the world (2005–2010). (Courtesy of Supaman89, Wikimedia Commons. http://en.wikipedia.org/wiki/File:Countriesbyfertilityrate.svg.)

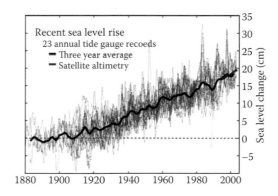

FIGURE 3.7
Recent global mean sea level rise (1880–2006). (Courtesy of Waldir, Wikimedia Commons. http://en.wikipedia.org/wiki/File:World_population_growth_(lin-log_scale).png.)

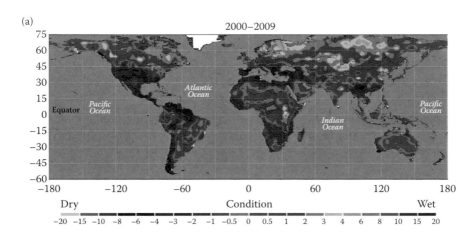

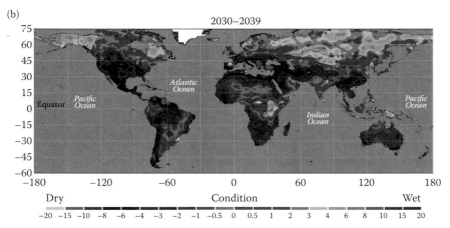

FIGURE 3.9
Color maps of computer-generated model drought predictions. (a) 2000–2009 (present: model verification). (b) 2030–2039. (c) 2060–2069. (d) 2090–2099. Note that the American Heartland and all of Europe surrounding the Mediterranean Sea are predicted to have severe drought. (From Dai, A., NCAR/CGD, University Corporation for Atmospheric Research. With permission.)

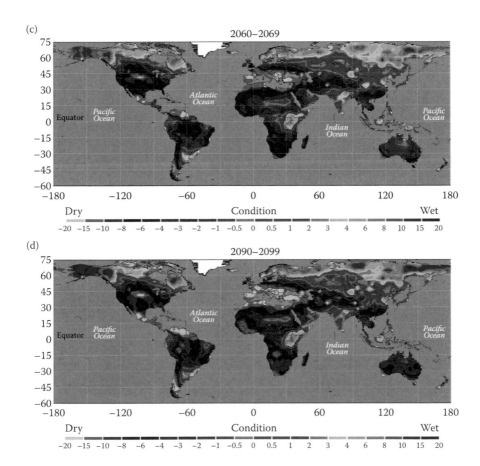

FIGURE 3.9
(Continued) Color maps of computer-generated model drought predictions. (a) 2000–2009 (present: model verification). (b) 2030–2039. (c) 2060–2069. (d) 2090–2099. Note that the American Heartland and all of Europe surrounding the Mediterranean Sea are predicted to have severe drought. (From Dai, A., NCAR/CGD, University Corporation for Atmospheric Research. With permission.)

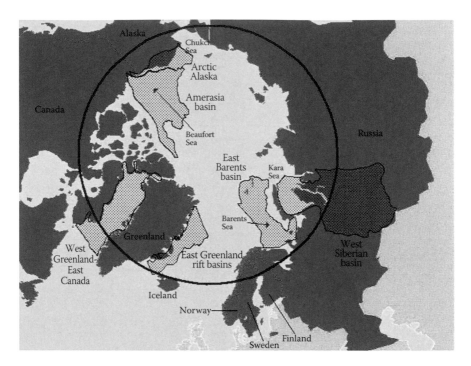

FIGURE 3.12
Oil reserves above the Arctic Circle. (Courtesy of the USGS.)

FIGURE 4.9
Aerial photograph of the STEPOG test facility at the Sandia NSTTS at Kirtland Air Force Base, NM. The heliostat mirrors are steerable. (Courtesy of NASA.)

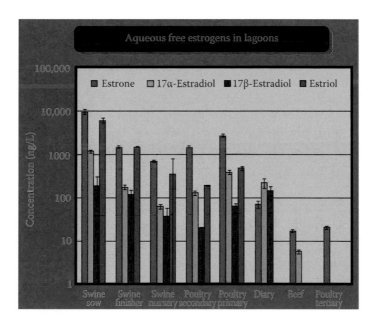

FIGURE 5.1
Estrogen levels in CAFO lagoons, EPA study. (Courtesy of Hutchins, S., Assessing potential for ground and surface water impacts from hormones in CAFOs, EPA Office of Research & Development CAFO Workshop, Chicago, Illinois, August 21, 2007. http://www.epa.gov/ncer/publications/workshop/pdf/hutchins_ord82007pdf.)

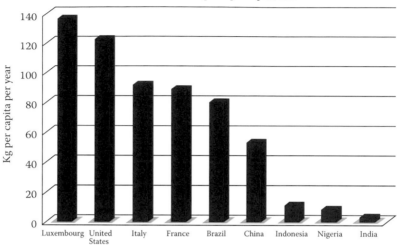

Luxembourg | United States | Italy | France | Brazil | China | Indonesia | Nigeria | India

FIGURE 5.2

Annual meat consumption per capita, 1999, for certain developed and developing countries. (Courtesy of Horrigan, L. et al., *Environ Health Perspect.* 110(5), 445–456, 2002. http://www.ncbi. nlm.nih.gov/pmc/issues/122660/.)

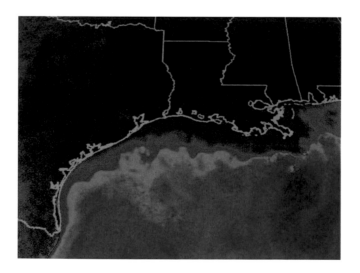

FIGURE 5.3

Gulf of Mexico dead zone, July 2010. Reds and oranges represent high concentrations of phytoplankton and river sediment. (Courtesy of NASA, Satellite image of eutrophic dead zone off the Mississippi River delta, 2012. www.nasaimages.org/luna/servlet/detail/NSVS~3 ~3~7277~107277:Mississippi-Dead-Zone.)

FIGURE 5.4
Algal bloom in the Baltic Sea, July 2010. (Courtesy of NASA. Satellite image of algal bloom in the Baltic Sea, 2012. http://earthobservatory.nasa.gov/images/2001184110526.LIA_HROM_Irg.jpg.)

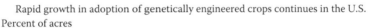

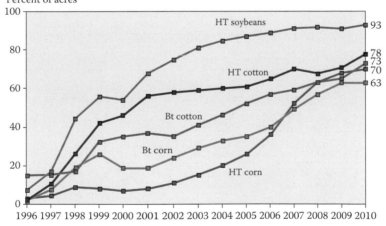

FIGURE 5.5
Increasing use of HT and insect tolerant (Bt) genetically modified (GM) crops in the United States. (Courtesy of USDA, *Agricultural Biotechnology: Adoption of Biotechnology and Its Production Impacts*, Economic Research Service, USDA, 2010. http://www.ers.usda.gov/data-products/adoption-of-genetically-engineered-crops-in-the-us/recent-trends-in-ge-adoption.aspx.)

FIGURE 5.6
Corn crops in a drought year: organic (left) versus conventional (right) farming was used. (From LaSalle, T. et al., *The Organic Green Revolution*, Rodale Institute, Emmaus, PA, 2008. With permission. http://www.rodaleinstitute.org/files/GreenRevUP.pdf.)

5

Sustainable Agriculture

5.1 Introduction

Current large-scale "industrial" farming practices constitute threats to human ecological sustainability, both in animal husbandry (meat and dairy production) and in agriculture (crop production). This chapter will examine the threats to sustainability posed by such farming operations and will then present some alternatives. The question of whether large-scale centralized agriculture and animal production is necessary to "feed the masses" will also be considered. Finally, we look at global food systems trends such as competition for cropland and the impact this has on food prices and world hunger.

5.2 Animal Husbandry: Concentrated Animal Feeding Operations

Most meat in the United States is produced by concentrated animal feeding operations (CAFOs). CAFOs make up only 5% of livestock operations but produce 50% of our food animals (Physicians for Social Responsibility 2011). Slaughterhouses and processing plants are also increasingly concentrated. As of 2005, four companies were responsible for the processing of over 80% of the beef and 64% of the pork in the United States (USDA 2011). There are an estimated 376,000 CAFOs in the United States (Duff and Davila 2005).

CAFOs place animals in crowded, unhealthy conditions where disease is easily spread. As a result, animals in these "factory farms" receive nontherapeutic doses of antibiotics as a preventive measure. It is also a common practice to administer hormone treatments to CAFO animals to promote faster growth. CAFOs yield vast, unregulated, sewage waste "lagoons" that release a powerful stench as well as pose a threat to local groundwater and surface water. The threat from CAFOs goes beyond animal cruelty or unpleasant conditions; CAFOs sicken or kill many humans each year and could create larger problems in the long run.

5.2.1 Bacteria from CAFOs

Unsanitary conditions in CAFOs and the industrial slaughterhouses they employ cause bacterial contamination in meat. In the United States, food-borne illness sickens 76 million people, causes 325,000 hospitalizations, and kills 5000 people annually (Mead et al. 1999). Not all food poisoning comes from meat, but it is a significant source.

Each year, salmonella and campylobacter from chicken and other food sources infect 3.4 million Americans, send 25,500 to hospitals, and kill about 500, according to Centers for Disease Control and Prevention (CDC) statistics (Consumer Reports 2010). The CDC (2012) reports that, between 1996 and 2010, the incidences of six key food-borne bacteria decreased by 23% overall, so some progress is being made. However, the incidence of salmonella (the most common cause of food-borne infection) continues to increase and, in 2010, was more than three times the national health objective target rate.

A recent (2010) Consumer Reports study found that 66% of chickens available in stores were contaminated with salmonella and/or campylobacter. Campylobacter was in 62% of the chickens, salmonella was in 14%, and both bacteria were in 9%. Only 34% of the birds were free of both pathogens. As will be discussed later, organic farming practices reduce the likelihood of meat contamination. In this study, store-brand organic chickens had no salmonella and were less likely than other types of chickens to harbor campylobacter (Consumer Reports 2010).

5.2.2 Antibiotic Resistance in Factory-Farmed Meat

CAFOs are heavy users of antibiotics. Antibiotics are used to fight the infections that would otherwise be rampant in crowded and unsanitary conditions, as well as to promote animal growth. In 2009, approximately 13.1 million kg or 28.8 million lb. of antibiotics was administered to animals (primarily in CAFOs) for nontherapeutic purposes. According to the same 2010 report by the Food and Drug Administration (FDA), approximately 80% of all antibiotics used in the United States were administered to food animals, compared with about 20% administered to humans (US Food and Drug Administration 2010). Overuse of antibiotics contributes to antibiotic resistance, and many scientists (Union of Concerned Scientists, Alliance for the Prudent Use of Antibiotics, among others) are concerned that this could lead to a pandemic of untreatable bacteria. In the Consumer Reports study cited in the section above, 68% of the salmonella and 60% of the campylobacter found in store-bought chickens showed resistance to one or more antibiotics (Consumer Reports 2010). Patients with antibiotic-resistant salmonella infections are up to 10.3 times more likely to die than the general population (Helms et al. 2002).

An example of a causal relationship between nontherapeutic use of antibiotics in food animals and the development of antibiotic resistance involves the antibiotic *avoparcin*. This drug has been used in poultry CAFOs but not in humans. Therefore, antibiotic resistance from avoparcin can be traced to animal use. The implication for human medicine is that bacteria that are resistant to avoparcin are *also* resistant to vancomycin, one of a few drugs remaining that can be used to treat methicillin-resistant *Staphylococcus aureus* (MRSA) in humans. In Europe, avoparcin was widely used in the 1990s in the poultry industry. During this time, vancomycin-resistant bacteria were commonly isolated from individuals who were not linked in any way with a hospital setting (Ebner 2007).

Antibiotic-resistant bacteria may be transmitted from animals to humans through meat, through water contamination, or through airborne particles (Barrett 2005).

In Denmark, nontherapeutic antibiotic use in food animals was banned in 1998. Virtually no antibiotics have been used in Denmark for this purpose since the end of 1999. In 2002, the World Health Organization conducted an extensive study of the impact of this decision on antibiotic resistance and human health. No negative impact was found on human health. The termination of the practice "dramatically reduced" the food animal reservoir of bacteria resistant to these drugs and therefore reduced the likelihood of bacteria developing resistance to several clinically important antibiotics in humans (World Health Organization 2003).

There are several databases available online tracking the development of antibiotic-resistant bacteria, some with specific reference to the genes involved.

- *Antibiotic Resistance Genes Database (ARDB):* Identifies which microbes are resistant to which drugs, with reference to the specific gene or genes supplying resistance. Each gene and resistance type is annotated with information including resistance profile, mechanism of action, ontology, and external links to sequence and protein databases. As of this writing, ARDB contains resistance information for 23,137 genes, 380 types, 249 antibiotics, 632 genomes, 1737 species, and 267 genera. ARDB is available at http://ardb.cbcb.umd.edu/ (Liu and Pop 2009).

- *Antibiotic Resistance Genes Online (ARGO):* A database on tetracycline, vancomycin, and β-lactam resistance genes. ARGO contains gene sequences conferring resistance to these classes of antibiotics. It is designed as a resource to enhance research on the prevalence and spread of antibiotic resistance genes and includes state-specific information. The database is affiliated with the Center for Biotechnology, Biological Sciences Group, Birla Institute of Technology and Science, Pilani, India. ARGO is available at http://www.argodb.org.

- *E-coli Resistance Database: Escherichia coli* accounts for 17.3% of clinical infections requiring hospitalization and is the second most common source of infection. Among outpatient infections, *E. coli* is the most common organism (38.6%). This database, affiliated with the Broad Institute, analyzes the genetic diversity of *E. coli*. This database is available at http://www.broadinstitute.org/annotation/genome/ escherichia_antibiotic_ resistance/MultiHome.html.

- *Pilot Comprehensive Antibiotic Resistance Database (CARD):* This pilot version of the database focuses primarily on MRSA and *Acinetobacter baumannii.* Molecular data are presented toward the development of an Antibiotic Resistance Ontology to guide annotation of resistance mechanisms, determinants, and targets. CARD is affiliated with McMaster University in Ontario, Canada. The database is available at http://arpcard.mcmaster.ca/.

One of the more common mechanisms for the acquisition of antibiotic resistance in bacteria is a mutation that modifies the target enzyme to which an antibiotic would normally bind. Resistance (of varying degrees) is conferred by the extent to which this modification disrupts the ability of the antibiotic to bind to the target (McLean et al. 2010).

Although such mutations occur at the individual level, they convey an advantage that allows resistant individuals to survive and reproduce at a greater rate than sensitive individuals, thus increasing the concentration of the mutated genotype in subsequent generations of the bacterial population. Population genetics are a useful framework for analyzing such changes in the gene pool of a bacterial population.

The mechanisms of antibiotic action and resistance are predictable using the mathematical framework of population genetics. A simple model can be built using estimates of the mutation rates (a) from sensitive to resistant and (b) from resistant to compensated resistant. The development of antibiotic resistance typically comes at some cost to fitness; compensated resistant bacteria are those that have recovered the fitness lost when developing antibiotic resistance (McLean et al. 2010).

Antibiotic-resistant bacteria will show greater relative fitness in the presence of antibiotics than antibiotic-sensitive bacteria. However, the same mutation that conferred resistance may also create other changes that reduce fitness in other ways. (Fitness refers to the ability of a genotype to be passed on to the next generation.) Therefore, the fitness of the three possible genotypes (sensitive, resistant, and compensated resistant) in the presence and absence of antibiotics (or combinations of antibiotics) must also be included in a predictive model for the evolution of antibiotic resistance in bacteria (McLean et al. 2010).

From the principles of population genetics, larger populations typically evolve more rapidly (Handel and Rozen 2009), another reason that large-scale CAFOs are at risk for the development of antibiotic resistance.

5.2.3 Anthelmintic Resistance in Farm Animal Parasites

Just as overuse of antibiotics promotes antibiotic resistance in farm animals, the global application of anthelmintics (drugs used to combat parasitic worms) is now promoting resistance to these drugs. Unfortunately, the standard protocol has become universal application of anthelmintics to livestock, even though parasitic worms are normally "overdispersed," meaning that a small percentage of hosts harbor most of the parasites. In horses, for example, this ratio is 20% of hosts harboring 80% of the worms (Kaplan 2009). This means that universal administration of anthelmintics is unnecessary, as well as creating serious unintended consequences.

Drug resistance in parasitic worms is most common in sheep and goats. In Australia, Brazil, and the United States, there are regions where half or more of farms are infested with drug-resistant worms. Cattle are affected in Argentina, Brazil, and New Zealand; horses harbor resistant worms in the United States and Europe (Kaplan and Vidyashankar 2012).

A University of Georgia study (2002–2006) in the southern United States found resistance to the anthelmintics *antibenzimidazole, levamisole, ivermectin,* and *moxidectin* on 98%, 54%, 76%, and 24%, respectively, of sheep and goat farms. Resistance to all three major classes of anthelmintics was detected on *half* the farms (Kaplan and Wolstenholme 2010).

The danger, of course, is that we are moving toward a scenario in which there is no way to kill parasitic worms, which could spread throughout numerous agricultural animals, particularly in the tight confines of a CAFO. Two possible solutions are increased investment in alternative drug discovery, since few alternatives exist, and more selective treatment.

Experts at the University of Georgia recommend targeted selective treatment (TST) of just those animals infected with worms, rather than global prophylactic treatment of all animals, and have piloted a program to assist farmers with TST. Since 2003, farmers in 47 states have received training and charts; the program has been very successful. Some 94% of farmers surveyed said that the program improved their ability to control parasites, 74% reported reduced parasite incidence, and 88% said they had saved money (Kaplan and Wolstenholme 2010).

An array of other "alternative" methods (in addition to TST) could also be considered, including animal-tree-crop integration, drug combination, move-and-dose methods, herbal medicine, and breeding for host resistance (Molento 2009). The only method that will not be sustainable in the long term is the current status quo.

5.2.4 Hormone Use in CAFOs and Endocrine Disruption

CAFOs use large quantities of hormones as growth promoters in food animals and dairy animals. Humans are exposed to these hormones through consumption of CAFO products and through the environment, including runoff into surface water and introduction into drinking water. Some 60% of biosolids are applied to agricultural fields to grow crops (Borch 2010), making them susceptible to runoff.

Synthetic steroid hormones, which are more resistant to biotransformation than endogenous steroids, are particularly persistent through various forms of transmission, including passage through soil and water (Kolodziej 2010). This is of concern because exogenous hormones (both natural and synthetic) act as endocrine disruptors in humans, with various health and developmental consequences.

5.2.4.1 Types of Hormones Used

Growth promoters commonly used in CAFOs include synthetic steroids as well as naturally occurring hormones. Approximately 80% (Raloff 2002) to 90% (Hutchins 2001) of feedlot cattle are administered with growth-enhancing hormones. They fall into three classes: estrogens, androgens, and progestins.

- The androgens include steroids such as trenbolone acetate, melengestrol acetate (USFDA 2002b), and zeranol (Kolodziej 2010), as well as testosterone propionate.
- The estrogens include estradiol, estradiol benzoate, estrone, and others.
- Progestins include progesterone.

According to a commission of the European Union on public health, the use of six growth hormones in beef production poses a potential risk to human health. Of these, three are naturally occurring (estradiol, progesterone, and testosterone) and three are synthetic (zeranol, trenbolone, and melengestrol) (European Union 1999).

- *Estradiol-17β* is the most active of the female sex hormones (estrogens). It affects many functions in human organs and systems, particularly those related to reproductive function. Estradiol is secreted in the early stages of embryogenesis and plays an active role in the normal development of female sex characteristics (EU 1999). Cattle commonly receive estradiol or estradiol benzoate in ear implants (Duff and Davila 2005). According to the US Environmental Protection Agency (EPA), cattle subjected to growth hormones generate urine with estradiol concentrations fivefold to sixfold greater than those not given hormones (Hutchins 2001). For cattle, the estradiol concentration in urine averages 13 ng/L (Hutchins 2001). Although poultry and swine operations typically do not administer hormones, the hormone concentration in runoff from these operations is still significant (Hutchins 2001).

- *Progesterone* is a natural steroid sex hormone synthesized by the corpus luteum of ovaries, particularly during the second half of the estrous cycle, and is regulated by polypeptides secreted by the brain. It plays a role in the implantation of an egg in the uterus, as well as fetal development (European Union 1999).
- *Testosterone* and dehydrotestosterone are the main sex hormones secreted by males. Testosterone is responsible for the early development and adult maintenance of male sex organs (European Union 1999).
- *Zeranol* is a natural mycoestrogen (fungal estrogen) derived from a compound produced by *Fusarium* molds. As with the animal estrogens, zeranol affects estrogen target organs and therefore can disturb reproductive patterns (European Union 1999).
- *Trenbolone* is a synthetic androgen with anabolic activity several times greater than that of testosterone. Trenbolone is metabolized in the bloodstream into 17β-trenbolone, its most active derivative. Trenbolone shares hormonal properties with testosterone (European Union 1999).
- *Melengestrol* is a synthetic progesterone analog that is about 30 times as active as progesterone. It interferes with the estrous cycle and is able to increase estrogen levels (European Union 1999).

Another growth hormone not discussed in the European study, because it is used in dairy rather than in meat production, is recombinant bovine growth hormone (rBGH). In large-scale dairy operations, it is common to inject dairy cows with this genetically engineered artificial growth hormone. This practice has been controversial since its inception in 1993. rBGH is manufactured under the brand name *Posalic* by Monsanto Corporation. It is also referred to as recombinant bovine somatotropin (rBST) (Sustainable Table 2011a). While the average dairy cow produced almost 5300 lb. of milk a year in 1950, today, a typical cow produces more than 18,000 lb. (Hallberg 2003).

One of the risks to human health from rBGH is the increase in concentrations of a hormone called insulinlike growth factor 1 (IGF-1) in milk from cows treated with rBGH or rBST. Numerous comprehensive studies "clearly demonstrate a significant increase in IGF-1 concentrations in milk from rBST-treated cows" (Hansen 1997). IGF-1 has been shown to play a role in the development of breast cancer (Rosen et al. 1991), colon cancer (Tricoli et al. 1986), and other cancers (Hoppener et al. 1988) in humans, at the levels found in IGF-1-treated milk.

5.2.4.2 Transmission of Endocrine Disruptors from CAFOs to Humans

Numerous studies have detected steroids at elevated concentrations in watersheds associated with CAFOs (Kolodziej 2010; Laessig and Snow 2011).

A 1999–2000 US Geological Survey (USGS) study detected steroid hormones in various US streams in median concentrations ranging from 9 ng/L (17β-estradiol) to 116 ng/L (testosterone) (Kolpin et al. 2002). In rivers, hormones from CAFOs have been identified up to 60 mi. away (Shore and Pruden 2009). To put this into perspective, feminization of male fish in response to estradiol begins at 1 ng/L (Sedlak 2011).

When animal waste from CAFOs is applied to agricultural fields as fertilizer, it is subject to runoff. It has been demonstrated that the hormones (particularly steroids) found in these biosolids contaminate the runoff (Yang et al. 2012). A recent study found that hormones (both endogenous and exogenous) can be observed in runoff more than 35 days after application and that androstenedione and progesterone were the most concentrated hormones in both biosolids and runoff (Borch 2010). According to the National Toxicology Program, androstenedione is a precursor to male and female sex hormones produced by the human body. It is currently designated as a con-

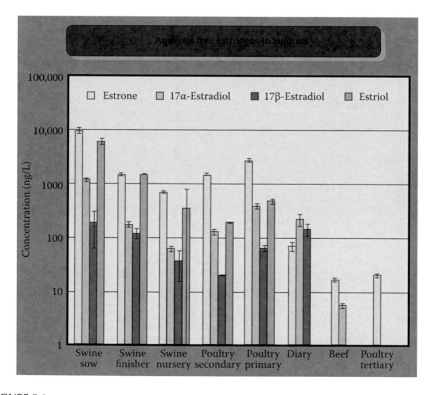

FIGURE 5.1
(See color insert.) Estrogen levels in CAFO lagoons, EPA study. (Courtesy of Hutchins, S., Assessing potential for ground and surface water impacts from hormones in CAFOs, EPA Office of Research & Development CAFO Workshop, Chicago, Illinois, August 21, 2007. http://www.epa.gov/ncer/publications/workshop/pdf/hutchins_ord82007pdf.)

trolled substance under federal law and has been shown in several studies to be carcinogenic (California Environmental Protection Agency 2010).

The appearance of endocrine-disrupting chemicals (EDCs) in groundwater is important not only because of the likelihood of transmission to drinking water but also because groundwater accounts for approximately 36% of the nation's irrigation water (Hutchins 2007). An EPA study found elevated concentrations of numerous EDCs in groundwater or "lagoons" associated with CAFOs. Figure 5.1 describes the estrogen levels in CAFO lagoons.

Another pathway for the introduction of EDCs into humans is, of course, through meat and dairy consumption, particularly consumption of products from hormone-treated animals. Concentrations of steroid hormones (estradiol, estrone, progesterone, and testosterone) have been found in fish, poultry, eggs, pork, cheese, milk, and milk products (Birkett and Lester 2003).

5.2.4.3 Actions of Endocrine Disruptors

Exogenous hormones, whether natural or synthetic, act as endocrine disruptors in humans. An environmental EDC was defined by the US EPA as "an exogenous agent that interferes with the production, release, transport, metabolism, binding, action or elimination of natural hormones in the body responsible for the maintenance of homeostasis and the regulation of developmental processes" (Kavlock et al. 1996).

Endocrine disruptors have been linked to reproductive changes in animals and humans, and also act as carcinogens. Because they are morphogens (affecting tissue development, not just cell development), exposure to even low doses of EDCs increases the susceptibility to cancer and developmental disorders, including changes in fetal development (Kajiwara and Imahori 2009).

Some of the specific effects of endocrine disruption that have been identified through research include changes in neuroendocrinology, behavior, thyroid function, and metabolism (Soto and Sonnenschein 2010), and the development and promotion of prostate cancer (Barrett-Connor 1990), pancreatic cancer (Fernández-del Castillo et al. 1990), and breast cancer (Lippman et al. 1977; Santner et al. 1993).

Postmenopausal women with high serum estrone levels have a significantly increased risk for estrogen receptor–positive breast cancer (Yasuo 2003). Estrone sulfate has been found to promote human breast cancer cell replication (Santner et al. 1993).

Estriol, an estrogenic steroid, can interfere with breast cancer treatment. It is capable of partially overcoming antiestrogen inhibition with the chemotherapy drug tamoxifen, even when antiestrogen is present in 1000-fold excess. Estradiol is also capable of overcoming antiestrogen inhibition to some degree and binds to as many sites as estriol (Lippman et al. 1977).

There is a huge amount of complexity involved in studying the effects of EDCs. Humans are exposed to numerous different environmental EDCs, in varying combinations and doses, making the specific contribution of any one EDC difficult to predict out of context. A systems approach would be needed to model the impact of EDCs in various scenarios (Soto and Sonnenschein 2010).

We do know that epidemiological trends related to EDCs began a significant upward movement in the latter half of the 20th century, coterminous with the widespread use of hormones as growth enhancers in agricultural operations (Soto and Sonnenschein 2010).

In addition to the impact of EDCs on humans, they pose a threat to biodiversity by interfering with the normal reproductive functions of numerous animal species. One example that has received extensive recent attention is the feminization of fish in the wild through exposure to EDCs entering waterways through runoff. As mentioned above, feminization of fish begins at 1 ng/L (Sedlak 2011). Concentrations as low as 25 ng/L have resulted in reproductive impairment as well as feminization of fish, yielding skewed populations (Duff and Davila 2005).

Other effects, according to the USGS, include the following:

- Masculinization of female and feminization of male fish, reptiles, birds, and mammals
- Abnormal thyroid function in birds
- Decreased hatching success in fish, birds, and turtles
- Altered immune function in birds, mammals, and fish
- Altered behavior such as overt aggression and reduced predator avoidance in fish
- Reproductive incompetence (e.g., infertility) in birds, fish, shellfish, and mammals (Larsen 2009)

There is ample evidence that endocrine systems of fish and wildlife have been affected by exposure to EDCs. The co-occurrence of exposure to EDCs and indicators of endocrine disruption (such as those listed above) have been documented in many countries, including the United States, United Kingdom, Denmark, South Africa, Germany, and others (Larsen 2009).

Some examples of endocrine disruption in fish have been found in recent USGS studies:

- A study of endocrine disruption in fish in Colorado demonstrated how a complex mixture of EDCs can have an additive effect on fish. In this study, 18% to 22% of fish downstream from a source of EDC contamination exhibited intersex characteristics (feminine traits in male fish and male traits in female fish) (Vajda et al. 2008).
- A Virginia study investigated the occurrence of intersex in male smallmouth bass in the Potomac River and its tributaries in Virginia

and West Virginia. The study reported a higher incidence of inter-sex in streams draining areas with intensive agricultural production (Blazer et al. 2007).

- A study of extensive fish kills in the Shenandoah River (Virginia) found that endocrine disruptors had damaged the immune systems of fish (Ripley et al. 2008).

Another interesting trend co-occurring with the widespread use of growth hormones in animal husbandry is increasingly early puberty in humans. The trend toward earlier puberty provides additional evidence of the effects of environmental EDCs. It is relevant not only because it shows that EDCs affect humans but also because there may be a link between early puberty and cancer, especially in females.

In 1950, a British pediatrician, Dr. James Tanner, introduced the Tanner scale, a system for mapping puberty. His benchmark for the earliest sign of puberty among girls was 11.5 years, and among boys, 11.2 years (Neil 2010). In 1997, a US study of 17,000 girls showed that the average age of puberty among girls had dropped to 9.9 years among white girls and 8.9 years among African-American girls (Herman-Giddens et al. 1996).

A recent study of about a thousand girls in Denmark found that the age of breast development (usually the first sign of puberty in girls) had dropped

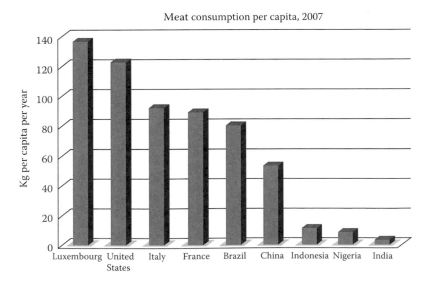

FIGURE 5.2
(See color insert.) Annual meat consumption per capita, 1999, for certain developed and developing countries. (Courtesy of Horrigan, L. et al., *Environ Health Perspect.* 110(5), 445–456, 2002. http://www.ncbi.nlm.nih.gov/pmc/issues/122660/.)

by a year, from 10.8 years in 1991 to 9.8 years in 2006. They were also having their first menses, on average, 3 months earlier (Aksglaede et al. 2009).

According to the National Cancer Institute (2006), one of the risk factors for breast cancer is "an early age at first menstrual period" (defined as before age of 12 years). A CDC study found that girls who start puberty before the age of 12 years are 51% more likely to die from ovarian cancer if they develop it than girls who go through puberty at 14 years or older (Robbins et al. 2009).

A direct link between eating meat and early puberty has recently been reported in a British study. The study found that 49% of girls eating more than 12 portions of meat a week at the age of 7 years had started their periods by age 12.5 years, compared to only 35% of those who ate less than four portions of meat a week. The researchers speculate that the increased amount of meat in the diet may account for the decline in the average age of puberty (Rogers et al. 2010). Figure 5.2 shows the annual meat consumption per capita in various countries around the world in 1999.

5.3 Industrial Agriculture

For the purposes of this chapter, "industrial agriculture" is defined as large-scale agriculture using synthetic chemicals, pesticides, and other methods not used in organic farming. The threats to human sustainability from industrial agriculture are many, primarily through damage to the environment and harm to human health from pesticide and fertilizer pollution. Other issues include loss of arable land through topsoil erosion and loss of biodiversity.

5.3.1 Pesticides and Human Health

Some 5 billion tons of pesticides a year are used worldwide (Wright and Welborn 2002). These pesticides affect human health by entering groundwater and drinking water through runoff, through meat from animals that have ingested pesticides, through ingestion in produce, and through the air (drift from crop dusting, etc.).

According to the National Institutes of Health, pesticides produce both short- and long-term effects on human health. Effects include increased cancer risks, as well as effects similar to those discussed above for endocrine disruptors, including interference with the reproductive system, immune system, and nervous system (Horrigan et al. 2002). The United Nations estimates that about 2 million poisonings and 10,000 deaths occur each year from pesticides; about three quarters of these occur in developing countries (Quijano et al. 1993).

TABLE 5.1

Pesticides and Associated Cancers

Type of Pesticide	Associated Type of Cancer
Phenoxyacetic acid herbicides	Non-Hodgkin's lymphoma, soft tissue sarcoma, prostate cancer
Organochlorine insecticides	Leukemia; non-Hodgkin's lymphoma; soft tissue sarcoma; pancreas, lung, breast cancers
Organophosphate insecticides	Non-Hodgkin's lymphoma, leukemia
Arsenical insecticides	Lung, skin cancers
Triazine herbicides	Ovarian cancer

Source: Horrigan, L. et al., *Environ Health Perspect.* 110(5), 445–456, 2002. http://www.ncbi.nlm. nih.gov/pmc/issues/122660/.)

5.3.1.1 Pesticides as Carcinogens

Table 5.1, from the National Institutes of Health, shows associations between various pesticides and types of cancers. These associations have been demonstrated in population-based studies.

Alachlor is an herbicide used for weed control in corn, soybean, sunflower, and cotton crops, controlling grasses and weeds (Secretariat of the Rotterdam Convention 2008). It has been linked to nasal, stomach, and thyroid tumors in rats (Krieger 2001).

Atrazine is an herbicide used to control weeds in corn and soybean fields (Wargo 1996). It has been linked to reproductive cancers (breast and prostate cancer) in both rodents and humans (Fan 2007).

5.3.1.2 Pesticides as Immune Suppressors

Pesticides can suppress the immune system in humans. An international study on pesticide use found, with all else held equal, a clear association between pesticide exposure and increased rates of disease, particularly those diseases to which immune-compromised individuals are susceptible. Increases were found in the incidence of respiratory, gastrointestinal, and acute inflammatory kidney infections, among other disorders (Repetto and Baliga 1996).

5.3.1.3 Pesticides as Endocrine Disruptors

Atrazine, a popular weed killer for corn and soybeans, acts as an endocrine disruptor in fish, amphibians, reptiles, and humans (Fan 2007). It has been shown to disrupt hormonal signaling in human cells and has been linked to abnormal birth weights (Suzawa et al. 2008). Atrazine also increases stress hormones in rats. These stress hormones are known to suppress the hormone signaling cascade needed for ovulation (Powers-Fraites et al. 2009).

Other pesticides found to act as endocrine disruptors in humans include (Lyons 1999) the following:

- *Chlordecone*, used to control insects on crops, including bananas and tobacco
- *Dicofol*, a nonsystemic organochlorine acaricide, used on cucumbers, tomatoes, lettuce, ornamentals, hops, apples, and strawberries
- *Endosulfan*, a contact and ingested organochlorine insecticide and acaricide, used on hops, canola, soft fruits, and ornamentals
- *Lindane* gamma-Hexachlorocyclohexane or "gamma-HCH," a contact, ingested, and fumigant organochlorine insecticide, used on many crops including sugar beet and canola
- *Methoxychlor*, an insecticide used on fruits, vegetables, forage crops, and livestock
- *Vinclozolin*, a dicarboximide fungicide used on canola, beans, peas, turf, and apple blossoms

5.3.2 Nitrate Pollution

Nitrates are a common ingredient in agricultural fertilizers. They are carried into rivers and streams through field runoff, where they are then carried to oceans as well as bodies of freshwater. Nitrogen also filters through soil easily and can enter wells and other sources of drinking water (Wisconsin DNR 2010).

5.3.2.1 Nitrates and Dead Zones

Nitrates fertilize phytoplankton (algae) that deplete oxygen as they bloom, die, and decompose, creating "dead zones" where no fish or other marine life can survive. Dead zones have been more common (and larger) recently, in part because streams are losing their ability to filter out excess nitrates before they reach the sea. In the Gulf of Mexico, the summer dead zone is in excess of 7700 mi.^{2} (20,000 km^2) (Bielo 2008). This is the second-largest dead zone in the world. Figure 5.3 shows the dead zone in the Gulf of Mexico in July 2010. Reds and oranges represent high concentrations of phytoplankton and river sediment.

The largest algal bloom recorded is in the Baltic Sea. In July 2010, it reached 377,000 km^2 in size (approximately the size of Germany), caused by a blue-green algae bloom that was promoted by a combination of warm weather and nitrate-rich fertilizers being washed into the Baltic (British Broadcasting Corporation 2010). Figure 5.4 Illustrates the massive algal bloom in the Baltic Sea in July 2010.

When not overwhelmed by nitrates, the bacteria that live in healthy streams can denitrify fertilizers, converting nitrate to nitrogen gas that is

FIGURE 5.3
(See color insert.) Gulf of Mexico dead zone, July 2010. Reds and oranges represent high concentrations of phytoplankton and river sediment. (Courtesy of NASA, Satellite image of eutrophic dead zone off the Mississippi River delta, 2012. www.nasaimages.org/luna/servlet/detail/NSVS~3 ~3~7277~107277:Mississippi-Dead-Zone.)

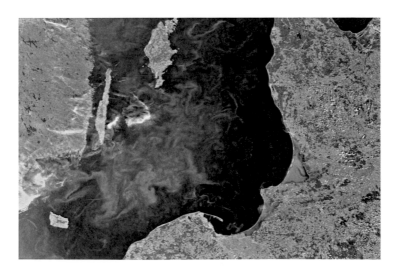

FIGURE 5.4
(See color insert.) Algal bloom in the Baltic Sea, July 2010. (Courtesy of NASA. Satellite image of algal bloom in the Baltic Sea, 2012. http://earthobservatory.nasa.gov/images/2001184110526. LIA_HROM_Irg.jpg.)

released into the atmosphere. Healthy streams can remove up to 43% of nitrogen. However, the typical US stream now denitrifies only 16% of nitrogen (Mulholland et al. 2008).

5.3.2.2 Nitrates and Human Health

Nitrogen pollution in groundwater can cause nitrogen poisoning in humans, especially in infants under 6 months of age. It reduces blood's capacity to carry oxygen. According to the Environmental Science Division of the Department of Energy:

> "Nitrates themselves are relatively nontoxic. However, when swallowed, they are converted to nitrites that can react with hemoglobin in the blood, oxidizing its divalent iron to the trivalent form and creating methemoglobin. This methemoglobin cannot bind oxygen, which decreases the capacity of the blood to transport oxygen so less oxygen is transported from the lungs to the body tissues, thus causing a condition known as methemoglobinemia" (Argonne National Laboratory 2005).

Early symptoms of methemoglobinemia can include irritability, lack of energy, headache, dizziness, vomiting, diarrhea, labored breathing, and a blue-gray or pale purple coloration to areas around the eyes, mouth, lips, hands, and feet (NHDES 2006). In infants, this can be fatal, known as "blue baby syndrome." The federal safety standard for nitrate concentration in drinking water is 10 mg/L (Wisconsin DNR 2010). Most cases of infant methemoglobinemia occur at 20 mg/L or higher (Argonne National Laboratory 2005).

Adults with heart or lung disease are particularly at risk for toxic effects (Wisconsin DNR 2010), as are pregnant women. According to the CDC, effects of methemoglobinemia in adults can include cyanosis, cardiac dysrhythmias, circulatory failure, and progressive central nervous system (CNS) effects ranging from mild dizziness and lethargy to coma and convulsions. Exposure of pregnant women to environmental nitrates and nitrites may increase the risk of anemia, threatened abortion/premature labor, and preeclampsia (Agency for Toxic Substances & Disease Registry 2011).

5.3.3 Topsoil Loss and Declining Crop Yields

> "Soil erosion is a major environmental threat to the sustainability and productive capacity of agriculture. During the last 40 years, nearly one-third of the world's arable land has been lost by erosion and continues to be lost at a rate of more than 10 million hectares per year" (Pimentel et al. 1995).

Arable land is being destroyed at an alarming rate because of nonsustainable farming practices. Because the average size of a farm has more than doubled in the past 50 years, farmers have adapted the layout of their fields to suit large-scale agricultural machines. They have removed grass strips, shelterbelts, and hedgerows, making soil more prone to erosion. Heavy industrial tractors, harvesters, and so forth also damage the soil ecosystem (Pimentel et al. 1995).

The loss of topsoil reduces crop yields, primarily through the loss of nutrients and the loss of the soil's ability to hold water. Various studies in the United States on corn yields have found reductions of 21% to 24% in severely eroded fields (Pimentel et al. 1995). The corn yield in moderately eroded soils (with about 5 in. of topsoil) was 13 bushels per acre more than the corn yield from severely eroded soils (with about 1.5 in. of topsoil) (Al-Kaisi 2001). Globally, yields for staple crops have dropped from 3% in the 1960s to 1% currently. This is the first time that crop yields are lower than the rate of population growth (HRH The Prince of Wales 2012).

Erosion is not the only way that topsoil is lost. Depletion of the soil's fertility also reduces crop yields, so soil productivity is lost even if the soil itself remains. Soil organic carbon is essential to soil health, both for structure and nutrient cycling (LaSalle et al. 2008). Some Midwestern soils that in the 1950s were composed of up to 20% carbon are now only 1%–2% carbon. Carbon loss, in turn, increases the propensity for soil erosion by making the soil less structurally sound (more friable) and more prone to compaction. Carbon loss also decreases the soil's ability to hold water (LaSalle and Hepperly 2008) and makes nutrients less available to crops. Chemical fertilization with nitrogen, a common practice in industrial agriculture, does *not* build up soil organic matter (LaSalle et al. 2008).

Soil organic content is a reliable predictor of crop yield potential (Mitchell and Entry 1998). A 2008 review of nearly 110 years of yield data from the world's oldest continuous cotton experiment found that "higher soil organic matter results in higher crop yields" (Mitchell et al. 2008).

Soil biota, killed by pesticides, herbicides, and soil compaction, are also important to soil health. For example, glomalin is a secretion from mycorrhizal fungi. Glomalin binds soil together so it is less likely to wash or blow away. Once the mycorrhizi are depleted or killed, a chemical fertilizer can supply essential elements but not the micronutrients captured by organic soil filled with air and water pockets (LaSalle et al. 2008).

One of the practices of large-scale industrial agriculture that contributes to topsoil loss is monocropping or monoculture, the practice of planting many acres of the same crop, year after year. Although this has its advantages in economy of scale and short-term productivity, it is destructive to the soil, depleting key nutrients and contributing, in turn, to soil compaction and erosion (DeJong-Hughes et al. 2001). Crop rotation, the use of cover crops, and mixing a variety of crops in proximity to one another help to prevent topsoil loss, as well as limiting pest cycles (LaSalle et al. 2008).

Long-term research at the Rodale Institute (the Farming Systems Trial, which has been operating for over 30 years) shows that the use of cover crops (legumes, grains, grasses, or mixtures) can provide all the nitrogen needed while reversing the loss of soil organic matter. Organic farms can gain about 1000 lb. of carbon per acre per year by using cover cropping and crop rotation. This is up to 10 times greater than the carbon gain from standard planting for corn or soybeans (LaSalle et al. 2008).

5.4 Loss of Genetic Diversity

Conventional industrial agriculture uses crops and animals in which the genetic variants (alleles) are nearly the same in every individual. While these species and varieties have been chosen for their productivity, lack of genetic variability reduces their ability to deal with new diseases and pests, as well as changes in environmental conditions such as drought or climate change (National Biological Information Infrastructure 2011a). The Irish potato famine of the 1840s was a tragic large-scale example of the result of a monoculture being struck by a new blight. If there had been numerous varieties of potatoes in widespread production, some of them might have been resistant.

According to the National Biological Information Infrastructure (2011a):

> "As farmers and agricultural scientists continue to address the world's ever-increasing food requirements, conservation of genetic diversity will play a pivotal role in the development of new varieties and in meeting new environmental challenges."

Since agriculture began about 12,000 years ago, about 7000 different species of plants have been raised as food crops (United Nations Convention on Biological Diversity 2008). Today, only 15 plant and 8 animal species are the basis for about 90% of all human food (Sustainable Table 2011b).

In the United States, 99% of all turkeys raised are Broad-Breasted Whites (Sustainable Table 2011b); 83% of dairy cows are Holsteins; 60% of beef cattle are Angus, Hereford, or Simmental; 75% of pigs are of only three breeds; and over 60% of sheep are concentrated in only four breeds. As a result of this concentration on certain breeds, 190 breeds of farm animals have gone extinct in the past 15 years, and some 1500 others are at risk of extinction (Sustainable Table 2011c).

Nearly 96% of the commercial vegetable varieties available in 1903 are now extinct, and over 1500 rice varieties in Indonesia have become extinct since modern rice varieties became common in farming there. Further, large corporations have developed virtual seed monopolies, reducing genetic diversity further. As of 2008, the four largest seed companies controlled 56% of the global seed market (Howard 2009).

This loss of genetic diversity is alarming, in part because such a large part of the human food supply could be wiped out by one new disease affecting a key crop or food animal. Research shows that mixed crops are more disease resistant, even without pesticides, than the same crops grown alone. For example, a study published in *Nature* in 2000 found that disease-susceptible rice varieties planted in mixtures with resistant varieties had 89% greater yield and disease was 94% less severe than when they were grown in monoculture (Zhu et al. 2000).

5.4.1 Responses to Loss of Genetic Diversity

The concern over loss of genetic diversity in agriculture has been so great that various groups have organized to preserve the seeds of food crops. In 1975, gardeners who valued plant varieties passed on by their grandparents (and that were not readily available anymore) saw the benefit of preserving what they called "heirloom" varieties. They founded the Seed Savers Exchange, a nonprofit, 501(c)(3), member supported organization. The mission of the Seed Savers Exchange (2011) is "to save North America's diverse, but endangered, garden heritage for future generations by building a network of people committed to collecting, conserving and sharing heirloom seeds and plants, while educating people about the value of genetic and cultural diversity."

More recently, scientists concerned about the potential loss of crops on a large scale (due to natural disaster, war, climate change, disease, etc.) founded a seed vault in the Arctic to store both conventional and heirloom seed varieties in a safe location. The Svalbard Global Seed Vault is dug into a mountainside on an island nearly a thousand kilometers north of mainland Norway. The facility is protected by thick rock and permafrost, so that the seed samples will remain frozen even if the electricity fails. The area is geologically stable, humidity levels are low, and there is no measurable radiation inside the mountain. It is located well above the sea level (130 m or 430 ft.), far above the point of any projected sea level rise. Since it was opened in February 2008, over half a million seed types have been collected at the Svalbard Vault. The Vault can store up to 4.5 million samples (each sample consisting of about 500 seeds) (Global Crop Diversity Trust 2011a).

5.5 Genetically Modified Organisms

The use of genetically modified organisms (GMOs) in agriculture has become increasingly more common. About 331 million acres of genetically engineered crops with herbicide tolerance and/or insect resistance traits were cultivated worldwide in 2009, a 7% increase over acreage in 2008. US acreage accounted for nearly half of all GMO cultivation worldwide in 2009. In the United States in 2010, genetically engineered herbicide-tolerant (HT) modified soybeans accounted for 93% of all soybean acreage; HT cotton was 78% of cotton acreage; and HT corn was 70% of all corn acreage. Insect-resistant GMO crops [such as those containing the gene from a soil bacterium *Bacillus thuringiensis* (Bt)] were nearly as common as HT crops (USDA 2010). Figure 5.5 shows the steady increase in genetically engineered crops in the United States in percentage of acres planted, from 1996 to 2010.

Although GMOs have numerous commercial benefits, they are associated with various risks and problems.

- Hidden allergens (a gene for a protein not normally found in a particular type of plant might have been spliced in) can harm some consumers.
- The spread of herbicide resistance. Many crops have been genetically modified to be resistant to the effects of the pesticide Roundup™. While this makes weed removal in fields relatively easy for farmers, this trait can spread to weeds near GMO fields (through cross-pollination), thus creating "super weeds" that cannot be killed with this pesticide.

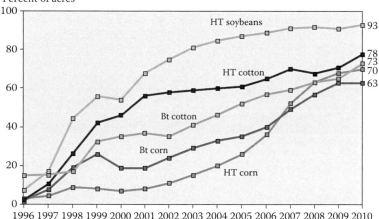

Rapid growth in adoption of genetically engineered crops continues in the U.S.

Data for each crop category include varieties with both HT and Bt (stacked) traits.
Sources: 1996–1999 data are from Fernandez-Cornejo and McBride (2002). Data for 2000–10 are available in the ERS data product. Adoption of Genetically Engineered Crops in the U.S., tables 1–3

FIGURE 5.5
(See color insert.) Increasing use of HT and insect tolerant (Bt) genetically modified (GM) crops in the United States. (Courtesy of USDA, *Agricultural Biotechnology: Adoption of Biotechnology and Its Production Impacts*, Economic Research Service, USDA, 2010. http://www.ers.usda.gov/data-products/adoption-of-genetically-engineered-crops-in-the-us/recent-trends-in-ge-adoption.aspx.)

- The use of pesticide-resistant crops makes crop rotation difficult or impossible, as "volunteers" remaining from a previous crop cannot be killed if they mix in with a cover crop.
- The method of gene splicing commonly used is by introducing new genes attached to antibiotic-resistant bacterial plasmids. There is a possibility that this antibiotic resistance could be transferred to other bacteria within the organism consuming GM plants, although this is probably dwarfed by the issues associated with massive non-therapeutic antibiotic use in CAFOs.

One of the concerns associated with GMOs, with all their unintended consequences, is that there is less and less choice for consumers wishing to avoid them. GM crops are widely known as a "leaky technology" (Montague 2009). Their use, and their ability to contaminate other crops, is now so widespread that even farmers who prefer not to grow GMOs may not have a choice. Wind-borne pollen allows genetically modified material to mix with that of other crops (or with weeds, as mentioned above) (Charles 2011).

> *As an economic sidebar*, this is a serious issue for farmers attempting to raise crops marketed as "non-GMO" or "organic." Contamination of non-GMO crops by GMO crops has been reported around the world. There were 39 cases of crop contamination in 23 countries in 2007 and more than 200 cases in 57 countries over the last 10 years (Gillam 2008). Further, because the companies that developed GMO crops hold patents on them, any farmer whose crops' genetic material has been contaminated by GMO crops is liable for damages for patent infringement (such cases have already been decided in favor of the seed companies) (Montague 2009).

In addition to cross-pollination by GMOs, it does happen that genetically engineered crops are confused (during transportation, for example) with non-GMO produce. In 2000, the first recall of a GMO product occurred when Kraft laboratory tests confirmed that they had accidentally used a genetically modified Bt corn to make taco shells under the Taco Bell brand name. The corn, called StarLink™ (developed by the pharmaceutical company Aventis), was not approved for human consumption and was produced only for livestock consumption or to be processed into ethanol. It contains a pest-repelling protein, Cry9C, that is toxic to insects and may be hard for humans to digest (Fulmer 2000). This incident demonstrates that such mix-ups can and do occur.

A new use of genetic engineering that shows promise in light of recent climate change (but which may yet yield to the law of unintended consequences) is an effort to genetically modify crops to be more drought resistant. For example, Monsanto's DroughtGard™ corn was field-tested in 2012 and will be on the market in the United States in 2013 (Monsanto 2012). It is the first crop genetically engineered for drought tolerance to be approved for commercial use.

The Union of Concerned Scientists issued a statement in December 2011 that new strains of crops alone will not solve the drought problem and that sustainable agriculture methods (such as those described in the section below) will have more impact over the long run (Gurian-Sherman 2011). In 2012, the Union of Concerned Scientists issued a report, *High and Dry: Why Genetic Engineering Is Not Solving Agriculture's Drought Problem in a Thirsty World*. This report described the gains from Monsanto's new strain of corn as "modest." Specifically: "Farmers are expected to plant [DroughtGard] on only about 15 percent of corn acres in the United States. If this corn reduces the yield normally lost during drought by 6 percent on 15 percent of corn acres, it would increase corn productivity nationwide by about 1 percent" (Gurian-Sherman 2012). Only time will tell what benefits and risks are associated with any genetically modified crop.

5.6 Sustainable Agriculture

Conventional industrial agriculture practices are not sustainable (or at least, not optimal) in the long run, due to destruction of arable land, damage to the environment, and harm to human health. The sustainable agriculture approach offers viable alternatives.

The *National Biological Information Infrastructure* (2011b) defines sustainable agriculture as a system to meet local food production needs that recycles resources (such as water and nutrients); needs little input (artificial fertilizers, pesticides); and has minimal environmental impact. Because sustainable farms produce foods without excessive use of pesticides and other chemicals, they produce foods that are healthier than their industrially produced counterparts (Sustainable Table 2011d). In his 2011 speech, "The Future of Food," the Prince of Wales succinctly defined sustainable agriculture as one that "does not exceed the carrying capacity of its local ecosystem and which recognizes that the soil is the planet's most valuable resource" (Prince of Wales 2012).

Organic farming is a subset of sustainable farming; organic farmers go a step further and use no pesticides or herbicides at all. Organic meats and produce are regulated by the US Department of Agriculture (USDA) in the United States and by other government agencies in other nations.

According to the USDA, "organic farming entails:

- Use of cover crops, green manures, animal manures and crop rotations to fertilize the soil, maximize biological activity and maintain long-term soil health.
- Use of biological control, crop rotations and other techniques to manage weeds, insects and diseases.

- An emphasis on biodiversity of the agricultural system and the surrounding environment.
- Using rotational grazing and mixed forage pastures for livestock operations and alternative health care for animal wellbeing.
- Reduction of external and off-farm inputs and elimination of synthetic pesticides and fertilizers and other materials, such as hormones and antibiotics.
- A focus on renewable resources, soil and water conservation, and management practices that restore, maintain and enhance ecological balance" (USDA 2006).

"Organic production is not simply the avoidance of conventional chemical inputs, nor is it the substitution of natural inputs for synthetic ones." Instead, organic farmers use age-old methods, such as those listed above, to promote overall system health. While an organic farm is not a closed system, the goal is to have minimal inputs and outputs (other than healthy crops) (USDA 2007).

According to the Rodale Institute, some of the advantages of sustainable agriculture include the following: competitive yields (see Section 5.7); improved soil quality, including nutrient content, as well as resistance to erosion and water loss; cost savings (on pesticides, herbicides, chemical fertilizers, etc.); energy savings, particularly in fossil fuel use; fewer CO_2 emissions; greater biodiversity, not only in plants and animals raised but also in terms of beneficial soil biota and nearby wildlife; water conservation; greater drought resistance and tolerance to weather and climate variations; denser nutrient content in food; and a reduced toxic load in the food produced (LaSalle et al. 2008).

5.7 Can Sustainable Agriculture Feed the World?

One of the most common objections to the idea of a paradigm shift from industrial agriculture to sustainable agriculture is that we could not "feed the masses" with the yield from sustainable agriculture. There is a perception that sustainable agriculture produces too little, at too high a cost, to be practical.

Aside from the fact that the hidden costs of damage to the environment and human health make industrial agriculture much more expensive than it purports to be, the argument that sustainable agriculture cannot feed the world ignores the facts.

> "Not only can organic agriculture feed the world..., it may be the only way we can solve the growing problem of hunger in developing countries" (LaSalle et al. 2008).

The estimated environmental and health care costs of the recommended use of pesticides in the United States alone are about $10 billion per year (Pimentel 2005). Excessive fertilizer use costs $2.5 billion a year in waste (National Academy of Sciences 2003). The estimated costs of public and environment health losses related to soil erosion caused by industrial agriculture is more than $45 billion annually (Pimentel et al. 1995).

The Rodale Institute Farming Systems Trial has compared organic and conventional farming systems since 1981, focusing on grains (corn, wheat, and oats) and soybeans. The trial found no statistically significant differences among soybean yields, comparing organic (animal-based and legume-based) and conventional (chemical) methods of adding nitrogen to the soil. It found an initial advantage for conventional methods regarding corn yields, but organic corn yields *equaled or exceeded* conventional corn yields after 5 years (Pimentel et al. 2005). Figure 5.6 illustrates the growth differences between corn crops in a drought year; organic (left) versus conventional (right) farming was used.

One of the most interesting findings of the Rodale Institute study was that organic corn yields far exceeded conventional corn yields in drought years: the organic-animal system had significantly higher corn yields during a severe drought (1511 kg/ha) than both the organic-legume (421 kg/ha) and the conventional system (1100 kg/ha) (Pimentel et al. 2005). This is probably because organically farmed soil is more effective at retaining moisture, as discussed in Section 5.3.3. Figure 5.7 is a bar graph illustrating average corn yields in kilograms per hectare in drought years (1988, 1994, 1995, 1997, and 1998).

FIGURE 5.6
(See color insert.) Corn crops in a drought year: organic (left) versus conventional (right) farming was used. (From LaSalle, T. et al., *The Organic Green Revolution*, Rodale Institute, Emmaus, PA, 2008. With permission. http://www.rodaleinstitute.org/files/GreenRevUP.pdf.)

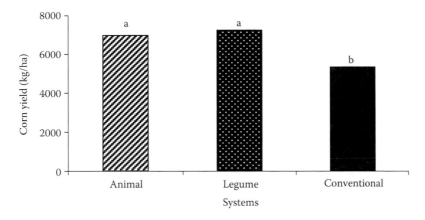

FIGURE 5.7
Average corn yields in drought years (1988, 1994, 1995, 1997, and 1998) for three different farming protocols (Rodale Institute Farming Systems Trial). Different letters above bars indicate statistical differences at the .05 level, using Duncan's multiple range test. (From Pimentel, D. et al., *Organic and Conventional Farming Systems: Environmental and Economic Issues*, Report 05-1, Rodale Institute, Emmaus, PA, 2005. With permission.)

There have been several other studies comparing sustainable agriculture with conventional agriculture, all with favorable findings for sustainable agriculture. The United National Environment Program conducted an analysis (2008) of 114 farming projects in 24 African countries and found that farming yields using organic practices outperformed industrial, chemical-intensive conventional farming and provided such benefits as improved soil fertility, better retention of water, and resistance to drought. In this African study, the yield increase was more than 100% (UNEP and UNCTAD 2008).

A 2007 study of agriculture in the developing world found that organic methods were two to three times more productive than conventional methods (Badgley et al. 2007). And a 2006 examination of yield data from 286 farms in 57 countries found that small farmers increased their crop yields by an average of 79% by using sustainable agriculture (Pretty et al. 2006). Not surprisingly, these studies also found that the more knowledgeable farmers were of organic and other sustainable farming practices, the better their yields become.

A 2001 evaluation of scientifically replicated research from seven state universities, the Rodale Institute, and the Michael Fields Agricultural Institute found that during 154 growing seasons, organically produced crops yielded 95% as much as crops grown using conventional methods (Liebhardt 2001).

In summary, "Yield data just by itself makes the case for a focused and persistent move to regenerative organic farming systems. When we also consider that organic systems are building the health of the soil, sequestering

CO_2, cleaning up the waterways, and returning more economic yield to the farmer, the argument for an Organic Green Revolution becomes overwhelming" (LaSalle et al. 2008).

5.8 Competition for Cropland

A recent global trend is increasing competition for arable cropland. There are two primary drivers of this competition: diversion of existing cropland into nonfood use, namely, the production of biofuels, and massive land purchases by foreign interests, usually in poorer and more vulnerable nations. Such competition is compounded by nonsustainable agricultural practices that damage and erode topsoil, making arable land more scarce. This trend toward competition for cropland, if not corrected, has the potential to push food prices higher and to result in an increased incidence of large-scale famines.

5.8.1 Biofuels and Food Prices

The diversion of existing cropland to the production of biofuels is both motivated by and complicated by the rising cost of fossil fuels, which makes food production and delivery more costly. In 2011, the New England Complex Systems Institute (NECSI) built a dynamic multivariable model that was borne out by actual food prices, unlike models limited only to supply and demand. They found that, while sharp peaks in price are attributable to investor speculation, the underlying cause of a steady upward trend in food prices since 2004 is increased demand for corn due to ethanol conversion (Lagi et al. 2011).

Figure 5.8 illustrates the results of a NECSI simulation of the *FAO [Food and Agriculture Organization of the United Nations] Food Price Index* (blue solid line); the *ethanol supply and demand model* (blue dashed line), where dominant supply shocks are due to the conversion of corn to ethanol so that price changes are proportional to ethanol production; and the results of the *speculator and ethanol model* (red dotted line), which adds speculator trend following and switching among investment markets, including commodities, equities, and bonds. The time period from 2004 to the first quarter of 2011 is simulated. Note the general upward trends.

The potential for food shortages, particularly among the poor, is not just speculative. Between 2006 and 2007, corn prices rose almost 70% in just 6 months, due to an increased demand for ethanol biofuel. In Mexico, this created a food crisis as many people could no longer afford corn tortillas, long a staple of their diet, as they reached prices equivalent to 20% of the minimum daily wage (Runge and Senauer 2007; Sauser 2007; Keleman and Rano 2008). See Chapter 4 for additional details on biofuels.

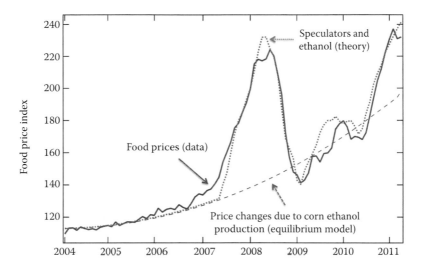

FIGURE 5.8

Food prices and model simulations. Shown are the *FAO Food Price Index* (blue solid line), the *ethanol supply and demand model* (blue dashed line), and an *ethanol model* (red dotted line), which adds speculator trend following and switching among investment markets, including commodities, equities, and bonds. (From Lagi, M. et al., The food crises: A quantitative model of food prices including speculators and ethanol conversion, 2011. http://arxiv.org/abs/1109.4859, With permission of the New England Complex Systems Institute.)

Note that the impact of climbing food prices is not felt equally among rich and poor nations. Lester Brown, in a 2011 article in *Foreign Policy*, stated:

> "For Americans, who spend less than one-tenth of their income in the supermarket, the soaring food prices we've seen so far this year are an annoyance, not a calamity. But for the planet's poorest 2 billion people, who spend 50 to 70 percent of their income on food, these soaring prices may mean going from two meals a day to one. Those who are barely hanging on to the lower rungs of the global economic ladder risk losing their grip entirely" (Brown 2011).

5.8.2 Land Grabs and Food Availability

Massive land purchases by foreign interests (known as "land grabs") widen the gap between wealthy nations and poor nations and leave poor nations open to the increased possibility of famine. When grain and soybean prices soared in 2008, affluent nations dependent upon imports started to buy or lease land in other countries.

According to the Lester Brown article (2011) in *Foreign Policy*, this trend was led by Saudi Arabia, South Korea, and China. Most of these land acquisitions were conducted without consultation with local populations, working with governments (primarily in Africa) willing to lease cropland to foreign interests for less than a dollar an acre per year.

In Liberia, 30% of the country's land was allocated to foreign investors between 2006 and 2011. In 2011, researchers estimated that 9% of South Sudan's land had been leased or bought by investors and that the percentage was growing (Provost 2012).

By the end of 2009, hundreds of "land grabs" had been negotiated. The World Bank reported in 2010 that nearly 140 million acres were involved. This is more than all the acres used to grow wheat and corn combined in the United States. Millions of people in the nations most affected, Ethiopia and Sudan, are already hungry, being fed by the UN World Food Program. Alarmingly, such deals typically include water rights, impacting another increasingly scarce commodity (Brown 2011).

There is serious concern among experts that the "land grab" trend will trigger widespread civil unrest and political instability (Provost 2012).

5.9 Chapter Summary

The costs of conventional or "industrial" agriculture are high and are not fully reflected in the price at the grocery store. The costs and dangers of conventional agriculture to human ecological sustainability include the following: damage to human health and the environment from hormones, pesticides, and herbicides; excessive use of antibiotics and their contribution to the development of antibiotic-resistant bacteria; and the role of agricultural antibiotic use in human and wildlife endocrine disruption. Nitrate pollution from conventional farms plays a significant role in environmental damage that threatens other human food sources (such as seafood) and has direct effects on human health. The loss of topsoil caused by conventional farming practices threatens a famine in the future as arable land is lost at alarming rates. The loss of genetic diversity in our food sources (both plant and animal) leaves us vulnerable to famines as we "put all our eggs in one basket"—a basket that may be knocked over by a new disease or environmental condition. The use of GMOs presents risks that have yet to be fully realized; if the time comes when we no longer want to use GMOs as a major food source, we may no longer have a choice, due to their growing genetic hegemony on the plant.

Sustainable agricultural practices, including organic farming, have the ability to solve many of these problems and save the hidden costs of conventional agriculture. Organic crop yields are competitive with conventional crop yields after an initial start-up period and are superior during drought conditions. Sustainable agriculture presents many advantages that could improve the chances of long-term human sustainability.

Regardless of the agricultural methods being employed, food prices are on a steep upward trend globally, with the greatest impact on the world's poorest and most vulnerable populations. Growing populations, the price of fossil fuels, and the demand for biofuels all impact food prices. This has resulted in some dangerous practices such as international "land grabs" that are likely to lead to increased hunger among the poorer nations, as well as potential unrest.

6

Unconventional Foods: Insects, Plankton, Fungi, and In Vitro Meat

6.1 Introduction

There are three unusual, yet nutritious, invertebrate food sources for humans and their farm animals that can contribute to future human sustainability if they are produced on an industrial level. These include insects, fungi, and plankton. Another unconventional, synthetic food source is from the tissue culture of vertebrate muscle stem cells.

The first source to be described is insects. Some insect species are easily raised industrially and have high food value. This "minilivestock" supplies all the basic nutrients of meats and whole grains. Insects are very efficient at converting plant biomass into their animal biomass. Certain phytophagous insects may prove suitable for industrial-scale mass production, and the United Nations (UN) Food and Agriculture Organization (FAO) is currently training farmers to raise edible insects at the Laos National University's Nabong Campus (KPL 2011). Even if insects or their larvae are not directly consumed by humans, they are a rich proteinaceous food source for farmed vertebrate animals (cattle, sheep, swine, goats, poultry, etc.).

Plant-feeding insects and their larvae are substantially free of transmissible endoparasites and are even eaten uncooked in some cultures. If eaten raw, carnivorous, carrion-eating, feces-eating, and blood-sucking insects can serve as vectors for a number of human parasites and pathogens. These include such benign insects as wild crickets and dung beetles (Olsen 1974).

There is a cultural dichotomy on the matter of human entomophagy (eating insects) between the developed, Western world and parts of Mexico, Central and South America, Africa, Japan, the Indian subcontinent, and other parts of Asia, Australia, and so forth. In Western Europe, the United States, and Canada, insects as human food are typically looked at with revulsion; they are generally used only as food for poultry, swine, freshwater fish, and certain zoo, laboratory, and household pet animals (DeFoliart 1999). In the regions and countries listed above, insects can range from a major portion of the diet to dessert treats. Why do you think this cultural dichotomy on entomophagy exists?

There is a long history of using insects as food. In the Bible, in *Leviticus* xi:21–23, God gave Moses His approval for the Israelites to eat grasshoppers, locusts, and beetles and also proscribed many other kinds of foods including "...all other flying creeping things, which have four feet, shall be an abomination unto you" (Lev. xi:23). Are bats, rats, mice, and lizards intended to be proscribed here? Also in the New Testament, *Matthew* iii:4 says that John the Baptist's meat in the wilderness was locusts with wild honey.

6.2 Nutritional Value of Insects

The consumption of insects as a human and animal food may be an important factor in future human nutritional sustainability (Goodyear 2011). Certain insects (adults and larvae) are rich in lipids, vitamins, proteins, and minerals. Unfortunately, mature insects have chitinous exoskeletons. About 10% of the mass of whole, dried, adult insects is chitin, a linear carbohydrate polymer that is also found in the exoskeletons of crustaceans, krill, certain protozoans, fungi, and certain algae. A section of a 2-D chitin polymer molecule is illustrated in Figure 6.1. It is composed of largely unbranched long chains of *N*-acetyl-D-glucosamine [also known as (aka) GlcNAc]. Chitin, $[C_8H_{13}O_5N]_n$, is only partially digestible by humans (Paoletti et al. 2009; Cohen-Kupiec and Chet 1998) and subtracts from the food value/mass of whole insects. The chitin is much less developed in the soft-bodied larvae of insects, such as dipteran maggots, beetle larvae, and the caterpillars of *Lepidoptera*; hence, they have higher nutrition per kilogram when eaten. Insect chitin can be removed in vitro by alkali (or enzymatic) extraction, improving the food value. When done, the true digestibility of protein concentrate from whole, dried, adult honeybees (*Apis mellifera*) was increased from 71.5% to 94.3%, the protein efficiency ratio (PER) from 1.50 to 2.47, and the net protein utilization (NPU) from 42.5 to 62.0 (DeFoliart 1992). We do not recommend using honeybee protein concentrate as a food; bees are more valuable to ecological sustainability as pollinators.

Animals that routinely eat foods containing chitin (insects, fungi) have chitinase enzymes in their digestive tracts. (Endochitinases cleave the chitin polymer anywhere in the polymer chain; exochitinases cleave chitin from the ends of the molecules.) Such animals include certain insect- and crustacean-eating fish and probably insectivorous mammals such as bats, shrews, skunks, voles, and so forth. Certain intestinal bacteria also can secrete chitinases to aid the digestion of insects. Complete digestion of chitin makes its sugar available for metabolism. Paoletti et al. (2009) found evidence of

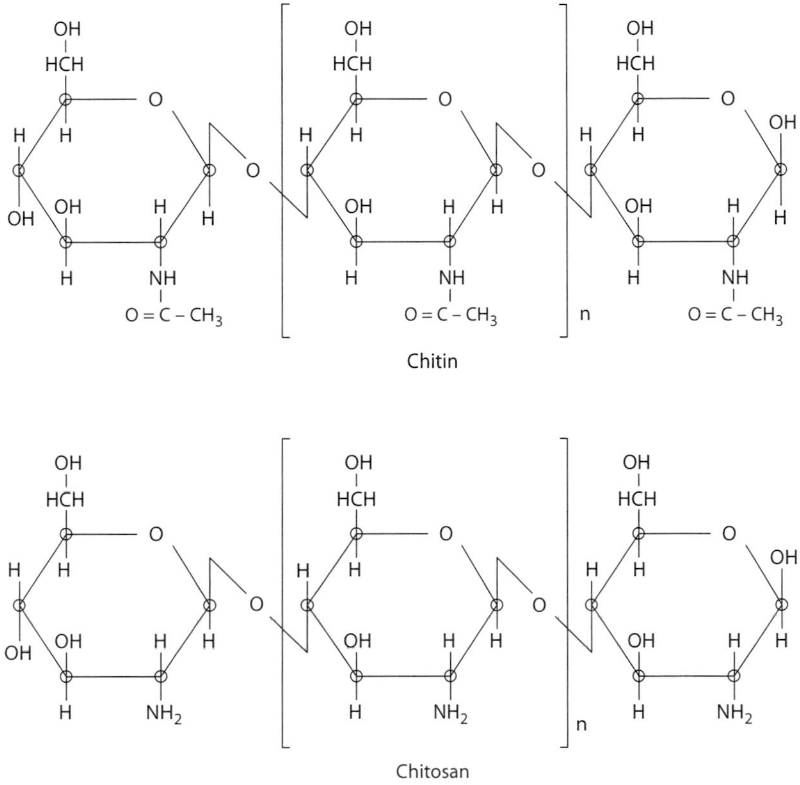

FIGURE 6.1
2-D molecular structure of the chitin and chitosan polymer molecules. Chitin is an oxygen-linked polymer of N-acetyl glucosamine. Chitosan is a polymer of oxygen-linked glucosamine molecules.

chitinase activity in acid human gastric fluid. One wonders if humans who routinely eat insects and fungi would have a higher acid chitinase activity.

In general, insect protein tends to be low in the sulfur-containing AAs, methionine and cysteine, but is high in lysine and threonine, one or both of which may be lacking in the wheat, rice, cassava, and corn-based diets found in the developing world (DeFoliart 1992). Table 6.1 gives the nutritional values of three African insects compared to beef and fish (Data from Parker 2006).

Table 6.2 lists the nutritional values of 10 common insects used for food (data from Nutrition 2000).

One would have to eat approximately 1000 small grasshoppers to obtain the amount of protein in a 12-ounce (340 g) steak (Goodyear 2011).

DeFoliart (1999) has reviewed in detail the entomophagous diets of the non-Western world by continent and country. Two points emerge: (1) Insects

TABLE 6.1

Nutritional Factors Associated with Certain Insects

Nutritional Value, per 100 g	Energy (kcal)	Protein (g)	Iron (mg)	Thiamine Vit. B$_1$ (mg)	Riboflavin Vit. B$_2$ (mg)	Niacin Nicotinic Acid (mg)
Termite (*Macrotermes subhyanlinus*)	613	14.2	0.75	0.13	1.15	0.95
Caterpillar (*Usata terpsichore*)	370	28.2	35.5	3.67	1.91	5.2
Weevil (*Rhynchophorus phoenicis*)	562	6.7	13.1	3.02	2.24	7.8
Beef (lean ground)	219	27.4	3.5	0.09	0.23	6.0
Fish (broiled cod)	170	28.5	1.0	0.08	0.11	3.0

TABLE 6.2

Nutritional Values for Certain Edible Insects, per 100 g Dry Weight

Insect	Protein (g)	Fat (g)	Carbohydrate	Calcium (mg)	Iron (mg)
Giant water bug	19.8	8.3	2.1	43.5	13.6
Red ant	13.9	3.5	2.9	47.8	5.7
Silkworm pupae	9.6	5.6	2.3	41.7	1.8
Dung beetle	17.2	4.3	0.2	30.9	7.7
Cricket (*Acheta domestica*)	12.9	5.5	5.1	75.8	9.5
Small grasshopper	20.6	6.1	3.9	35.2	5.0
Large grasshopper	14.3	3.3	2.2	27.5	3.0
June beetle	13.4	1.4	2.9	22.6	6.0
Meal worm (*Tenebrio molitor*)	20.3	12.7	—	—	—
Mopane worm[a]	48	7.7	25	16	12.7

[a] Mopane worm data from Goodscience. 2011. Nutritional value of mopane worms. 4 pp. Web paper. Available at: http://informalscientist.com/tag/nutritional-value-of-mopane-worms/, accessed May 13, 2011. Mopane "worms" are the large larva (caterpillars) of the South African Emperor moth (*Gonimbrasia belina*, family Saturnidae).

are widely eaten (both raw and cooked), in some cases forming a significant caloric fraction of human diets. (2) Most of the insects eaten are gathered in their habitats and not farmed.

6.3 Can Insects Be Farmed?

If Western humans were to overcome their phobias of entomophagy and rely on insects to any degree or form as a direct food source, means of raising various insect species at industrial levels would need to be developed. That is, insects would have to be factory-farmed. Two species that are currently farmed in the United States for reptile, bird, and fish food are described below.

1. *Meal worms* (*Tenebrio molitor*) are relatively easy to raise. They are beetle larvae having human food potential. They require a dry, high-protein food, such as a chick starter mash bran cereal, or possibly distiller's solids residue. *Tenebrio* larvae can get their water from slices of potato, cabbage hearts, apple slices, and so forth in their brood box. Their life cycle takes approximately 3 to 6 months, that is, eggs to worms to pupae to adult egg-laying beetles. Current practice for raising mealworms uses large plastic boxes, such as used for storage of woolens. The tops of the boxes are given screened openings for ventilation and to keep the beetles in and other critters out. A variety of foods can be used; the larvae are not fussy. Examples are as follows: (1) wheat bran plus dried skim milk (3:1 ratio); (2) chick starter mash (nonmedicated) in 4 layers 1/4 in. thick, separated by burlap or newspapers; and (3) crushed oat or wheat kernels plus whole wheat flour plus wheat germ or powdered milk plus brewer's yeast in a 10:10:1:1 ratio. Other foods may include oatmeal, cornmeal, oat bran, ground-up dry dog or cat food, leftover or stale low-sugar cereals, and so forth. The optimum rearing temperature is around 80°F with a relative humidity of approximately 70%. Every few weeks, the dead beetles, shed worm exoskeletons, and pupa cases are sifted out. Dried-up and eaten potatoes and vegetable remains may contain beetle eggs, so they are put in a new colony with fresh moisture food (unless they have mold, in which case they are disposed of). Once the worms are big enough, they are sifted out (harvested). Old bedding and worm waste (frass) is disposed of in a garden compost pile. The brooding box is washed out thoroughly and dried, and then mealworms are added, with fresh food and moisture food, to grow, pupate, and breed. Detailed procedures for raising mealworms can be found in Sialis (2011), Newsletter (1996) (also recipes), and Beckham (2011).

2. *Crickets: Achetus domesticus,* the house cricket, is another easily raised food insect species. It is commercially available as food for raising fish, snakes, mantises, and so forth. As with mealworms, a cricket life-cycle colony can be established in a plastic storage box (26 × 14 × 16 in. deep). Unlike mealworms, crickets do like water—it can be supplied from a small chick waterer (an inverted Mason jar screwed into a plastic trough base). Plastic souring pads are placed in the trough of the waterer to keep the crickets from drowning. Fiber egg "flats" that are stacked in the rearing box to within 4 in. of the top give the crickets places to hide and socialize. Mated females need to oviposit in a damp sand nesting material, so one or two pint containers of

sand are placed near the waterer. Once a good density of eggs is found 1–2 in. below the sand surface, the original lid is placed on the nesting container, and it is placed on top of the heat pad on top of the breeding box. The first instar crickets hatch in approximately 7–10 days, and then these tiny nymphs can be placed in a fresh breeding box to grow.

Food for the crickets can consist of alfalfa pellets plus dry cat food rolled in a mixture of skim milk powder or a calcium supplement powder (10:1 mixture). This food is placed in a dish on top of the egg cartons. Fresh fruit and vegetable scraps can be fed but must be removed before they mold or rot.

Crickets thrive at temperatures between 80°F and 90°F. A thermostatically controlled cage heater or an adjustable heating pad might be used. Detailed instructions on cricket raising can be found in Kaplan (2009) and also Crickets (2011). Commercial production of the common black field cricket, *Gryllus assimilis,* may also be practical.

It is generally better to eat insects in their larval stages to avoid the thick chitin of adults. An exception is adult and nymph crickets and grasshoppers. With these orthopterous insects, it is best to remove their wings and legs to maximize the protein-to-chitin mass ratio before cooking and eating. Mealworms in their later molt stages also tend to be more chitinous. However, their chitinous surface area scales roughly as the 2/3 power of their mass. In other words, one gets a higher ratio of soft flesh (muscle, fat) to chitin mass as the insect's size increases.

Other rapidly growing insects, such as certain species of flies, might also be produced industrially. To maximize the ratio of protein to chitin, the maggots would be harvested and processed.

Farmed insects may never constitute a major food source, but in areas of the world where entomophagy is accepted, they may provide an important dietary protein supplement for both humans and farm animals.

6.4 Plankton as a Source of Human Food

Oceanic plankton can be classified as either phytoplankton (containing chlorophyll) or zooplankton. All planktons are small, ranging from microscopic to some zooplanktons (e.g., Antarctic krill) that are as much as 60 mm in length. Phytoplanktons (e.g., diatoms) live by converting dissolved CO_2 to sugars and other metabolites through photosynthesis. They provide basic food for zooplankton, which in turn are the basic food stock of small fish, baleen whales, and even the giant whale shark, a filter feeder. The food values of various planktons were summarized by Swamy (1974). His data are presented in Table 6.3.

TABLE 6.3

Nutritional Components of Certain Plankton as Percentage of Dry Weight

Plankton (*Phylum*)	Protein %	Fat %	Carbohydrate %	P_2O_5 %	Nitrogen %	Ash %
Copepods (*Arthropoda*)	70.9–77.0	4.6–19.2	0–4.4	0.9–2.6	11.1–12.0	4.2–6.4
Sagittae (*Chaetognatha*)	69.6	1.9	13.9	3.6	10.9	16.3
Diatoms (*Heterokontophyta*)	24.0–48.1	2.0–10.4	30.7	0.9–3.7	3.8–7.5	30.4–59.0
Dinoflagellates (*Dinoflagellata*)	40.9–66.2	2.4–6.0	5.9–36.1	0.7–2.9	6.4–10.3	12.2–26.5

Source: Swamy, P.K., *Seafood Export Journal* 64, 23–26, 1974.

Copepods may be the dominant members of the zooplankton; they are a significant food organism for small fish, baleen whales, the whale shark, seabirds, and other planktonic crustaceans such as krill. They typically measure 1 to 2 mm in length and have large antennae and a teardrop-shaped body. Some polar ocean copepods reach 10 mm in length. Copepods have a single, median, compound eye, generally bright red. Copepod species are also are found in wet, freshwater, terrestrial habitats (wet forests, bogs, springs, ponds, puddles, damp moss, stream beds, etc.).

Cocccolithophores are single-cell phytoplankton of the phylum Haptophyta. They are spherical eukaryotic cells, approximately 15–100 μm in diameter, enclosed by many calcareous plates called coccoliths, which are about 2–25 μm across. The coccoliths are composed of calcium carbonate, not silicates. Under a scanning electron microscope, the coccoliths look somewhat like two-hole buttons.

Sagittae are in the phylum Chaetognatha. Adult *Sagitta lyra* are typically 42 mm long with a tail taking up 15%–17% of their length. *Sagittae* are found in the warmer oceanic regions between 40°N and 40°S. Most *Chaetognaths* are transparent and torpedo shaped; they range in length from 2 to 120 mm and are approximately 8 times their diameter in length. They have two compound eyes, each consisting of a number of pigment-cup ocelli fused together. *Chaetognaths* swim in short bursts using a dorsoventral undulating motion of their flattish tails, which can be from 15% to 45% of their body length. There are more than 120 modern *Chaetognath* species assigned to over 20 genera.

Diatoms are a major group of algae and are one of the most common types of unicellular phytoplankton. Diatom cells are characteristically encased in a unique cell wall made from a pair of hydrated silicon dioxide (silica) shells, called a frustule. Diatoms live in both freshwater and marine environments and range from approximately 2 to 200 μm in diameter. The frustules of various diatom species have a wide range of shapes, symmetries, and designs (Brodie 2004).

Not all planktons are suitable for food. *Dinoflagellates* are a large phylum of flagellate protists, found as marine plankton as well as in freshwater habitats. About half of all dinoflagellates do photosynthesis, and these make up the second largest group of eukaryotic marine algae compared to the diatoms. *Dinoflagellates* sometimes bloom in concentrations of more than 10^6 cells/mL. Some species (*Gonyaulax* sp.) color the water reddish brown, producing a "red tide," and they produce a neurotoxin (saxitoxin) that can accumulate in filter feeders such as oysters, clams, and mussels, which can poison humans who eat these toxic mollusks (EFSA 2009; ISSHA 2010; PSP 2012; Paralytic 2012).

Under the scanning electron microscope, the dinoflagellate *Gonyaulax polygramma* has an interesting exterior morphology: the "top" *epitheca* (shell) resembles an inverted funnel sitting on a bowl-shaped bottom, the *hypotheca*. Where the top and bottom join, there is an "equatorial" groove, the *cingulum*. The top, the *epitheca*, has longitudinal ridges for reinforcement. The bottom, the *hypotheca*, has two downward-pointing spines on its bottom. The *cingulum* contains one cilium, and a vertical groove in the *hypotheca* contains another cilium, providing mobility for the cell. Equatorial diameters of *Gonyaulax* sp. range from 30 to 66 µm, and the distance from the tip of the *epitheca* to the bottom of the *hypotheca* is 26–75 µm (Identifying 2011).

Swamy (1974) estimated that one has to filter approximately 5 million gal. (18,927 m^3) of seawater to recover 1 lb. (0.454 kg) of plankton. Presumably, he was talking about microplanktons in a "normal" concentration, too small to effectively net. With the use of enhanced sonar to locate plankton blooms, this ratio might be improved to less than 1 million gal. to 1 kg of planktons recovered. Pumping seawater from a depth of 200 m or more is energy intensive and hence would be very expensive for a very small protein return. Also consider the possible effects on the fisheries whose fish are dependent on zooplanktons for food if we start harvesting Arctic or Antarctic planktons en masse. Harvesting planktons for human food without regulation does not appear to be a good idea. Running a sustainable fishery is probably a better path to food sustainability.

Krill are small, shrimp-like marine crustaceans of the order Euphausiacea. They are found in greatest concentrations in cold Antarctic and Arctic waters; the estimated biomass of Antarctic krill, *Euphausia surperba*, is from 125 to 725 million t (1.25–7.25 × 10^{11} kg) (Krill 2012; Webster: krill fishery 2012). Of this mass, over half is eaten by whales, seals, penguins, fish, and squid each year and is replaced by robust yearly reproductive growth. Most adult krill are about 1–2 cm long as adults; the Antarctic *Euphausia surperba* adults are approximately 6 cm (2 3/8 in.), weigh up to 2 g, and live up to 6 years. They are classified as megaplanktons. *Euphausia* feed primarily on phytoplanktons (diatoms) but are generally omnivorous, preying on smaller zooplanktons as well (Webster: krill 2012; Nicol and Endo 1997).

Because of their larger size, krill can be harvested with nets; about 0.1 M t of krill are taken yearly in the Antarctic waters. Nicol and Endo (1997) reported that a total of 528,201 t of Antarctic krill (*Euphausia superba*) was taken in 1982, 93% of which by the former Union of Soviet Socialist Republics (USSR). The USSR caught 275,495 t in 1991, and the Ukraine and Russia reported a total of 199,029 t in 1992, and then the catch fell abruptly to 9036 t in 1993. In 2007, the annual Antarctic krill catch was approximately 100,000 t; the nation taking the largest krill catch was Japan, followed by South Korea, Ukraine, and Poland (Webster: krill 2012).

Krill have been used as a food source (*okiami*) in Japan for several centuries. They taste salty and somewhat stronger than shrimp. The chitinous exoskeletons of food krill must be peeled before consumption because they contain fluorides [e.g., $Ca_5(PO_4)_3F$], up to 1500 mg/kg, which are toxic in high concentrations. (Chitin is the polymer *poly-N-acetyl-D-glucosamine*.) The EC has set the upper limit for fluoride in animal feed, including feed for farmed fish, at 150 mg/kg. Krill oil is now used as a patent dietary supplement for a variety of human maladies. It contains the omega-3 oils docosahexaenoic acid (DHA) and eicosapentaenoic acid (EPA). A large percentage of the world krill catch is frozen and used for animal food and fish bait, and also as food for farmed fish (Suontama 2006).

Determination of the optimum, sustainable harvest of krill and other planktons is a complex issue. Obviously, planktons play a pivotal role as a food (energy source) in marine ecosystems and fisheries. Phytoplanktons capture solar energy and dissolved CO_2 and convert them to useful sugars, proteins, and so forth. They are consumed by generally larger zooplanktons, which in turn are food for fish, whales, penguins, humans, and so forth. We suspect that as with other fisheries, there will be a tipping point if excess zooplanktons are harvested. This will adversely affect fisheries that man depends on for food. As an example, Weier (1999) stated, "Disturbance of an ecosystem resulting in a decline in the krill population can have far-reaching effects. During a coccolithophore bloom in the Bering Sea in 1998, for instance, the diatom concentration dropped in the affected area. Krill cannot feed on the smaller coccolithophores, and consequently the krill population (mainly *E. pacifica*) in that region declined sharply. This in turn affected other species: the shearwater [a seabird] population dropped, and the incident was even thought to have been a reason for salmon not returning to the rivers of western Alaska that season." Note that this was a natural decrement on the northern krill population, not caused by a man-made overharvest. There is evidently a high sensitivity of certain fish population densities to krill population variations.

Oceanic phytoplankton blooms appear to be becoming more intense worldwide. This may be caused by larger deepwater upwelling currents carrying nutrients such as dissolved nitrates, phosphates, silicates, carbonates, magnesium, iron ions, and so forth that stimulate rapid phytoplankton growth in the warmer coastal waters in spring and summer, producing a population far in excess of what zooplanktons can normally keep in check (SERC 2011; Biello 2010b). These excess phytoplanktons die, sink to the ocean bottom, and decay. This decay process depletes local dissolved O_2 concentration, causing the zooplanktons to also die off, depriving small fish of food and limiting their population. Fewer small fish are then available to feed larger fish; hence, their population, too, can collapse. The increased frequency of upwelling currents may be due to global warming in the oceans, more intense oceanic storms, and other unknown factors.

James et al. (2003) considered the interactions between fish larvae, zooplankton, and phytoplankton using mathematical models. They showed the theoretical conditions under which fish larvae feeding on zooplankton could cause a phytoplankton bloom and population collapse.

Another scenario that was witnessed off the coast of Spain is that local overfishing of sardines, a small zooplanktonic predator fish, allowed the zooplankton population they normally hold in check to bloom. The excess zooplanktons depleted the local phytoplanktons and then died. The microorganisms feeding on their decaying biomass rendered the water low in oxygen, and sulfate-reducing bacteria that fed on their dead biomass secreted toxic hydrogen sulfide gas (H_2S), which discouraged the return of fish and zooplanktons, creating a temporary ecological "dead zone." [The lethal concentration of H_2S gas for 50% of humans (LC_{50}) is about 800 ppm, but severe peripheral neural damage begins at 1/10 that LC_{50}. Fish are vulnerable, too, when they swim in a sea of dissolved H_2S.]

Clearly, our harvesting of top-of-the-food chain zooplankton such as krill must be done along with appropriate ecological sampling of other plankton densities, fish populations, and so forth, to insure that the krill fishery remains viable and sustainable and that a tipping point is not being approached for other marine animals dependent on the phytoplankton and zooplankton ecosystems. (See Section 4.1.5: *Ecosystem Interactions* and Section 4.1.6: *Ecosystem Monitoring* in Nicol and Endo 1997.) Ecosystem sampling will be expensive and probably will not be done at a level required to feed valid data into a krill fishery management model. Thus, harvesting zooplankton for food appears to potentially be a major threat to our ocean ecosystems' sustainability. Time will tell.

6.5 Fungi: Food and More

6.5.1 Introduction

In this section, we introduce the kingdom Fungi and describe its enormous taxonomic diversity and the properties of many of its members, both beneficial and harmful to human sustainability. There may be over 1.5 million species of fungi, and about 10^5 species have been taxonomically specified. However, the taxonomy of fungi is in flux; older taxonomies were based on physical appearance and habitat; revisions are occurring because of DNA genome typing. Fungi belong to the domain Eukarya and kingdom Fungi, and they can be classified into six divisions:

Ascomycota

Basidiomycota

Chytridiomycota

Glomeromycota

Microsporidia

Zygomycota

DNA testing has also led to revisions in the branching fungal phylogeny (Palaeos 2008; Wang et al. 2009). Even with modern DNA testing, there are many fungi *incertae sedis* (*Latin* for uncertain dwelling place), which defy phylogenetic classification. Although fungi are eukaryotes, their cell walls are composed of glucans and chitin; fungal cell walls do not contain cellulose, as found in plant cells.

Fungi may be classified as beneficial to human sustainability if they are edible, or useful, such as the yeast *Saccharomyces* spp. used in fermentation (beer, wine, soy sauce), raising bread, and so forth; those species that are used in cheese-making; those that can produce antibiotics (e.g., penicillin); and those that can act as biological pesticides. Also in the list of useful fungi are those used in industrial processes to manufacture chemicals including citric, gluconic, lactic, and malic acids, as well as industrial catalysts such as lipases used in biological detergents, cellulases used in making cellulosic ethanol, invertases, proteases, and xylanases. Some beneficial fungi can be used to control insect pests. For example, the fungus *Beauveria bassiana* will parasitize and kill crop-eating grasshoppers. Other fungi used as biological insecticides are *Metarhizium* spp., *Hirsutella* spp., *Paecilomyces (Isaria)* spp., and *Lecanicillium muscarium* (Costa 2009; Deshpande 1999).

Harmful fungi are poisonous when ingested, infect living plants and animals, or spoil food. Because of the many types of fungi, their biology is complex (sex, reproduction, nutrition, habitats), and we refer the interested reader to texts on fungal biology, for example, Deacon (2005), Alexopoulos et al. (1996), and Jennings and Lysek (1996).

6.5.2 Edible Fungi

6.5.2.1 Introduction

Most of the edible fungi are mushrooms, although the mycelia of a fungus have been processed into nutritious vegetarian food, Quorn (see below). Certain mushrooms, truffles, puff balls, and bracket fungi are also edible and nutritious. They contain vitamins, minerals, and little fats, and their proteins contain all 10 amino acids (AAs) essential for human nutrition (essential AAs cannot be synthesized by the human body in adequate quantities to meet human nutrition) (Northrop and Connor 2009). Mushrooms also contain the biopolymers chitin and chitosan (part of their cell walls), which are largely indigestible by humans.

6.5.2.2 Quorn

Quorn is the first commercially produced, processed mycoprotein food derived from the filamentous fungi *Fusarium venenatum,* strain PTA-2684, which is grown industrially in large vats. It is marketed largely in the United Kingdom and Europe and has recently become available in the United States. It is sold as a health food and as a vegetarian alternative to meat. Potato protein or egg white (albumin) is used as a binder to give it texture. Quorn was first developed in the United Kingdom in 1985 from a joint venture between Rank Hovis McDougall Ltd. (RHM) and Imperial Chemical Industries (ICI) Ltd. Quorn was first distributed in the United Kingdom in 1994, entered the European market in the later 1990s, and was sold in the United States after 2002 (Mycoprotein 2012; Quorn 2011b).

The parasitic soil mold *Fusarium venenatum* is grown in oxygenated water in large, sterile fermentation tanks; glucose, vitamins, and minerals are added to improve the food value of the product. The product mycoprotein is heat-treated to break down excessive levels of RNA; this is to lower the excessive purine content. The product is dried and mixed with a binder; it is then textured, flavored (beef, chicken, pork, mushroom, etc.), colored, cut to size, cooked, and then frozen. Some of the Quorn products available in the United States include the following: Classic Burger, Cheese Burger, Southwestern Chik'n Wing, Chik'n Patties, Naked Chik'n Cutlet, Garlic & Herb Chik'n Cutlet, Cranberry & Goat Cheese Chik'n Cutlet, Chik'n Nuggets, Gruyere Chik'n Cutlet, and Turk'y Roast (Quorn 2011a). The Quorn website also lists 10 recipes for Quorn products. Quorn products are said to naturally be low in fat, high in protein, easy to prepare, and flavorful, and have zero cholesterol.

The marketing of Quorn products is not without controversy. Some people have reported symptoms of food allergy after having eaten Quorn products, generally nausea and diarrhea. This led to competitors [the American Mushroom Institute and Gardenburger (a soy product)] and the Center for Science in the Public Interest (CSPI) to file various suits and complaints with advertising and marketing standards watchdog groups in the European Union and United States. CSPI claimed that in 2003, Quorn sickened 4.5% of eaters and should be removed from stores (Warner 2005). This claim was argued to be specious by the manufacturer's figures, which claimed that there were 1 in 146,000 adverse reactions (0.0007%) to Quorn. CSPI's claims were also refuted by a professor of nutrition at the University of Pittsburgh, as well as a writer for the Fox News channel (Milloy 2002). CSPI is continuing to try to get the Food and Drug Administration (FDA) to classify Quorn food products as potentially dangerous allergens to some people by citing undocumented statistics and anecdotal evidence of allergic sensitivity to Quorn in letters to the FDA. In a recent letter, CSPI (2011) claimed that 5% of sensitive individuals reacted adversely to Quorn, compared to 3% for shell-fish, 2% each for milk and peanuts, and 1% for gluten. [In fact, 0.3%–1% of the general US population is genetically gluten intolerant and reacts with a wide spectrum of symptoms from no apparent reaction to severe, life-threatening allergic symptoms (Arora 2011; Greco 1995).] CSPI suggested that Quorn products should carry a prominent allergic warning label, and they also urged that the FDA "...revoke the GRAS status of mycoprotein and get this dangerous product off grocery store shelves." (The FDA's GRAS desig-nation stands for "generally recognized as safe.") We note that Quorn Chik'n Tenders and Beef-style Grounds are gluten-free but are manufactured in a facility that also processes gluten-containing products.

Evidently, the mold *Fusarium venenatum* is easy to grow in batches and has simple nutritional needs, which makes it ideal for large-scale, industrial production of Quorn mycoprotein food products.

6.5.2.3 Edible Mushrooms

Edible mushrooms are the fleshy and edible fruiting bodies of a number of species of the fungal kingdom. Evidence of humans using mushrooms as food goes back approximately 13,000 years to archeological sites in Chile. The pre-Christian era Romans and Chinese also consumed mushrooms (Boa 2004). Edibility of mushrooms can be defined in terms of an absence of toxic effects and the presence of desirable taste, aroma, and texture. Edible mush-rooms can be divided into two categories: those gathered in their wild habi-tat and factory farm-raised mushrooms. Wild mushrooms, unless gathered by an expert, present a risk to the consumer because many toxic mushrooms in certain stages of growth resemble their edible relatives.

The top 10 edible mushroom– and truffle-producing countries are listed in Table 6.4.

Figure 6.2 shows a plot of world mushroom and truffle production, drawn from FAOSTAT data cited above. The continuous curve was fit by eye. The data points are plotted to the closest 125,000 t.

The species listed below are commonly harvested from the wild and are consumed with a variety of foods (steaks, salads, stews, gravies, etc.) or just grilled. They include the following:

Agaricus bisporus, the white button mushrooms; when mature, the por-tobello mushroom *A. bisporus is raised commercially.*

Boletus edulis, or edible boletus, native to Europe.

Cantharellus cibarius, the chanterelle. Found worldwide.

Cantharellus tubiformus, the tube chanterelle or yellow-leg.

Clitocybe nuda, the blewit.

TABLE 6.4

Mushroom Production Data for 2008

Country	Output in Tonnes (2008)	% of Total World Output
China	1,608,219	45.9
United States	363,560	10.4
Netherlands	240,000	6.86
Poland	180,000	5.15
France	150,450	4.30
Spain	131,974	3.77
Italy	100,000	2.86
Canada	86,946	2.49
Ireland	75,000	2.14
Japan	67,000	1.92
Top 10 total (2008)	3,003,149	–
World (2009)	6,941,858	–

Source: Data from UN Food and Agriculture Organization statistics web site (FAOSTAT).

Note: World production was 857,987 t in 1971.

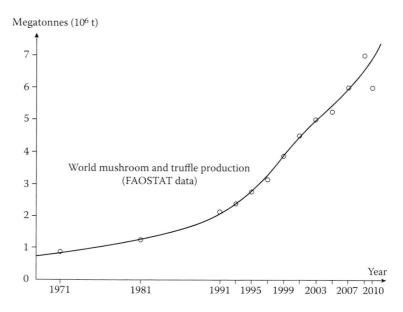

FIGURE 6.2

History of world mushroom and truffle production, 1917–2010. (Plotted from FAOSTAT data.)

Cortinarius caperatus, the gypsy mushroom.

Craterellus cornicopioides, the horn of plenty or trompette du mort.

Grifola frondosa, the "hen of the woods" or "sheep's head"; in Japan, known as the *maitake*.

Gyromitra esculenta, the false morel; deadly raw, must be parboiled to detoxify.

Hericium erinaceus, a tooth fungus, aka "lion's mane mushroom."

Hydnum repandum, the "sweet tooth fungus," "hedgehog mushroom," "urchin of the woods."

Lactarius deliciosus, the saffron milk cap.

Morchella sp., the morel family, which belongs to the *ascomycete* group of fungi. Can be confused with the toxic *Gyromitra esculenta*.

Tricholoma matsutake, the famous Japanese *matsutake* mushroom.

Tuber sp., the well-known truffles. Some have been domesticated: for example, *T. borchii, T. brumale, T. indicum, T. macrosporum, T. mesentericum,* and *T. uncinatum*.

In addition, young wild puffballs (*Lycoperdon* sp. and *Calvatia* sp.) are quite edible, and so are the oyster mushroom, *Pleurotus ostreatus,* and the sulfur shelf (aka bracket fungus), *Laetiporus suphureus,* which grows in clusters on dead trees. Note that it is possible to confuse immature poison *Amanita phalloides* mushrooms with immature puffballs.

The nutritional value of mushrooms makes them a useful food for human sustainability. Mushrooms are excellent sources of thiamin, riboflavin (vitamin B_2), niacin (vitamin B_3), pantothenic acid (vitamin B_5), vitamin B_6, vitamin C, vitamin D, folate, iron, phosphorus, and selenium, which helps protect the body's cells from free radical damage and also helps the thyroid gland to function properly. There is no vitamin A, B_{12}, E, K, or folic acid, however (Nature's Dry 2009). Mushrooms generally are very low in lipids, high in potassium, and low in sodium ions; they also contain zinc, copper, and manganese ions. Their proteins generally contain 18 AAs, including the following: alanine (Ala), arginine (Arg), aspartic acid (Asp), cysteine (Cys), glutamic acid (Glu), glycine (Gly), histidine (His), isoleucine (Ile), leucine (Leu), lysine (Lys), methionine (Met), phenylalanine (Phe), proline (Pro), serine (Ser), threonine (Thr), tryptophan (Trp), tyrosine (Tyr), and valine (Val). Not present are glutamine (Gln) and asparagine (Asn) (Vegan 2011; Nature's Dry 2009). The 10 essential AAs for humans (Arg, His, Ile, Leu, Lys, Met, Phe, Thr, Trp, and Val) are present, however, in mushroom proteins (Northrop and Connor 2009).

6.5.2.4 Mushroom Growth Media

Commercial mushroom cultivation is done in the dark in caves or temperature-, light-, and humidity-controlled barns. The spores are started on a growth medium having two parts: a lower compost layer, which is covered by a second, top layer known as casing soil. The casing soil assists in maintaining a desirable moisture level, CO_2/O_2 ratio, and neutral pH. Sphagnum peat (partially decomposed sphagnum moss) is widely used as casing soil by most mushroom growers. In 2003, J.W. Stamp (2003) patented a non-sphagnum peat casing soil formula for mushroom growth. Stamp's recipe called for approximately 6.25 parts by weight of sedge peat or coconut fiber (shredded coconut coir) peat and approximately 3.75 parts by weight of sugarcane mill mud, adding water to adjust the moisture content to approximately 75% and adjusting the pH to between 6.7 and 7.2 using gypsum and/or lime. (Sugarcane mill mud is the material washed from sugarcane mills; it contains cane washings, lime, cane juice impurities, sugar, and fine bagasse.)

6.5.2.5 Poisonous Fungi

The toxicity of certain mushroom species varies from mild to deadly. Poisonous mushrooms have been used in historically famous assassinations, and notable persons have been killed by accidentally eating poisonous mushrooms. A majority of the accidental poisonings have resulted from mistaken identity when gathering wild mushrooms. The mushroom that accounts for the majority of fatal poisonings worldwide is the well-known *A. phalloides* ("death cap"), which contains the toxins α- and β-*amanitin* and *phallotoxin*, among others. *A. phalloides* is native to Europe and the United Kingdom. α-Amanitin and the related β-amanitin act to inhibit RNA polymerase II, an enzyme critical for the biosynthesis of messenger RNA (mRNA), micro RNAs, and small nuclear RNAs (snRNAs). Without mRNA, essential intracellular protein synthesis stops, cell metabolism halts, and the cell dies. The liver and kidneys are the organs most seriously affected. The toxic effects of the amanitins and phallotoxin are not expressed for 6 to 12 h. Then there is acute gastrointestinal (GI) distress (vomiting, diarrhea) that lasts approximately 24 h, followed by acute liver damage. Mortality from *A. phalloides* toxins is about 10%–15%. A liver transplant may be indicated.

Other poison mushrooms are noted for their psychogenic effects. Their hallucinogenic agents include psilocybin (converted in the body to psilocin) and ibotenic acid (metabolized into muscimol). Some fungal species of the genus *Psilocybe* contain psilocybin, as do some other mushrooms of the genera *Panaeolus, Copelandia, Conocybe,* and *Gymnopilus. Amanita muscaria, A. pantherina,* and *A. gemmata* contain ibotenic acid.

The plant fungus *Claviceps purpurea* infects rye plants and produces ergot alkaloids (e.g., ergotamine), which, when ingested in bread, can produce dangerous circulatory and psychogenic effects (St. Anthony's fire).

In Table 6.5, we summarize a number of investigated fungal toxins and their effects; most are from mushrooms.

TABLE 6.5

Mushroom and Fungal Toxins, Their Sources and Symptoms

Toxins	Type of Mushroom	Effects on Humans
Isoxazole derivatives (muscimol, ibotenic acid, etc.)	*Amanita muscaria, A. pantherina, A. gemmata, A. multisquamosa, A. frostiana, A. crenulata, A. strobiformis, Tricholoma muscarium*	Confusion, visual distortion, delusions, hallucinations, convulsions, drowsiness, and coma that can last 24+ h; nausea and vomiting also common
Amanitin (amatoxins), including alpha-amatoxin, which inhibits RNA polymerase II and protein synthesis; also beta-amatoxin	*A. phalloides* ("death cap"), *A. virosa* ("destroying angel"), *A. verna* ("fool's mushroom"), *A. ocreata, A. bisporigera, Conocybe filaris, Galerina autumnalis* ("autumn skullcap"), *G. marginata, G. venenata*	Mortality rate is 50% 1. Latency period of 6–24 h, during which toxins damage kidneys and liver, but the victim experiences no discomfort 2. 24 h of violent vomiting, bloody diarrhea, and severe abdominal cramps 3. 24 h of false "recovery" 4. Relapse with renal and hepatic failure or internal bleeding (often fatal)
Allenic norleucine (2-amino-4,5-hexadienoic acid)	*Amanita smithiana, A. proxima, A. pseudoporphyria*	GI distress, anxiety, chills, cramps, disorientation, renal failure, sweating, weakness, oliguria, polyurea, thirst
Muscarine	*Inocybe* sp., *Clitocybe dealbata, Omphalatus* sp. ("jack-o'-lantern" mushrooms), certain red-pored *Boletus* sp.	Excessive glandular secretions (sweat, salivation, tears, etc.), visual disturbances, irregular pulse, decreased blood pressure, difficulty breathing (can lead to respiratory failure), severe vomiting and diarrhea
Gyromitrin: Gyromitrin's hydrolysis product is monomethylhydrazine, a colorless, volatile, highly toxic, carcinogenic compound, sometimes used as rocket fuel	*Gyromitra* sp., including *G. esculenta, G. ambigua, G. infula, G. montanum, G. gigas, G. fastigiata, G. californica, G. sphaerospora,* and related Ascomycetes such as some species of *Helvella, Verpa,* and *Cudonia*	Headaches, abdominal distress, severe diarrhea and vomiting. Renal, hepatic, and red blood cell damage may occur, possibly resulting in death. Increased future risk of cancer

(continued)

TABLE 6.5 (Continued)

Mushroom and Fungal Toxins, Their Sources and Symptoms

Toxins	Type of Mushroom	Effects on Humans
Orellanine	*Cortinarius orellanus, C. rubellus, C. splendens, C. atrovirens, C. venenosus, C. gentillus*	Onset may be delayed up to 3 weeks. Nausea, vomiting, lethargy, anorexia, frequent urination, severe thirst, headache, feeling cold and shivering, renal failure
Psilocybin, psilocin, and other indole derivatives (serotonin agonists)	*Psylocybe* sp. including *P. cyanescens, P. stuntzii; P. cubensis, P. semilanceata; Panaeolous* sp. including *P. cyanescens and P. subbalteatus; Gymnopilus* sp. including *G. spectabilis*	Psychedelic effects; heightened color perception, emotional effects such as ecstasy or anxiety, hallucinations or delusions
Involutin	*Paxillus involutus*	Abdominal pain, nausea, vomiting, diarrhea; immune complex-mediated hemolytic anemia with hemoglobinuria, oliguria, anuria, and renal failure
Coprine: causes symptoms when alcohol is ingested: blocks acetaldehyde dehydrogenase	*Coprinopsis atramentaria, C. insignis, C. quadrifidus, C. variegatus*	Tachycardia, heart palpitations, tingling arms and legs, warmth and flushing, and sometimes, headache, heavy limbs, and salivation

Sources: NAMA, Mushroom poisoning syndromes, North American Mycological Association, 2012. www.namyco.org/toxicology/poison_syndromes. html; Habal, R. and J. Martinez, Mushroom toxicity. Medscape, July 27, 2011. http://emedicine.medscape.com/article/167398-overview#a010; Fischer, D., A detailed look at America's poisonous mushrooms, 2011. http://americanmushrooms.com/toxicms.htm.

6.5.2.6   Harmful Fungi

A number of fungal species work against human sustainability by infecting grain crops (wheat, rice, rye, etc.) and decreasing yields, and in the case of rye, rendering the grain and its products poisonous. Wheat is infected by *Puccinia graminis*, the black stem wheat rust; European barberry, *Berberis vulgaris*, is an intermediate host for this destructive fungus (Stokstad 2007). *Fusarium solani f. sp. glycine* attacks the roots of soybean plants under cool, wet soil conditions, causing "sudden death" syndrome (Huffstutter 2011). The rice blast fungus, *Magnaporthe grisea* (aka *Magnaporthe oryzae*), is a complex of similar species that principally infect rice species but also can attack other cereals including wheat, rye, barley, and pearl millet, causing blast or blight disease. Each year, *M. grisea* destroys enough rice that potentially could feed more than 60 million people! It is known to occur in approximately 85 countries (Scardaci 2003; Sesma and Osbourn 2004). Researchers at the University of Exeter (United Kingdom) found a single *M. grisea* gene responsible for its infectivity. Deletion of the MgAPT2 gene from the fungal genome, in effect, disarmed it. In addition, the gene deletion triggered no defense mechanism in rice plants (Sample 2006).

Humans are also parasitized by fungi, as anyone who has had athlete's foot disease can attest. *Candida* sp. parasitizes moist human skin and epithelial tissues, including genital tissues (jock itch, vulvovaginitis), sweaty toes (athlete's foot), bladder, throat, esophagus, and so forth. Other fungi infecting humans include *Microsporum* spp., *Trichophyton* sp., *Aspergillus* sp. (lungs), *Epidermophyton* sp., and ringworm fungi (dermatophytes) (Bennett 1996).

The cold-loving fungus, *Geomyces destructans*, is implicated in the deadly white-nose syndrome (WNS) threatening hibernating bats in Eastern North America. This cold-loving fungus infects hibernating bats in their caves and leads to their deaths (USGS 2010b; Flory et al. 2012; WNS 2012; Blehert et al. 2011). One hypothesis for the sudden incidence of WNS (first observed in winter, 2006) is that some other infective factor (such as an immune deficiency virus), or a combination of factors, has compromised the infected bats' immune systems, leading to a fatal level of *G. destructans* proliferation. Another possibility is that the US fungus strain itself may have mutated to a form that evades the bats' immune systems. Curiously, *G. destructans* has been found in bats throughout Europe but is generally not fatal to these animals. One theory is that the European bats have coevolved with the fungus and may be more immunologically resistant to it (Wibbelt et al. 2010; Puechmaille et al. 2011). Bats are ecologically important because they eat moths, flies, and mosquitoes; the latter spread Eastern equine encephalitis and West Nile disease viruses, and southern species can spread malaria. Moths spread other plant pests (e.g., gypsy moth caterpillars eat oak tree leaves).

6.5.3 Antibiotic Fungi

The plant-dwelling fungus *Muscador albus* is in the order Xylariales. It has the ability to produce a mixture of volatile organic compounds (VOCs) including alcohols and esters, which can kill other molds and bacterial pathogens like *Listeria* sp., *Salmonella* sp., and many plant pathogens (endophytes). It was isolated from the bark of a Honduran cinnamon tree. For example, *M. albus* can inhibit the growth of the gray mold, *Botrytis cinerea*, on table grapes. In a greenhouse study, Riga et al. (2008) showed that harmful, plant root–eating nematodes could be killed or paralyzed by being exposed to the VOCs from *M. albus*. Lacey et al. (2008) reported that the fumigation ability of *M. albus'* VOCs could control potato tuber moth in stored potatoes. Still another application for the VOCs of *M. albus* was studied by Mercier and Jiménez (2007). In a laboratory study, they grew the molds *Cladosporium cladosporoides*, *Aspergillus niger*, and *Stachybotrys chartarum* on wet drywall to reproduce mold damage to buildings following water immersion. They then fumigated the samples with grain cultures of *M. albus*. *C. cladosporoides* was eliminated after 48 h of fumigation, while 96 h was required to reduce the cultures of *A. niger* and *S. chartarum* by a factor of approximately 10^{-5}. Mercier and Jiménez reported that the three most concentrated VOCs from *M. albus* in their testing were isobutyric acid [25 µg/L (m/v)], 2-methyl-1-butanol (10 µg/L), and isobutanol (µg/L; m/v) (all peak concentrations).

Probably the best-known examples of antibiotics derived from mold are the penicillins. Penicillins G and V were developed from *Penicillin* sp. molds and are effective against Gram-positive *Streptococcus* sp., *Neisseria meningitidis*, many anaerobes, spirochetes, and others but are largely ineffective against Gram-negative *Staphylococcus aureus*. Figure 6.3 illustrates the structure of Penicillin G and several of the other "R" side molecules made synthetically to confer new antibiotic properties to the basic molecule. *T* is a thiazolidine ring, and β is a beta-lactam ring. Both penicillin G and V have no resistance to bacterial penicillinase; the methicillin molecule is resistant to penicillinase and therefore effective against varieties of *S. aureus*, except methicillin-resistant *Staphylococcus aureus* (MERSA). Amoxicillin is not penicillinase resistant and is effective against *Listeria monocytogenes*, *Proteus mirabilis*, and 70% of strains of *Escherichia coli*. Piperacillin is not penicillinase resistant and is effective against *Pseudomonas* sp., *Enterobacter* sp., and many *Klebsiella*. Immune hypersensitivity reactions are the most common adverse side effects of the various penicillins.

Another important antibiotic-producing fungus is *Cephalosporium acremonium*. Figure 6.4 illustrates the basic cephalosporin (*cephem*) nucleus and the R-groups that make the first-generation antibiotic Keflex. The cephalosporins use many side groups (R_1 and R_2), and the many antibiotics they create can be found in Table 45-2 in Hardman et al. (1996). Keflex is effective against most strains of *S. aureus* and *Streptococcus* sp. It is not effective against *Listeria*

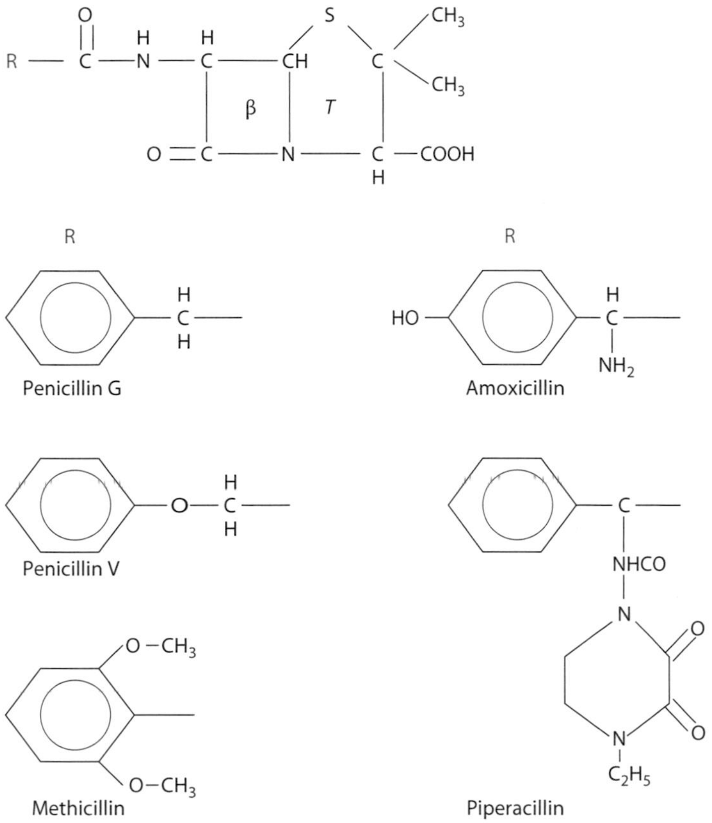

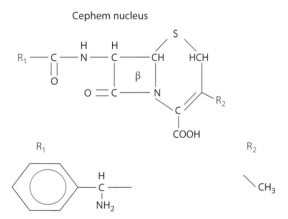

FIGURE 6.3
Penicillin family molecules.

FIGURE 6.4
Cephem molecule and side groups.

or *Enterococcus* sp., however. Hardman et al. list a total of 19 R_1/R_2 antibi-otic variants for the *Cephem* nucleus. These were developed to win the Red Queen contests with mutating and evolving bacterial pathogens.

Kurtzman (2005) reviewed the medical uses of fungal chitin and chitosan polymers. Chitin and chitosan make up the cell walls of fungal cells. Humans lack the chitinase enzymes required to break these polymers down in the digestive process, so they pass through the GI tract largely as undigested fiber. By processing the waste from packaging mushrooms (butts, culls), it is possible to purify chitin and chitosan. Chitosan is chitin that has lost its acetyl groups through hydrolysis. Figure 6.1 illustrates the 2-D structures of chitin and chitosan. Kurtzman cited a paper in which chitosan purified from wastes from *Ganoderma tsugae* mushrooms was used successfully as a wound dress-ing. The US health food industry has used chitosan as a putative cholesterol reducer, the theory being that the polymer chelates in some manner the low density lipid (LDL) "bad" cholesterol in the gut and carries it out in the stool so it is not absorbed by the body. Kurtzman also commented that the butts and culls of the mushroom *Coprinus comatus* can be processed to extract the dietary supplement glucosamine. Glucosamine (plus chondroitin) is used as an aid for joint cartilage regeneration to mitigate osteoarthritis symptoms.

6.5.4 Fuel Synthesis by Fungi

Gary Strobel at the Department of Plant Sciences at Montana State University and his coworkers have studied an endophytic fungus, *Gliocladium roseum,* and found that it emits a collection of volatile hydrocarbons and hydrocar-bon derivatives when cultured on an oatmeal-based agar under microaero-philic conditions. The hydrocarbon series of *G. roseum* contains a number of compounds normally found in petrodiesel fuel, so the VOCs were called "mycodiesel" by Strobel et al. (2010). Some of these mycodiesel components were found to be the following: *octane, 1-octene, 2-methyl heptane, hexadecane, 4-methyl undecane, 3-methyl nonane,* and *1,3-dimethyl benzene*. These VOCs were found after 18 days of culture in the approximate concentration of 80 ppmv (parts per million by volume). Other endophyte fungi also produce mycodiesel VOCs. Banerjee et al. (2010) reported that *Myrothecium inunda-tum* produced 32 identified mycodiesel VOCs, including *sesquiterpene 5, (1-ethylpropyl)-cyclohexane, acetic acid, 3-methyl-1-butanol, 1-methyl-1,4-cyclo-hexane, 2-nonanone, monoterpene 2,* and so forth, which were the most abun-dant. Singh et al. (2011) also found that the endophytic fungus *Phomopsis* sp. also produced mycodiesel VOCs.

An obvious objective, given fungal species that can produce diesel-like VOCs, is to modify their genomes to increase their VOC production to a point where it will be economically practical to grow these fungi, harvest their VOCs, separate them, and use them in mycodiesel fuel production. A good deal of genomic and chemical research and development will be required before mycodiesel is competitive with biodiesel.

Another future use of genetically-modified (GM) fungi is to make cellulosic ethanol fuel. GM fungi are used to break down the lignocellulose in waste biomass to its component sugars (glucose, xylose), which can then be fermented to ethanol by GM yeasts (*Saccharomyces* sp.) or bacteria (*Clostridium thermocellum* and *Thermophilus saccharolyticum*). See Section 4.2.4 for a more detailed description of cellulosic ethanol production.

6.5.5 Mushroom Farms

That mushrooms can be raised in factories is indeed fortunate. Varieties such as the button, Swiss brown, and portobello are raised en masse in controlled conditions of lighting, temperature, and humidity. Special growth media made from sphagnum moss and also proprietary formulations are used, described in Section 6.5.2.4.

Also, as described above, the mycelia of *Fusarium venenatum* are grown in industrial bioreactor tanks and then processed to make the mycoprotein food Quorn.

Mushrooms and Quorn-like products have good nutrition and more popular appeal as food in the United States and Western Europe than processed insects. As available fish and mammal-based food supplies become scarcer with time and population growth, and thus more expensive, we predict that a larger fraction of our dietary protein and minerals will come from farmed fungi.

6.6 Food from Tissue Culture Using Animal Stem Cells

Using animal muscle cells grown industrially for food, in vitro, is not a new idea. The "Chicken Little" tissue culture provided food for an overpopulated future world in L. Ron Hubbard and C.M. Kornbluth's 1952 sci-fi novel, *The Space Merchants*. As with the 21st-century fungus food Quorn, it should be possible to use continuous-flow or batch processing to grow and harvest animal muscle cells of various types (cardiac, striated, smooth) from appropriate committed stem cell cultures (Northrop and Connor 2009). Bovine, chicken, ovine, piscine, porcine, or turkey committed stem cells might be used. The cells would form not differentiated organs but rather sheets of undifferentiated muscle cells (myocytes) that could be peeled off moving substrate belts used for their growth. The harvested cells would be washed, compressed, textured, flavored, colored, cooked, and so forth, and then frozen. One would not obtain Porterhouse steaks or lamb chops per se. Would a future beef-based, in vitro cutlet be more appealing to humans than Quorn's *Classic Burger*? Importantly, would it be cost-competitive with the Quorn product now on the market? Or with real beef hamburger? Note that production of

1 lb. of real beef requires an average of 1800 gal. of water, plus 6.6 lb. of grain (e.g., corn), plus the water and fuel to grow, cultivate, harvest, and process the grain, approximately 36 lb. of roughage or grass (plus its irrigation water), and an additional 19 gal. of water for processing (Featherman 2011).

Below are some of the challenges the mass production of in vitro meat faces:

The development by genetic engineering of *committed* (or *satellite) muscle stem cells* that can be caused to divide into muscle cells and fuse onto the surface of a continuously moving conveyor belt matrix from which the layers of mature muscle cells can be separated and harvested at the bioreactor's output.

The development of an inexpensive, effective growth medium is necessary to supply the growing muscle cells. See, for example, the complicated composition of an early growth medium (Medium No. 199) of Morgan, Morton, and Parker given in Table VII in a work by Parker (1950). This medium contained all of the normal AAs plus the nucleic acid bases adenine, guanine, thymine, and uracil (no cytosine), plus many vitamins. (Morgan et al. used their medium to culture 11-day chick embryos.) The medium used for in vitro meat production will need to contain glucose, AAs, nucleic acid bases, sodium, potassium, magnesium, calcium, iron, magnesium, and chloride, sulfate, phosphate, and nitrate ions, as well as pH buffers. It could also contain cell growth factors such as *insulin-like growth factor 2* (IGF-2), *fibroblast growth factor,* and *platelet-derived growth factor* (PDGF) (Doumit et al. 1993). Hormones such as insulin and animal growth hormone (GH) probably should not be used in the growth medium because if they contaminate the myocyte product, human health will be adversely affected. Antibiotics should also not be used because they could contaminate the product and put human health at risk through the development of resistant bacteria. Cooking the harvested myocytes may denature GHs and antibiotics and render them harmless.

The growth medium must be capable of being regenerated and recycled, and the medium must be able to be sterilized [by heat and/or ultraviolet (UV) radiation] before reuse to keep the bioreactors and growing cells free of bacteria and fungi. Once present, microorganisms will outgrow the slowly growing mammalian muscle cells, so it is critical to maintain sterility. Muscle cell metabolic waste products must be removed (including CO_2), the pH adjusted, and any components used up in growing a batch of cells (e.g. glucose, AAs, ions) must be added to the regenerated medium. Ethylene oxide gas sterilization could be used on the bioreactor between batches. Perhaps natural antimicrobial peptides (AMPs), such as human α-defensins (or other AMPs), could be used in the growth medium; humans are evidently not allergic to these endogenous 18–25 AA proteins (Northrop 2011; Section 6.4.6).

Yet to be determined is the effect of pulsed DC electrical stimulation on the growing myocytes, causing them to contract. Would such periodic contraction accelerate their growth, slow it, or have no effect? Would it firm up their texture?

Myocytes can be grown in suspension (batch processing) or on a moving matrix in growth medium. Which will be more cost effective? The matrix could be an inert matrix that could be recycled or one made of an organic tissue that could be harvested with the myocytes.

Once harvested, the mature myocytes must be separated from the matrix and washed free of the growth medium, compressed, and processed into the form of the in vitro meat end product. The cells must be inspected for bacterial and viral contamination, and their chemical purity must be verified.

Part of the flavor in cooked real meat comes from fat in the meat. A small amount of natural animal fat could be added to the in vitro meat after harvesting, or some adipocytes could be cultured along with the myocytes.

What would be the most palatable forms of in vitro *meat*? Cutlets? Sausages? Cubes for stewing? Bricks of compressed myocytes that can be sliced or ground into burgers? Sushi? Market research will need to be done.

Two US patents have already been issued on in vitro meat production: (1) *Method for Producing Tissue Engineered Meat for Consumption*, US Patent No. 6,835,390 B1, issued to Jon Vein, December 28, 2004, and (2) *Industrial Production of Meat Using Cell Culture Methods*, US Patent No. 7,270,829, issued to W.F. van Eelen (Amsterdam, the Netherlands), September 18, 2007. These patents spell out in detail the procedures that might be used to grow in vitro muscle cells, but neither patent shows figures of a process flow chart or of a growth apparatus/bioreactor.

To successfully grow mammal or fish muscle or liver cells in large, industrial-sized batches will require much future research on GM muscle stem cells, mass tissue culture methods, cell adhesion, growth media, how to maintain sterility, and so forth. This in turn will require capital investment and economic and altruistic motivation (Specter 2011; Global Envision 2011; Benjaminson et al. 2002). We predict that by 2032, in vitro meat products will become common and market competitive with natural animal meat and fish, and Quorn products. However, there still will be a market for some natural meat specialties, such as bacon, caviar, and sweetbreads, as well as an occasional real filet mignon and lobster. A 2008 study on the economic feasibility of in vitro meat production (Exmoor Pharma Concepts 2008) predicted the production costs of in vitro meat myocytes grown free in tissue culture media and meat grown in a bioreactor on a 3-D matrix. They predicted (in 2008) that large quantities of in vitro meat *could be produced* for approximately Euro 3300–3500 per tonne, depending on the culture method. This translates to Euro 1.50–1.59 per pound in 2008 Euros.

One obvious positive impact of the rise in consumption of in vitro meats will be a decreased demand for meat animals, a decrease in factory farms and CAFOs, and an increase in available arable land to grow grain crops for humans. Newly available grazing land might be planted in crops used for the production of cellulosic ethanol, such as *Miscanthus* sp. In vitro meat consumption will also reduce GHG emissions from cattle.

The total, worldwide GHG emissions due to the production and marketing of *all* livestock products have been calculated to be approximately 32.6 Gt CO_2 equivalent (includes methane), which is over 50% of anthropogenic GHG emissions (Goodland and Anhang 2009)! In 2005, cattle alone may have accounted for 8.8 Gt CO_2 equivalent (21% of anthropogenic GHGs) (Calverd 2005, cited by Goodland and Anhang). And we worry about emissions from fossil fuels.

6.7 Chapter Summary

Human sustainability can be enhanced by producing food that exploits the protein, minerals, fats, and so forth in insects, zooplanktons, and fungi. To become a significant fraction of the world's food, insects will have to be farmed, not just caught wild. One can imagine netting vast swarms of desert locusts for processing into some sort of insect protein cake that can serve as human or livestock food. Our present technologies for raising crickets and mealworms can easily be expanded to industrial scale. In North America and Western Europe, the "yuk factor" precludes extensive use of cooked, adult crickets or mealworm larvae without processing. We are more likely to see this protein presented as a dried, flavorless meal or as flavored patties (*similar to* tofu or Quorn).

Planktons also offer a relatively easily acquired biomass. Just as the food value of adult insects goes up when their indigestible chitin is removed, the calcified exoskeletons of planktons also have to be removed in processing. Extensive harvesting (hence depletion) of apex planktons such as krill may have untoward effects on other ocean ecosystem members, including anchovies and sardines, and even predatory fish higher on the food chain, such as tuna, salmon, and cod.

Expect to see a growing use of insect and plankton food supplements in Eastern Asian countries in the near future. Americans will probably prefer mushroom and fungal mycelia-based foods over eating insect-derived protein. Their farm animals will not know the difference.

There will be a rise in the consumption of in vitro meats once the production technology is perfected and they reach cost parity with real meats, perhaps by 2025. Both in vitro meats and Quorn-type fungal foods require complicated growth media, a significant component of which will be sugars. Sugarcane, corn syrup, and sugar beet sugar will be used, adding to the competition for sugars from the food and fermentation industries.

7

Complex Economic Systems and Sustainability

7.1 Introduction to Economic Systems

While this book has focused mostly on the general relationships between human sustainability and the ecological, environmental, food, energy, water, and social systems with which we interact, our actions also significantly affect, and are affected by, the very important, large, complex nonlinear systems (CNLSs) that probably receive the most public attention today, that is, economic systems (ESs). ESs are driven by human behavior in response to many factors, and all ESs are ultimately affected parametrically by human population growth, resource availability, and climate and ecosystem changes. Some of the factors that adversely directly affect ESs are shortages in fossil fuels (FFs) and shortages in human and animal foods (due to weather, shortages of water, loss of pollinators, bad farming practices, diseases, crop diversion to fuel plants, and shortages in raw materials needed in manufacturing) (Economic Situation 2010). From the principle of supply and demand (S&D; see Section 7.2), shortages generally result in increased prices of goods, services, and food.

There are three reasons that ESs must be viewed as complex: (1) Their models contain many nodes and directed branches that have nonlinear, time-variable and parametric gains. (2) They are noisy (look at the daily prices of commodities on the stock market). (3) Human behavior is an integral component of their dynamics. Human behavior on the individual level, in small groups, and *en masse* is modulated by interpersonal and intergroup communications, information flow, and news. Human agents make the decisions to invest; buy; sell; adjust prices; raise and lower taxes; print money; sell bonds; invest in government-sponsored job programs; and grow, manufacture, catch, plant, harvest, mine, refine, stockpile, and hoard goods—based on information from surveys, from economic reports in the media, on the Internet, and from interpersonal contacts, all of which influence individual, group, and mass behavior. The results of polls and economic reports also affect human agent behavior.

Take, for example, the hard-to-quantify issue of *consumer confidence*. Confident consumers have little hesitation to purchase expensive capital goods (automobiles, boats, houses, furniture, kitchen appliances, etc.), invest their money in markets, and go on expensive vacations. They have job security and good cash flow. They are confident that they can pay later. The quantification of human attitudes and behaviors on economic issues presents a daunting challenge to ES modelers.

To sample a taste of the complexity of ES graphs, see the colorful nodal diagrams that appeared in the article by Harford et al. (2011) in the *NY Times Magazine* entitled "The Art of Economic Complexity." The MIT Media Lab (2011) and the *Center for International Development* at Harvard University created an *Economic Complexity Observatory* (ECO) (2011) in October 2011. "The goal of the ECO is to develop new tools that can help visualize and make sense of large volumes of data that are relevant for macroeconomic development decision making." Also see the seminal papers on the network view of economic development, the building blocks of economic complexity, and the history of economic complexity over a 42-year period by Hildago and Hausmann (2008, 2009) and Hidalgo (2009).

ESs have taken the forefront of public (and media) attention in 2008–2012 in the United States because of unemployment, the monotonic increases in energy costs, and their linkage with worldwide increases in population growth and prices for food, durable manufactured goods, transportation, real estate, and so forth. Yes, the prices of these items fluctuate up and down, but their long-term trend is up.

Other factors affecting ESs include the rise in US and global unemployment, partly caused by the steady loss of manufacturing jobs to overseas facilities, and the October 2008 lending crisis, which morphed into the 2009 US recession—going into 2012—as well as government attempts to break the recession by targeted stimulatory spending and lending. The human agents that run economies appear to have situational rather than sustainable values. This behavior has contributed to the US's slow economic recovery following the events of 2008–2010.

We have heard complaints (generally political) that the US government's economic stimulus spending under President Obama has been in vain because US unemployment is still high (8.3%, August 2012). Do you think this is because not enough stimulus money was spent, because it was spent in the wrong places to affect employment, because there are built-in time delays in any ES (in outputs resulting from various inputs), all of the preceding, and/or some other reasons? ESs are every bit as complex as biological systems. They include hard-to-quantify factors governing human behavior such as consumer confidence, prosperity, and flourishing (Jackson 2011).

Some good news released on March 13, 2012 showed that Connecticut's unemployment rate dropped to 8.0% for the first time in nearly 3 years (it peaked at 9.4% in August 2010). The nationwide unemployment rate was 8.3% in August 2012 (Economist 2012f; Haar 2012). The bad news is that the prices of food and fuels are steadily increasing (Lagi et al. 2012; Seetharaman 2012).

A number of time-variable input variables as well as time-variable branch gains make ESs of all kinds *nonstationary* and exacerbate our difficulty in modeling them. These time-variable input parameters include but are not limited to the following:

(1) The steadily increasing world population, which puts strains on eco-systems, resources, markets, and job availability.

(2) Limits on resources (e.g., water and food) that fluctuate with plan-etary weather, which in turn is affected by such factors as seasons; ocean currents such as *El Niño, La Niña,* and the North Atlantic Oscillation; sunspots; greenhouse gasses produced by power gener-ation and agriculture; and so forth. Finite, nonrenewable, FF energy resources are becoming exhausted, and FFs in reserves are more expensive to extract (i.e., oil, natural gas, and coal).

(3) The gradual exhaustion of easily mined metal ores and mineral deposits and the increased energy cost of extracting these materials.

(4) Emotional human behavior in attempting to manage ESs. ("No new taxes"; "Cut government spending"; "Tax the rich"; etc.)

In summary, we view ESs to be nonstationary, noisy, CNLSs.

Anyone who watched CNN news and commentary on television in the last half of 2008 might recall that the United States's and the world's eco-nomic problems were their "Issue #1." Unquestionably, ESs are complex. To understand them and predict their behavior, we can try to construct valid mathematical models of them in the form of nonlinear signal flow graphs. To do this, we must define the nodes (input variables, states, and output vari-ables) and unidirectional branches (or edges) describing quantitative causal relationships between the variables.

It was remarked above that ESs are more complex because humans run economies. A major source of ES complexity comes from human behavior in response to information (Arthur et al. 1997; Arthur 1999; Bowles and Gintis 2002; Easley and Kleinberg 2010). Human psychology, group dynamics, and communications figure largely in economic decisions. The human emotions of confidence, greed, fear, anxiety, and panic are hard to quantify, but their effects are seen reflected in stock market trends, commodities trading, the derivative markets, the loan industry, and real-estate markets. Also, eco-nomic complexity arises from the nonstationary and noisy behavior of eco-nomic variables affected by climate and stochastic events such as tsunamis, earthquakes, and volcanic eruptions.

Pryor (1996) described the *structural complexity* of ESs in three ways: (1) An eco-nomic process or system is more complex if it requires an increase in infor-mation for its effective operation. (2) ES complexity increases if, in its model, there is an increase in the density of paths between nodes. (3) Increased complexity also occurs when there is an increase in the number of heterogeneous *modules* in it (see Section 2.5 in this book). Thus, a metropolitan area's economy will be made more complex by population growth, the inclusion of more heterogeneous ethnic groups, and the spatial segregation of ethnic groups. Because the exchange of financial information is an integral part of both macroeconomic and microeconomic systems, the increased use of cell phones, PDAs, and the Internet has contributed to the increase of structural economic complexity. Electronic communications have eliminated destabilizing delays in the transmission of economic information. Delays still appear, however, in the time required to collect and analyze data and present reports. More sophisticated, targeted TV and Internet advertising has also affected demand-side economics.

Like many other classes of CNLSs, macroeconomic systems can be partitioned into interconnected modules or subsystems (Harford et al. 2011). We have the world ES (WES), which can be subdivided into regional ESs (RESs), including but not limited to the US RES, the EU RES, the Chinese RES, the Russian RES, and so forth. Clearly, these regional macroeconomic models are all functionally linked—increasingly so with globalization and the Internet. The countries in the EU RES are linked not only by trade but also by a common currency, the Euro. Financial information and money can be exchanged at the click of a mouse. Often, each RES can be further subdivided into smaller modules in order to facilitate analysis, for example, individual countries, such as Greece in the EU RES.

Perhaps one of the more important economic modules or subsystems in any national economy concerns energy (sources, sinks, reserves, production costs, delays in production, etc.); another module deals with agriculture (food production, marketing and distribution, and growth of plants for biodiesel and ethanol); others include health care, manufacturing, recycling, and stock markets. The energy module appears to be highly connected to the agriculture module (growing operations, transportation) and the manufacturing and mining modules. Weather certainly affects food production (hence prices) and also alternate electrical energy production (wind turbines, solar cells, hydropower). It is clear that in developed countries, energy is also heavily used for personal comfort. While a new nuclear energy plant is safer than its predecessors, and has a near-zero carbon footprint to operate, it is expensive to run safely, and its construction has a huge carbon footprint.

Economics as an academic discipline has been largely Balkanized; that is, a number of "schools of economic thought" and approaches to economic analysis have arisen since the seminal work of Adam Smith in 1776, *The Wealth of Nations*. In addition, economics is generally studied on two levels of scale: *macroeconomics* and *microeconomics*. Both levels are complex systems, and their behaviors are quite interdependent. Foster (2004) cited *neoclassical, evolutionary, post-Keynesian, Chicago,* and *neo-Austrian* economic schools of thought. There is also a *thermodynamic* (second law) approach to economic analysis [also known as (aka) *econophysics*] (Kafri 2008; Raine et al. 2006; Saslow 1999; Georgescu-Roegen 1971). (Over 20 schools of economic thought, past and present, are described in Competing 2012.)

John Foster (2004) made a strong case for viewing ESs as complex, dissipative structures that import free energy (in a thermodynamic sense) and export *entropy* in a way that enables them to self-organize their structures and functions. These are open systems that absorb information from their environments and create stores of knowledge that facilitate their growth. They are, in fact, *complex adaptive systems* (CASs) (cf. Section 2.1.5 in this text). Foster made the point that adaptation, hence evolution, of an ES cannot occur if there is a very high degree of connectivity (edges) between its nodes—there is no room for learned plasticity in the edge structure that can affect behavior. Foster viewed complexity in his paper simplistically as the "...connective structure of a system."

Why are there so many approaches to modeling and analysis of ESs? The answer is simple: ESs are very complex, and most economists treat the facets of economics they are attracted to and are familiar with. By this, we do not mean their analyses are necessarily invalid, only limited in scope and effectiveness. The "big picture" must be considered in dealing with any CNLS. For example, in the 19th century, economists developed the *quantity theory of money* (QTM; inflation or deflation could be controlled by varying the quantity of money in circulation inversely with the level of prices). In 1936, economist John Maynard Keynes argued that the effect of circulating money on the prices of goods was virtually nil, illustrating that the sensitivity of prices to circulating money was small, invalidating the QTM for most economists. Keynes maintained that government budgetary and tax policy, and the direct control of investment, had higher sensitivities with respect to the prices of goods. Then in the 1960s, economists Milton Friedman and Anna Schwartz refuted Keynes' approach and reestablished the validity of the QTM. Their ship floated until the 1990s, when other economic models focusing on growth and development surged. And so contemporary economists have also included the areas of public finance (taxation), labor, industrial organization, international economics, agriculture, information, and law into their modeling and analyses. In other words, economists have now justifiably expanded the scope of their models to include more variables and parameters.

Horgan (2008) commented that "economics [economic analysis] keeps lurching faddishly from one approach to another rather than converging on a single paradigm the way that more successful scientific fields such as nuclear physics or molecular biology do. The obvious reason is that economies are fantastically complicated in comparison to atomic nuclei or galaxies or *E. coli*." In our opinion, to be successful, studies of ESs must be interdisciplinary, drawing on complex systems theory, chaos theory (Horgan's *chaoplexy*), dynamic modeling, as well as mathematical approaches adapted from ecology and evolutionary biology, and very importantly, psychology (after all, fallible humans operate economies). Perhaps economics should be viewed more from a social science viewpoint. This approach is, in fact, taken in the text *Complex Adaptive Systems: An Introduction to Computational Models of Social Life* by Miller and Page (2008). The role of CASs in economics is treated in this interesting text, but unfortunately, there is a dearth of mathematical detail in describing various models.

7.2 Basic Economics; Steady-State S&D

One of the most fundamental concepts in *neoclassical economic* (NCE) theory are the "laws" of *supply, demand,* and *supply and demand.* These "laws" describe steady-state (SS) or equilibrium relations and are usually analyzed in the framework of *microeconomics.* NCE models rely on the assumption that there is no trading at all unless and until all prices reach equilibrium, at which point all sellers and buyers simply exchange a good at a certain price. In an NCE model, there is no excess demand or shortage of goods, labor, or services. Also, unemployment is not considered! Probably the most severe criticism of NCE models is that they do not consider money; it does not and cannot appear. There is no capital accumulation, earnings, savings, taxation, and so forth. The price **P** in an S&D model is considered to be only a label (McCauley and Küffner 2004).

We shall review the generalities of NCE systems and discuss how their principles might guide us in formulating an SS mathematical model for an ES. First, below are some economics terms that will be used:

Good (n): A good is any object or service that increases the utility, directly or indirectly, of the consumer. A good is manufactured, grown, crafted, and so forth and generally sold for profit. A good is *supplied* to meet a general *demand.*

Price: The cost a consumer must pay for goods or services.

Demand: The (average) quantity of a certain good or service desired by consumers.

Service: Work done for others as an occupation or business, for example, income tax preparation, mowing grass, and so forth.

Supply: How much of a good or service the market can supply at a given price.

Production function: An equation that expresses the fact that a producer's output depends on the quantity of raw material inputs it employs. Inputs can be combined in different proportions to produce a given level of output.

Utility function: An equation that attempts to model the pleasure or satisfaction households (basic microeconomic consuming units) derive from consumption. It depends on the products purchased and how they are consumed. Utility functions provide a general description of the household's preferences between all of the paired alternatives it might be presented with.

These terms are related functionally and can vary parametrically. *S&D curves* are used in *microeconomics* to illustrate how the *supply of a good,* $P_S(Q)$, has a common, SS solution (price P_o) with the *consumer demand for that good,* $P_D(Q)$. Various production and *utility functions* taken with certain assumptions about human behavior lead to S&D *curves* such as those illustrated in Figure 7.1. Note that not all S&D curves look alike. In general, most *supply curves* (**S**) have positive slopes, while most *demand curves* have negative slopes. For example, the *demand curve,* $P_D(Q)$, represents the *quantity of a good or service* (**Q**) consumers are willing and able to purchase at various *prices* (**P**). $P_D(Q)$ is also a function of geographic location (not much demand for air conditioners in Greenland), the consumer population density, demographics, seasonal need (e.g., snow shovels, seed corn), and so forth. The *supply curve,* $P_S(Q)$, illustrates the SS relationship between the market price and the *amount of a good produced* (**Q**). Suppliers are likely to produce more of a good at higher prices; they try to maximize their profit.

In the simple, SS demand curve (**D**) illustrated in Figure 7.1, the price (**P**) of a good is seen to be a decreasing function of the quantity willing to be

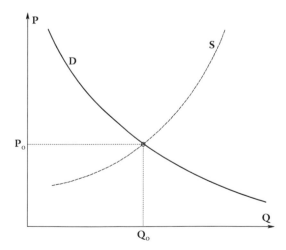

FIGURE 7.1
Generic, static, microeconomic supply (**S**) and demand (**D**) curves. **P** = price of a good, **Q** = quantity of a good. See text for discussion. (From Northrop, R.B., *Introduction to Complexity and Complex Systems,* CRC Press. Boca Raton, FL. 2011. With permission.)

purchased by consumers (Q), *ceteris paribus* (Latin: *all other factors remaining constant*). As the price increases, so does the *opportunity cost* for that good. The SS supply curve (**S**) in Figure 7.1 generally has a positive slope, meaning that producers will supply more goods at a higher price because selling a higher quantity at a higher price increases their profit, *ceteris paribus*.

The *SS S&D equilibrium* is at ($\mathbf{P_o}$, $\mathbf{Q_o}$) in Figure 7.1. This intersection illustrates the *law of S&D* for a good. The $\mathbf{P_S}(\mathbf{Q})$ and $\mathbf{P_D}(\mathbf{Q})$ curves have a unique intersection where the *supply price* of a good, P, equals the *demand price* ($\mathbf{P_o}$). At this point, the amount of a good being supplied equals, in theory, the amount demanded by the consumers. S&D curves represent SS analysis and are valid only in the short term. The economic models describing sustainability effects are nonstationary in the long term because of the steadily rising world population, and gradually dwindling resources. Nonstationary and nonlinear ESs are best modeled and studied using dynamic models based on large sets of nonlinear ODEs.

A challenge in understanding SS S&D is to consider what happens to prices when other factors cause a shift in the SS demand curve, $\mathbf{P_D} = f(\mathbf{Q})$, while the supply curve, $\mathbf{P_S} = g(\mathbf{Q})$, remains fixed. Such a scenario is illustrated generally in Figure 7.2. Here the demand curve shifts to the right, from **D** to **D'**. As a result of generally increasing consumer demand, a new equilibrium, ($\mathbf{Q'}$, $\mathbf{P'}$), is established. Along with the increased demand, the price increases by $\mathbf{\Delta P} = (\mathbf{P'} - \mathbf{P_o}) > 0$, and the number of goods sold also increases by $\mathbf{\Delta Q} = (\mathbf{Q'} - \mathbf{Q_o}) > 0$. The converse relation holds as well; if **D'** shifts to the left, then both $\mathbf{\Delta P}$ and $\mathbf{\Delta Q}$ are negative.

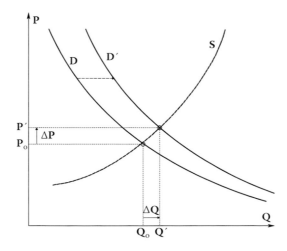

FIGURE 7.2
Effect of demand shift on **P** and **Q**. See text for discussion. (From Northrop, R.B., *Introduction to Complexity and Complex Systems*, CRC Press. Boca Raton, FL. 2011. With permission.)

In some cases, the supply curve can be nearly vertical; that is, the quantity supplied to the market is fixed (or nearly fixed) by the manufacturer, regardless of market price. (A short-term example of a nearly horizontal supply curve is oil production regulated by the Organization of Petroleum Exporting Countries (OPEC), land availability in a region, or diamond production regulated by De Beers.) A nearly vertical supply curve, S_L, is shown in Figure 7.3. When demand increases (shifts to the right), $\Delta P > 0$.

A supply curve can also be double valued. Figure 7.3 shows such a curve (S_C) that has been observed in the labor market and the crude oil market (Samuelson and Nordhaus 2001). After the 1973 oil crisis, many OPEC countries *decreased* their production of oil even though prices increased. This was due to *human behavior*: Why sell a valuable asset? Save it to market later at a higher price (in this case, the seller is hoarding or warehousing a commodity).

The *price elasticity of S&D* is an important concept in static S&D theory. For example, if a vendor decides to increase the price of his good, how will this affect his sales revenue? Will the increased unit price offset the likely decrease in sales volume? Or, if a government imposes the tax on a good, thereby effectively increasing the price to consumers, how will this affect the quantity demanded?

One way to define *elasticity* is the percentage change in one variable divided by the percentage change in another variable. Thus, elasticity is basically a system sensitivity as defined in systems engineering (see Glossary). For example, if the price of a widget is raised from $1.00 to $1.05, and the quantity supplied rises from 100 to 102 units, the supply slope at a

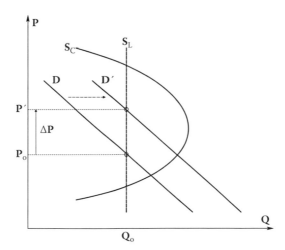

FIGURE 7.3

S&D curves with a vertical, constant good supply (S_L) and a double-valued-good supply curve (S_C). This can be the result of producer warehousing. See text for discussion. (From Northrop, R.B., *Introduction to Complexity and Complex Systems*, CRC Press. Boca Raton, FL. 2011. With permission.)

point is 2/0.05 = 40 units per dollar (see Figure 7.4). Because the elasticity is defined in terms of percentages, the quantity of goods sold increased by 2%, and the price increase was 5%, so the price elasticity is 2/5 = 0.4. Note that for a perfectly inelastic supply, the supply curve is vertical, as shown by S_L in Figure 7.3.

The *income elasticity of demand* (**IED**) tells us how the demand for a good will change if purchaser income is increased (or decreased). (Figure 7.1 illustrates a generic demand curve.) For example, how much would the demand for a giant plasma TV screen change if the average purchaser income increased by 10%? Assuming it is positive, the increase in demand would be reflected by a positive (up and right) shift of the demand curve, resulting in a positive ΔP and ΔQ. This is shown in Figure 7.2. Thus, $\mathbf{IED} \equiv (\Delta Q/Q_o)/(\Delta P/P_o)$ is seen to be an economic sensitivity.

The *cross elasticity of demand* (**CED**) is another important differential *sensitivity* used in economic models. Here we calculate the percent $\mathbf{Q}$ of a good demanded in response to a percent change in the price of another competing or substitute (alternative) good, or the price of a complement good that must be used with the good in question (e.g., a set of loudspeakers used with a sound system). Thus, the sensitivity $\mathbf{CED} \equiv (\Delta Q/Q_1)/(\Delta P/P_c) = (\partial Q/\partial P)(P_c/Q_1)$. An example of **CED** is the decrease in the demand for a certain brand of SUV by 20% when the price of gasoline increases by 30%. Here, **CED** = −0.667.

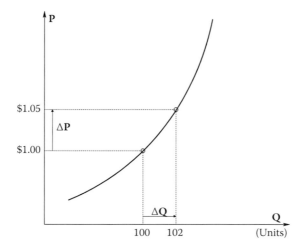

FIGURE 7.4
Illustration of supply elasticity for a good. See text for discussion. (From Northrop, R.B., *Introduction to Complexity and Complex Systems*, CRC Press. Boca Raton, FL. 2011. With permission.)

7.2.1 Forrester's Views

The challenge of formulating quantitative, dynamic models of complex ESs (CESs) was recognized over 56 years ago by economist/mathematician Jay W. Forrester (2003) at the Massachusetts Institute of Technology. In 1956, Forrester commented on the factors governing the time-dependent interplay of money, materials, and information flow in ESs and industrial organizations. The first point he made was that CESs contain causal feedback loops. He wrote: "The flows of money, materials, and information feed one another around closed re-entering paths." He went on to state that such systems often develop bounded oscillations (limit cycles), a well-known property of CNLSs having feedback paths and time delays. Forrester noted some other factors that operate in CESs:

The human factor of resistance to change, including habit, inertia, prejudices, traditions, and so forth. (Quantifying these properties lies more in the realm of psychology than economics because they involve human behavior.)

Accumulation (of materials or cash): Accumulation (saving, investing, hoarding, stockpiling, warehousing) also involves human decision making.

Delays (transport lags): In 1956, Forrester cited transport lags in the following ESs: (1) the delay between actual sales and accounting reports submitted to corporate decision makers; (2) the time required to process orders for goods, including building inventories of components; (3) the delay between the decision to make a good and its production (i.e., manufacturing throughput time); (4) the length of time from the decision to plant a crop to the harvest of the mature crop; (5) the time between collection of taxes and disbursement of governmental funds; (6) the length of time from the decision to build new factories, oil wells, refineries, pipelines, wind turbine "farms," solar cell farms, nuclear plants, power distribution lines, and so forth; (7) other delay generators in a model ES, including mail delays, freight shipment transit times, and time required for changes in human behavior in response to social or economic changes (generally a tipping point is seen); (8) the $800B economic stimulus package appropriated by Congress in early 2009, which appeared to have built-in bureaucratic delays in its distribution to target agencies, causing lags in job creation and slowing economic recovery.

Quantizing, in which the periodic availability of financial information (e.g., monthly and quarterly economic reports) builds lags into decision making.) This is actually a discrete sampled input and is more appropriately called *sampling*. (Quantizing describes the information loss that takes place in analog-to-digital signal conversion.)

Policy and decision-making criteria: Forrester argues that these criteria
have first-order effects on "amplification characteristics of the system" (i.e., path gains).

In commenting on the interrelationship of goods, money, information, and
labor, Forrester argued that previous models (before and up to 1956) had
generally neglected to adequately interrelate these variables. At this writing,
evidently this is still largely true, or we would be able to understand and
manage the US economy more competently than insisting on tax cuts on
one hand while attempting to implement expensive wars, health care programs, and multibillion-dollar corporate bail-out loans on the other. Giving
tax rebates in 2008 and arguing for a reduction on federal fuel taxes was
a short-term palliative for a currently sinking US economy, as well as an
ineffective political gesture.

ES behavior depends on human behavior in response to information. It is
evident that such behaviors must be included in any comprehensive dynamic
model of the US economy. As we stated above, the challenge remains: How
do you model human emotionally driven behavior such as fear, greed, dis-
satisfaction, satisfaction, confidence/pessimism, happiness/paranoia, and
so forth in ESs, and how they spread among a population? Forrester commented: "I believe that many of the characteristics of a proper model of the
national economy depend on deeply ingrained mental attitudes, which may
change with time constants no shorter than one or two generations of the
population." That is, one's (micro)economic attitudes and behavior can be
influenced by one's parent's attitudes or the attitudes of one's peers or ethnic
group. Modern attitudes that include factors such as the encouragement of
personal deficit spending using credit cards can be shaped by advertising:
"Buy and enjoy now, pay later" (if you can).

Forrester concluded his prescient paper by discussing the factors he felt
should be considered in ES dynamic modeling. He noted: "Almost every
characteristic that one examines in the economic system is highly nonlinear." (We should add "...and time-variable," as well.) Without saying so
explicitly, he argued that ESs are CNLSs. He called for the use of sets of nonlinear ODEs as modeling tools. All this happened over 56 years ago, on the
threshold of the computer revolution and at the dawn of organized complex
system thinking.

Some texts that have appeared since 1992 that deal specifically with
the dynamic modeling of ESs include, but are not limited to, the following: Brock and Malliaris (1992), *Differential Equations, Stability and Chaos in
Dynamic Economics;* Ruth and Hannon (1997), *Modeling Dynamic Economic
Systems;* Neck (2003), *Modeling and Control of Economic Systems;* Barnett et
al. (2004), *Economic Complexity: Non-Linear Dynamics, Multi-Agents Economies,
and Learning;* Zhang (2005), *Differential Equations, Bifurcations and Chaos
in Economics;* Cuaresma et al. (2009), *Dynamic Systems, Economic Growth,
and the Environment (Dynamic Modeling and Econometrics in Economics and*

Finance); Weber (2010), *Demographic Change and Economic Growth: Simulations on Growth Models (Contributions to Economics)*; and Plasmans et al. (2010), *Dynamic Modeling of Monetary and Fiscal Cooperation Among Nations (Dynamic Modeling and Econometrics in Economics and Finance)*. We note that Forrester's call is finally being heeded. Hopefully, the new-era economists have the mathematical backgrounds needed to effectively extract information from such texts.

7.2.2 Dynamic Models of ESs

Forrester's seminal (1956) paper called for the dynamic modeling of ESs. It was noted that in general, NCE theory ignores processes that take time to occur. This includes, for example, delays on the supply side, including, for example, the time for a crop to mature, the time to build a new wind farm and put it online, or the time to develop a new fuel-efficient automobile engine. It is also clear that the dynamic path of the economy cannot be ignored. Economists generally do not consider time when analyzing S&D or any other key variables.

NCE does use mathematical models in static analysis; these models use simultaneous, linear, *algebraic equations* and linear algebra (matrix methods) to reach solutions. Forrester argued that ESs must be modeled using sets of *nonlinear differential equations* and that such dynamic models will behave chaotically, ideally following limit-cycle attractors similar to real-world ES behavior. We note that in extrapolating from models to the real world, economic variables are likely to be in disequilibrium—even in the absence of external transient inputs. The conditions that NCEs have proven to apply at equilibrium will thus be irrelevant in actual ESs. It is clear that SS economic analysis cannot be used as a simplified proxy for dynamic analysis. The real question is whether we can sufficiently control such unstable, chaotic systems. Can we constrain their instability (chaos) within acceptable bounds while producing desired outcomes?

Some of the time-dependent processes that ought to be included in dynamic microeconomic models include the following: (1) The *functional lifetime* (FL) of a good. (Does FL follow Poisson statistics or some other distribution? Think of automobile batteries as an example.) What affects the human motivation to replace a failed good? Or do we buy a new, improved model? (2) Planned obsolescence in design. (We see this in the automobile, appliance, and clothing industries.) (3) The role of an advertising campaign in shifting the demand curve. The "new model is better" marketing strategy is applied to all kinds of durable goods ranging from air conditioners and automobiles to computers, furniture, and windows. A 21st-century advertising approach is "greenness"; buyers must be convinced that the improved energy efficiency of a good is worth the capital outlay for future operational savings. (4) Population growth is important because it presents an ever-increasing need for tax-based resources such as health care, unemployment compensation, and municipal services (water, sewer, power, refuse collection, schools, etc.), as well as providing an increasing source of labor and consumers.

There has been a slow trend toward dynamic modeling of ESs in the past 50 years. Wisely, some modelers have started with simpler model structures (three ODEs) (Abta et al. 2008), while others have constructed more complicated economic models (about 100 ODEs with more than 80 parameters) and then have had to reduce their model to a more realistic architecture (Olenev 2007). Olenev's model was faced with many unknown path parameters that had to be estimated, and there were unknown initial conditions on many model (node) variables that also had to be determined. Eight nonlinear ODEs were described in his paper (volume restrictions prevented inclusion of more detail). Both Olenev's and Abta's models were interesting because they described the flow of capital.

The relatively simple model of Abta et al. (2008) was used to study limit cycles in a business cycle model. Their three ODEs are summarized below:

$$Y = \alpha \, [I(Y) - \delta_1 K - (\beta_1 + \beta_2)R - l_1 Y] \tag{7.1}$$

$$K = I \, [Y(t - \tau)] - (\delta + \delta_1)K - \beta_1 \, R \tag{7.2}$$

$$R = \beta \, [l_2 Y - \beta_3 R - M] \tag{7.3}$$

where Y = gross product, K = capital stock, R = interest rate, τ = delay for new capital to be installed, α = adjustment coefficient in goods market, β = adjustment coefficient in money market, $I(*)$ = linear investment function, $S(*)$ = linear savings function, M = constant money supply. (We have used the notation from the Abta et al. paper.)

Representation of this simple ES model by a signal flow graph (SFG) (cf. Figure 7.5) shows that it has three integrators and eight nodes. Abta et al. used the "Kaldor-type investment function":

$$I(Y) = \frac{\exp(Y)}{1 + \exp(Y)} \tag{7.4}$$

Note that if $0 \le Y \ll 1$, $I(Y) \cong 1/2(1 + Y) \cong 1/2$, and for $Y \gg 1$, $I(Y) \to 1$. Thus, the Kaldor function saturates at 1 for $Y \gg 1$. If $Y \ll 1$, $I(Y)$ is linear, and the model is, in fact, a linear, cubic system. Inspection shows that the complicated linear SFG has five negative-feedback loops; one has a delay. Their linear loop gains are $A_{Ln1}(s) = -(\delta + \delta_1)/s$, $A_{Ln2}(s) = -\alpha\delta_1(1/4)e^{-s\tau}/s^2$, $A_{Ln3}(s) = -\beta\beta_3/s$, $A_{Ln4}(s) = -\alpha l_1/s$, and $A_{Ln5}(s) = -\beta(\beta_1 + \beta_2)l_2/s^2$. There are two positive-feedback loops with linear loop gains: $A_{Lp1}(s) = +\alpha(1/4)/s$ and $A_{Lp2}(s) = +\alpha\beta(\beta_1 + \beta_2)l_2/s^2$.

Abta et al. found that when $\alpha = 3$, $\beta = 2$, $\delta = 0.1$, $\delta_1 = 0.5$, $M = 0.05$, $l_1 = 0.2$, $l_2 = 0.1$, $\beta_1 = \beta_2 = \beta_3 = 0.2$, the system exhibited SS, bounded, limit-cycle oscillations for $\tau \ge 1.7975$. (Initial conditions (ICs) and parameter units were not specified.)

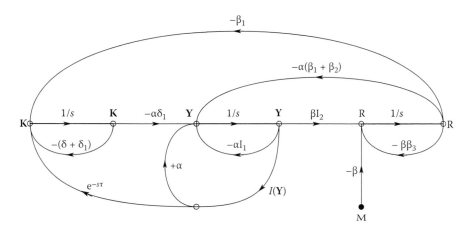

FIGURE 7.5
Nonlinear signal flow graph describing the three-state ES model of Abta et al. (2008). The SFG describes Equations 7.1 through 7.3. $I(\mathbf{Y})$ is the nonlinear, Kaldor-type investment function. See text for discussion. (From Northrop, R.B., *Introduction to Complexity and Complex Systems*, CRC Press. Boca Raton, FL. 2011. With permission.)

The results of this paper were not remarkable [otherwise, linear, cubic, feedback systems with delay(s) can easily be unstable]; what was of interest, however, is the fact the authors examined the stability of a simple ES model known to have a closed attractor for a sufficiently large delay in capital investment, **Y**.

Olenev's (2007) dynamic model for a regional economy used ODEs on the following eight variables (only the eight X variables are shown in his paper): $Q_V^X(t)$ = *shadow products in sector X*, $W_X(t)$ = *stock of open ("white") money of economic agent X*, B^X = *shadow incomes from sector X*, Z^X = *debts of agent X to bank system B*, Q_X^L = *stock for final product X of timber industry complex directed to household markets, L*, p_X^L = *consumer price index (CPI) on product X*, s_L^X = *open wage sector X*, W^G = *stock of money in the regional consolidated budget*. Production sectors X, Y, and Z were assumed. Additional variables are obtained by substituting Y or Z for X in the preceding list of variables. Space prevents us from writing the Olenev system's ODEs in detail. Olenev stated, "The main purpose of the paper is to illustrate the method, rather than present a state-of-the-art analysis."

Another approach to dynamic modeling of ESs is through the use on *agent-based models* (ABMs). [See the excellent reviews on ABMs by Macal and North (2006a, 2006b).] Instead of using nonlinear ODEs, as in Equations 7.1 through 7.4, the ABM uses interconnected *agent models* to mimic human behaviors in the ES. Many different kinds of agents are incorporated into the economic ABM. For example, some agents are *consumers* (they decide whether to buy, hoard, or save their money). Others are *manufacturers* (they decide whether to manufacture and/or inventory goods, lay off workers). Still others can be *marketers/distributors*, others can be *suppliers of raw materials, investors*, and so forth. Each agent is a self-contained, discrete entity with a designed set of characteristics and rules governing its input/output (I/O) behaviors and decision-making capability. In the ABM, agents interact with other agents and have the ability to recognize the traits of other agents. An agent may be goal directed, having goals to achieve in its behaviors (e.g., maximize manufacturing profit). It also is autonomous, is self-directed, and has the ability to learn and adapt its behaviors, based on experience. This is quite a programming challenge for the designers of economic ABMs. Because of the complexity of agent-based, dynamic economic models, they are generally simulated on enhanced, multiprocessor computers, not PCs [see Tesfatsion (2002, 2005a, 2005b, 2009) and Holland and Miller (1991) to explore more material on economic agent-based modeling]. How would you craft the attributes of a stock trader agent? Agent-based modeling is examined in more detail in Section 7.3.

7.2.3 What We Should Know about Economic Complexity

To be a competent economist in the 21st century, it appears that one should be a polymath and have broad interdisciplinary skills that embrace classical economic theory, mathematics (including the theory of complex systems, differential calculus, chaos theory, and the dynamics of nonlinear systems), and also, importantly, psychology. Quite a requirement, but then competent biomedical engineers should know differential and integral calculus, ODEs, engineering systems analysis, physiology, as well as introductory anatomy, cell biology, biochemistry, complex systems theory, and so forth.

To form a meaningful dynamic model of an ES, one must be able to identify (define) inputs, outputs, and internal variables (node parameters), and equally importantly, the branch (edge) functions relating the variables. One should also have measured relevant system sensitivities. In short, one should be able to build a valid dynamic model of a selected ES and then run simulations. All models must be verified using known data before they can be used predictively.

Wisely, Durlauf (1997) addressed the need for economists to have a background in complex systems theory. He used the example of a stock market having many traders with idiosyncratic beliefs about the future behavior of stock prices. These traders react to common information and exchange information rapidly through electronic media, but ultimately, stock prices are determined by a large number of decentralized buy-and-sell decisions (a scenario that can be simulated by ABMs). Durlauf commented that there is feedback between the aggregate characteristic of the ES under analysis and the individual (microeconomic) effectors that comprise that environment. "Movements in stock prices influence the beliefs of individual traders and in turn influence their subsequent decisions." Information flow was seen to be critical.

Durlauf noted that *conformity effects,* in which *an individual's perceived benefit from a choice increases with the percentage of his or her friends who make the same choice*, must figure in economic dynamic models. Conformity effects are an example of autocatalytic, positive-feedback processes between (human) economic agents. Another example of positive feedback, first noted by Forrester, is *economic attitude* passed on from parents to offspring in a family. Such familial (group) feedback can determine education and job choices, as well as spending/saving patterns. Such mundane things as the choice of automobile make to purchase (e.g., Ford or Chevy) can also be influenced by family biases.

Durlauf offered two messages for economic policymakers: First, interdependence among various component subsystems can create multiple types of internally consistent overall behavior. As a result, ES behavior can become stuck in an undesirable SS. "Such undesirable steady-states may include high levels of social pathologies or inferior technology choices." Second, "...the consequences of policies will depend critically on the nature of the interdependencies." He pointed out that the effects of different policies may be highly nonlinear, producing unexpected, chaotic results. He stated: "At a minimum, detailed empirical studies which underlie conventional policy analysis should prove to be even more valuable in complex environments." This suggests that to manipulate complex ESs, one must know estimates of the system's relevant sensitivities.

Lansing (2003) was of the opinion that complexity theory has an impact on economics: There is "...a shift from equilibrium models constructed with differential equations to nonlinear dynamics, as researchers recognize that economies, like ecosystems, may never settle down into an equilibrium" (a sort of economic homeostasis). This trend was noted by Arthur (1999), who argued: "...complexity economics is not a temporary adjunct to static economic theory, but theory at a more general level, out-of-equilibrium level. The approach is making itself felt in every area of economics: game theory, the theory of money and finance, learning in the economy, economic history, the evolution of trading networks, the stability of the economy, and political economy" (see also Arthur et al. 1997; Kaufman 1955).

Krugman (1994) urged the use of the metaphoric phrase "complex landscapes" in dynamic economic analysis. *Landscape* in this context refers to the N-dimensional phase plane behavior of an N-state ES as it seeks equilibrium, given a transient input, or as behavior for an initial value problem (IVP). (Note that a three-state system has a 3D landscape surface—similar to a contour map.) Krugman pointed out that complex nonlinear ESs can have *many* point attractors, "basins of attraction," and closed attractors for each variable. Hence, a complex landscape can exist for a CES. Krugman also remarked: "The most provocative claim of the prophets of complexity is that complex systems often exhibit spontaneous properties of self-organization, in at least two senses: starting from disordered initial conditions they tend to move to highly ordered behavior, and at least in a statistical sense this behavior exhibits surprisingly simple regularities." Self-organization or learning behavior in CESs should not be unexpected. After all, humans with brains operate them, and brains are adaptive CNLSs that can learn.

An assessment of economic complexity was treated in terms of complexity theory and measures in a paper by Lopes et al. (2008). Lopes et al. gave 12 algebraic indicators of connectedness (hence complexity) based on the large **A** matrices describing the sets of *N linear, simultaneous equations* describing SS, input–output relations in ESs associated with the Organization for Economic Co-operation and Development (OECD) member countries. For example, one of the simpler measures given to quantify complexity in ESs is the *inverse determinant measure*, IDET: IDET = $1/|\mathbf{I} - \mathbf{A}|$, where **I** is the $(N \times N)$ unit matrix. Another simple measure they applied is the *mean intermediate coefficients total per sector*, MIPS: MIPS $\equiv \mathbf{i}^T \mathbf{A} \, \mathbf{i}/\mathbf{n}$, where $\mathbf{i}^T$ is the transpose of a unit vector, **i**, of appropriate dimension, and **n** is the number of sectors.

They compared the results of their 12 measures applied to standard sets of economic data from 9 countries in the 1970s and the 1990s. Not unexpectedly, they found that large economies (United States and Japan) are more "intensely connected" and thus more "complex" than small ones (Netherlands, Denmark). The Lopes et al. paper is significant because it used quantitative measures of assessing complexity in ESs based on linear algebraic models of the economies.

Raine et al. (2006) wrote an interesting paper in which they speculated on why both biological and socioeconomic systems expand their structures (and populations) with the result that they use increasing amounts of "free energy" and associated materials. In biological systems, *free energy* is the thermodynamic energy released by breaking (or making) chemical bonds plus the energy from solar radiation. In socioeconomic systems, energy can be from exothermic chemical reactions (combustion), nuclear reactors, wind, solar power, tides, and so forth. This energy is primarily converted to and distributed as transmitted electricity. (It is certainly not "free.") They address this natural increase in complexity from the viewpoint of a modified *second law of thermodynamics*.

The traditional *second law of thermodynamics* states that any real process in an isolated system can proceed only in a direction that results in an *entropy increase*. Schneider and Kay (1994) (cited by Raine et al. 2006) reformulated the second law to apply to living systems: "The thermodynamic principle which governs the behaviour of systems is that, as they are moved away from equilibrium, they will utilize all avenues to counter the applied gradients. As the applied gradients increase, so does the system's ability to oppose further degradation." They asserted that *ecosystems* develop in ways that systematically increase their ability to degrade incoming solar and chemical free energy. According to Raine et al., "This reformulated second law is appropriate for the analysis of spatially fixed and open systems such as plants and forests [e.g., ecosystems]. Such systems are subject to two opposing gradients: the thermodynamic degradation gradient and the incoming solar radiation gradient." Solar radiation provides the free energy required for these systems to oppose thermodynamic degradation inherent in the entropy law.

Raine et al. (2006) observed that biological systems, ecological systems, and ESs are never near classical thermodynamic equilibrium. The ability to do self-organization requires structures (and algorithms) that throughput free energy in a way that resists the thermodynamic gradient. These self-organized processes in complex systems emerge and adapt through the creation of endogenous feedback paths, subsystems, or modules in order to maximize the efficiency of energy utilization and resist the degradation process. These authors maintained that it is the role of knowledge (information) that differentiates economic evolution (development) from biological evolution. They cited three ways that CESs differ from their biological counterparts: (1) CESs share knowledge (information transmission). (2) Knowledge interactions imply that experimental mechanisms [research and development (R&D)] may contribute to ES evolution (applied science can create information). (3) CESs growth limits (i.e., a quasi-SS) are not reached as easily as in biosystems because of the continual innovation created by new knowledge (e.g., the better use of materials in manufacturing, the use of new materials to make new goods). Economic growth also results from technical improvements in the physical means to process energy. The accumulation of knowledge (as "free energy") in ESs is part of their evolutionary process.

Raine et al. concluded with these thoughts: "Economic systems are characterized by the explicit use of knowledge (information) in harnessing energy, and consequently creating value [of goods and services]." "The reformatted second law suggests that knowledge structures be considered as unique complements that allow socioeconomic systems to utilize more energy than other biological and ecological species. Knowledge coevolves with energy using structures, facilitating economic growth through the use of functional and organizational rules under the governance of social institutions."

While the modified second law of thermodynamics provides an interesting viewpoint on socioeconomic systems, we view the major contribution of Raine et al. to be the viewpoint that knowledge and information are critical parameters in ESs. Only humans can create, process, and store abstract information, and human agents run ESs.

We submit that the viewpoint of Raine et al. is valid for ES but fails in the consideration of biological systems. Genomic information is critical in directing life processes and programming homeostatic systems. It is the equivalent of their "knowledge" in CESs. Genomic information is passed on (or communicated) generationally, and its alteration is necessary for evolution to occur.

7.2.4 Tipping Points in ESs; Recession, Inflation, and Stagflation

There are several examples of tipping points in ESs: One is the *onset of a recession/depression*. Human behavior figures largely in these scenarios. In a recession, there is a contraction of the business cycle; the economy stagnates—there is high unemployment, there is little circulating cash, factories close, stock markets plummet, and so forth. A sustained, severe recession may morph into a depression with massive unemployment and collapse of markets. A rule of thumb is that a recession occurs when the nation's *gross domestic product* (GDP) growth is negative for two or more consecutive quarters. Recessions are the result of the falling demand for goods; buyers lack the available funds and motivation and confidence to purchase certain goods. Prices generally fall but not enough to stimulate turnover. Lenders lack cash to loan and are unwilling to extend poorly secured credit. US unemployment soars as manufacturers cut back domestic operations due to a lack of demand for goods or move overseas to find cheaper labor. The morphing of a recession into a depression involves autocatalytic or positive-feedback behavior. [For an in-depth analysis of many of the factors that contribute to stock market crashes, see the interesting text by Sornette (2003), *Why Stock Markets Crash*.]

The US Bureau of Economic Research (Bureau of Economic Analysis) prepared a bar graph of Historic Recessions in the US economy; percent change in the GDP by quarter, annual rate, was plotted versus year by quarters from 1970 to the first quarter of 2011. Seven recession events (REs) where the percent change in GDP went negative were identified: (1) from 4th quarter 1969 to 1971, (2) 2nd q. 1973–1975, (3) 4th q. 1979–4th q. 1980, (4) 1st q. 1981–1983, (5) 1st q. 1990–2nd q. 1991, (6) 4th q. 2000–2002, and (7) 2nd q. 2007–2nd q. 2009 (Isidore 2011). Recessions (3), (4), and (7) were more negative than –5% GDP. Inspection of the graph shows that each RE is followed by a brisk recovery to positive percent change in GDP by quarter, and then a dip, and another positive peak followed by the next RE. While this pattern is relatively consistent, the periods of the REs are not regular. If the pattern of the RE events repeats itself, there should be a recovery around 2013, followed by another RE sometime in 2014–2015.

To counteract recession, Keynesian economists may advocate government deficit spending (e.g., construction and research on alternate energy sources to create jobs) to catalyze economic growth. Supply-side economists may suggest tax cuts to promote business capital investment, and laissez-faire economists favor a Darwinian, do-nothing approach by government in which the markets sort themselves out; the weak fail (a capitalist survival of the fittest), and the workers suffer. The populist economic approach is for government to implement lower-bracket and midbracket tax relief and simultaneously provide subsidies for manufacturing, banks, and agriculture to stimulate the economy. Such subsidies are ultimately paid for by all taxpayers. Their object is to "prime the economic pump." The built-in delays of government fund allocation and setting up spending programs can make it appear that government efforts are ineffective; people expect instant results—there is no such thing in ESs.

One tipping point for a recessionary (autocatalytic) spiral is generally excessive debt, private and public. Nearly everyone uses credit cards, and credit card interest rates are usurious. Information in the form of advertising—"keeping up with the Joneses," variable-rate mortgages, "no cash down payment plans," and so forth—encourages buyers to accrue debt. At some point, some individuals no longer can meet their combined tax, interest, health care, and cost-of-living obligations and have to declare bankruptcy and/or default on their mortgages. Few have been willing to "live within their means"; after all, the government does not.

Another symptom of economic pathology is *inflation*. Inflation is seen as a rapid rise in the level of prices for goods and services, as well as a decline in the real value of money (i.e., a loss in its purchasing power). The adverse effects on the economy from inflation can include the hoarding of consumer durables by households in the form of cash, precious metals, canned goods, flour, and so forth as stores of wealth. Investment and saving are discouraged by uncertainty and fear about the future. One trigger for inflation is known to be a high growth rate of the money supply, exceeding that of the economy. This happened in pre-World War II (WWII) Germany. Typically, inflation is quantified by calculating a *price index,* such as the *US CPI.* The annual US CPI fell from 3.8% in 2008 to −0.4% in 2009, a sign of recession (BLS 2010). The CPI was 7.6% in 1978, jumped to 11.3%, 13.5%, and 10.3% in 1979, 1980, and 1981, respectively (a brief inflationary surge), and then fell to 6.2% in 1982. Observed from a microeconomic viewpoint, inflation has affected certain foodstuffs more than other durable goods. For example, the rise in the prices of beef, chicken, certain cereals, and corn syrup can be traced in part to the decrease in the supply of corn grown for food use. A significant fraction of corn grown in the United States now goes into the production of ethanol for a gasoline additive, rather than directly feeding farm animals or humans.

Many of the theories on the causes of inflation seem to converge on it being a monetary phenomenon. In the Keynesian view of the cause of inflation, money is considered to be transparent to real forces in the economy, and economic pressures express themselves as increases of prices seen as visible inflation. The monetarist viewpoint considers inflation to be a purely monetary phenomenon; the total amount of spending in an economy is primarily determined by the total amount of money in existence. According to the Austrian school of economic theory, inflation is a state-induced increase in the money supply. Rising prices are the consequences of this increase. Other economist splinter groups have devised other theories and explanations for inflation.

Because the causes of inflation are complex, varied and debatable, a variety of approaches have been used to try to control it. The US Federal Reserve Bank manipulates the prime interest rate. A raised interest rate and a slow growth of the money supply are the traditional tools to dampen inflation. Ideally, a 2% to 3% per year rate is sought. Keynesian economists stress reducing demand in general by increased sales taxes or reduced government spending to reduce demand. Government wage and price controls were successful in WWII in conjunction with rationing (to prevent hoarding). Supply-side economists have proposed fighting inflation by fixing the exchange rate between the domestic currency and some stable, reference currency or commodity (e.g. the Euro, the Swiss franc, or gold).

So is there a crisp tipping point for inflation, or does it tend to increase monotonically if left uncorrected? The "wage–price spiral" is a well-known 20th-century phenomenon. At present, if wages become too high, the manufacturer or service provider can move overseas to find a pool of cheaper labor, leaving behind unemployed workers that tend to dampen the inflationary spiral.

Stagflation is yet another pathological macroeconomic condition. Stagflation is simultaneous inflation and economic stagnation. (Stagflation is defined as low economic growth coupled with high unemployment and high prices for goods and food.) The stagnation can result from an unfavorable *supply shock*, such as a sharp increase in the price per barrel of light, sweet crude oil. Increased energy and/or material costs can slow production of goods and cause their prices to be raised to meet continued demand. Other commodities can trigger supply shock, for example, a corn crop failure (due to weather) or the failure of the Peruvian anchovy fishery (due to overfishing or global warming affecting ocean currents). (Anchovies are a major source of fertilizer and animal food protein.) If at the same time, central banks use an excessively stimulatory monetary policy to counteract a perceived recession (low interest rates), the money supply increases, and a runaway, inflationary wage–price spiral may occur.

The classic cure for stagflation appears to be to restore, or find alternatives to, the interrupted money supply while raising lending interest rates to dampen the inflation. Growth can also be stimulated by the government reducing taxes. Note that steady world population growth means increased competition for key commodities such as FFs, hence a steady increase in their prices. While not strictly a supply shock, limited FF resources will have a dampening effect on all kinds of production; it will also affect manufacturer confidence.

7.3 Introduction to ABMs and Simulations of Economic and Other Complex Systems

An *agent*, as used in an ABM or *multiagent simulation* (MAS), is generally a software module that is a component in an ABM. An agent inputs information from adjacent agents and its environment, and according to simple designed-in rules, it outputs actions and/or information. An ABM consists of an assembly of dynamically interacting, rule-based agents. They can respond to externally introduced information as well as information from other nearby agents and system parameters. An ABM can often exhibit complex, emergent behavior.

The agents in a MAS can have several important characteristics:

(1) *Autonomy:* The agents are at least partially autonomous.
(2) *Local connection:* No one agent has a full global view of the entire system; it is generally connected locally.
(3) An agent can be programmed to exhibit *proactive behavior,* directed at achieving a goal. (For example, in an economic MAS, an agent can be made to be a hoarder.)
(4) *Decentralization:* There is no designated controlling agent; every agent is its own "decider."

Gilbert and Terna (1999) commented: "There is no one best way of building agents for a multi-agent system. Different architectures (that is, designs) have merits depending on the purpose of the simulation." They go on to say that one of the simplest effective designs for an agent is in a *production system* (PS). The three basic components of a PS agent are (1) a set of rules, (2) a rule interpreter, and (3) a working memory. The rules have two parts: a condition that specifies when (and if) a rule is executed and an action part that determines the consequences of the rule's execution. The interpreter considers each rule in turn, executes those for which the conditions are met, and repeats the cycle indefinitely. Obviously, not every rule fires on each cycle of the interpreter because the agent's inputs may have changed. The MAS's agents' memories can change, but the agents' rules do not. In an adaptive MAS, the agent rules can be made to change or evolve, to pursue some goal(s).

Readers interested in pursuing agent-based modeling and MAS in-depth should consult the tutorial papers by Angus-Monash (2011), Axelrod and Tesfatsion (2009), Macal and North (2006a, 2006b), Bonabeau (2002), and Gilbert and Terna (1999). Also see the texts by Gilbert (2007), Tesfatsion and Judd (2006), and Wooldridge (2002).

ABMs are used in MASs of a broad selection of CNLSs. There are many applications for MASs, for example, to investigate the behavior of microeconomic and macroeconomic systems, terrorist network structures, animal group behavior, behavior of immune system cell trafficking, behavior of the Internet, vehicular traffic flows, air traffic control, wars, social segregation, stock market crashes, disaster responses, and so forth. For example, see the 2005 paper by Heppenstall et al. on their ABM for petrol price setting in West Yorkshire in the United Kingdom. This simple, economic ABM used interconnected, like-agent models for petrol station behavior. Ormerod et al. (2001) wrote an interesting paper entitled "An ABM of the Extinction Patterns of Capitalism's Largest Firms." Their ABM contained many agents, all interconnected. Model rules specified how the interconnections are updated, how the fitness of each agent is measured, how an agent becomes extinct, and how extinct agents are replaced. The overall properties of the ABM emerged from the interactions between the agents. They stated: "The empirical relationship between the frequency and size of extinctions of capitalism's largest firms is described well by a power law. This power law is very similar to that which describes the extinction of biological species."

See also the papers by Farmer and Foley (2009) and Buchanan (2009) on the need for ABMs in economics. *Agent-based Computational Economics* (ACE) is an officially designated special interest group of the *Society for Computational Economics.* See, for example, ACE at the Iowa State University website: http://www2.econ.iastate.edu/tesfatsi/ace.htm (accessed on August 2, 2011). Altreva offers *Adaptive Modeler v. 1.2.8* simulation software for ABM applied to stock trading (http://altreva.com, accessed on August 2, 2011). They state: "Instead of optimizing one or a few trading rules by back-testing them over and over on the same historical data, *Adaptive Modeler* lets a multitude of trading strategies compete [through agents] and evolve on a virtual market in real time." "*Adaptive Modeler* automates the process of creating new trading rules to adapt to market changes..."

There are a large and growing number of software applications for ABM/MAS, some proprietary, some open-source, and some free. An extensive *comparison of agent-based modeling software* can be found online (ABM Software 2010; Tools 2010). (Too many programs and toolkits are given to cite in detail here.) Also see the review and evaluation of ABS platforms by Railsback et al. (2006).

In any modeling formalism, model verification and validation is imperative. A model's ability to correctly simulate system behavior for novel inputs and ICs can be based only on its competence in simulating known, past scenarios. This is certainly true for ABMs, as well as large dynamic models based on sets of nonlinear ODEs.

7.4 Economic Challenges to Human Sustainability

As the world population has grown, the economies of areas, countries, and the world have been affected. Demand-caused food shortages have been exacerbated by climate change (floods, drought), soil exhaustion by continuous single cropping, loss of arable land by drought-caused wind erosion, insect and rodent infestations, plant viruses, loss of pollinators, and use of arable land to grow fuel crops (e.g., corn for ethanol, soy for biodiesel) instead of food. These factors have all contributed to the gradual inflation of food prices, hence the rise in the cost of living. See the papers by Lagi et al. (2011, 2012) for a detailed modeling study of contemporary food prices and the factors that affect them.

Short-sighted human actions motivated by short-term financial gain have contributed to food shortages and food price inflation. These include the following:

(1) The overuse of insecticides, causing insecticide-resistant insect pest strains to emerge, and the loss of beneficial insect pollinators.

(2) The overuse of aquifer water, causing aquifer depletion and salination.

(3) The overuse of chemical fertilizers to support monocropping and to increase present-year crop yields; this causes pollution of potable water and plankton blooms in estuaries from runoff, eventually depleting fisheries.

(4) The overuse of antibiotics on cattle and poultry to allow their raising in crowded conditions on more efficient "factory farms" (CAFOs) has led to drug-resistant bacterial strains that infect the animals as well as humans (cf. Chapter 5).

Food shortages inevitably lead to rising food prices and, in some areas, actual food shortages, and then food riots, famine, and eventually, human and domestic animal deaths from pestilence and starvation. Human migrations out of afflicted areas have stressed ecosystems and carrying capacities of new, marginal homelands. We have seen this occur in central East Africa.

The slow exhaustion of FF energy sources has also contributed to inflation: the costs of land, air, and sea transportation of humans, food, raw materials, and goods have steadily grown. The price per kilowatt-hour of electricity in the United States has steadily risen with the cost of fuel and the demand (most US electricity is still generated by burning coal, natural gas, or oil). About 19% of US electric energy came from nuclear plants, and approximately 13% came from renewable (hydro, wind) sources in 2011 (USEIA 2012). The United States produced 134 GW of electric energy from renewable sources in 2010, while China produced 263 GW (most of China's from hydropower) (REN21 2011). Thus, the cost of keeping warm in winter has risen along with the cost of air conditioning, refrigerating food, pumping water from wells, and industrial uses of electric power such as aluminum production, metal refining, oil refining, and so forth. Eventually, the cost per mile of running an electric automobile whose batteries are recharged from the grid will approach that for one run by 87-octane gasohol (E10). This is one type of road transportation parity. New, more efficient mass transportation networks will have to be developed. Who will pay for these networks? Who can pay?

Data from the US Energy Information Administration (EIA) illustrates how the average yearly cost of all US grades of conventional retail gasoline has risen abruptly from 2002 to 2012 (Figure 7.6), as have No. 2 diesel prices (Figure 7.7). Figure 7.8 from the US EIA shows how US field production of crude oil has risen steadily from about 1920 to a peak at around 1970, and then declined until around 2010, when it began an upward trend. The upward trend is partly from oil-shale oil, oil from new offshore wells, as well as fracking old wells. We are now well into our petroleum reserves.

Water shortages, caused by increased human consumption and decreased rainfall in certain areas, have limited agricultural production and use by industries and mining operations. As freshwater aquifers have been depleted, attention has turned to desalination of saline and brackish waters, an expensive process requiring large energy inputs. Decreasing precipitation needed to recharge aquifers is correlated to global warming events. Thus, there will be a steadily rising cost for freshwater linked to population growth and climate change.

The bottom line is that there will be a *slow*, monotonic inflation of the cost of living, ultimately driven by population growth and the *slow* depletion of water, food, energy, and natural resources. Living standards in the developed countries will *slowly* drop, in spite of technology and short-term economic "fixes" by governments.

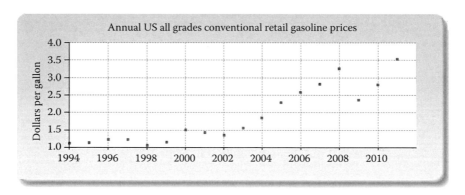

FIGURE 7.6
US retail gasoline prices (1994–2011). (Courtesy of US EIA.)

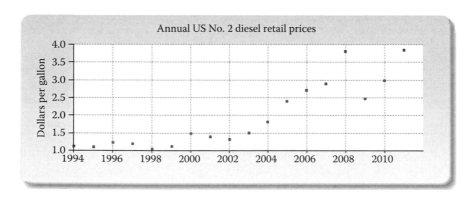

FIGURE 7.7
US No. 2 diesel retail prices (1994–2011). (Courtesy of US EIA.)

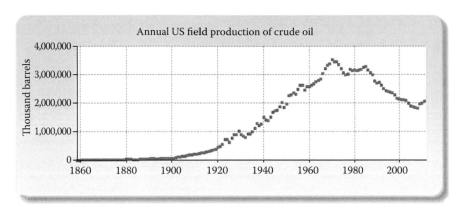

FIGURE 7.8
Annual US field production of crude oil. (Public domain figure—US EIA.)

Stagflation will occur worldwide (inflation of the costs of energy, food, water, health care, transportation, etc., while at the same time, rising unemployment and falling GDPs).

In his interesting book, *Prosperity Without Growth*, Tim Jackson (2011) asked the very meaningful question, "How—and for how long—is continued growth possible without coming up against the ecological limits of a finite planet?" Concern over limits to growth goes back at least to the writings of Thomas Malthus in the late 18th century.

Jackson argued that an important component of prosperity is the ability to participate meaningfully in the life of society. He outlined the factors influencing subjective well-being (happiness), including partner/spouse and family relationships (47%), health (24%), a nice place to live (8%), money and financial situation (7%), religious/spiritual life (6%), community and friends (5%), work fulfillment (2%), and other (1%) (Jackson 2011, Figure 3.1). Prosperity also contains the capabilities for *flourishing*. The type of socioeconomic homeostasis that Jackson advocates requires a number of factors: The labor content must be reduced; manufacturing goods must have increased durability and reparability; organic agriculture must be used; new infrastructures must be sustainable, maintainable, and repairable; and financial behavior will have to depend less on monetary expansion and more on prudent, long-term, stable investments.

Jackson addressed the complex problems of how humans can flourish, achieve greater social cohesion, find higher levels of well-being, and still reduce their adverse material impact on world ecosystems, that is, live sustainably. He made the case that increasing prosperity is not the same thing as economic growth per se (e.g., increasing *per capita* GDP). However, a high GDP is a *necessary* condition for flourishing human societies; it is closely correlated with certain basic social entitlements, including health care and education.

In a commentary in Jackson's book, Herman E. Daly considered the problem of how big a nation's economy can be before it overwhelms and destroys the ecosystems that sustain it. What is its optimum scale relative to its associated ecosystems? Daly wrote: "There is much evidence that some countries have passed this optimal scale, and entered an era of uneconomic growth that accumulates illth [sic] faster than it adds to wealth. Once growth becomes uneconomic at the margin, it begins to make us poorer." [This is another example of the *principle of diminishing marginal returns,* seen by Joseph A. Tainter (1988) to be one of the root causes of the collapse of complex societies.] *Per capita* GDP (pGDP) is widely used by economists as a measure of economic growth, hence economic health. It is a functional, a number calculated from the annual marketed flow of final goods and services. As a widely used economic measure, pGDP ignores *economic throughput*—the important "metabolic flow" of energy and materiel from environmental (ecological) sources into an economy and back out to environmental sinks as waste (think heat, polluted water, heavy metals, chlorinated hydrocarbons, deforested land, airborne chemicals including CO_2 and VOCs, etc.).

Perhaps economists should devise a new functional to replace traditional pGDP that is based on the whole activity of an economy and its influence on the people, environment, and ecosystems. The new functional should not only contain the traditional sum of *consumption, gross investment, government spending,* and *(exports – imports)* but also include consideration of the following factors: *unpaid work* (including housework, in-home child care, elder care, volunteer work, and community service); *distribution of wealth* in society; *quality of life* [e.g., including environmental degradation, leisure time, cultural events (operas, concerts, fairs, etc.), *increased life expectancy, increased traffic congestion, air and water pollution, pollution-related diseases, crime,* etc.]; and *quality of goods produced* (e.g., increased fuel efficiency in new cars, safety in new cars, increased energy efficiency in home heating, etc.). All this information should be distilled into one number; call it the *prosperity index* (PI).

7.5 Chapter Summary

This chapter has introduced and described classical static (S&D) and dynamic modeling of CESs. In 1956, Forrester (2003) urged that the dynamics of CESs be modeled by sets of nonlinear ODEs. The formulation of these ODEs promises to be every bit as challenging as writing the differential equations describing complex physiological systems such as the human immune system. They will contain transport lags, thresholds, and saturation terms, as well as parametric branch gain determination.

However, the largest problem in formulating accurate dynamic models of CESs is seen to be how to model human economic behavior, on both microeconomic and macroeconomic scales. We live in an era of rapid communications: cell phones with cameras and GPS systems, the Internet, PDAs, Twitter, Skype, Facebook, cable news, online financial analysis, and so forth. Thus, the transport delays in the communication branches are negligible. Attitudes, motivations, decisions (buy, sell, manufacture, do nothing, etc.), fear, and anger (e.g., at subprime lenders) must be cobbled into mathematical relationships in new dynamic models of CESs. Delays still exist in the execution of government programs designed to fight recession, however, and in the changes of public attitudes on economic issues. Some people blame the incumbent and former presidents for unemployment, high food prices, and the national debt, an example of the single-cause mentality applied to very CNLSs.

The use of ABMs was introduced as a means of incorporating information-modulated, decision-making behavior into ES modeling. ABMs were shown to have application in biology as well, in such uses as modeling human immune system cell trafficking.

Human sustainability is reflected in the economic changes we see around us. Inflation plus unemployment morphs into stagflation. High food prices, energy shortages, and people out of work translate into a lower living standard and lower GDP, not surprising when we must slice a finite resource pie into many smaller pieces because of population growth. The unemployed pay little if any taxes, placing a burden on the government financing of health, education, welfare, and other social programs. Governments have to borrow more money and/or issue bonds to operate and then must pay interest on these loans. One remedy is to downsize government spending (i.e., implement austerity in military spending, social programs, and government-sponsored R&D). This, in turn, presents problems when a government wants to provide essential services for a growing population who have grown to expect them. Thus, a dilemma appears for the role of government in regulating macroeconomic and microeconomic sustainability. The challenge we all now face: What are the best approaches to do this? Why does it take several years for a country's ES to respond to corrective measures? Does having a common currency (the Euro) help or hinder an individual country's economic well-being? We note that two of the healthiest economies in Europe at present (Poland, Switzerland) have not adopted the Euro and, by revaluing their currencies, have more leverage on their GDPs and so forth even though they trade with EU countries. Is this a case for modular economies being more robust?

Finally, we considered the problems inherent in implementing prosperity without growth.

8

Application of Complex Systems Thinking to Solve Ecological Sustainability Problems

---

8.1 Introduction

An important purpose for formulating a detailed mathematical model of a complex, nonlinear system (CNLS) in ecological sustainability analysis is to be able to simulate (and verify) its behavior under a variety of initial conditions and inputs in order to anticipate unintended behaviors, including tipping points and limit-cycle oscillations.

To be able to anticipate and predict quantitatively (or even just qualitatively) the behavior of a CNLS that we are introduced to for the first time, we need to be able to describe and understand the relationships between its states in qualitative terms and then mathematically in quantitative algebraic and differential (or difference) equations. This mathematical description allows us to model it dynamically and formulate a directed graph structure (relevant nodes and branches) with ODEs or difference equations (DEs) relating node parameters to branch (internode) transmissions. Computer simulations can then be run under different conditions. We need to be able to estimate the nonlinear relationships between parameters, and the effects of their initial conditions. One of the universal attributes of CNLSs is that one generally does not know *all* of the relevant signals (variables) and the relationships between them—we can only approximate or estimate these uncertain data. Thus, all mathematical models of complex systems are characterized by an aura of uncertainty, whether they be weather, economic, ecological, energy, food, social, or other system representations.

In many examples of CNLSs, we can identify the system's major *independent variables* (inputs) and define sets of *dependent variables* (outputs), but we have little idea how a given input will functionally and quantitatively affect the set of outputs. Inputs can interact in a nonlinear manner (superposition is not present), and tipping points and chaotic behavior may occur. Initial conditions can also affect input/output behavior.

More challenging is the need to establish branch gains relating interior (unobservable) states. Often, these gains have to be estimated and guessed at. Branch gains often are functions of node variables, which further contribute to system complexity. This property in certain physiological and biochemical systems is called *parametric regulation* (Northrop 2010, Section 11.3). Measurement of some of a real-world, multiple-input, multiple-output (MIMO) CNLS's sensitivities to parameter changes, and its cross-coupling gains, is helpful in constructing a valid, verified model that will enable the in silico exploration of its properties.

Below, we examine two little-known but nevertheless effective late 20th-century approaches to first encounters with general CNLSs.

8.2 Dörner's Approaches to Tackling Complex Problems

Dietrich Dörner's 1997 book, *The Logic of Failure: Recognizing and Avoiding Error in Complex Situations,* stressed the importance of complex systems thinking in problem solving. Dörner, a cognitive psychology professor at the University of Bamberg, addressed the dark side of the law of unintended consequences (LUC) in CNLSs and its causes.

Things can go wrong (the LUC), according to Dörner, because we focus on just one element of a CNLS. We tend to apply corrective measures too aggressively or too timidly (especially when the measures are expensive to implement); we ignore basic premises, overgeneralize, follow blind alleys, overlook potential "side effects," and narrowly extrapolate from the moment, basing our predictions on the future of those variables and parameters that most attract our attention. We also tend to incorrectly compensate for system time lags between inputs and reactions to them. In short, we tend to oversimplify problems. Dörner stated: "An individual's reality model can be right or wrong, complete or incomplete. As a rule it will be both incomplete and wrong, and one would do well to keep that probability in mind."

Dörner identified four human behavioral trends that contribute to failures when working with complex systems: (1) The slowness of our thinking—we streamline the process of problem solving to save time and energy. (2) We wish to feel confident and competent in our problem-solving abilities—so we try to repeat past successes using the same methods, even though the system has changed. (3) We have an inability to absorb and retain large amounts of information (in our heads) quickly—Dörner was of the opinion that we prefer static mental models, which cannot capture a dynamic, ever-changing process. (4) We have a tendency to focus on immediately pressing problems—we are captives of the moment. We ignore the future problems our solutions may create. "They may not happen." We add our fifth and sixth reasons for getting adverse, unintended consequences: (5) Political decision-makers try to look good and appear competent (the goal is to be reelected); they often make this a priority over actually accomplishing the goal. Then when things go wrong, a scapegoat is found (generally from the other party), or a specious, single-cause hypothesis is created to explain the failure. (6) Our responses are rate sensitive to our informational inputs. If something adverse happens slowly enough, we tend to ignore it (perhaps it will go away). Things that happen suddenly (e.g., 9/11, throughway bridge collapses, power grid failures, levee breaches) trigger massive responses, even overresponses, that often have not been thought through. [This is the social action rate sensitivity law (SARSL) principle we described in Section 2.2.2.]

It is clear that fact-based, critical thinking leads to better decision outcomes than hunch-based, guess-based, political-based, or faith-based thinking. The complex system's "steersman" should also "look outside the window" to constantly observe how his actions affect all the system's visible states. In economic systems, this view should include marketing research, including a review of the conditions that shape consumer preferences—something the "big three" auto makers in Detroit had clearly failed at, at least until the last decade. The big three evidently sampled only the opinions and preferences from their boardrooms. Clearly, *system sensitivities should be constantly monitored* to generate feedback to the inputs, in order to prevent the LUC from exerting itself.

Dörner's book used a real situation (i.e., Chernobyl) and two simulated research scenarios using human subjects: a fictitious African country, *Tanaland*, and a fictitious community, *Greenvale*, were used to verify his thesis on complex systems operation. Dörner showed the various behaviors individuals exhibit when challenged with ambiguity in decision-making scenarios and when facing overwhelming complexity. In some instances, to avoid coping with complexity, subjects focused tightly on a small area in which they were comfortable or allowed themselves to become distracted by small items. Some individuals were willing to change; others jumped right in without any situational analysis and became confused by unintended consequences and started creating bogus hypotheses on why they were experiencing problems.

Following his research, Dörner identified seven common problems people have when dealing with CNLSs: (1) failure to state and prioritize specific goals; (2) failure to reprioritize as events change; (3) failure to anticipate "side effects" and long-term consequences; (4) failure to gather the right amount of system information (neither to rush ahead with no detailed plan nor to excessively overplan); (5) failure to realize that actions often have *delayed consequences*, leading to overcorrection (and possible instability) when the results of an action (input) do not occur at once; (6) failure to construct suitably complex models of the system/situation; and (7) failure to monitor progress and reevaluate input actions.

Clearly, one needs a detailed foreknowledge about the organization of a CNLS and its sensitivities before one attempts to manipulate it in a meaningful manner.

8.3 Frederic Vester's "Paper Computer"

Another German scientist, Frederic Vester (b. 1925, d. 2003), developed a set of procedures, including a software analysis package, *Sensitivity Model Prof. Vester* (Vester 2004; Malik mzsg 2011), for characterizing a CNLSs being analyzed and modeled for the first time. Vester's approach is summarized below, beginning with his "paper computer."

As part of his integrated approach to dealing with certain complex systems and situations, Vester developed a paradigm that he called the "paper computer." This procedure is a heuristic method of making estimates of the relationships and sensitivities between the (nodal) variables of a CNLS being characterized for the first time. Vester listed five steps in the paper computer process, which we paraphrase here:

1. The systems analyst first picks a suitable CNLS to analyze.

2. Next, he/she makes a list of the relevant variables between which he/she wishes to establish causal relations. These can be obvious system inputs and outputs, as well as known "internal" variables. (Vester suggested picking between $N = 15$ and 30 variables, but about 50 might be a more appropriate upper bound when the analysis is done on a computer.)

3. Now an *impact matrix* is made on a piece of graph paper (or on a computer). The formulation of the impact matrix is generally subjective; it depends on the interpretations and judgments of the person(s) evaluating the system. To make the impact matrix, one lists the N variables along the x-axis and the same N variables along the y-axis. Then a 2-bit scale (0, 1, 2, 3) is used to estimate the impact (influence) of each variable (x_k) on each other variable ($x_j, j \neq k$). Obviously 0 = no impact; 1 = small impact (a big change in the chosen variable makes a small change in the target variable, i.e., a low sensitivity); 2 = medium impact (a small change causes a small change, but a large change causes a medium to big change in the target variable); and 3 = high impact (a small change in the variable causes a large change in the target variable). The impact matrix diagonal is all zeros. Note that there are $N(N-1)$ boxes to hold the impact numbers (2-bit sensitivity estimates).

   Table 8.1 illustrates an impact matrix for a hypothetical, relatively simple, $N = 8$ system we created as an example. Figure 8.1a illustrates the hypothetical system's connectivity with a directed graph with eight nodes, and in Figure 8.1b, the Vester diagraph for this hypothetical system. (In this example, the directed graph was made up before the impact matrix was constructed.)

4. Having established all the obvious cause-and-effect relationships for the system for each of the N variables in the impact matrix, one

TABLE 8.1

Impact Matrix for Hypothetical $N = 8$ State System

State	x_1	x_2	x_3	x_4	x_5	x_6	x_7	x_8	AS	PS	AS/PS	L_k	θ_k
x_1	0	1	0	2	3	2	0	0	8	3	2.67	8.54	69.4°
x_2	0	0	0	0	1	0	0	0	1	1	1	1.41	45°
x_3	1	0	0	0	0	0	0	3	4	3	1.333	5.0	53.1°
x_4	0	0	0	0	0	0	0	1	1	3	0.333	3.16	18.4°
x_5	0	0	1	1	0	0	2	0	4	6	0.667	7.21	33.7°
x_6	0	0	2	0	0	0	0	0	2	5	0.40	5.39	21.8°
x_7	2	0	0	0	0	0	0	1	3	2	1.50	3.51	56.3°
x_8	0	0	0	0	3	3	3	0	6	5	1.20	7.81	50.2°
PS	3	1	3	3	6	5	2	5	—	—	—	—	—

Note: AS = active sum, PS = passive sum. Length L_k is the impact vector magnitude, $L_k = \sqrt{AS_k^2 + PS_k^2}$. θ_k = impact vector angle = $\tan^{-1}(AS_k/PS_k)$. See the eight impact vectors ($L_k \angle \theta_k$) plotted on the diagraph in Figure 8.1b. We consider x_5 and x_6 to be system outputs, x_1 and x_8 system inputs, and the internal nodes are $x_2, x_3, x_4,$ and x_7 in this example.

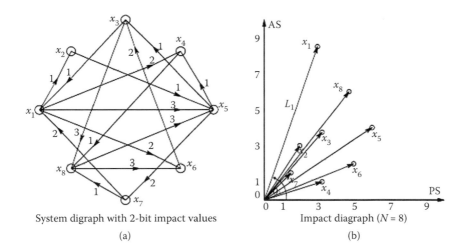

System digraph with 2-bit impact values

(a)

Impact diagraph ($N = 8$)

(b)

FIGURE 8.1
(a) Eight-node digraph with 2-bit impact (branch) values. (b) Eight-vector, Vester-type impact diagraph. (From Northrop, R.B., *Introduction to Complexity and Complex Systems*, CRC Press. Boca Raton, FL. 2011. With permission.)

adds up the impact numbers in their vertical columns and also in their horizontal rows. Vester called each row sum an *active sum*, and each column sum is called a *passive sum*. Thus, each variable is characterized by a pair of numbers that are an indicator of the strength of a given variable in influencing the whole system's performance and also of the degree to which that variable responds to the whole system's $N - 1$ nodal variables.

5. The final operation in the paper computer paradigm is to plot the
   2-D vector descriptors for each variable on an x–y graph that Vester
   called a *diagraph*. (Note that a diagraph is not the same thing as a
   digraph.) The *passive sum numbers* are plotted on the positive x-axis,
   and the *active sum numbers* along the positive y-axis. A pair of active
   and passive numbers determines the location of that variable in 2-D,
   impact vector space in the first quadrant.

   Vester identified eight regions in the diagraph's 2-D space (see
   Figure 8.2). Points lying in *region 1* (the top left corner) belong to
   active variables that exert a lot of influence on the system as a whole
   but do not receive much impact from the $N - 1$ other variables. The
   region 1 variables are helpful "levers" for change in the whole system. They are, in effect, inputs. *Region 2* (high active and passive
   numbers) variables are called *critical variables*. They receive much
   impact and also exert much influence on the other variables (on the
   average). They may behave like switches that can shift the whole

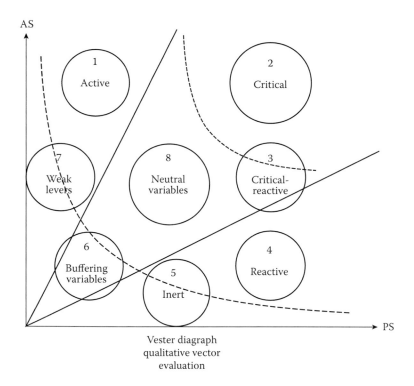

Vester diagraph
qualitative vector
evaluation

FIGURE 8.2
Eight vector evaluation ("influence") regions of a Vester diagraph. (From Northrop, R.B.,
Introduction to Complexity and Complex Systems, CRC Press. Boca Raton, FL. 2011. With
permission.)

system up and down the performance scale. They are stronger than *active variables*. In *region 3*, the vectors for critical *reactive variables* are located that exert some influence on the system and are highly sensitive to the system's status. *Reactive variables* are located in *region 4*; they can serve as indicators but not for steering the system. *Inert variables* are found in *region 5*. In *region 6* are the *buffering variables* that are weakly connected to the system. Vester calls the variables in *region 7 "weak levers."* Finally, *region 8* is in the center of the diagraph. In it are *neutral variables* that may serve for the self-regulation of the system. Note that the numbered regions in the diagraph have fuzzy boundaries. We have found by working several examples that it is possible for system outputs to appear in Vester's regions 8 and 6. This is not surprising, however, with 2-bit impact numbers (0,...,3) and eight regions (3-bits) defined in the diagraph.

The entire diagraph with its N points can provide an insight on how to manipulate the complex system. Its effectiveness depends entirely on your skill in assigning 2-bit impact numbers between the variables. The paper computer process is used as one component of an iterative systems analysis paradigm developed by Vester called the *Sensitivity Model*. More about this is discussed below.

8.4 Sensitivity Model of Vester

The purpose of Vester's Sensitivity Model software is to enable effective management of certain CNLSs while minimizing unintended consequences (see Figure 8.3). He recommended it for application in the following: corporate strategic planning; technology assessment; developmental aid projects; examination of economic sectors; city, regional, and environmental planning; traffic planning; insurance and risk management; financial services; and research and training. I suggest that the sustainability-related areas of agriculture, water resource management, fisheries management, ecosystems management, energy resource planning, epidemiology, waste management and resource recovery, and so forth be added to this applications list. On the Vester Web site (accessed March 14, 2011), some 66 licensees and users of his software are listed. Not surprisingly, 62 are in German-speaking countries (Germany, Austria, Switzerland); however, one licensee is Danish, one in Namibian, one is in Taiwan, and one is the European Air Control. No licensees were cited in English-speaking countries. Perhaps one reason Vester's scholarly works and software are relatively unknown in the United States is that he did not write in English, and his writings do not refer to the methodological developments of systems thinking in the past 30 years in the English

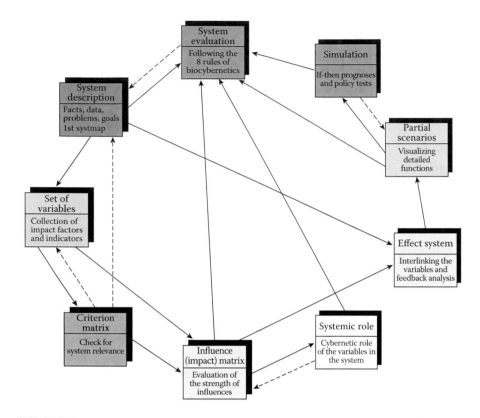

FIGURE 8.3

Vester's *Sensitivity Model* paradigm. Note the recursive structure. There are nine software modules in which information is processed, presumably replicating how evolutionary management works in nature. See text for description. (From Northrop, R.B., *Introduction to Complexity and Complex Systems*, CRC Press. Boca Raton, FL. 2011. With permission.)

literature. In our opinion, Vester's most significant work was his 1999 book: *Die Kunst vernetzt zu Denken: Ideen und Werkzeuge für einen neuen Umgang mit Komplexität* (*The Art of Network Thinking: Ideas and Tools for a New Way of Dealing with Complexity*). This text was reviewed in English by Ulrich (2005) and Business Bestseller (2000).

The central theme of the book was described in its preface (translated from German by Ulrich):

> Do we have the right approach to complexity: do we really understand what it is? Man's attempt to learn how to deal with complexity more efficiently by means of storing and evaluating ever more information with the help of electronic data processing is proving increasingly to be the wrong approach. We are certainly able to accumulate an immense amount of knowledge, yet this does not help us to understand better

the world we are living in; quite the contrary, this flood of information merely exacerbates our lack of understanding and serves to make us feel insecure... Man should not become the slave of complexity but its master.

One reason Vester's book is important is that it gives ordinary researchers, professionals, managers, and decision-makers (as opposed to trained complex systems scientists) a new sense of competence in dealing with certain of the complex systems issues in 21st-century sustainability studies. The other reason is that it describes his unique Sensitivity Model software.

8.5 Can We Learn From Our Mistakes?

A foolish consistency is the hobgoblin of little minds.

R.W. Emerson

As we have noted above, in dealing with CNLSs with the best intentions, one is liable to make mistakes, that is, take an action that leads to poor or inappropriate results. When one repeats the same action and expects a different result, one might be accused of insanity or, at best, naive behavior. Clearly, the manipulator(s) of the CNLS needs to "take the lid off" the system and reexamine in detail its behavior under a variety of branch parameters, inputs, and initial conditions. Generally, a more detailed model must be constructed—more variables and more sensitivities need to be measured or estimated in order to generate valid predictions. Above all, the manipulator(s) must not fall victim to the "single-cause mentality" or be of the "not in my box" mindset.

Nowhere is this need more obvious than in economics. Certain parameters of the US economy have been manipulated by the US government and the private sector to try to prevent the recession of November 2008 from morphing into a full-blown depression with soaring unemployment and market stagnation. In early 2010, there have been indicators that the hundreds of billions of dollars the government has spent on bank bailouts and rescue for certain automobile industries have done some good toward restoring US economic health and consumer confidence; however, high average national unemployment continues to lag these good indicators, perhaps because of built-in delays in the complex, nonlinear, nonstationary US economic system. [On January 27, 2012, the mean US unemployment fraction was down from over 9% to about 8.5%. As of August 2012, it is 8.3% (Economist 2012f).]

Have economists learned enough from dealing with the "great depression" and other more recent "mini-recessions" to advise appropriate government actions that will speed US (and global) economic recoveries? Time will tell.

D.J. Smith (2004) published online a very insightful essay, "Systems Thinking: The Knowledge Structures and the Cognitive Process." We recommend its reading. We were impressed with his "5th Law" (a systems' equivalent to the "Peter Principle"): "Systems controllers rapidly advance to the level of systems complexity at which their systems competence starts to break down." Smith's paper echoes many of the thoughts found in Dörner's book; Smith gives many interesting examples of problems arising in complex systems brought about by human incompetence in dealing with CNLSs.

8.6 Chapter Summary

The insightful analytical approaches of Dörner and Vester described above offer a beginning at describing and understanding a new complex system we are presented with and must characterize. Vester offered a heuristic, first-step approach at clarifying the inner structures of a complex system's graphical model (network).

Picture a 3-D glass spherical volume filled with a large, directed graph modeling a CNLS. The graph consists of a very large number of nodes (vertices) and branches (edges). The locations of the nodes in the volume appear random in some parts of the graph and regularly spaced in others. Some of the nodes are found in densely connected clusters because they describe related variables, subsystems, or modules in the CNLS. From outside the sphere, you can see the nodes and the branches connecting them very well on the periphery of the volume facing you, and one can identify them quantitatively. Deeper into the volume, the nodes and branches are not so visible, but one can estimate their values. However, at the center of the 3-D graph, one is hard-pressed to even count the nodes, and the connections of the branches are mostly unknown. The central structure of the graph is hidden by the outer layers and is very uncertain. To appreciate the detailed topology of the graph, we need to be able to rotate the graph in 3-D space so the viewer can see all the features hidden in a 2-D view.

Ordinarily, the node locations of a large graph in 2-D or 3-D space are arbitrary. However, in describing food networks (who eats whom), ecologists have arranged predator and prey species in arrays on separate levels and have also coded them by color. These ecological graphs (called networks or webs by ecologists) have appeared in many articles (Feldman 2008; FoodWeb3D 2004; Carafa et al. 2007). Ecological networks (graphs) produced using the FoodWeb3D software by R.J. Williams (Yoon et al. 2004) can be rotated in 3-D on your computer screen, revealing the "hidden details" caused by the high density of nodes and branches seen in 2-D. An online example can be found in Spruce Budworm (2011). Node/edge networks have also been used to characterize systems such as the following: high school friendships;

interdisciplinary research collaborations; online social networks (Friendster); protein interaction networks; human sexual contacts (HIV epidemiology); the Internet; characters in literature (*Les Miserables*); transportation networks: roads; transportation networks: airlines; politics; and of course, sustainability (Feldman 2008).

One class of complex system that appears to get little attention in terms of visualizing its characteristics in the form of node/edge networks is economic systems. Angus-Monash (2011) made the case for agent-based modeling (ABM) of economic systems (cf. Section 7.3 in this text). Quoting his comments on the current methods of dealing with economic complexity:

"In today's high-tech age, one naturally assumes that US President Barack Obama's economic team and its international counterparts are using sophisticated quantitative computer models to guide us out of the current economic crisis. They are not."

"The best models they have are of two types, both with fatal flaws. Type one is econometric: empirical statistical models that are fitted to past data. These successfully forecast a few quarters ahead as long as things stay more or less the same, but fail in the face of great change. Type two goes by the name of 'dynamic stochastic general equilibrium.' These models assume a perfect world, and by their very nature rule out crises of the type we are experiencing now."

Angus-Monash (2011) lamented that papers using ABM of economic systems count for less than 0.03% of the top economic research publications. Research papers using ABM in economics seem to be limited to specialized journals such as the *Journal of Economic Dynamics and Control*, the *Journal of Artificial Societies and Social Simulation, Computational Economics*, and the *Journal of Economic Behavior and Organization*.

Humans are addicted to linear thinking; we generally avoid having to deal with complex systems and seek oversimplified, single-cause hypotheses as to why CNLSs go bad. For the human race to be sustainable on this planet, we must learn how to deal with complexity—the complex social, environmental, and economic systems we create and the natural complexity inherent in ecosystems and in all living systems. We must think "outside the box," consider facts, and eschew dogma.

.

9

What Will Happen to Us?
FAQs on Sustainability

Reality requires verifiable facts and data; it also requires a balance between pessimism and optimism.

9.1 Introduction

Prognostications on human ecological sustainability are difficult to make because the governing variables are related in a complex, nonlinear, nonstationary manner. Also, changes in variables often occur slowly, dulling our perception of them. For example, we can take the record of atmospheric CO_2 concentration ($[CO_2]$) over time and use various numerical extrapolation programs to predict what the $[CO_2]$ will be in 2050, or even in 2100. There is a certain mathematical purity in doing this, which inspires confidence in the results, but this confidence is necessarily based on the past (and present) $[CO_2]$ records, plus estimates of future emissions and the conditions that affect them. The actual future value of $[CO_2]$ in 2050 will be influenced by, among other things, the world population growth, our future rate of combustion of *all* carbon-containing fuels, the types of fuels themselves, and the global phytomass (CO_2 uptake for photosynthesis and release by decay). Fuel combustion, in turn, depends on the population, the economies of countries, and their rates of development, whether the world's vast oil, natural gas (NG), and coal reserves are successfully exploited, whether societies will learn to live with less energy in general, and whether renewable energy sources (solar, wind, hydro, tidal, wave) continue to be developed. Further fossil fuel (FF) prospecting, energy conservation, and the development of non-FF sources are all connected to the economies of countries. Economies around the world at present are in flux (look at the United States; the United Kingdom; the European Union including Greece, Italy, Spain, etc.). Inflation of FF energy prices means less FFs will be burned, particularly in countries having recessions; their standards of living will fall.

The same uncertainty surrounds extrapolations looking forward in time on future food sources. A food supply is dependent on many factors, not the least of which is climate change producing droughts; competition for a crop (e.g., maize, soybeans) for energy use; crop infestations by insects, fungi, rodents, and so forth; pollution; overharvesting fisheries; overharvesting plankton; natural disasters (hail, floods, tsunamis, etc.); and increased demand by growing populations. On the plus side for future food resources is the future development of safe, genetically modified (GM) food crops, animals, and fish that are engineered to be more resilient to environmental challenges such as drought and increased environmental temperatures. We have the technology.

As we have already discussed, a very important factor affecting our future sustainability is population growth. This is actually less difficult to predict, based on the birth and death rates in a region. Birth rates tend to decrease as the standard of living grows, and they also fall in countries with extreme poverty. Miscarriages and infant mortality rise in undeveloped, economically stressed regions. Higher populations mean more demands on limited resources, causing inflation and a lowered standard of living. The Earth's human population must eventually stabilize at a sustainable value, or else the distant future will indeed be grim. (See the interesting papers on population growth in the July 29, 2011, issue of the journal *Science*.)

One of the very challenging complex systems affecting our sustainability is the economy of a country. One can argue that if economists really understood the incredible complexity of economic systems, we would not now see certain cities, states, and countries with double-digit unemployment, zero growth, and high *per capita* poverty, and on the edge of bankruptcy. Knowledgeable economists could advise legislators and executives on courses of regulatory actions that would lead to reduced unemployment, no symptoms of recession, and robust gross domestic product (GDP) growth. They would also be able to predict how long certain corrective actions would take to be effective. Then there is the chilling possibility that none of the actions taken by governments or by the private sector will be effective at rectifying world economic malaise.

To make a simile, today, the science of economics is at the same level of understanding economic systems as neuroscience was in the 1930s of understanding the human central nervous system (CNS; a complex adaptive system), when prefrontal lobotomies (aka leucotomies) were used to try to cure or palliate certain forms of mental illness (Swayze 1995).

9.2 Will Technology Sustain Us?

9.2.1 Food

We predict that in spite of certain untoward effects of early GM plants, new, safer, genetically engineered plants and animals will be developed and will significantly increase world food productivity. A limiting factor in food production will be water shortages in areas that formerly had just enough (e.g. Texas, Oklahoma, Iowa, Indiana, Nebraska). Future rain and snowfall water sources will not keep up with the drawdown of aquifers for agriculture and cattle ranching. Our food production will have to compete for water sources with energy production. Water is used in oil production from oil shale and oil sand, for fracking, as well as in cooling and steam production. Corn grown for ethanol production requires the same amount of water per acre, cultivation, and fertilizer as corn grown for food. With future changes in rainfall patterns, some cropland will become ineffective; other land will become arable. Certain crops (e.g., corn, soybeans) will be made drought resistant by genetic engineering, with the insertion of drought-resistance genes into their genomes from plants such as sunflowers.

Fisheries will continue to be depleted because of overharvest, pollution, and habitat change due to global warming (changes in established ocean currents, salinity, and ocean acidification). Overharvesting macroplanktons such as krill will also adversely impact fisheries.

The factors affecting our future food insecurity were addressed by HRH the Prince of Wales (2012) in his insightful essay, *On the Future of Food*. Basically, the Prince of Wales makes the case for changing from the factory farming sources that place short-term corporate profits ahead of health, environmental, and ecological concerns to many smaller, organic, ecologically friendly, sustainable farms.

New sources of food for humans and their farm animals will be exploited—insect and plankton biomass, as well as a number of edible fungi. Muscle cells grown in vitro in mass tissue culture reactors may possibly compete with natural meat and other new food sources.

Food prices will steadily increase regardless of sources. Researchers at the New England Complex Systems Institute (NECSI) have developed a predictive model for the United Nations Food and Agriculture Organization (UNFAO) Food Price Index (FPI) that tracks the real data very closely, beginning at an FPI of 110 in 2004 to an FPI of 210 in 2012. Their model makes use of inputs from market speculation and conversion of food (corn) to ethanol and produces the observed, short-term "bubbles" in food prices very well. They showed that by the beginning of 2013, their model predicts that a bubble will increase prices above that of the price peak bubbles in 2008 and 2011. The 2011 peak reached an FPI of approximately 240. [The FPI was approximately 110 in 2004 (Lagi et al. 2012).] These results are not encouraging for sustainability. Current NECSI research is examining the effects of the 2012 Midwestern drought on US food prices.

Another threat to sustainable food supplies is the effect of coastal eutrophication on fisheries. This is seen very clearly in the waters of the Chesapeake Bay, which was once a tidal ecosystem rich in fish, oysters, clams, and blue-claw crabs. There has been a gradual increase in phytoplankton blooms in the bay. The large plankton blooms are influenced by the unnaturally high concentrations of nitrates and phosphates (from agricultural runoff from farms bordering the bay and also from sewage effluent from surrounding private septic systems and overflowing municipal systems) and by higher-than-normal water temperatures. When the phytoplanktons die, they sink and are consumed by bacteria that deplete the oxygen concentration in the bottom water. Certain species of these bacteria also release toxic hydrogen sulfide, which poisons fish and bottom-dwelling clams, oysters, and crabs. Many will die. The loss of many filter-feeding mollusks means the bay waters will not be naturally filtered and cleaned. If these sources of undesired nutrients are not abated, the Chesapeake Bay ecosystems may slowly degrade. They will no longer be able to support the large, economically viable seafood industries in the surrounding states. On the present trajectory, the bay will end up as a "dead estuary" with a moribund fishing industry and few recreational boaters and tourists. There is a trade-off in regional food sustainability: high yields from crops bordering the bay versus high yields from estuarine fisheries. Perhaps the best way out of this dilemma is to replace the profligate use of chemical fertilizers with organic farming strategies. The crop yields may be slightly reduced, but the fisheries yields would remain robust, giving a net total food gain. People will be willing to pay more for organic foods and fish caught from clean water. It remains to be seen whether governments can act together to abate the sources of pollution to the bay "common."

There will be more attention paid to foods from unconventional sources, including insects, planktons, fungi, and muscle cell tissue culture (see Chapter 6.) A little-known, nutritious seed that may see more cultivation in the northern hemisphere is quinoa. Grown and consumed in the Andes for over 5000 years, quinoas are the seeds of a broadleaf plant of the *Amaranthaceae* family, *Chenopodium quinoa*. Quinoas are conventionally harvested by hand because the seeds mature over the entire growing season. Like grain crops in post–Ice Age Europe and Asia Minor, selective breeding could narrow the period over which the quinoa seeds mature, making farming and harvesting more practical. Quinoa is highly nutritious. Cooked quinoa contains, per 100 g, 120 cal; 21.3 g carbohydrates; 17.6 g starch; 2.8 g dietary fiber; 1.9 g total fat; 0 g cholesterol; 4.4 g protein [21 amino acids (AAs)]; vitamins B_1, B_2, B_6, B_9, and E; copper; iron; magnesium; manganese; phosphorus; potassium; selenium; sodium; and zinc. Quinoa is gluten-free (Nutrition Facts 2012).

There will be a slow, steady increase in the number of small, organic farms in the United States. This increase in small organic farming operations will be a reaction to the increasing unsustainability of industrial agriculture. For example, monocropping will lead to one or more spectacular crop failures, which will promote a resurgence in the use of "heirloom" crops and food animals, as well as increased use of some of the next-generation GM crops. An increase in food poisoning outbreaks (due to overuse of antibiotics, pesticides, and contamination of crops with CAFO wastes) will also lead to greater consumer demand for organic food. Increasing drought conditions (which most climate models predict) will raise the cost of industrially farmed crops, making water-efficient organic crops relatively more affordable.

Water shortages will also affect meat costs. Beef prices in particular will rise precipitously, as beef is far more water intensive to raise than any other food animal or crop. It takes approximately 1799 gal. of water to produce 1 lb. of beef, compared with 468 gal. per pound of chicken (Mekonnen and Hoekstra 2012). If drought trends persist, consumers will probably see beef as a luxury item in the future, not unlike lobster today.

The rising prices of food and a desire for healthier produce will encourage a resurgence in home gardening similar to the "victory garden" phenomenon during World War II. "Backyard chickens" for home egg production are already a popular trend in US cities.

Topsoil will continue to be degraded and eroded, further encouraging the development of small, organic farms on properties not large enough to support monolithic industrial farms. The loss of arable land will also contribute to more "landgrabs" in developing countries, possibly causing unrest in those regions. Unrest will make this strategy less and less economically appealing, causing wealthy nations to turn increasingly to local, sustainable, agricultural solutions.

In short, we will see a series of actions and reactions eventually "moving the needle" toward more sustainable farming practices.

Big factory farms will continue, but their nongreen operating costs (fertilizers, herbicides, insecticides) will increase, making them less profitable. There will be more farmers' markets used to purchase fresh, in-season, organic produce, rather than supermarkets. People will pay more for produce free from chemicals. Using next-generation GM plants, we believe that organic farming methods can, indeed, eventually "feed the world" (Estabrook 2012).

9.2.2 Water

As we have seen, areas of drought will develop as the result of the Earth's surface receiving, and the surface and atmosphere retaining, more solar energy. This will translate into regional water shortages adversely affecting agriculture, manufacturing, oil and NG production, and life in cities (cf. Section 3.4).

What can be done? One word: *conservation.* One option is to collect local rain runoff (from roofs, parking lots, etc.); another is to recycle domestic "gray water" for irrigation and toilet waste disposal. There are a number of methods that have been devised to desalinate seawater; all require expensive energy, however. One author (RBN) has an old-fashioned rain barrel under the gable roof to catch rainwater to water the garden. (This is water that the deep well pump does not have to pump.)

A clever means of saving water stored in reservoirs, lakes, and ponds coming into wider use is the reduction of surface evaporation by covering the water surface with polar, hydrophilic molecules that block the escape of water vapor. Factors that increase evaporation are solar radiation, wind, high air temperature, and low humidity (McJannet et al. 2008). One treatment that has been used worldwide is the patented *WaterSavr* powder, composed of powdered hydrated lime [calcium hydroxide, $Ca(OH)_2$], plus hexadecanol and octadecanol (at ca. 5% each), which when applied forms a polar, mono-molecular layer on the water surface. The powder is self-spreading, and the molecules arrange themselves in a tight monolayer. If the monolayer is inter-rupted by wind, waves, boat wakes, and so forth, it will spontaneously reform (WaterSavr 2012). The powder is generally applied at a density of approxi-mately 1 kg/ha (1 ha = 10^4 m^2 = 2.741 acres). The WaterSavr powder has been demonstrated in tests in Owens Lake (CA, USA), Turkey, India, Spain, and Australia (Coliban 2006) to reduce surface evaporation up to 38%. The pow-der is fully biodegradable in 48–72 h (ChemistryViews 2010) and is nontoxic. However, it requires a still water surface for effective use. Depending on wind conditions, the powder needs to be reapplied about every 2–3 days. It was Au \$18/kg in 2012 in Australia (Monolayers 2012). Flexible Solutions claims that a 30% evaporation reduction on a 1000-acre or 400 ha reservoir with 120 in. or 3048 mm annual evaporation saves approximately 3.7 Mm^3 (9.78 × 10^8 US gal.) of water, enough to supply 8000 homes for a year. (This implies treat-ment with WaterSavr throughout the year.) WaterSavr is a patented product of Flexible Solutions International Ltd. (US Patent No. 6,303,133).

Another evaporation-retarding substance used in Australia is *Hydrotect*, an emulsion of 60% water and 40% aliphatic alcohols. Like WaterSavr powder, Hydrotect is claimed to be nontoxic and biodegradable and is suitable for application to drinking water in liquid form (Monolayers 2012). Presumably, Hydrotect can be applied as a spray. A study done in Burkina Faso, Africa, using Hydrotect at 1–3 kg/ha/day showed evaporation reduction of up to 50% for small ponds (0.5 ha) and 25%–35% in larger water storage volumes. In 2003, Hydrotect was delivered in 190 kg drums at a cost of Au \$5.00/kg. It is recommended to be applied at a rate of 1.5 kg/ha. Hydrotect is made in Australia by Swift & Co. Ltd (Monolayers 2012; GHD 2003).

There appear to be negligible effects on fish and invertebrates living in ponds treated with monolayers. However, saturated oxygen concentration in certain treated ponds has been shown to decrease by 10%–15% (Wixson 1966, cited in McJannet et al. 2008). In a 2008 Libyan test of WaterSavr, 16.42% of water was saved in a treated, open-top tank; the other identical, nontreated tank was the control. Curiously, the treated tank experienced a heavy algal bloom (Ikweiri et al. 2008).

Two other surface treatments used to inhibit water evaporation are stearyl alcohol and CIBA PAM, a water-soluble polyacrylamide resin. In a Cuban study of stearyl alcohol effectiveness, an evaporation reduction of 49.5% was reported (Poloni et al. 2009). One liter of PAM for water treatment is priced at US\$ 37 by Sigma–Aldritch Co., and cationic hydrophilic PAM powder is priced at US\$ 1000 – 1500/tonne by eb88.com E-business (1/25/13).

9.2.3 Energy

There will be a slow reduction in the production of FFs in the last half of the 21st century. It will be slow because petroleum reserves will be exploited (tar sands, oil shale, deepwater oil), as well as the new NG deposits being discovered. These reserves will help the world's active FF supplies to decrease less rapidly, slowing our transition to noncarbon, alternative energy sources (AESs). The world's "peak petroleum" may be approached asymptotically for decades to come. Unfortunately, the new FF sources have high production costs, including a high need for freshwater. The world demand for oil will continue to increase, largely due to the industrialization of Brazil, Russia, India, and China and the voracious oil consumption of the United States. Global demand will exceed supply, driving petroleum prices up and up in the mid-21st century. It will be this inflation of petroleum prices that will trigger a more rapid investment in AESs in the United States.

Public fear of nuclear meltdowns, exacerbated by the four Fukushima Daiichi reactor failures, may be reversed if China's new, pilot, pebble-bed reactor works without incident. The continuing burning of CO_2-producing FFs will accelerate the global warming syndrome, in spite of our token efforts to capture CO_2 from point sources. There are all these automobiles and animals (including 7 billion humans) exhaling CO_2, and decaying organic matter worldwide, which no one seems to worry about very much.

As the world population grows, so will the demand for electric energy. In addition, production costs for oil shale oil, tar sands oil, and deepwater oil will be high. This will be reflected in inflated prices for oil-based fuels and oil-based goods (e.g., various plastics and petrochemicals), so we can expect their prices to rise in the future, along with the price of coal. However, feedstock for plastic production (e.g., ethylene) can be obtained from syngas derived from plant biomass. This "green" ethylene can save some FFs for their energy uses.

Fuel cells (FCs), powered by hydrogen, methane, or methanol, will gain more use, and so will flow batteries. We also predict a steady, worldwide rise in wind turbine power and more power from inexpensive solar cells. Means of storing energy from intermittent power sources will also grow (e.g., pumped hydro, compressed air, thermal, batteries, flywheels), increasing the utility of wind, tidal, wave, and solar sources.

The US IEA (2012) stated that in 2007, about 69% of electric energy worldwide was generated from FFs (42% coal, 21% NG, 6% oil). Hydroelectric energy accounted for 16%, nuclear 14%, and nonhydro renewables only 2%! See the IEA *World Energy Outlook 2011 Factsheet* (2011) for predictions on how total anthropogenic CO_2 emissions will vary in the future, and the shifts predicted on how FF and renewable energy consumption will vary looking forward to 2025. The US EIA (2011) predicted that the world would consume 770 quadrillion × (10^{15}) Btu of energy in 2034 (compared to 505 quadrillion Btu in 2008), an amazing 52.5% increase in 27 years. (Note that 770 × 10^{15} Btu = 770 PBtu = $8.12 × 10^{20}$ J = 812 EJ = $812 × 10^{18}$ J.)

NG is going to displace coal and oil FF energy in many applications because it is the least expensive and lowest-polluting FF. David Rotman, in a recent article in *Technology Review*, observed that NG used in US power plants now accounts for as much electricity as coal! This parity with coal has happened in the last few months in 2012. There are two major reasons for this downward shift in the use of "king coal": Electricity from new power plants using NG is approximately \$0.04/kWh, while new coal plants produce at approximately \$0.062/kWh. Also, this recent, increased use of NG for electric power generation is estimated to save 400 million metric tonnes (Mt) of carbon emissions annually. In addition, electric power from new nuclear plants will cost approximately \$0.105/kWh. New solar grid power (with NG backup) will be most expensive, approximately \$0.11/kWh (Rotman 2012).

9.2.4 Electric Vehicles

Electric vehicles (EVs) can be propelled by direct-current (AC) motors, including brushless DC motors, or alternating-current (AC) polyphase induction motors (PIMs). Certain DC and brushless motors make use of powerful, rare-earth, permanent magnets (PMs). The rare earths (lanthanum, cerium, dysprosium, neodymium, samarium, etc.) are used to make superstrength PMs used in DC motors and AC generators (alternators). Most of these elements are currently found in ores in China, who controls their export (Economist 2012d). AC PIMs do not require expensive, rare-earth, PMs and can be made with torque–speed curves having high, zero-speed torque. The DC power from onboard batteries or an FC is inverted and transformed to make audio-frequency (400–800 Hz) AC to run the PIM propulsion systems. DC motors and generators using electromagnetic field excitation (rather than expensive PMs) have been around for years (Fitzgerald and Kingsley 1952). DC generators use commutators and brushes and can be categorized as shunt field or compound excitation (both series field coil and shunt field coil). DC motors can use series field windings, shunt field windings, or compound windings to generate the desired torque–speed curves.

EVs using batteries must charge their batteries either from an onboard, FF-powered generator or from electric power from the grid. Most grid power today (about 70%) is not renewable; it comes from burning FFs, and hence, it creates more greenhouse gasses (GHGs) than does an equivalent amount of mechanical work done by a conventional, onboard diesel or gasoline engine–powered vehicle. This is because of the losses in generating high-voltage, 60 Hz AC; the transmission losses for high-voltage AC; losses in transforming grid AC voltage to low voltage DC; losses in charging the batteries; and losses in the DC propulsion motors and drive train. These losses (inefficiencies) accumulate to approximately 79%, (i.e. the typical battery EV has an *overall efficiency* of about 21%) and must be added to the total energy cost of operating a battery-powered EV. Petrol-fueled vehicles have overall efficiencies ranging from approximately 10% to 40% (Ridley 2006).

Another lossy, EV process is the *wireless* charging of EV batteries. This is accomplished through the use of an AC magnetic field generated by a coil external to the EV, coupled closely to a fixed coil under the vehicle's body panels. This creates what is, in effect, a lossy air-core transformer. The transformer output is rectified and conditioned to charge the EV's batteries. Coupling audio-frequency AC power through a transformer with a large core air gap is very inefficient, even at high audio frequencies. While wireless charging may appeal to those challenged by inserting an AC power plug into the vehicle, any convenience of wireless charging is well offset by the power loss in the process, further lowering the charging (and vehicle) energy efficiency. Wireless charging is a bad idea.

EVs can also derive their electric power from H_2- or CH_4-based FCs. For this to occur, there will need to be a network of "gas stations" (literally) where FC-powered vehicles can replenish their hydrogen or methane. A large capital investment will be needed to build such gas stations and set up gas distribution systems to them. We already have an abundance of methane, but hydrogen can also be extracted from some NG or made by electrolysis of water (also expensive). This is a "chicken or egg" situation: investors will not want to finance "gas stations" unless there will be a market for the H_2 or CH_4, that is, many FC EVs that use these gasses are being built.

EVs powered by energy stored in flywheels or compressed air have poor efficiencies and probably will not be used.

9.2.5 Anthropogenic GHGs

The rise in the atmospheric concentration of the GHG CO_2 has been well documented (NOAA 2011). The physics of GHGs and their correlation with the global warming phenomenon are also well known. What is not certain from numerous modeling studies is the actual extent of future global warming, hence the level of its adverse effects on human sustainability.

Carbon capture and sequestration (CCS) may gain some momentum in the future. ("Carbon" means CO_2 in this context.) The rate of CCS will lag the total CO_2 output rate of new and future FF-burning power plants, especially in developing countries. Future CCS will slow the future rate of increase of atmospheric [CO_2], but not reverse it, and will have little impact on mitigating global climate change, a process already well underway.

Atmospheric [CO_2] will continue to climb, in spite of our point-source CCS efforts. The major reasons for the limited future application of CCS are its capital and operating costs, borne by the consumers of electricity and taxpayers, and the problem of what to do with the CO_2, once sequestered.

Recall that much atmospheric CO_2 comes from diffuse (nonfixed) sources that include the following: over 7 billion humans breathing, their farm animals, wild animals, decaying vegetation, volcanoes, forest fires, motor vehicles, airplanes, ships, and so forth. Technologies are being developed to permit direct extraction and sequestration of CO_2 from the atmosphere. The costs of future atmospheric CCS plants will be borne by taxpayers. Do not expect to see many operational in the future. And once we have dealt with trying to reduce the total atmospheric [CO_2], how will we deal with the GHG methane? Atmospheric methane comes from the following: decaying vegetation, in NG from deep oil wells, melting methane clathrate deposits in tundra and under the Arctic sea bottom, the many ruminant farm animals, the bacterial decay of manure, and so forth. If it is acclaimed as an imminent global warming threat, there will be political pressure to have it scrubbed directly from the atmosphere, again at some great taxpayer cost. At least captured methane can be burned directly as fuel (turned into heat energy and CO_2), used to make other fuels such as methanol or dimethyl ether (DME), or used in an FC to directly make DC electric power and CO_2. Atmospheric methane is 21–23 times as effective a GHG as is CO_2, but it has a far shorter half-life in the atmosphere than does CO_2.

9.3 FAQs Concerning Sustainability

- *What are the threats to human sustainability?* Threats to human sustainability can be grouped into those *stochastic events* we have no control over and slow, *anthropogenic threats* we have marginal control over. Under *stochastic events,* we have the following: hurricanes, earthquakes, tsunamis, massive volcanic eruptions, solar flares, asteroid strike, tornadoes, and so forth. (These events all have low probabilities or low frequencies of occurrence.) Under slow, *anthropogenic threats,* we have the following: population growth and rise in atmospheric [CO_2] from burning fossil and other organic fuels. High [CO_2] has several adverse effects: It lowers ocean pH, threatening marine ecosystems and fisheries. It also triggers global weather change, including droughts leading to food shortages and shortages of potable water. Exhaustion of FFs will threaten energy supplies and raise the cost of feedstock for plastics and other industrial chemicals. FF exhaustion will also increase the cost of transportation. Expect inflation.

- *What will be the effects of planetary warming on ecological sustainability?* The dire conditions brought about by future, severe planetary warming have been well documented. They include, but are not limited to, the following: changes in weather patterns, including increased areas of drought affecting crops and freshwater supplies; melting polar ice caps, glaciers, and ice flows; and interruption of ocean currents carrying equatorial heat to northern landmasses (a negative feedback to northern hemisphere warming). There will be more atmospheric water vapor (AWV), hence more precipitation in global regions at >50° N and S latitudes, as well as changes in regional crops that can be grown optimally. There will also be an increase in invasive southern plant and animal species into ecosystems in northern latitudes (land and sea); the size (mass) of most animal, fish, and plant species will decrease noticeably with habitat temperature rise, affecting food sustainability

(Sheridan and Bickford 2011). Tropical diseases, especially those with insect vectors (e.g., malaria, sleeping sickness, plus diseases such as elephantiasis and schistosomiasis caused by parasites), and fungal diseases affecting both plants and humans will spread north and south from equatorial regions. CO_2-caused acidification of the oceans and water temperature increase will decrease plankton and coral growth in ocean ecosystems—ultimately impacting fisheries. Warming Arctic oceans are releasing methane into the atmosphere from melting methane clathrate crystals on the ocean bottom; methane is 21–23 times as effective a GHG as is CO_2. Warming of the Southern Ocean will contribute significantly to global sea level rise through the melting of the Antarctic landmass ice cap (Liu and Curry 2010).

- *Are there any benefits from global warming?* Higher ambient temperatures will mean lower rates of consumption of FFs for domestic heating in temperate regions. (This will be offset in developed and developing countries by higher electric energy consumption rates for air conditioning.) Higher atmospheric [CO_2] will mean plants and green algae will grow faster, consuming more CO_2 and emitting O_2. (Faster-growing plants require more water. Increased green algae densities cause eutrophication, hence ocean "dead zones" adversely impacting coastal fisheries.)

- *What will happen to food supplies around the world?* Supplies will decrease per capita, and food prices will continue to climb as global warming becomes more severe and the world population continues to increase. This is not just from supply-and-demand effects; the production and distribution costs of food will also rise because of increasing energy costs (Feed 2011; Food Insecurity 2010; Lagi et al. 2012). Available arable land will decrease from soil loss, and crop sizes will decrease due to soil nutrient depletion from monocropping and chemical farming. In underdeveloped countries, land owned by small farmers for subsistence food growth is being bought up by agribusinesses for raising energy and food crops for export (factory farming). This creates local food shortages and inflation in food prices. The increasing use of organic farming will mitigate the rising costs of produce in the United States.

- *What technological changes can be made to improve food supplies?* Grain food plants (rice maize, wheat barley, oats, etc.) can be bred selectively for higher yields. Second-generation GM plants can also be "tuned" to their environments. For example, drought-resistant genes from sunflowers will be added to corn and other crops, making them drought- and heat-tolerant xerophytes (Schenkelaars 2007).

- *What will happen to our FF resources?* Prices will steadily rise. These increases will be driven by simple supply and demand, by increased production and transportation costs, and also by market speculation driven by international events (e.g., revolutions, threats of war, and natural disasters). FF production rates will slowly decline as supplies dwindle, accompanied by a rise in the population-driven demand for energy in developing countries, for example, Brazil, Russia, India and China (BRIC). Higher FF prices will be reflected in the increase in costs of food, manufactured goods, and transportation. More NG will be used as a primary energy source (its combustion produces less CO_2 than an equivalent amount of coal or fuel oil), and it is cheaper than coal (Rotman 2012).

- *The return of the Dark Ages for mankind in North America and Western Europe by 2099?* No, we have science, technology, and limited religious hegemony. By 2050, the standard of living in developed countries may fall to early-20th-century levels partly because of the high cost of energy, water, and food to an increasing world population. (One estimated world population figure in 2050 is about 9.1 billion.)

- *Will procreation laws be enacted in the future?* This will probably never happen. It is much better to make free reproductive education and contraceptives available in developing and underdeveloped countries to dampen population growth rates.

- *Will one-car families dominate in the United States?* Very likely; also zero-car families in cities. New car designs will use engines that run on alternate fuels: H_2 and methane FCs, DME, biodiesel, and so forth, when the cost of alternate fuels reaches parity with FFs and grid electricity. Car rentals will become more common.

- *Will there be less flying and more public ground transportation?* Definitely: Who will pay for expanded public ground transportation expansion? "Road trip parity" will be reached when the cost of personal travel (including fuel, tolls, and lodging) in automobiles exceeds the cost of public transportation for the same trip. Road trip parity will make it easier to convince taxpayers that investment in efficient public transportation is worthwhile.

- *Will there be water rationing?* In many places, it is now not rationed and will be in the future. The city of San Antonio, TX, already has water use restrictions due to the unprecedented drop in the Edwards Aquifer caused by drought and overconsumption (due to population growth, agriculture, lawn watering, fracking, etc.).

- *How effective will point-source CO_2 capture and sequestration be in mitigating global warming?* We feel it will be too little, too late, and too expensive in providing a small, delayed return (only a slight decrease in the rate of global warming). Other factors such as cyclical increases in the solar energy flux hitting the Earth must be accounted for, as well as cloud albedo influenced by certain pollutants. *All* anthropogenic CO_2 emissions (including respiration) must be considered. No one will want to pay to clean up the atmospheric carbon common, when other carbon emitters are not, cannot, or will not. (See David Victor's book: *Global Warming Gridlock*, 2011.)

- *Will there be resource wars?* Probably: in Asia and Africa over land and ocean FF deposits, for example, in Nigeria and the South China Sea. Also, conflicts may arise because of burgeoning regional populations competing for water from rivers and arable land, as well as for gas, oil, and water pipeline rights. Global warming will cause declining spring river flows due to shrinking winter snows. Thus, there will be less water for hydroelectric generation and irrigation.

- *Will SARSL rule until it is too late?* Probably. The steady state is known; certain changes leading to uncertain outcomes are perceived as threatening.

- *Will there be a strong mobilization toward green, AESs (solar, wind, tide, wave, hydro)?* Only if the cost of FFs becomes too high, grid parity is reached, and quick money can be made by building AESs.

- *Which non-FF fuels will dominate in 2050?* CH_3OH (MtOH), CH_3CH_2OH (cellulosic EtOH), CH_4, H_2, biodiesel, and H_3COCH_3 (DME). H_2, CH_4, and MtOH can be used in *FCs*, MtOH and DME in diesels.

- *Will US automotive CAFE standards be raised?* Eventually. Only when the US auto industry and FF energy companies stop lobbying against it and public demand for fuel-efficient vehicles reaches congress and the White House at a volume they will pay attention to. It will cost US auto manufacturers money to redesign certain inefficient internal combustion engines.

- *Which will be the dominant future biodiesel source?* Nonfood seed oils, palm oil, DME from biogenic methane.

- *Will rising FF prices trigger a mass movement to sustainable energy sources?* Yes, eventually there will be a SARSL tipping point. However, the continuing discovery and exploitation of undersea NG and oil deposits will push this tipping point way down the road.

- *How far will the US standard of living fall?* We predict to a pre-1930 level by 2050. (See Chakravarty and Majumder 2005; Makoka and Kaplan 2005; Standard 2003; Economic Situation 2010.)

- *What must happen in the United States for it to maintain its standard of living and GDP growth (prosperity)?* Americans must respond effectively to seven major 21st-century challenges: (1) that posed by globalization; (2) rapid advances in information technology; (3) the huge and growing financial deficits of federal, state, and local governments and banks; (4) our profligate pattern of energy consumption; (5) jobs must be created (Friedman and Mandelbaum 2011; Smil 2011b); (6) ideologue politicians must relearn the art of compromise or find new jobs; and (7) new sources of energy must be developed, both from nonconventional FFs (oil shale, tar sands, and NG) and from noncarbon sources (solar, wind, tide, wave, etc.).

- *How can these seven challenges be met?* (1) and (2) By education and technical training. (3) By agreeing politically on meaningful legislative actions to regulate government spending and restructure taxation, and appreciating that economic systems are very complex and have built-in delays in response to input actions. (4) US and state legislatures must agree to take steps to stimulate private investment in alternate energy sources and aggressively encourage energy conservation by offering tax incentives for buying and owning vehicles with high fuel miles-per-gallon (MPG) ratings, installing building insulation, and building using passive solar designs. Enhanced public transportation networks must be built. Research and development (R&D) and production of alternate fuels and energy sources must be expedited. (5) Job creation is a multidimensional problem: public and private investment in small businesses using high technology will be required. New markets must be created. (6) Vote out nonproductive incumbents. Increase public feedback to congress. Create citizen lobby groups with clout to compete with big business' lobbyists. (7) The development and production of new FF energy sources will be expensive, adding to fuel prices. A huge capital investment will also be required to produce meaningful levels of noncarbon energy. Thus, all forms of mid- to late 21st-century energy will be much more expensive than their mid-20th century counterparts, dampening GDP growth.

- *What can be done to get people to realize there is a problem and take concerted actions?* Public education is needed on sustainability (not just "greenness"); start very early. How long does the rubber band of fuel prices need to get stretched before it snaps and the SARSL threshold is reached? Perhaps US Department of Energy (DOE) commercials on TV exhorting the use of sustainable energy will help. (Or will this be a political conflict of interest?) We already have seen TV commercials by energy companies urging the consumption of "clean coal" and how benign fracking is in recovering NG. ["Clean coal" is an oxymoron. Anthracite (hard) coal does burn cleaner than soft coal, but the "cleanest" FF is NG, or methane from hydrates.] The US government should also do commercials on energy conservation, explaining how it can be done (e.g., insulation, lights out, drive less, etc.).

- *What will be the effect of China's "one-child policy" on that country's future economic growth?* A greater percentage of China's population will become "over 65." By 2050, this shift in the population's age distribution will have two dampening effects on China's economy: (1) The money spent on national health care and welfare will increase. (2) There will be fewer young workers available to drive a vibrant economy.

Other developed countries in which the total fertility rate (TFR) has approached 2 will have the same problems, but to a lesser degree.

- *What can we do as individuals to mitigate the threats to our sustainability?* Conserve: energy, water, food. Drive high-MPG vehicles, including hybrids and those powered by compressed natural gas (CNG) (the cost of fuel will insure this). Use fluorescent lamps. Turn off power to unused lights and systems. Insulate dwellings and workplaces. Collect rainfall for watering gardens and animals. Filter municipal water in the home (water quality will deteriorate). Have a local source of potable water that does not require electricity to pump. Where possible, plant organic "victory gardens." Plant high-yield, safe, GM crops that are drought resistant. Harvest and preserve family or communal food.

In the 19th century and earlier, meat was preserved by drying, smoking, and salting. Vegetables and fruits were preserved by drying or canning in mason jars. Raise food animals such as chickens, rabbits, goats, ducks, and geese that can be fed table scraps and that can forage. Expect future power "brownouts"; have alternate sources of lighting and heating. Where possible, install an individual home wind turbine and/or solar panels for alternate power; sell some excess electric power back to the grid.

9.4 Chapter Summary

It is very challenging to try to predict what will happen to our sustainability in the future, given a collection of interacting, complex, nonlinear, time-variable systems that determine its outcome. However, persistent trends do seem to emerge in many cases that tempt extrapolation. Most of our opinions in this chapter are debatable; this is good. We would like our readers to examine the facts and trends in issues affecting human sustainability and form their own opinions. Eschew politicization.

The major issues surrounding all aspects of ecological sustainability were seen to be the following:

(1) Population growth and the consequence of the reduction of TFR causing a population's age distribution to shift to the more aged end of the distribution.

(2) A slow reduction in the available potable water from all sources. Part of this shortage will come from climate change, part from excessive domestic use by a growing population for agriculture, and part from FF production, including NG production.

(3) A slow rise in the cost of food.

(4) A slow rise in the cost of electric energy from all sources.

(5) A gradual rise in the cost of all natural resources (wood, minerals, FFs).

(6) A slow, global, rise in air and water pollution. This will be from human waste and industrial wastes, including the GHGs CO_2 and CH_4. This pollution will continue to negatively impact ecosystems, fisheries, agriculture, and human health.

(7) Human ecological sustainability will also be negatively impacted by an increase in pathogenic fungi, viruses, and bacteria, some spread by insects and rodents. The continual use (overuse) of antibiotics and pesticides in factory farms, in animal feedlots, and domestically is causing pathogen mutations that are antibiotic resistant and insects unaffected by certain chronically used insecticides.

Glossary

A. Abbreviations and Acronyms Used in This Book

AA: Amino acid
ABM: Agent-based model
ABS: Agent-based simulation
AC: Alternating current
ACS: Adaptive complex system
AD: *Anno Domini:* the years following the birth of Christ
AMP: Antimicrobial peptide (protein)
AR: Antireflective (coating)
AWV: Atmospheric water vapor
BC: The years before the birth of Christ
BCE: The years before the Christian era
BP: Before present
BRIC: The developing countries: Brazil, Russia, India, and China
BSE: Bovine spongiform encephalopathy
Bt (or **BT**): *Bacillus thuringiensis* (an insecticidal protein originally derived from a bacterium of that name)
Btu: British thermal unit, heat energy equal to 1054.35 J or 252.0 g-cal
*ca.***:** Abbreviation for *circa* (Latin)*: approximately, or about*
CAFE: Acronym for "corporate automobile fuel efficiency"
CAFO: Acronym for "concentrated animal feeding operation"
CC: Carbon capture
CCS: Carbon capture and storage (or sequestration)
CDC: U.S. Centers for Disease Control and Prevention
CE: Years in the Christian era
CES: Complex economic system
CIS: Commonwealth of Independent States
CLAW: Crude look at the whole (in analyzing a complex system)
CNG: Compressed natural gas
CNL: Complex nonlinear
CNLS: Complex nonlinear system
CS: Complex system
DC: Direct current
DE: Difference equation
DG: Distributed generators

DME: Dimethyl ether
EC: European community
EM: Electromagnetic
ES: Economic system
EtOH: Ethanol
EU: European Union
exa-: $\times 10^{18}$ multiplier
FAO: Food and Agricultural Organization (of the United Nations)
FC: Fuel cell
FF: Fossil fuel
FFL: Feed-forward loop [reaction architecture term coined by Alon (2007)]
FFP: Feed-forward path
FrF: Fracking fluid
GDP: Gross domestic product
GHG: Greenhouse gas
GI: Gastrointestinal
giga-: $\times 10^{9}$ multiplier (also, in the United States, billion)
GM: Genetically modified
GMOs: Genetically modified organisms
Gt: gigatonne = 10^{9} tonnes = 10^{12} kg
GW: Global warming
HAWT: Horizontal axis wind turbine
hCNS: Human central nervous system
hIS: Human immune system
IC: Internal combustion (engine); initial condition
ICs: Initial conditions (of an ordinary differential equation)
I/O: Input/output
IPCC: Intergovernmental Panel on Climate Change
IR: Infrared (radiation)
IV: Initial value
IVP: (In solving sets of ordinary differential equations): initial value problem
KE: Kinetic energy
KERS: Kinetic energy recovery system
LD$_{50}$: The dose of a poison that will kill (statistically) one-half of the animals receiving it
LFG: Landfill gas (methane + CO_2)
LIA: Little Ice Age
LIR: Long-wave infrared radiation
LTI: Linear time-invariant (system)
LUC: Law of unintended consequences
L-V: Lotka–Volterra
MA: Mass action
MAHB: Acronym for Millennium Assessment of Human Behavior
MAS: Multiagent simulation
MetOH: Methanol

MIMO: (In systems theory): acronym for multiple-input, multiple-output system
MPa: Megapascals (1 MPa = 145.04 psi)
MRSA: Methicillin-resistant *Staphylococcus aureus*
Mt: Million metric tons (tonnes)
NFB: Negative feedback
NFF: Non–fossil fuel
NG: Natural gas
NIMBY: Not in my backyard
NIMFY: Not in my front yard
NL: Nonlinear
NLCS: Nonlinear complex system
OCV: Open-circuit voltage
OECD: The Organization for Economic Cooperation and Development
ODE: Ordinary differential equation
ORNL: Oak Ridge National Laboratory
OTC: Over the counter (medicine)
OWCCAT: Oscillating water column air turbine
OWE: Ocean wave energy
Pa: Pascal (unit of pressure)
PAH: Polyaromatic hydrocarbons
PC: Parametric control
pd: Population density (number per volume or area)
PDF: Probability density function
peta-: $\times 10^{15}$ multiplier
PETN: Pentaerythritol tetranitrate (an explosive)
PFB: Positive feedback
pGDP: Per capita gross domestic product
PM: Permanent magnet
ppmv: Parts per million by volume
PV: Photovoltaic
RR: Railroad
RV: Random variable
SARSL: Social action rate sensitivity law
SFG: Signal flow graph
SISO: (In systems theory): acronym for single-input, single-output system
SoS: Systems of systems
sp: Species
SRV: Stationary random variable
SS: Steady state
SST: Sea surface temperature
STEG: Solar thermal electric generation
STP: Standard temperature and pressure for gasses (e.g., 1 atm and 0°C)
SV: State variable
tera-: $\times 10^{12}$ multiplier (also, in the United States, *trillion*)

TFR: Total fertility rate
tonne: 1000 kg (abbreviated: **t**)
UN: United Nations
USD: United States dollars
USGS: United States Geological Service
VAR: Volt-amperes, reactive (in AC power)
VOC: Volatile organic compound (in smoke or exhaust)
V&V: Verification and validation (of a model)
WMD: Weapon of mass destruction
WV: Water vapor

B. Glossary Terms

absorbance: In spectrophotometry, the absorbance is defined as $A(\lambda) \equiv -\log_{10}[T(\lambda)]$. The absorbance is also called the optical density (OD). $A(\lambda)$ ranges from 0 to large positive numbers. $T(\lambda)$ is the optical *transmittance* (see below).

adjacency matrix, A (in graph theory): This is an $N \times N$ matrix, where N is the number of vertices (nodes or states) in the graph. If an edge (path) exists from some vertex **p** to some vertex **q**, then the element $m_{p,q} = 1$, else it is 0.

agent [in complex adaptive system (CAS) modeling]: An agent is an entity or a computer simulation of an entity that inputs information and outputs behavior and information. Agents are autonomous decision-making units with diverse characteristics. A network of information flow can exist between agents. Agents can have "memories." Some agents are adaptive and can learn. Computer models of human agents figure in *agent-based models* (ABMs) of political, economic, contagion, terrorism, and social complex nonlinear systems (CNLSs), most of which can be treated as CASs. In economic ABMs, human agent models can be programmed to produce, consume, buy, sell, bid, and so forth. They can be designed to be influenced by the behavior of other, adjacent agents. ABMs can exhibit emergent, collective behaviors. Dynamic ABMs have been used to model bacterial chemotaxis (Emonet et al. 2005) and have application in hIS modeling (immune cell trafficking). Electric power companies have used ABMs to better understand the complexities of electric power generation, transmission, and usage. The following species of agents were used in power simulations: customer, physical generator, fuel, transmission line, demand, generation company, transmission company, regulatory (power market + independent system operator (ISO) + real-time dispatch = administrative), and random event (lightning, failures, etc.) agents. See the ABM tutorial papers by Macal and North (2006) and Bonabeau (2002).

agent-based model (ABM): An ABM is a dynamic system of interacting, autonomous entities. It consists of the following: (1) a set of user-defined agents; (2) a set of agent relationships; and (3) a framework for simulating agent behaviors and interactions (i.e., ABM software, e.g., *Swarm, Repast S, Ascape, MASON, MATLAB, DIAS, NetLogo, StarLogo, SeSAm, Cormas, VOMAS*, etc.).

angiosperm: Any plant of the class Angiospermae, characterized by having seeds enclosed in an ovary; a flowering plant.

Anthropocene epoch: Our present epoch, the Holocene, has been renamed the *Anthropocene* (the recent age of man), where the activities of mankind have altered Earth systems, by atmospheric chemist Paul Crutzen. Crutzen suggested that the Anthropocene epoch began in the late 18th century when ice cores show that atmospheric CO_2 levels began to rise monotonically. Other scientists put the beginning of the Anthropocene epoch in the middle of the 20th century, when the rates of both population growth and energy consumption accelerated rapidly (Kolbert 2011; Economist 2011e).

Ashby's law of requisite variety [LRV; in complex systems (CSs)]: "The survival of a system depends on its ability to generate at least as much variety (entropy) within its boundaries as exists in the form of threatening disturbances from its environment." In essence, this means that "...a control[ler or regulatory] system has to be more complicated than the system it is controlling [regulating]" (Smith 2004). Another formulation of Ashby's law: "A model system or controller can only model or control something to the extent that it has sufficient interval variety to represent it. For example, in order to make a choice between two alternatives, the controller must be able to represent at least two possibilities, and thus one distinction" (Heylighen and Joslyn 2001).

atmospheric time constant of a gas (also, atmospheric lifetime): The time in years it takes the atmospheric concentration of a certain gas to decay to 0.368 $(1/e)$ of its initial concentration, given zero input of that gas to the atmosphere at $t = 0$. See **half-life**.

attractor (in phase plane plots of CNLS behavior): An attractor can be a point, a curve, a closed path or a set of closed paths in the phase plane to which the phase trajectories of the CNLS converge as the system reaches the steady state (SS).

In the case of the closed path attractor, the system's trajectories can form a *stable limit cycle* (LC), that is, one in which the system ends oscillating in the SS, given any set of initial conditions (x_o, y_o). (There are also *semistable LCs*. An *unstable LC* is better called a "repellor" than an attractor.) One example of a *strange attractor* is a pair of connected LC orbits (A and B) that the system follows in the phase plane, switching from A to B and back again in a seemingly chaotic yet deterministic manner, a behavior noted in the three coupled *Lorenz equations.*

autocatalytic reaction (in chemical dynamics): Those reactions in which at least one of the *products* (on the right-hand side of the chemical equation) is a *reactant* (on the left-hand side of the chemical equation). An example of a simple autocatalytic reaction is $\mathbf{A} + \mathbf{B} \underset{k_r}{\overset{kf}{\Longleftrightarrow}} 2\mathbf{B}$. An autocatalytic reaction system uses positive feedback, and its mass-action (MA) equation(s) is (are) nonlinear (NL). The autocatalytic MA ordinary differential equation (ODE) is $\dot{b} = k_f ab - k_r b^2$, where a and b are the running concentrations of $\mathbf{A}$ and $\mathbf{B}$, respectively. When the simple autocatalytic ODE above is viewed as an initial value problem (IVP), with initial concentration of $\mathbf{A} = a_o$, and the initial concentration of $\mathbf{B} = b_o$, assuming $k_f \gg k_r$, the running concentrations of $\mathbf{A}$ and $\mathbf{B}$ can be shown to be given by the following:

$$a(t) = (a_o + b_o)/\{1 + (b_o/a_o)\exp[+(a_o + b_o)k_f t]\} \text{ and}$$
$$b(t) = (a_o + b_o)/\{1 + (a_o/b_o)\exp[-(a_o + b_o)k_f t]\} \tag{GL1}$$

Note that $a(\infty) \to 0$ and $b(\infty) \to (a_o + b_o)$. Both the $a(t)$ and $b(t)$ curves are sigmoid. One example of an autocatalytic reaction is in the formation of bovine spongiform encephalopathy (BSE) prion protein ($\mathrm{PrP^{Sc}}$) from normal prion protein ($\mathrm{PrP^C}$). The reaction is $\mathrm{PrP^C} + \mathrm{PrP^{Sc}} \Rightarrow 2\ \mathrm{PrP^{Sc}}$.

average gross domestic product (GDP) per capita: This number tells us how big each person's share of the GDP would be if we were to divide the total GDP by the population of the country; we take the value of all goods and services produced within a country's borders, adjust for inflation, and divide by the total population. GDP *per capita* does not take into account the following factors: (1) unpaid work including housework, in-home child care, elder care, volunteer work, and community service; (2) distribution of wealth; (3) changes in the quality of life, including environmental degradation, more leisure time, increased life expectancy, increased traffic congestion, and so forth; and (4) changes in the quality of goods, for example, increased fuel efficiency in new cars, increased energy efficiency in home heating, and so forth (Standard 2003).

average geodesic distance (of a graph): The average distance D of each node to any other defines the average geodesic distance of a graph:

$$D = (1/m)\sum_{i=j}^{n}\sum_{j=1}^{n} d(i, j) \tag{GL2}$$

where n is the total number of nodes, $d(i, j)$ is the shortest path distance between i and j, and $m =$ total number of edges in the graph.

azeotrope: A mixture of two or more liquids in such a ratio that the liquids cannot be separated by simple distillation. The ethanol and water azeotrope has 95.6% ethanol and 4.4% water, and both components boil at 78.1°C at 1 atm.

bagasse: The dried, crushed stalks of sugarcane after the sugar juice has been extracted. Used as a renewable, biomass fuel.

bar (pressure unit): 1 bar = 100 kPa = 14.5 psi = 1 atm (1 Pa = N/m²).

barrel: 1 bbl. US liquid = 31.5 US gal. liquid = 119.24 L. 1 US petroleum bbl. = 42 US gal. = 0.159 L.

Beer's law (in spectrophotometry): Describes how the intensity I_λ of light at a given wavelength λ is attenuated with distance as it travels through an absorbing medium. Ideally, for solutions $dI_\lambda/dx = -\alpha I_\lambda C$, where C is the molar concentration of the infrared (IR)–absorbing substance, $\alpha(\lambda)$ is the *molar absorption coefficient*, L is the total optical path length, and $I_{\lambda 0}$ is the intensity of the input light. Integrating, we find that $(I_\lambda/I_{\lambda 0}) = \exp[-\alpha(\lambda)CL]$. If an air path is involved with absorbing gasses, Beer's law can be written as $(I_\lambda/I_{\lambda 0}) = \exp[-(k_{\lambda 1} + k_{\lambda 2} + k_{\lambda 3} +...) L]$. The $k_{\lambda j}$ are the *extinction coefficients* of the component, IR-absorbing gasses at wavelengths λ, k_λ, and $\alpha(\lambda)$ are functions of temperature; $k_{\lambda j}$ is also a function of partial pressure of gas j.

bifurcation (of a phase portrait): A bifurcation occurs when a small, smooth change made to a parameter (input, initial condition, gain, etc.) in a continuous CNLS causes a sudden topological change in the system's long-term dynamical behavior. In the phase plane, a new attractor appears. Under some conditions, the cause of a bifurcation is called a *tipping point*. See also **Hopf bifurcation**.

billion: 1×10^9 (a giga-).

birth rate: The *crude birth rate* is the number of childbirths per 1000 people (male and female), per year, in a population. (In 2009, the world average birth rate was 19.95 per year per 1000 total population.) See **fertility rate**.

bitumen (aka asphalt): Found in tar sands. A thick, multicomponent, viscous form of crude oil, so heavy that it will not flow unless heated and/or diluted with lighter hydrocarbon solvents. At room temperatures, its rheology is much like cold molasses.

blackbody (BB): In thermodynamics, an ideally black surface that absorbs all incident energy and also is the best emitting surface for radiated energy. See below.

blackbody radiation: W_λ describes the distribution of radiated power per square meter from a hot, *ideal* radiating object (BB) as a function of wavelength. The radiated watts/(m²λ) is given by Planck's relation:

$$W_\lambda = \frac{c_1 \lambda^{-5}}{\exp[c_2/\lambda T]-1} \text{watts}/(\text{m}^2 \times \text{nm}) \tag{GL3}$$

where W_λ is the BB's spectral emittance in watts/(m^2 nm) at λ nm, λ is the radiation wavelength in nanometers (1 nm = 10^{-9} m), $c_1 = 3.740 \times 10^{20}$, $c_2 = 1.4385 \times 10^7$, T is the Kelvin temperature, $c_1 = hc/k$, $c_2 = 2\pi c^2 h$, h = Planck's constant, k = Boltzmann's constant, and c = speed of light (Sears 1949).

Note that the maximum of the W_λ curve at BB Kelvin temperature T can be shown to occur at $\lambda_{pk} = 2.8971 \times 10^6/T$ nm. Thus, the hotter a BB object is, the more power it emits at shorter wavelengths. Using the Stefan–Boltzmann law, we can find the total radiant emittance of an ideal BB at T Kelvin. By integrating $W_\lambda d\lambda$, we obtain the total power per square meter emitted by the ideal BB. This can be shown to be $W_{bb} = \sigma T^4$ W/m^2, where $\sigma = 5.672 \times 10^{-8}$ (Sears 1949, Chapter 12). The average BB temperature of the Earth is 288 K; thus, its BB emissivity peaks at $\lambda = 2.8971 \times 10^6/288 = 1.01 \times 10^4$ nm. The 288 K Earth ideal BB emits $W_{bb} = 390.2$ W/m^2.

butterfly effect (in CNLSs): A term given to the *tipping-point* behavior of a CNLS. Often misused in a flawed analogy where the feeble air currents caused by the gentle flapping of a butterfly's wings may lead to a hurricane or tornado at some distant location. The analogy fails to take into consideration the real-world dissipative effect of action at a distance and that some system parameter must cross a threshold before the CNLS switches behavioral modes.

CAFE (Corporate Average Fuel Efficiency) Standards: A law enacted in 1975 by the US government, setting minimum fleet mileage standards for automobiles. At first, CAFE required car companies to double fuel efficiency of their auto fleets from 13.5 mpg (1975) to 27.5 mpg (1985). The US Energy Security and Independence Act of 2007 raised the fuel efficiency standards to 35 mpg by 2020.

Campbell's law: "The more any quantitative social indicator is used for social decision-making, the more subject it will be to corruption pressures and the more apt it will be to distort and corrupt the social processes it is intended to monitor" (Campbell 1976). Educational achievement tests are cited as subject to Campbell's law, especially in the context of the *No Child Left Behind Act*.

canola: A cultivar of the rapeseed plant *(Brassica campestris L.)*. Their seeds are used to produce edible oil that is fit for human consumption because it contains lower concentrations of toxic *erucic acid* [$CH_3(CH_2)_7CH = CH(CH_2)_{11}COOH$] than common rapeseed oils. Canola was originally bred in Canada from rapeseed plants in the 1970s. Canola is actually an acronym for "Canadian oil, low acid." Canola oil is low in saturated fats, is high in monounsaturated fat, and has established heart health benefits. About 42% of canola seed is oil; the remaining pulp is a high-quality cattle food. One needs 22.68 kg of canola seed to make approximately 10 L of oil. Canola oil is used in foods, cosmetics, industrial lubricants, and biodiesel fuel. Canola plants have now been genetically engineered to carry genes giving them resistance to Monsanto's Roundup herbicide. GM canola plants are grown in Canada and Australia. Eighty percent of wild rapeseed plants in North Dakota have been found with transgenes from Canadian GM canola, making them resistant to chemical control (Nature News 2010, Beckie et al. 2011), an example of the law of unintended consequences.

carbon dioxide: CO_2. Molecular weight = 44.0095. 1 Gt of CO_2 = (12/44) Gt carbon = 0.267 GtC. Critical pressure (P_c) = 73.8 bar. Critical temperature (T_c) = 304.2 K. Triple point (TP) at T_t = 216.6 K, P_t = 5.185 bar. CO_2 is a liquid for $T > T_t$ and $P > P_t$. One tonne of CO_2 contains 12/44 = 0.2727 tonne of carbon (C); 32/44 is oxygen.

carrying capacity (in sustainable living): The number (or density) of human inhabitants that can be supported [in terms of food (calories), water (potable and irrigation), and energy (for cooking, heating, lighting)] *in a particular area,* without degrading the physical, social, and economic environments at present or in the future. When a local population's numbers or density exceeds the area's carrying capacity, further growth is considered unsustainable. The carrying capacity's thresholds for adequate food, water, and fuel are often not well defined (fuzzy). The *area* cited above can be a specific city (e.g., Mumbai), a country (e.g. Japan), or a geographical area (e.g., North Africa).

carrying capacity (of an ecosystem): The population densities of nearly all prokaryotes, plants, and animals are in a complex, dynamic equilibrium with other prokaryotes, plants, and animals in their ecosystem. If factors such as disease, drought, climate change, loss of food, change in habitat by invasive species, predation, hunting, fishing, and so forth exceed a tipping point, the ecosystem's carrying capacity for a population can decrease rapidly beyond its normal equilibrium range to a low level, allowing a population to crash or even become extinct, for examples, the Atlantic cod fishery, passenger pigeons, blue whales, polar bears, and so forth. Hardin (1977) defined the carrying capacity of a particular area as "...the maximum number of a species that random changes, without degradation can be supported indefinitely by a particular habitat, allowing for seasonal and random changes, without degradation of the environment and without diminishing carrying capacity in the future."

catalyst: A substance whose surface speeds the rate of a chemical reaction but itself remains unchanged chemically is called a catalyst. Precious metals are often used as catalysts, for example, platinum and gold; however, some reactions can be catalyzed at high temperatures by cerium (see below), copper, iron, or even carbon nanotubes. Many catalysts are given rough or porous surfaces to increase their effectiveness. Catalysts can be "poisoned" by certain chemicals that bind to their surfaces and decrease their catalytic activity.

cellulolytic enzymes: Enzymes that break down cellulose, lignin, and xylan plant polymers.

cellulose: A linear polymer chain of glucose molecules, 100 to 10^4 units in length, covalently bonded by oxygen atoms. Many chains are cross-linked together by hydrogen bonding. The basic cellulose chain is $[C_6H_{10}O_5]_N$, $100 < N < 10^4$. Cellulose is found in the cell walls of plants.

centrifugal force: When a tethered point mass rotates in a circular path around the center of the circle, a radially acting, centrifugal force F tries to make the mass fly outward. It can be shown that $F = M R \omega^2$ newtons, where M is the mass in kilograms, R is the radius in meters, and $\omega =$ the angular velocity of the mass around the center in radians per second. Centrifugal force is a consideration in the design of high-velocity, energy storage flywheels. (Note 1 N = 0.2248 lb. force.)

ceria: Cerium (IV) oxide, also known as (aka) ceric oxide, cerium oxide, or cerium dioxide, is an oxide of the tetravalent rare earth metal, cerium. Formula: CeO_2, molecular weight: 172.115 g/mol. Ceria has many industrial uses, including that of a catalyst. It is used for catalytic converters in automotive applications. Ceria can give up oxygen without decomposing. In association with platinum, ceria can effectively reduce NO_x emissions as well as oxidize toxic CO to CO_2. A reduced amount of expensive platinum is required in catalytic converters when ceria is used with it as a cocatalyst. Ceria can be used to thermochemically convert $CO_2 + H_2O$ to syngas (Solarbenzin 2011).

cetane number (CN): The CN is a measure of the combustion quality of diesel fuel during compression ignition. The CN is actually a measure of the fuel's ignition delay: the time between fuel injection and the start of combustion. *Cetane* is an unbranched, open-chain, alkane molecule that ignites very easily under compression; it was assigned a CN of 100. Alpha-methyl naphthalene has a CN = 0. Most diesel engines run well with fuels with CNs between 40 and 55. Dimethyl ether, a nonfossil diesel fuel, has a high CN of 55.

ceteris paribus: *Latin:* other things being equal.

chaoplexy: A fusion of chaos theory with complexity mathematics; a term coined by John Horgan (2008).

chaos (in CNLS): Certain NL dynamic systems described by differential equations exhibit tipping-point behavior. That is, their dynamics are highly sensitive to minute changes in initial conditions, input variables, and/or signal rate parameters. These abrupt, seemingly stochastic changes in behavior have been called chaos; however, they are in fact deterministic and are the result of dynamic interactions in the system. The initiation of turbulence in fluid flow is an example of chaos. Certain simple, NL ODEs such as the van der Pol equation and the Lorenz equations exhibit chaos in their behavior in response to small parameter changes.

chaos theory (in CNLS): Chaos theory attempts to describe the behavior of NL dynamical systems that under certain conditions (of inputs and initial conditions) exhibit the phenomenon known as *chaos*. Such chaotic behavior is, in fact, deterministic. However, its occurrence may at first appear random. Under appropriate conditions, a chaotic CNLS may exhibit periodic LC oscillations in its states or unbounded responses. Certain chaotic CNLSs can exhibit *tipping-point behavior* where a combination of initial conditions and/or inputs can trigger an abrupt transition from nonoscillatory, SS behavior to *LCs*, and vice versa. CSs exist on a spectrum ranging from equilibrium to complete chaos. A system in equilibrium does not have the internal dynamics to enable it to respond to changes in its environment and will die (or is not alive, e.g., viruses, crystals). A system in chaos ceases to function as an organized system. The most productive state to be in is at the *edge of chaos*, where there is maximum variety and creativity, leading to new possibilities of behavior.

circuit rank (in graph theory; aka the cyclomatic number): The circuit rank of a graph is the minimum number r of edges that must be removed from a graph to make it cycle-free (i.e., to remove all feedback loops). $r = e - n + c$, where e is the number of edges in the graph, n is the number of nodes in the graph, and p is the number of disjoint partitions the graph divides into (also given as "separate components").

clathrate compound: A chemical substance consisting of a molecular lattice of one molecular species containing or physically trapping a second type of molecule.

clathrate hydrate: A crystalline, water-based solid in which small, nonpolar molecules are trapped inside "cages" of hydrogen-bonded water molecules. Clathrate hydrates are not chemical compounds as the sequestered gas molecules are never chemically bonded to the lattice. Low-molecular-weight gasses such as O_2, H_2, N_2, CO_2, CH_4, H_2S, Ar, and so forth will form clathrate hydrates. Methane hydrates are a potential fossil fuel (FF) source, and deep-sea CO_2 hydrates offer a possible means of sequestering carbon.

clique (in graph theory): A clique is a *subgraph* in which every possible pair of nodes is directly connected and the clique is not contained in any other clique. See also **modularity**.

Club of Rome (CoR): A global, nonprofit, nongovernmental organization (NGO) think tank founded in 1968. The CoR states that its mission is "to act as a global catalyst for change through the identification and analysis of the crucial problems facing humanity and the communication of such problems to the most important public and private decision makers as well as to the general public." The CoR is concerned with sustainability: In 1972, it sponsored the prescient and well-known book, *The Limits to Growth*, by D.H. and D.I. Meadows, J. Randers, and W.W. Behrens III. Most of the warnings and predictions in *The Limits to Growth* remain surprisingly valid today, in 2012. See www.clubofrome. org/ and Johnson (2011) for more details about CoR.

clustering coefficient (CC; property of a graph node): The ratio between the number of edges between the graph nodes that the node in question is connected to and the number of all possible edges between them. The CC ranges between 0.0 and 1.0.

cogen or co-gen power plant: A power plant that can use two or more fuels, e.g., oil and natural gas.

compartment (in physiology and pharmacokinetics): A well-defined volume in a living organism over which the concentration of a chemical species (e.g. of a drug, hormone, neurotransmitter, etc.) may assumed to be well mixed and constant. For example, blood in the circulatory system is one compartment for potassium ions [K^+], the cerebrospinal fluid is another, extracellular tissue space is another, the hepatic portal circulation is another, and so forth. Pharmacokinetic systems describing the dynamics of drug distribution in the body make use of compartments.

complete graph: A simple graph having n nodes (vertices) in which every pair of nodes is connected with one unique edge. It has been shown that a complete graph with n vertices has $K_n = n(n-1)/2$ edges. Thus, $K_1 = 0$, $K_2 = 1$, $K_3 = 3$, $K_4 = 6$, $K_5 = 10$, $K_6 = 15$, $K_7 = 21,\ldots$, $K_{10} = 45$, $K_{11} = 55$, $K_{12} = 66,\ldots$, $K_{100} = 4950,\ldots$, $K_{1000} = 4.9950$ E5,$\ldots$, $K_{1E4} = 4.9995$ E7,$\ldots$, $K_{1E5} = 5.0000$ E9, and so forth. The number of ODEs or difference equations (DEs) describing an n-node system is roughly proportional to K_n. An n-graph with *modularity* will have significantly fewer edges than K_n.

complex adaptive system (CAS): A CAS has the capacity to change its parameters in order to optimize its performance, a form of learning. The term *complex adaptive systems* was coined at the Santa Fe Institute by J.H. Holland, M. Gell-Mann, and colleagues. Examples of *CASs* include the brain, the immune system, ant colonies, manufacturing businesses, and so forth. CASs generally develop *robustness* as they adapt. Holland's definition of a CAS is as follows: "A Complex Adaptive System (CAS) is a dynamic network of many agents (which represent cells, species, individuals, firms, nations) acting in parallel, constantly acting and reacting to what the other agents are doing. The control of a CAS tends to be highly dispersed and decentralized. If there is to be any coherent behavior in the system, it has to arise from competition and cooperation among the agents themselves. The overall behavior of the system is the result of a huge number of decisions made every moment by many individual agents" (Holland 1995).

A key feature of all CASs is that their behavior patterns as a whole are not determined by centralized authorities but by the collective results of NL interactions among independent entities or agents.

CAS system behavior generally exhibits *resilience* and *robustness*. The human brain is a CAS.

complexity (from the Latin word complexus, meaning twisted together or entwined): Broadly stated, complexity is a subjective measure of the difficulty in describing and modeling a system (thing or process). In this respect, complexity lies in the eyes of the beholder. In other words, complexity is a global characteristic of a system; it represents the gap between component knowledge and knowledge of overall behavior. There is a lack of high predictability of the system's output.

Complexity is relative, graded, and system dependent. A necessary (but not sufficient) property of a CS is that it is composed of many components that are interconnected in intricate ways, some NL and/or time variable. The complexity of a system can also be correlated with stochastic parameters and unspecified connectivity parameters. One measure of a system's complexity is the amount of *information* necessary to describe the system. A CS may exhibit hierarchy and self-organization resulting from the dynamic interaction of its parts. It is also sensitive to initial conditions on its states. CSs can be subdivided into fixed (time-invariant) and dynamic systems. A major challenge in objectively analyzing and describing a CS model is the choice of mathematical algorithms to use.

CSs include, but are not limited to, biochemical pathways; cells; organs [the liver, the immune system, the central nervous system (CNS)], physiological systems, organisms, ecosystems, economic systems, weather, political parties, governments, corporate management, and so forth.

connectance (in graph theory): Connectance has been defined as:

$$\text{Conn}[\%] \equiv \frac{2E}{V(V-1)} \qquad \text{(GL4)}$$

where V = number of graph vertices (nodes), and E = number of graph edges (Bonchev 2004).

connected component (in graph theory): A connected component of an undirected graph is a subgraph in which any two vertices are connected to each other by paths, and to which no more vertices or edges can be added while preserving its connectivity. See also **strongly connected component**.

connected graph (in graph theory): A graph is *connected* if there is a *path* connecting every pair of nodes. A graph that is not connected can be divided into *connected components* (disjoint connected subgraphs). See **complete graph**.

connectivity (in graph theory): A graph is said to be *connected* if it is possible to establish a path from any vertex (node) to any other vertex of the graph. Otherwise, the graph is *disconnected*. A graph is *totally disconnected* if there is no path connecting any pair of vertices.

consumer confidence [also consumer confidence index (CCI)]: Consumer confidence is an economic indicator that estimates the degree of optimism that people feel about the state of the US economy and their personal financial situation. It determines their propensity to spend for goods and services. CCI has been determined monthly by the NGO, the Conference Board, since 1967. It intends to assess the overall confidence, relative financial health, and spending power of the average US consumer. Five thousand US households (a low sample number) are surveyed, providing index information for the United States as a whole and for each of the country's nine census regions. Five questions are asked on each of the following topics: (1) current business conditions, (2) business conditions expected for the next 6 months, (3) current employment conditions, (4) employment conditions expected for the next 6 months, and (5) expected total family income for the next 6 months. The *Consumer Confidence Average Index* (CCAI; range +150 to −300) is a monthly indicator that combines data from three national polls on consumer confidence: (1) the rescaled average of the Conference Board CCI, (2) the Reuters–University of Michigan Consumer Sentiment Index, and (3) the ABC News Consumer Comfort Index. The CCAI is produced and published by the StateOfEconomy.com. The CCAI was +100 in January 2004, was fairly constant around +50 up to January 2007, began to decrease during 2008–2009, and reached −275 by December 2008–January 2009.

controller (in feedback systems): A controller is a dynamic subsystem of a feedback control system that causes an output to follow a variable input. The departure of the controlled output from the desired value is called the *error.* In a simple feedback controller, the error causes the controller to generate a corrected output. A controller can also act as a *regulator,* forcing a feedback system output to remain constant in spite of external disturbances and internal system noise.

Coriolis force: A mass m moving in a linear path with velocity $\mathbf{v}$ on a platform rotating with angular velocity ω is acted on by a Coriolis force, which can be shown to be given by the vector cross-product $\mathbf{F}_c = -2m\Omega \times \mathbf{v}$, where m is the mass of the moving element, and Ω is the angular velocity vector, which has a magnitude equal to the rotation rate ω and is directed perpendicular to the plane of rotation by the right-hand screw rule. It is collinear with the axis of rotation of the rotating platform. The × symbol represents the vector cross-product operator. The vector cross-product between two vectors, $\mathbf{A}$ and $\mathbf{B}$, separated by an angle θ is a vector $\mathbf{C} = \mathbf{A} \times \mathbf{B}$, whose magnitude is $C = AB\sin\theta$ and whose direction is perpendicular to the plane containing A and B and is directed by a right-hand screw rotated in the direction of $\mathbf{A}$ rotated into fixed $\mathbf{B}$.

Note that

- If $\mathbf{v}$ is parallel to the rotation axis (RA), $\mathbf{F}_c = 0$.
- If $\mathbf{v}$ is directed straight in toward the RA, $\mathbf{F}_c$ is in the direction of the rotation (tangential to it).

- If **v** is directed straight outward from the RA, $\mathbf{F}_c$ is against the direction of rotation.
- If **v** is in the direction of rotation (tangential to it), $\mathbf{F}_c$ is outward from the RA.
- If **v** is against the direction of rotation (tangential to it), $\mathbf{F}_c$ is inward toward the RA.

On the surface of a sphere rotating at ω radians per second, the object velocity **v** can be resolved into three vector components, $\mathbf{v}_e$, $\mathbf{v}_n$, and $\mathbf{v}_u$, where $\mathbf{v}_e$ is the east vector component, $\mathbf{v}_n$ is the north component, and $\mathbf{v}_u$ is the vertical (upward) component. The velocity vectors $\mathbf{v}_n$ and $\mathbf{v}_e$ are tangential to the Earth's surface; $\mathbf{v}_u$ is perpendicular. See Figure GL1; all angles and vectors in the figure are positive. Thus,

$$\mathbf{\Omega} \equiv \omega \begin{pmatrix} 0 \\ \cos\varphi \\ \sin\varphi \end{pmatrix} \text{ and } v = \begin{pmatrix} v_e \\ v_n \\ v_u \end{pmatrix}. \text{ Thus,}$$

$$\mathbf{F}_c = 2m\mathbf{\Omega} \times v = 2\omega \begin{pmatrix} v_n \sin\varphi - v_u \cos\varphi \\ -v_e \sin\varphi \\ v_e \cos\varphi \end{pmatrix} \tag{GL5}$$

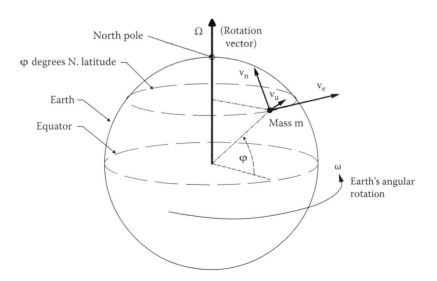

FIGURE GL1

Diagram showing *Coriolis velocity vectors* at N. latitude φ on a rotating Earth. See the Glossary for analysis.

When considering the Coriolis effect on winds and water, $\mathbf{v}_u$ is small, and the vertical component of the Coriolis force is negligible compared to gravity. Thus, only the horizontal velocity components matter, and we can write

$$v = \begin{pmatrix} v_e \\ v_n \end{pmatrix} \text{ and } \mathbf{F}_c = m \begin{pmatrix} v_n \\ -v_e \end{pmatrix} (2\omega \sin \varphi) \tag{GL6}$$

From the vector matrix equations above, we can see that if $v_n = 0$ and ω and $\varphi > 0$, a mass point moving due east experiences a Coriolis force accelerating it due south. Similarly, setting $v_e = 0$, a mass element moving due north is acted on by a Coriolis force moving it due east.

Coulombic efficiency (of battery): The coulombic efficiency, η_c, of a battery is the ratio of the number of charges that enter the battery during charging to the number of charges that can be extracted from the battery during discharging. Ideally, η_c is unity. The losses that reduce η_c are due to secondary reactions in the battery, such as electrolysis of water and other redox reactions. η_c's generally are around 95% or higher.

Crambe: A plant genus of *Brassicaceae* (brassicas) native to Europe, southwest and central Asia, and Eastern Africa. Three major species of *Crambe* are *C. maratima* (seakale), grown as a leaf vegetable; *C. cordifolia*, which is grown as an herbaceous border perennial; and *C. abyssinica*, which is grown for the oil from its seeds. This light, inedible oil has similar characteristics to whale oil and can be used to replace light petroleum oil. *C. abyssinica* is a hardy perennial that is drought resistant and tolerates temperatures as low as 24°F. It costs less to plant, fertilize, and grow than the oil crops soybeans and canola (Treehugger 2009). *C. abyssinica* has been genetically engineered to increase its seed oil yield (Li et al. 2009).

critical point (CP; of a gas): See **supercritical fluid**.

cycle (in graph theory): A cycle is a closed walk or path having at least three nodes, in which no edge is repeated. An engineer will recognize a cycle as a feedback loop.

cyclomatic complexity (in computer science): A software metric. CC = $\mathbf{M}$ is computed using a graph that describes the control flow in a program. The graph's nodes (vertices) correspond to the commands of the program. A directed edge connects two nodes if the second command might be executed immediately after the first command. Cyclomatic complexity directly measures the number of linearly independent paths through a program's source code. $\mathbf{M}$ is given by

$$\mathbf{M} \equiv \mathbf{E} + 2\mathbf{P} - \mathbf{N} \tag{GL7}$$

where **E** = number of edges in the graph, **N** = number of nodes in the graph, and **P** = the number of connected components. An alternative definition is $\mathbf{M} \equiv \mathbf{E} + \mathbf{P} - \mathbf{E}$ (this has also been called "circuit rank"). Yet another way of calculating **M** uses the number of closed loops **C** in the graph: $\mathbf{M} \equiv \mathbf{C} + 1$ (see McCabe 1976; Crawford 1992).

cyclomatic number r (in graph theory): See **circuit rank**.

degeneracy (in biological and other systems): A feature of complex and complex adaptive systems that confers robustness or resilience on the system. It is the ability of structurally different elements (subsystems, modules) to perform the same function, allowing a damaged or noisy system to maintain its same function(s).

degree (in graph theory): A nonnegative integer. The degree, $d_G(v)$, of a vertex (node) v in a graph G is the number of edges incident to v; loops are counted twice. If $d_G(v) = 0$, node v is isolated in graph G.

degree distribution (in graph theory): The degree distribution of a graph or network is the probability distribution, $P(k)$, of node degrees over the N nodes in the whole graph or network. $P(k)$ is the fraction of the nodes in the graph/network with degree k. (k is an integer ≥ 0.) That is, $P(k) = n_k/N$, where n_k of N nodes have degree k.

deuteron (in atomic physics): A deuteron consists of a *deuterium atom* (heavy hydrogen) nucleus with one proton + one neutron, without its one orbiting electron. It is an ion with a net positive charge.

diagraph (in Vester's analysis of CSs): A 2-D plot of vectors derived from the "impact matrix" of two-bit, internodal "influences" in the graphical model of a CS. The diagraph indicates approximately the impact of nodes on each other and allows tentative identification of inputs, outputs, and internal (intermediate) nodes. (See Section 8.3.) A diagraph *is not* a digraph (see below).

dielectric absorption (in capacitors): A property of certain capacitor dielectric materials. The voltage across an ideal capacitor when completely discharged remains at zero volts. Because of energy storage in the dielectric, the voltage across a capacitor with dialectic absorption goes to zero when shorted, but when open-circuited, V_C rises to a direct current (DC) level that is a certain percentage of the initially applied DC voltage. This rebound voltage may be less than 1%–2% for capacitors with polymer film dielectrics. Electrolytic filter capacitors and supercapacitors can have as much as 15%–25% rebound voltage. This NL behavior makes electrolytic and supercapacitors totally unsuitable for conducting alternating current (AC) signals and sample-and-hold applications (Kundert 2008).

digraph (in graph theory): A graph with directed branches between its nodes, such as a *signal flow graph* (SFG) or a food web. The directed branches indicate the direction of signal (or information) flow between pairs of nodes (source and target) and specify a mathematical operation performed on the signal in the source node (j), which is then added linearly to the signal in the target node (k).

dimethyl ether (DME): $H_3C–O–CH_3$. A clean-burning fuel with characteristics similar to liquefied petroleum gas (LPG). It has applications as a household fuel (heating, cooking) and as a diesel engine fuel. DME can be made from any carbonaceous feedstock, including natural gas (NG), methane from landfills and clathrates, coal, or biomass. It is noncarcinogenic, nonteratogenic, nonmutagenic, and virtually nontoxic. Exhaust from DME-fuelled piston engines is very low in NO_x and SO_x pollutants and also has no soot. Some physical properties: *Critical temperature and pressure,* +127°C and 57.3 bar. *Boiling point,* −24.9°C. *Vapor pressure at 20°C, 5.1* kg/m^2. *Liquid density at 20°C, 668* kg/m^3. *Lower heating value, 28.43 MJ/kg. CN, 55–60. Ignition temperature at 1 atm,* +235°C. *Heat of vaporization at 20°C, 410 kJ/kg.* China is the leading world producer of DME made from coal gas (Larson and Yang 2004; Semelsberger et al. 2006).

diminishing marginal returns (in economics): This "law" describes how the marginal production of a good increases as the result of increasing a *production factor* (such as the number of workers), reaches a peak, and then decreases with further addition of workers. An example from agricultural production: 0.5 kg of seed sewn produces 0.5**N** kg of seed at harvest. One kilogram of seed sewn in a plot yields **N** kg of seeds at harvest. Sewing 2 kg seed yields 1.5**N** kg at harvest; a third kilogram of seed sewn yields 1.75**N** kg at harvest, and so forth. (The reason for decreasing seed yield can be overcrowding of plants or finite nutrients in the field, *ceteris paribus.*)

See Figure GL2 for an illustration of a generic marginal production curve. The change in the output, **ΔO**, is plotted on the vertical axis versus the input **I** on the horizontal axis (**ΔO** = dO/dI). **I** can be increasing investments in additional workers, physical inputs (resources), technology, or more generally, complexity. "After

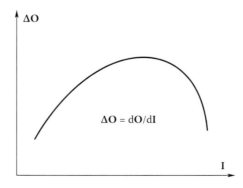

FIGURE GL2
Generic *marginal production curve.* See the text for analysis.

a certain point, increased investments in complexity fail to yield proportionately increasing returns. Marginal returns decline and marginal costs rise. Complexity as a strategy becomes increasingly costly, and yields decreasing marginal returns" (Tainter 1988, Chapter 4). The curve is clearly double valued.

directional sensitivity function: The solid angle over which a photonic or radiation sensor will respond to incident rays, $S(\theta, \varphi)$. By definition, $S(0, 0) = 1.0$ for on-axis input radiation. $S(\theta, \varphi)$ is usually assumed to have symmetrical, circular symmetry, so only $S(\theta)$ need be considered. The shape of $S(\theta)$ for a photosensor can fall off gradually with θ, for example, with a Gaussian-like curve, or for a radiation sensor with a shielded body, be near unity for $-a \le \theta \le a$ and be zero for $|\theta| > a$.

distance (in graph theory): In a graph, d_{pq} is the smallest number of edges connecting nodes **p** and **q**.

distance matrix D (in graph theory): A symmetric $N \times N$ matrix, an element of which, d_{pq}, is the shortest path length between nodes **p** and **q**. If there is no path between **p** and **q**, then $d_{pq} = \infty$. If the shortest path between **p** and **q** contains three edges, then $d_{pq} = 3$.

distributed robustness (DR; in CNLSs): In DR, many parts of a system contribute to overall system function, but all of these parts have different roles. When one part fails or is changed through mutations, the system can compensate for this failure, but not because a redundant (backup) part takes over for the failed part (Wagner 2005). DR is related to complex adaptive behavior.

distance (in graph theory): The distance d_G (u,v) between two vertices (nodes), u and v, in a graph **G** is the (integer) number of edges in a shortest *path* connecting them. (This is also known as the *geodesic distance*.) If u and v are identical, d_G $(u,v) = 0$. When u and v are unreachable from each other, d_G $(u,v) = \infty$.

distance matrix (in graph theory): A symmetrical, $N \times N$ matrix containing the *distances,* taken pairwise, of a set of points in a graph. Its elements are nonnegative real numbers, given N points in euclidean space. A distance matrix describes the costs or distances between the vertices (nodes) and can be thought of as a weighted form of an *adjacency matrix.* The number of pairs of points, $N \times (N - 1)/2$, is the number of *independent elements* in the $N \times N$ distance matrix.

dynamic braking (of electric drive vehicles): A process in which the DC drive motors of an electric vehicle (EV) are disconnected from the line voltage supplying power to the motor, and a resistive load is placed across the motor's armature. The motor now acts as a DC generator; its mechanical input power is supplied by the rotating tires on the road. The generator's load torque slows the vehicle as it supplies electric power to the load resistor, dissipating mechanical input power as V_o^2/R heat. The motors' field windings must retain excitation during dynamic braking. Varying the size of the load resistor controls the braking torque. Some of the generator's output power can be used to charge the EV's batteries. Dynamic braking can also be used to regulate wind turbine speed.

ecology (adj. ecological) and ecosystems (n): *Ecology* is commonly viewed as an interdisciplinary branch of life science, the study of ecosystems. *Ecosystems* are hierarchical systems that are seen to be organized into a graded series of interacting parts and modules (e.g., of species, habitats). They can be global, regional, or highly localized (e.g., a particular rainforest, swamp, or desert). Ecosystems are characterized by their energy inputs and the biodiversity within them. Ecosystems must necessarily contain extensive biophysical feedback pathways between living (biotic) and nonliving (abiotic) components of the planet, including their solar energy input. These feedback loops regulate and sustain local communities of plants, animals, fungi and bacteria, continental climate systems, and planetary biogeochemical cycles. Ecosystems sustain every life-supporting function on the planet, including climate regulation, water filtration, soil formation, food, fibers, organic medicines, ocean currents, and so forth. Humans are a major planetary ecosystem component; their actions influence nearly all planetary ecosystem components, biotic and abiotic.

ecological debt, ecological reserve: Ecological debt occurs when human activities exceed the carrying capacities of one or more ecosystems supporting human life (e.g., fisheries); that is, they are unsustainable. *Ecological reserve* occurs when human activities are sustainable; the supporting ecosystems in question can regenerate (energy in > energy out).

ecological sustainability: The capacity of ecosystems to maintain their essential functions and processes and retain their full biodiversity over the long term, including a "natural" extinction rate. *Ecological sustainability* is one of three interdependent components of human sustainability, with social and economic being the other components. Human actions and inactions can alter ecological sustainability.

economic allocation [in complex economic systems (CESs)]: Allocation is the channeling of the various factors (components) of production into the different types of production required to generate the particular mix of products (goods) that households (markets) demand. A market will be allocatively efficient if it is producing the right goods for the right people at the right price, that is, its prices are equal to its marginal costs in a perfectly competitive market.

economic development (ED): Broadly viewed, the increase in the *standard of living* in a nation's population with sustained growth from a simple, low-income economy to a modern, high-income economy. It involves social and technological progress. It implies a change in the way goods and services are produced, not just an increase in a country's GDP, which is the aggregate value added by the economic activity within a country's borders. ED typically involves improvements in a variety of indicators such as literacy rates, life expectancy, and poverty. It also is concerned with hard-to-quantify aspects such as leisure time, environmental quality, freedom, social justice, and so forth.

economic distribution (in CESs): Distribution describes exactly who gets what goods that have been created. Considered are the particular mix of persons who consume a product by virtue of the incomes they have earned for the factors of production they have contributed to in the prior production of the goods currently being consumed.

edge (in graph theory): A line connecting two vertices (aka nodes), called *end vertices* (aka end points). An edge is also called a (unidirectional) branch in *SFGs*.

El Niño event (in climatology): Unusually warm water temperatures in the equatorial Pacific Ocean. See www.pmel.noaa.gov/tao/elnino/el-nino-story.html (accessed August 15, 2011).

emergence (in CASs): The unpredicted appearance of new characteristics, organization, structures, or behaviors in the course of biological or social evolution. Unexpected emergent characteristics are the result of dynamic interactions between the components of a CS. Biological complexity may emerge as a response to evolutionary pressure on the encoding of structural features that lead to differential survival.

energy (in physics): The *work* that a physical system is capable of doing in changing from a specified reference state to another state. Resting (potential) energy, kinetic energy (KE), stored energy, and energy losses (as from friction or heat loss) are involved. Mechanical energy can be stored by compressing a spring or a gas, giving a flywheel KE, or in the form of heat: 1 g-cal = 4.18605 J, 1 Btu = 251.996 g-cal = 778.26 ft.-lb. Energy can also be stored by pumping water up to a high reservoir, giving it potential energy to turn turbines. Electrical energy can be stored as a charge on a capacitor or in a magnetic field. Energy has the same dimensions as work, ML^2T^{-2}.

energy conversion efficiency: The energy conversion efficiency η of a machine or system is the ratio between the useful energy or power output P_{out} and the input energy or power P_{in}. Generally, $P_{out} < P_{in}$ because of inherent friction or heat losses in producing the useful output. A simple example is an incandescent lamp: The input power is electric; $P_{in} = V_{in} \times I_{in}$ watts. The useful output power is visible light power (photons). The nonuseful (waste) power output is in the form of heat (IR photon radiation). (See **blackbody radiation**.) The η of an incandescent bulb is approximately 5%; the η of a fluorescent lamp can be over 25%. The natural photochemical process of photosynthesis can approach $\eta = 6\%$. A water turbine converting water energy to shaft mechanical energy can have η approaching 90%; similarly, a wind turbine can approach 59% efficiency.

enthalpy (in thermodynamics): A measure of the total energy of a thermodynamic system, in the International System of Units (SI units), joules. One formulation of a system's enthalpy **H** is **H** = **U** + **pV**, where **U** is the internal energy of the system, equivalent to the energy required to create the system, and the **pV** term is equivalent to the energy that would be required to "make room" for the system if the pressure of the environment remained constant.

enthalpy of fusion (latent heat of fusion): The heat energy that must be removed from a liquid in order to change it to a solid (i.e., freeze it). The same amount of heat must be supplied to the solid to change it back to a liquid. Units: kilojoules/mole (kJ/mol). For example, the heat of fusion of water is 79.72 cal/g or 333.55 kJ/kg.

enthalpy of vaporization (heat of vaporization): Usually measured at a substance's boiling temperature, it is the heat required to vaporize a molar quantity of a substance. Units: kJ/mol. The same amount of energy must be removed from a molar quantity of the same substance in vapor state to convert it to a liquid; this is called the enthalpy of condensation or heat of condensation. The heat of vaporization of water is 40.65 kJ/mol or 2257 kJ/kg.

entropy (in thermodynamics, ecology, economic systems, and information theory): In thermodynamic systems, entropy is defined by the equation

$$S_2 - S_1 \equiv \int_2^1 \frac{dQ}{T} \geq 0 \text{ (a thermodynamic system going from state 1 to state 2)}$$

(GL8)

where S_k is the entropy of a system at state k; its units are joules/K. T is the system's Kelvin temperature, and dQ is a differential change system energy (joules, ergs, calories, Btu, etc.) going from state 1 to state 2. Note that no real physical process is possible in which the entropy decreases. Entropy also can be thought of as a measure of thermodynamic disorder that is maximum at thermodynamic equilibrium, where a system has uniform temperature, pressure, composition, and so forth at all points in its volume.

The *order* that living cells produce as they metabolize, grow, and reproduce is more than compensated for by the *disorder* they create in their surroundings as a result of their lives. Cells preserve their internal order by using stored information (their genomes) to direct their processing of free energy and returning to their surroundings an equal amount of energy as heat and entropy. Unlike inanimate physical systems, living and economical systems make use of stored information. Some free energy must be used to reproduce or transmit this information to successive generations.

In information theory, entropy is a measure of the amount of information that is missing in a transmitted message. Informational entropy was defined by Claude Shannon at Bell Labs in the late 1940s.

ergodic (in noisy signals and systems): A noisy signal is assumed to be ergodic when one sample record taken from an ensemble of records is assumed to be an equivalent of any other record taken from the ensemble. Thus, *time averages* can be used in lieu of ensemble or probability averaging to calculate statistics for the signal (e.g., mean, mean-square, autocorrelation, etc.).

eutrophication: A process whereby an excess of anthropogenic plant growth substances such as nitrates and phosphates are discharged into freshwater rivers, lakes, ponds, and marshes, or directly into the oceans. These substances stimulate the overgrowth of aquatic phytoplankton and microalgae such as cyanobacteria. Eutrophication is generally caused by human activities such as the runoff of fertilizers used on crops farmed near waterways and lakes, the discharge of untreated (or treated) sewage effluent, and runoff from heavily grazed pastures and feedlots. It can also have natural causes, including the concentration of normal nutrients in lake water caused by evaporation due to climate change. Dead microalgae fall to the lake bottom and decompose, absorbing oxygen and releasing the greenhouse gasses (GHGs) CO_2, CH_4, and also H_2S, which inhibits the growth of zooplanktons and fish. The entire ecosystem of the lake is changed and made less species diverse and, in general, less viable. Marine eutrophication recently occurred on a large scale (20,000 km^2) off the Chinese beaches in Qingdao on the Yellow Sea (July 2011). The alga blanketing the beaches was *Enteromorpha prolifera* (Algae 2011).

evapotranspiration (ET): The combined water loss rate from both terrestrial plants and the soil around them.

evolutionary landscape (in evolutionary theory): A multidimensional parameter space that organisms occupy. The dependent parameter is usually the fitness of the organism (plotted on the ordinate or y-axis). When a landscape is "flat," it signifies that changes in the immediate mutational neighborhood are of negligible effect on the fitness function. This seldom-seen situation can be interpreted to mean that the organism is not under any significant selection pressure. A poorly adapted organism exhibits a rough landscape with many peaks and troughs. Although many parameters affect fitness, we humans are most comfortable with 3-D plots (x, y, z), or even 4-D plots ($\mathbf{r}$, x, y, z; $\mathbf{r}$ is the radius of a vector from the origin). See **fitness landscape**.

Fermat's spiral: In polar coordinates (r, θ), Fermat's spiral is given by the equation $r = a\,\theta^{1/2}$. It is a type of Archimedean spiral. Another form of describing Fermat's spiral devised by Vogel to describe the position of sunflower florets is

$$r = c\sqrt{(\theta / 137.508°)}$$

where c is a scaling factor and $137.508°$ is the golden angle that is approximated by the ratios of Fibonacci numbers (Vogel 1979).

fertility rate [aka total fertility rate (TFR)]: Basically, a country or region's birth rate: the number of children born in a given population per total number of childbearing women in that population, per year. Census data are often used in calculating TFR.

Fischer–Tropsch (F-T) process: A set of catalyzed chemical reactions that can convert *syngas* (hydrogen and carbon monoxide) to a mixture of liquid hydrocarbons (alkanes or paraffins). Certain alkanes can be used for diesel fuel. The net production of alkanes can be described by

$$\underset{\text{(Syngas)}}{(2n+1)\text{H}_2} + n\text{CO} \xrightarrow{\text{(Cat)}} \underset{\text{(Alkane)}}{\text{C}_n\text{H}_{(2n+2)}} + n\text{H}_2\text{O} \qquad \text{(GL9)}$$

where n is a positive integer (chain length), $n = 1, 2, 3,....$

The F-T process produces a spectrum of alkanes described by the Anderson–Schulz–Flory distribution,

$$W_n = n(1 - \alpha)^2 \alpha^{n-1} \qquad \text{(GL10)}$$

where W_n is the weight fraction of alkanes containing n carbon atoms. α is the chain growth probability ($0 < \alpha < 1$), the probability that a molecule will continue reacting to form a longer chain. In general, α is determined by the catalyst and process parameters used. The alkanes produced include methane ($n = 1$), ethane ($n = 2$), propane ($n = 3$), butane ($n = 4$), pentane ($n = 5$), and so forth. Ethane can be separated and catalytically converted to ethylene ($\text{H}_2\text{C} = \text{CH}_2$), a feedstock for many industrially produced petrochemicals such as polyvinyl chloride, styrene, polystyrene, styrene–butadiene for tires and footwear, and so forth. Recent research on catalysts promises to make the F-T process selective for C_2–C_4 alkanes (Torres Galvis et al. 2012).

fitness (in evolutionary biology; definitions may vary): Fitness is a measure of a species' reproductive success in its *niche*. Those individuals who leave the largest number of mature, breedable offspring are the fittest. Fitness can be attained in several ways: (1) selection for longevity (survival); (2) selection for mating success; (3) selection for fecundity or family size (e.g., large litters, several breeding seasons);

(4) robustness to adverse conditions in the niche (e.g., sudden changes in temperature, light, salinity, pO_2, and so forth, and escape from predators); (5) ability to find food/energy sources; (6) ability to pass on advantageous, learned behavior to the next (F1) generation, and so forth; (7) any combination of the preceding ways.

If fitness is the capability of an individual with a certain genotype to reproduce, one measure of fitness is the proportion of an individual's genes found in all of the genes of the F1 generation. The *absolute fitness* is given by $f_{abs} \equiv N_{after}/N_{before}$, where N_{before} is the number of individuals with a particular genotype, and N_{after} is the number of individuals with that genotype after selection. Absolute fitness can also be calculated as the product of the proportion of a certain genotype surviving times the average fecundity, and it is equivalent to the *reproductive success* of a genotype. A fitness $f_{abs} > 1$ indicates that the frequency of that genotype in the population increases, while $f_{abs} < 1$ means that it decreases. If differences in individual genotypes affect fitness, then clearly, the frequencies of the genotypes will change over successive generations. In *natural selection*, genotypes with higher fitness become more prevalent.

Relative fitness is given by $f_{rel} \equiv \overline{N_{F1}} / \overline{N_{compF1}}$, where $\overline{N_{F1}}$ is the average number of surviving progeny of a certain genotype in the F1 generation, and $\overline{N_{compF1}}$ is the average number of surviving progeny of all competing genotypes in the F1 generation.

The fitness of nonbiological CNLSs can also be defined. In a complex manufacturing process, fitness might be made a function of the rate of production of a good, R, and the number of manufacturing errors per unit, E. Thus, we might use $F = R/E$ as a fitness measure.

fitness landscape (FL; in evolutionary biology; aka adaptive landscape): A poorly defined but much-used term in evolutionary biology. One version of the *FL* is that it is a hypersurface whose height is proportional to a species' reproductive rate (or fitness). The fitness can be plotted as a function of one or more environmental (niche) variables such as temperature, salinity, availability of a certain energy source (food, light), density of another species that competes for resources, and so forth. A population tends to move to the peaks in its FL over a number of generations.

Another view of the FL plots fitness in a population as a function of the frequency of different forms of specific genes (Alleles). This produces a contoured landscape in which the peaks represent local optima corresponding to a particular genotype. The niche is assumed to be constant.

flow battery: A form of rechargeable battery (secondary cell) in which an electrolyte containing one or more dissolved electroactive species flows through an electrochemical cell that converts chemical energy directly to electrical energy. Flow batteries can be rapidly "recharged" by pumping new electrolyte into the cells while simultaneously recovering the spent electrolyte for chemical re-energization.

fly ash (FA; from coal-burning power plants): The light mineral residue from the combustion of coal. Its composition varies widely, depending on the type of coal used and the degree of combustion. FA leaves the burner in the exhaust stack and is generally trapped with electrostatic precipitators. FA contains a large percentage of amorphous and crystalline silicon dioxide (SiO_2), calcium oxide (CaO) (unslaked lime), Fe_2O_3, and Al_2O_3. Depending on the grade of coal used, various oxides of toxic metals can also be present at ppm levels: arsenic, beryllium, boron, cadmium, chromium (including hexavalent Cr = chromium-6), cobalt, lead, manganese, mercury, molybdenum, selenium, strontium, thallium, and vanadium. In addition, FA toxins can include dioxins, furans, and polyaromatic hydrocarbon (PAH) compounds. FA particles are generally spherical and range from 0.5 to 100 µm in diameter. The most common industrial uses of FA are portland cement and grout; embankments and structural fill; waste stabilization and solidification; road subbase; mineral filler in asphaltic concrete; loose application on rivers to melt ice, and so forth (Kennedy 2008). FA is generally disposed of and stored in landfills as a slurry. In 2005, coal-fired power plants in the United States produced approximately 71.1 million tons of FA, of which 29.1 million tons were recycled in various applications.

flywheel energy source: The rotational KE stored in a spinning flywheel is $E = 1/2\,J\,\omega^2$ joules. Assume this flywheel is supplying mechanical energy to an electrical generator running at constant angular velocity $\dot\theta$. The mechanical input power to the generator is $P_{in} = \tau\dot\theta$ watts, where τ is the generator's shaft load torque in newton-meters. (Assume the motor is coupled to the flywheel through a lossless transmission.) In general, $\omega \gg \dot\theta$. The energy input to the motor over a short time Δt is just $\Delta E = \tau\dot\theta\Delta t$. This energy comes from the flywheel's stored KE, causing it to slow to ω_1. Thus, we can write

$$1/2\,J\omega_0^2 - \tau\dot\theta\Delta t = 1/2\,J\omega_1^2 \text{ joules} \tag{GL11}$$

We solve the quadratic equation for $\Delta\omega = \omega_0 - \omega_1$ of the flywheel. This gives $\Delta\omega/\omega_0 \cong \dfrac{1/2\tau\dot\theta\Delta t}{(1/2\,J\omega_0^2)}$, assuming $1/2\,J\omega_0^2\tau\dot\theta\Delta t$. Another way of posing the conversion of flywheel KE to generator energy is to write

$$1/2\,J\omega_0^2 - t\times(\tau\dot\theta) = 1/2\,J\omega^2 \tag{GL12}$$

The flywheel angular velocity at time t is simply $\omega = \omega_o \sqrt{1 - \omega_o^2(\tau\dot\theta)t/(1/2J\omega_o^2)}$ rad/s, until $t = 1/2J/(\tau\dot\theta)$. ω_o is the initial flywheel angular velocity. The initial transmission gear ratio is $n = \omega_o/\dot\theta = 20,000/1800 = 0.0900$, assuming the flywheel is spinning at 20,000 rpm and the motor at 1800 rpm.

Centrifugal forces act on rapidly spinning flywheels, exerting stresses that act to tear them apart. There is no simple, universal formula for the maximum safe speed for a flywheel because of the wide range of mechanical designs and materials used. For example, one formula given for the maximum safe speed is (Oberg et al. 1976)

$$N_{max} = \frac{CAMEK}{D} \tag{GL13}$$

where N = safe maximum speed in rpm, C = constant = 1 for wheels driven by constant speed motors, A = spoke constant = 1.08 for 8 spokes, M = 2.45 for cast steel of s = 60,000 psi tensile strength, E = joint efficiency (1.0 for solid rim), K = rim thickness constant (2340 for rim thickness equal to 20% of outside diameter), and D = outside diameter of wheel in feet.

Another formula approximating the bursting velocity of a flywheel is

$$V_{Tmax} = \sqrt{10s} \text{ ft./s (maximum tangential velocity)} \tag{GL14}$$

From this, we can write

$$N_{max} = \left[\sqrt{10s}/R\right](60/2\pi) \text{ rpm} \tag{GL15}$$

where s = the psi tensile strength of the flywheel rim. Note that the maximum angular velocity of the flywheel is just $\omega_{max} = V_{Tmax}/R$ rad/s; R is the outer radius of the flywheel in feet.

Four Horsemen of the Apocalypse: From the last book of the New Testament in the Bible, *Revelations* 6:1–8. A scroll in God's right hand is sealed with seven seals. When the first four seals are opened by The Lamb (Christ), four horses and horsemen are described: The first horse is white, and its rider is referred to as *Conquest*. The second horse is red, and its rider represents *War*. The third seal reveals a black horse whose rider is interpreted as *Famine*. The fourth seal reveals a pale horse whose rider is named *Death*. Another interpretation of the white and red horsemen, which perhaps is more appropriate for a work on sustainability, is that the white horse's rider represents *War* with all its horrors, and the red horseman represents *Pestilence and Disease*. Various other interpretations of *Revelations* 6:1–8 exist.

fractal: A rough or fragmented geometric shape that can be subdivided into parts, each of which is generally a reduced size copy of the whole. The fine structure of a fractal exists at arbitrarily small scales; it is too irregular to be easily described by traditional euclidean geometric language. Fractals have simple, recursive, mathematical definitions. Approximate fractals occur in nature: for example, snowflakes, blood vessels, cauliflower, and so forth. The term was coined by mathematician Benoit Mandelbrot in 1975, derived from the Latin *fractus,* meaning "broken" or "fractured." Broadband fractal antennas have been designed for ultrahigh-frequency (UHF) radio systems and may prove effective at IR wavelengths.

free energy (in chemistry, physics, and economics; aka Gibbs free energy, ΔG joules): "Also called the available energy, or energy set free. In chemical reaction systems, ΔG is the chemical potential that is minimized when a system reaches equilibrium at constant pressure and temperature. If a reaction is spontaneous, $\Delta G < 0$ and energy is given off, usually in the form of heat and/or photons. A nonspontaneous reaction requires energy input (heat, photons, etc.) and $\Delta G > 0$. At equilibrium, $\Delta G = 0$. In general, in chemical thermodynamics, $\Delta G = \Delta H - T\Delta S$, where ΔH = change in enthalpy (joules) and ΔS = change in entropy (joules/Kelvin). For example, when hydrogen is burned in oxygen, the water vapor is formed with a release of -54.64 kcal/mol (heat and photons). When CO_2 is formed by burning carbon in oxygen, $\Delta G = -94.26$ kcal/mol.

functional (mathematics): A functional is traditionally a map from a multidimensional vector space to the field underlying the vector space, which is usually the real numbers (scalars). A functional takes for its input argument a vector and returns a scalar, for example, a graph complexity measure that uses the number of nodes, the number of branches, and the number of closed paths to calculate a complexity number. The percent connectedness function is one such measure.

gain (in linear systems analysis): Gain is a term derived from electronics; it refers to the amplification factor of an amplifier. More generally, gain is the frequency-domain transfer function (or SS frequency response) between an input signal, V_{in}, and an output signal, V_o. As a frequency response, the gain can be written as a complex number (vector): $H(j\omega) = V_o/V_{in}$. Gain of a linear system can also be expressed in terms of the Laplace complex variable, s; $H(s) = \dfrac{V_o}{V_{in}}(s)$. $H(s)$ will generally be a rational polynomial in s.

genetic algorithm (GA; in computing): A search technique used in computing (and artificial neural network optimization) to find exact or approximate solutions to optimization and search problems. GAs use techniques inspired by evolutionary biology such as inheritance, mutation, selection, and recombination (crossover), as well as synaptic strength in neural networks. GAs have been applied to problems in bioinformatics, chemistry, computer science, economics, engineering, manufacturing, mathematics, physics, and many other fields. A problem-dependent *fitness function* is used to measure the quality of the solution calculated iteratively by the GA. The GA initializes a population of solutions randomly and then improves it by repetitive application of mutation, crossover, inversion, and selection operators. The generational process continues until a terminating condition has been reached using the fitness function.

glucosamine: ($C_6H_{11}O_5$–NH_2) A ubiquitous hexose sugar that is a prominent biochemical precursor in the biosynthesis of glycosylated proteins and lipids. It is also the building block of the glycan polymers chitosan and chitin, found in arthropod exoskeletons and certain fungal cell walls. It is used as a human dietary supplement.

glycan: A polysaccharide or oligosaccharide, usually consisting of O-glycosidic linkages of monosaccharides. Cellulose and chitin are glycans.

good (n.; in economics): In economics, a good is any object or service that increases utility, directly or indirectly, of the consumer. A good is manufactured, grown, crafted, and so forth and generally sold for profit. A good is *supplied* to meet a general *demand*.

graph (in graph theory and systems analysis): A set of objects called *nodes, vertices,* or *points,* connected by *branches,* also called *edges, links,* or *lines.* A graph can be used to model a system. Graphs can be *undirected,* in which a line from node **A** to node **C** is the same as a line from node **C** to node **A**. In a *directed graph* (aka *digraph*), all branches leave certain nodes and are directed to other nodes. This directivity implies unidirectional flow of information, signal, mass, and so forth from the originating (source) nodes to a sink node. A *mixed graph* contains both directed and undirected edges. A *simple graph* has no multiple edges or loops. An *oriented graph* is one in which all its nodes and branches are numbered, and arbitrary directions are assigned to the branches (i.e., the branches are directed). [For an extensive glossary of terms used in graph theory, see Albert (2005), Caldwell (1995), and also en.wikipedia.org/wiki/Glossary_of graph_theory.]

gray water: Domestic waste water from dishwashing, bathtubs, showers, sinks, and laundry, excluding toilet waste. It contains soap, detergents, and some organic material. After treatment, it is generally suitable for irrigation and toilet flushing.

"green collar" jobs: Those involved with renewable energy technologies.

grid parity (for an energy source): The condition where electric power from a source [e.g., photovoltaic (PV) panels, wind turbines, fuel cells, etc.]

is equal to or cheaper than power from the grid, whatever its source (hydro, coal, oil, diesel, etc.).

gross domestic product (GDP; in economic systems): GDP is a measure of national income and output for a country's economy. GDP is the total market value of all final goods and services produced within the country in a year. It can be calculated from GDP = $C + I + G + (X - M)$, where C = consumption, I = gross investment, G = government spending, $(X - M)$ = (exports – imports). See **Average GDP per capita**.

gymnosperm: Any plant of the class *Gymnospermae,* which includes the coniferous trees and other plants not having seeds enclosed in an ovary.

Haber process: A multistep, nitrogen fixation reaction where first CH_4 + steam is catalytically converted to carbon monoxide + hydrogen gas. Second, $N_2 + CO + H_2$ is catalytically reacted to form $N_2 + H_2 + CO_2$ at 500°C. This gas is compressed, and $H_2O + CO_2$ is separated from $N_2 + H_2$. The actual *Haber process* is where the $H_2 + N_2$ gasses are finally converted catalytically to ammonia gas, NH_3, which is used as fertilizer, a refrigerant gas, or a chemical reagent to make explosives. The net reaction is $N_2(g) + 3H_2(g) \xrightarrow{\text{(Catalyst)}} 2NH_3(g)$, ($\Delta H$ = -92.22 kJ/mol). The Haber reaction occurs at approximately 200 atm and between 350°C and 550°C over an iron + KOH, or ruthenium catalyst (Clark 2002). About 100 million tonnes of nitrogen fertilizer is produced per year by the Haber process, mostly in the form of anhydrous ammonia, ammonium nitrate, and urea. About 3%–5% of the world's NG production goes into the Haber process. The fertilizers produced are responsible for food sustaining approximately one-third of the Earth's population. [For more details on the Haber process, see Smil (2001).]

half-life (in biochemistry): The time it takes the concentration of a drug, hormone, cytokine, and so forth in a compartment to fall to one-half its concentration at $t = 0$, given zero further input. $T_{1/2}$ is not the same as the time constant, τ. $T_{1/2}$ can also be used to describe the loss rate of an atmospheric gas concentration, given no further input at $t = 0$.

Hamiltonian path (in graph theory): A Hamiltonian path is a path in an undirected graph that visits each node only once. A Hamiltonian cycle is a closed path in a graph that visits each node only once and also returns to the starting node. A complete graph with more than two nodes is Hamiltonian.

heat (in thermodynamics): Thermal energy, Q: One kilogram-calorie (kg-cal) is the quantity of heat that must be supplied to 1 kg of water in order to raise its temperature 1°C. Other units of heat are the British thermal unit (Btu; 1 Btu = 0.252 kg-cal). The mechanical work equivalent of heat is 1 kg-cal = 4186 J. Supplying heat to a material raises the temperature of the material according to $\Delta T = Q/(mc)$, where Q is the SI quantity of heat in joules, m is the mass of the substance in kilograms, and c is the *specific heat capacity* of the substance in joules per kilogram Kelvin.

heat of combustion (in chemical thermodynamics)*:* The thermal energy Q released as heat when a compound undergoes complete combustion with pure oxygen under standard conditions in a bomb calorimeter. Its units can be energy/mole of fuel (J/mol), energy/mass of fuel (J/kg), or energy/volume of fuel (J/m³). For example, H_2 gas has a heat of combustion of 141.8 MJ/kg, and petrodiesel has 44.8 MJ/kg.

heat of fusion: The heat taken up by a material when it undergoes a phase change from solid to liquid. The same heat is released when a given liquid mass of the material solidifies. The heat of fusion of water is 80 cal/g, or 4182 J/kg.

hectare: A metric unit of area equal to 2.471 acres. 1 hectare = 10^4 m^2.

Hill function (used in modeling dynamic biochemical systems): The nth order Hill function is used to model parametric variation of rate constants in CNLS models. They provide one means of introducing parametric feedback. Rate-constant saturation is modeled by the Hill function:

$$K_S = \frac{K_{So}[M]^n}{\beta + [M]^n} \tag{GL16}$$

where K_{So} and β are positive constants, $[M]$ is the concentration of the regulating variable, M, and n is an integer ≥ 1. $K_S \to K_{So}$ for $[M]^n \gg \beta$.

Rate-constant suppression by the concentration of M is modeled by the decreasing Hill function:

$$K_N = \frac{K_{So}}{\beta + [M]^n} \tag{GL17}$$

K_N is a decreasing function of $[M]^n$. $K_N \to 0$ as $[M]^n \gg \beta$.

homeostasis (in physiology): The dynamic maintenance of SS conditions in the internal compartment (cytosol) of a cell by the programmed, regulated expenditure of chemical energy. Cellular homeostasis stabilizes the interior of a cell against chemical and physical changes in its external environment as well as from chemical and physical changes arising within the cell from metabolic processes, that is, homeostasis confers resilience. Homeostasis requires the dynamic equilibria of many biochemical systems and is generally the result of the action of multiple chemical negative feedback control systems.

An example of homeostasis is the maintenance of the concentration of sodium ions in the cytosol of a cell within "normal" bounds against extracellular changes in sodium concentration. Sodium "leaks" into the cell down concentration and membrane potential gradients. Sodium homeostasis in cells is effected by sodium "pumps", specialized proteins fixed in the cell membrane that have the capability of moving sodium ions from the cell's cytosol to the extracellular fluid at the expense of metabolic energy. The sodium ion concentration in a cell determines the osmotic pressure across its cell membrane, hence the amount of water that leaks in or out of the cell. Even water passage through a cell membrane is regulated by transmembrane *aquaporin proteins* as a part of osmotic homeostasis. Molecular signals that activate transmembrane pumps and gates are generally regulated by the expression of genes by regulated transcription factors (TFs).

Multicellular organisms practice physiological homeostasis. Regulation is generally effected by the action of multicellular organs, such as the kidneys, pancreas, and liver, under hormonal and CNS regulatory signals. Homeostasis at all levels of scale leads to organismal resilience.

Hopf bifurcation (in NL dynamical systems): A local bifurcation in which a fixed point of a dynamical system loses stability as a pair of complex-conjugate eigenvalues of the linearization around the fixed point cross the imaginary axis of the complex (root) plane. Hopf bifurcations occur in the Hodgkin–Huxley model for nerve impulse generation (Northrop 2001), the Lorenz system, and the Oregonator and Brusselator chemical oscillator models. In a CNLS's phase plane, a bifurcation point is one where the phase plane attractor changes as the result of one or more changes in the system's parameters. See also **Tipping point**.

hub (in a graph): A node in a network that has an above-average number of interactions with other nodes in that graph. (In a directed graph, these interactions can be by both input and output edges.)

Hubbert curve (in ecosystem modeling): It is used as a model for the *rate of increase, r,* of a resource or population over time. It is a logistic distribution curve with the general form

$$r(x) = \frac{4ke^{-ax}}{(1+e^{-ax})^2} = \frac{2k}{[1+\cosh(ax)]} \tag{GL18}$$

where $x = t - t_o$. t_o is the year at which the peak rate occurs.

Plotted, the Hubbert curve is a smooth, Gaussian-looking function. Its peak value, k, occurs at $x = 0$ and a sets the curve's width.

Real-world plots of a resource's depletion rate versus time are often fit with an asymmetrical or skewed Hubbert curve model. US oil production, for example, has a very long "tail" following its peak around 1970 due to the development of tar sands and oil shale deposits in the late 20th and 21st centuries.

human development index (HDI): A complicated statistical measure used to rank countries by their level of "human development." It generates a number that can be used to separate developed, developing, and underdeveloped countries, making these distinctions less fuzzy. The HDI statistic uses data from *life expectancy, education*, and *per capita GDP* (as an indicator of standard of living). HDI statistics can also be applied to states, cities, and villages. The HDI is calculated as

$$HDI \equiv (LEI \times EI \times II)^{1/3}$$

where LEI is the *life expectancy index*, LEI $\equiv$ (LE – 20)/(83.4 – 20); LE is the mean life expectancy; EI is the *education index*; and II is the *income index*.

$$EI \equiv \frac{\sqrt{MYSI \times EYSI}}{0.951} \tag{GL19}$$

where the *mean years of schooling index* = MYSI $\equiv$ MYS/13.2, the *expected years of schooling index* = EYSI $\equiv$ EYS/20.6, MYS = mean years of schooling (years that a 25-year-old person or older has spent in schools), and EYS = expected years of schooling (years that a 5-year-old will spend with his/her education in his/her whole life).

$$\text{The income index} = II \equiv \frac{\ln(GNIpc) - \ln(100)}{\ln(107{,}721) - \ln(100)},$$

where GNIpc = gross national income at purchasing power parity, per capita.

An HDI of over 0.8 is considered to be "high human development." Norway has an HDI = 0.938, the United States 0.902, Switzerland 0.874, Hong Kong (not PRC) 0.862, and so forth. Connecticut led the 50 states with an (2008) HDI = 6.30.

human impact equation: In an attempt to make the concept of the "human impact on the environment" more quantitative, Ehrlich and Holdren (1971) presented a formula in order to quantify and compare the impact of population groups (PGs). This is the $I = P \times A \times T$ relation. I is the impact number for a given PG (e.g., world population, North American population, Australia/New Zealand population, etc.). P is the population being considered. A is affluence, figured from the average consumption per capita of food, energy, water, manufactured goods, and so forth. The GDP per capita for the PG is often used to make assigning a number to A simpler. GDP per capita measures production, and it is often assumed that consumption increases when production increases. The **T** parameter is technology. It represents how resource intensive the production of affluence is: how much environmental impact is involved in creating, transporting, and disposing of goods, services, and amenities used by the population. Technology has the power to reduce **T**, as well as increase it. If human impact **I** on climate change were being considered, **T** might be GHG emissions per unit of GDP. Clearly, the **IPAT** equation is a gross oversimplification of a very complex group of systems; however, it does permit comparisons between the impacts of various PGs in areas such as climate change, fisheries, or freshwater resources, and so forth.

human poverty index: A numerical indication of the standard of living in a country, developed by the United Nations (UN). The smaller the HPI, the higher the living standard in a country. For developing countries, the HPI-1 is given by

$$\text{HPI} - 1 \equiv \left[1/3 \left(P_1^\alpha + P_2^\alpha + P_3^\alpha \right) \right]^{1/\alpha} \tag{GL20}$$

where P_1 = probability at birth of not surviving to age 40 ($\times$ 100), P_2 = adult illiteracy rate (%), and P_3 = unweighted average of population without sustainable access to an improved water source plus children under weight for their age. $\alpha = 3$.

HPI-2 for selected Organization for Economic Co-operation and Development (OECD) countries is given by

$$\text{HPI} - 2 \equiv \left[1/4 \left(P_1^\alpha + P_2^\alpha + P_3^\alpha + P_4^\alpha \right) \right]^{1/\alpha} \tag{GL21}$$

where P_1 = probability at birth of not surviving to age 60 ($\times$ 100), P_2 = % of adults lacking functional literacy skills, P_3 = % of population below income poverty line (50% of median adjusted household disposable income), and P_4 = % rate of long-term unemployment (lasting $\geq$ 12 months). $\alpha \equiv 3$.

In a recent UN report (2007–2008), Sweden ranked #1 with HPI-2 = 6.3, Switzerland was ranked #7 with 10.7, and the United States ranked #17 with 15.4 (cf. Wikipedia: Human_Poverty_Index).

hydrothermal carbonization (HTC): Wet vegetative material (e.g., crop waste, sewage sludge, forestry waste, microalgae, etc.) is heated anaerobically in a pressure vessel to approximately 200°C for several hours. Complicated, spontaneous, chemical reactions occur that produce "biocoal" or "biochar", a product rich in reduced carbon-containing molecules, plus water (no CO_2 is emitted). The biochar can be stored and used for combustible fuel or as a soil-conditioning material. The process is "green"; the CO_2 given off during combustion of biochar comes from plants that recently grew and took up CO_2.

incidence matrix (in graph theory): A graph is characterized by a matrix of E (edges) by V (vertices), where [edge, vertex] contains the edges' data (simplest case: 1 if connected, 0 if not connected).

increasing returns (to scale; in complex economics): A phenomenon cor-related with positive feedback loops in a CES. Goods or profits are seen to increase at a rate larger than the scale of a microeconomic system. That is, goods or profit are proportional to S^k, where S is the size (scale) of the microeconomic system and the exponent $k > 1$. If $k < 1$, the system exhibits decreasing returns to scale; if $k = 1$, there are constant returns to scale.

inflation (in economic systems): A general rise in the prices of goods and services over a time. It is due to a decline in the purchasing power of money. Inflation is quantified by calculating the *inflation rate* = % change in a price index over time. Price indices include the consumer price index (CPI), cost-of living index, commodity price index, and so forth.

infrastructure (in social systems): A society's *infrastructure* is a collective noun denoting its highways; bridges; tunnels; airports; public trans-portation systems; electric power supply and distribution; water sup-ply; waste removal; communications (newspapers, Internet, TV, radio, telephone, cell phone, telecom cable, fiber optic, etc.); dams; inland waterways levees and flood control; and so forth. Infrastructure is a mixture of publicly and privately supported works that make our society work smoothly.

insolation: The *solar radiation energy* received on a given earth surface area (at a specific latitude, longitude, and time of day), recorded during a given time. Also called solar irradiation. Unit generally MJ/m^2. Also units of Wh/m^2 are used (watt-hours per square meter). If this energy is divided by the measuring time in hours, it then becomes a power density called *irradiance,* in W/m^2.

integrated gasification combined cycle (IGCC): A complicated industrial process that makes *syngas* (containing CO, CO_2, CH_4, H_2) from coal. IGCC plants have very high capital costs, approximately $3600/kW. Because CO_2 is part of the output stream, IGCC plants are suitable for carbon capture and storage (CCS). The syngas is used to run gas turbines driving alternators (190 MW)*, and "waste heat" is recov-ered from both the gasification process and the turbines and used to make steam to make additional electrical power (120 MW)*. SO_x is scrubbed from the syngas and stack exhaust, and CO_2 is captured. (*In an IGCC plant that uses 2500 tonnes/day of coal.)

inverter: An electronic or mechanoelectric system that converts DC from sources such as solar cells (SCs), wind turbines, batteries, or fuel cells to AC power.

irradiance: *Radiant flux* on a surface per unit area; unit, W/m^2.

issue (as used in this book): "A matter of wide public concern" [definition from Morris (1973)].

joule: The SI unit of energy or work. $1 J \equiv 1$ N-m $= 0.239006$ g-cal $= 0.000948451$ Btu. A joule is a relatively small energy unit, for example, 1 kWh $= 3.60 \times 10^6$ J.

Kermack–McKendrick models for epidemics: Mathematical compartmental models using ODEs to describe the dynamics of the spread of an infectious disease. Some models include infection latent period; recovery; deaths; conferred immunity; as well as the use of antivirals, antibiotics, and quarantines.

kerogen: A mixture of organic chemical compounds found in certain shale rock deposits. Kerogen is formed naturally as the first step in producing crude oil. The high temperatures and pressures underground accelerate the breakdown of algae, diatoms, plankton, spores, pollen, and other organic material trapped in the layers of shale rock. The large biopolymers from proteins and carbohydrates (lignin, keratin, xylan, etc.) break down and are rearranged to form *geopolymers*, the precursors for kerogen. In the formation of geopolymers, hydrogen, oxygen, nitrogen, and sulfur are lost in functional groups, and *aromatization* occurs (the formation of cyclic, PAHs; see Section 4.2.8). Kerogen from the Green River oil shale deposits in the western United States contains elements in the proportions of carbon (215): hydrogen (330): oxygen (12): nitrogen (5): sulfur (1). Labile kerogen can be processed to obtain heavy hydrocarbons (oils) and gasses, and inert kerogen forms graphite (carbon).

Kolmogorov complexity (KC; in computer science and genomics; aka algorithmic entropy, algorithmic complexity, algorithmic information content, descriptive complexity, Kolmogorov–Chaitin complexity, program-size complexity, or stochastic complexity): Consider a string of data [e.g., a finite binary string of 0s and 1s, a finite coding region of DNA (groups of three bases called codons, each of which codes one of 20 amino acids [AAs] in a protein peptide sequence), or words in a paragraph of English text]. The NIST (2006) definition of KC is as follows: "The minimum number of bits into which a string can be compressed without losing information. This is defined with respect to a fixed, but universal decompression scheme, given by a universal Turing machine." Funes (2008) stated: "One problem with AIC is its uncomputability. The fact that no program can be known in advance to halt at some point makes it impossible to know the AIC of a given string." "The KC of a stochastic process can be proved to approach its entropy..." Hutter (2008) gave the history of KC and some interesting inequalities describing the KC of strings.

Kuznets curve: A graph indicating that economic inequality increases over time while a country is developing, and then after a certain average income is attained, economic inequality begins to decrease (named after economist Simon Kuznets). For example, a simple form of Kuznets curve is an inverted parabola, described by the equation

$$\Delta = -k (I - I_o)^2 + \Delta_{max} \qquad (GL22)$$

where I is the per capita income, Δ is the *economic inequality*, I_o is the per capita income giving maximum inequality, Δ_{max}, and k is a positive constant.

Other Kuznets-type curves are seen in the graphs of *declining marginal returns:* "In manufacturing, diminishing returns set in when investment in the form of additional inputs causes a decline in the rate of productivity." "Investment in sociopolitical complexity as a problem-solving response often reaches a point of declining marginal returns" (Tainter 1988, Chapter 4).

La Niña event (in climatology): An oceanographic event where there are unusually cold water temperatures in the equatorial Pacific Ocean. (See **El Niño.**) See www.pmel.noaa.gov/tao/elnino/la-nina-story. html (accessed August 15, 2011).

law of requisite variety (LRV; in feedback systems): Ashby's *LRV* states in essence that a feedback system's controller always has to be more complicated than the system it is controlling.

law of unintended consequences (in CSs; actually, more of a maxim): A property of the behavior of CSs in which an action, inaction, or input can lead to an unanticipated, unintended "side effect," as well as a desired action. The side effect can be beneficial, neutral, or bad.

length of path (in graph theory): Consider nodes j and k in a directed graph: The path length L_{jk} is the number of concatenated branches between nodes j and k that do not pass through any intermediate node more than once. There may be more than one distinct path between nodes j and k. The *path gain* is the product of the branch gains in L_{jk}.

lignin: A complex organic polymer of variable composition found in the cell walls of plants and the red algae, *Calliathron*. In paper production by sulfite pulping, lignin is removed from wood pulp as lignosulfonates, which have several uses: (1) as dispersants in high-performance cement applications, water treatment formulations, and textile dyes; (2) as additives in specialty oil field applications and agricultural chemicals; (3) as raw materials for several chemicals such as vanillin, dimethyl sulfoxide (DMSO), ethanol, xylitol sugar, and humic acid; and (4) as an environmentally sustainable dust suppression agent for dirt roads. Lignin extracted from shrubby willow has been used to produce expanded polyurethane foam. In 1998, a German company, Tecnaro, developed a process for turning lignin into a substance called Arboform, which behaves identically to plastic for injection molding. When the Arboform plastic item is discarded, it can be burned for its energy. Components of the lignin polymer include paracoumaryl alcohol, coniferyl alcohol, and sinapyl alcohol.

limit cycle (LC; in NL, dynamic, CSs): Under certain initial conditions and/ or inputs, an NL, dynamic CS may exhibit bounded, self-sustained oscillations of its states. LCs can further be characterized as *stable, unstable,* or *conditionally stable.* The kth state's stable LCs exhibit a closed path in the $(\dot{x}_k, x_k)$ plane (Ogata 1970). Stable LCs in enzymatic biochemical systems may form the basis for *biological clocks.*

linear system (LS): An LS obeys *all* of the following properties; an *NL system does not obey one or more of the following properties:* $x(t)$ is the input, $y(t)$ is the output, and $h(t)$ is the LS's impulse response, or weighting function.

$$x_1 \to \{LS\} \to y_1 = x_1 \otimes h \quad real\ convolution \tag{GL23}$$

$$\frac{if}{then\ a_2 x_2} \xrightarrow{} LS \xrightarrow{\begin{array}{c} y_2 = x_2 \otimes h \\ y_2' = a_2(x_2 \otimes h) \end{array}} \quad scaling \tag{GL24}$$

$$\xrightarrow{a_1 x_1 + a_2 x_2} LS \xrightarrow{y = a_1 y_1 + a_2 y_2} \quad superposition \tag{GL25}$$

$$\begin{array}{c} x_1(t - t_1) \\ + x_2(t - t_2) \end{array} LS \begin{array}{c} y = y_1(t - t_1) \\ y_2(t - t_2) \end{array} \quad shift\ invariance \tag{GL26}$$

LMC complexity measure (in physical systems): LMC stands for the authors who first described the LMC measure: Lopez-Ruiz, H.L. Mancini, and X. Calbet (1995, *Phys Lett A* 209: 321). The LMC measure is based on probability theory and is applied to CSs in physics and chemistry. [See Section 9.4 in Northrop (2011).]

loop (in graphs): A collection of branches in a graph (oriented or unoriented) that form a *closed path.* Loops can be *feed-forward* or *feedback* in topology. A loop in an *SFG* is a directed path that originates at and terminates on the same node, and along which no node is encountered more than once. The product of the branch transmissions in a loop is called the *loop gain.*

loop gain [in a single-input, single-output (SISO) linear time invariant system (LTIS)]: The loop gain, $A_L(s)$, in a linear feedback system is the *net gain around a feedback loop,* starting at one node and finishing at that node. The zeros of the *return difference* transfer function, $F(s) \equiv 1 - A_L(s)$, are the closed-loop system's poles, which determine its stability and, in part, its transient response. A system with *negative feedback* has a loop gain with a net minus sign: for example, $A_L(s) = - [K_p K_c (s + a)]/[s (s + b)]$.

Lyapunov exponent (LE; in phase plane plots of CNLS behavior): The LE characterizes the rate of separation of a pair of close trajectories originating at the same time from closely spaced points. The trajectories can be converging on a common attractor or diverging. Consider two closely spaced points in 2-D phase space at $t = t_o$, P_0 on the first (reference trajectory) and P_0' on the second trajectory. Each point will generate a trajectory determined by the system's dynamics and its initial conditions. t seconds later, the vector distance between $P(t)$ on the first trajectory, and the corresponding point on the second trajectory, is $\rho(t)$.

The LE λ is defined by the equation, $|\rho(t)| = \exp(\lambda t)|\rho_0|$. [Absolute values are used because ρ_0 and $\rho(t)$ are 2-D vectors.] Here, ρ_0 is the initial, small separation at $t = t_o$ between the points P_0 and P_0' on adjacent trajectories, and $\rho(t)$ is the separation between corresponding points on the trajectories at $t = t_o + t$. Often, the *maximal LE* (MLE) is used to characterize the phase plane behavior of CNLSs. The MLE is defined by

$$\lambda = \lim_{\substack{\tau \to \infty \\ \rho_0 \to 0}} (1/t) \ln \frac{|\rho(t)|}{|\rho_0|} \qquad \text{(GL27)}$$

In general, if $\lambda > 0$, then the trajectories diverge from each other rapidly. If $\lambda = 0$, the system is a stable oscillator. If $\lambda < 0$, then the system is focal [trajectories converge on an attractor in the SS ($t \to 0$)]. Measuring the LE is of significance only at the attractor.

macroeconomics (in economic systems): A branch of economics that deals with the structure, modeling, and behavior of a global, national, or regional economy *as a whole*. Macroeconomic models seek to describe the relationships between factors such as national income, overall employment/unemployment, investments, debt, savings, international trade and finance, GDP, and so forth.

marasmus: A form of severe protein + carbohydrate malnutrition in which body weight may be reduced to less than 80% of the average. It generally affects infants less than 1 year old.

Marburg virus: The Marburg genus of viruses is in the family *Filoviridae* along with the Ebola virus. *The Lake Victoria Marburgvirus* species causes *Marburg hemorrhagic fever* in humans. The natural reservoir for this virus is believed to be the Egyptian fruit bat, as well as monkeys (grivets) from Uganda. *Marburgvirus* is highly contagious, being spread through bodily fluids. Treatment is only palliative. The fatality rate is from 23% to 90%. *Marburgvirus* was named after the German city of Marburg, where the first outbreak occurred in the northern hemisphere in 1967. The virus was introduced by Ugandan grivet monkeys being used to develop polio vaccines.

mass action (MA; in chemical systems): The basis for writing dynamic models for chemical reactions (also used in population dynamic models). A group of coupled chemical reactions described by MA kinetics can be expressed as a set of first-order ODEs. These ODEs are generally NL and can have time-varying coefficients. The *law of mass action* is based on the assumptions that the reacting and product molecular species are well mixed and free to move in a *closed compartment* (i.e., they are not bound to a surface—like a mitochondrial membrane or a cell surface), and there are no diffusion barriers. Furthermore, it is assumed that the reacting molecules and ions are in weak concentrations and move randomly due to thermal energy and momentum exchanges due to collisions with themselves and each other. The *law of mass action* is based on the probabilities that molecular species will collide <u>and</u> react, that is, chemical bonds will be broken and made. Basically it states that the *rate* at which a chemical is produced in a reaction (at a constant temperature) is proportional to the product of the concentrations of the reactants.

   For example, if two reactants, **A** and **B**, combine and form a product, **C**: **A** + **B** → **C** → **D**, then by MA, the rate of appearance and disappearance of **C** is given by the simple, first-order, NL ODE: $[\dot{C}] = k_1[A][B] - k_2[C]$. Also, it is easy to see that both the reactants behave as $[\dot{A}] = [\dot{B}] = -k_1[A][B]$. In these simple ODEs, brackets [*] denote a concentration, such as moles per liter; k_1 is the reaction rate constant; and k_2 is the loss rate constant for **C**. Now examine a reaction in which two molecules of **B** must combine with one of **A** to make **C**:

$$\mathbf{A} + 2\mathbf{B} \xrightarrow{k_1} \mathbf{C} \xrightarrow{k_2} \mathbf{D} \qquad\qquad\qquad \text{(GL28)}$$

MA shows that the net rate of increase of the molar concentration of the product, **C**, is given by the ODE:

$$[\dot{C}] = k_1[A][B]^2 - k_2[C] \qquad\qquad\qquad \text{(GL29)}$$

Note that the concentration of **B** is squared in the ODE. More examples of the application of MA kinetics can be found in Chapter 4 in Northrop and Connor (2009).

methanogen: A prokaryote microorganism belonging to the domain *Archaea*. Methanogens are anaerobic; the hydrogenotrophic methanogens use CO_2 as a source of carbon and H_2 as a reducing agent to produce methane, which produces an electrochemical gradient across certain membranes, used to make ATP by chemiosmosis. There are about 36 strains of methanogens.

microbiome, human: The over 10^{14} bacteria that live in and on the adult human body. Four phyla of bacteria are normally dominant: Actinobacteria, Bacteroidetes, Firmicutes, and Proteobacteria. The total mass of the adult human microbiome is approximately 1 kg. Normally, these bacteria live commensally.

microeconomics: A branch of economics that studies how individuals, households, or firms make decisions to allocate limited resources. It considers SS supply and demand in markets where goods and services are being bought and sold.

million: 1×10^6 (also, mega-).

Miscanthus: A genus of about 15 species of a tall perennial grass. *Miscanthus* is native to tropical and subtropical regions of Africa and southern Asia. One species, *M. sinensis*, is found in China and Japan. *M. giganteus*, a sterile hybrid between *M. sinensis* and *M. sacchariflorus*, has been grown in Europe as a biofuel crop since the early 1980s. *M. giganteus* grows densely to a height of approximately 3.5 m. When harvested annually, it yields approximately 8–15 t/ha (3–6 t/acre), has a low moisture content (15–20%) and low ash, and is well suited for use directly as a biofuel (after drying). *Miscanthus* spp. are also well suited for feedstock in cellulosic ethanol production and HTC. *Miscanthus* rhizomes have a low tolerance for frost (Bioenergy 2011).

modularity (in graphs and biology): Modularity is independent of scale. Anatomic modularity generally involves parts of organs, for example, the pancreatic α-, β-, γ-, and δ-cells, the adrenal cortical cells versus the adrenal medullary cells. At the intracellular level, ribosomes, mitochondria, and chloroplasts are seen as functional and structural modules. At the molecular level, we can identify complex reaction networks such as the Krebs cycle, which has limited interfaces with other reaction networks. Anatomic modularity is expressed in the evolution of organ systems and a segmented body plan. Modules allow evolutionary flexibility; specific features may undergo changes during development without substantially altering the functionality of the entire organism. Each module is thus free to evolve, as long as the metabolic pathways at the edges between modules are relatively fixed. Each organism may have an optimal level of modularity. (See **segmentation**.) For a discussion of modularity in graphs, see Chapter 4 in Northrop (2010).

module (in a living organism): A part of an organism that is integrated with respect to a certain kind of process (natural variation, function, development, etc.) and relatively autonomous with respect to other parts (and modules) of the organism (Wagner et al. 2007). A *functional module* is composed of features that act together in performing some discrete physiological function that is semiautonomous in relation to other functional modules.

module (in a graph or network): Broadly defined as a subnetwork of a graph, the nodes of which have more connections to other nodes within the module than to external nodes.

motif (in graph theory): A pattern of interconnections occurring either in an undirected or a directed graph **G** at a number significantly higher than found in randomized versions of the graph, that is, in graphs with the same number of nodes, links, and degree distribution as the original one, but where the links are distributed at random. The pattern **M** is usually taken as a *subgraph* of **G**.

myoblast: A muscle precursor cell. Myoblasts remain in a proliferative state until they are signaled to differentiate into myocytes. The myoblasts that do not form muscle fibers dedifferentiate back into *satellite cells*. The satellite cells are a type of stem cell that are generally situated between the *sarcolemma* and the *endomysium* (the connective tissue complex that separates the muscle fascicles into individual fibers). *In vivo,* they serve to regenerate damaged muscles.

myocyte: A muscle cell that contains myofibrils (long chains of sarcomeres), the contractile units of the cell. Specialized myocytes include those in cardiac, skeletal, and smooth muscles. The myocyte contraction process involves a sequence of six major steps: (1) Either neural or external electrical stimulation causes specific ion channels on the cell membrane to open transiently. Sodium ions rush into the cell, driving its resting transmembrane potential positive, depolarizing the cell. (2) Simultaneously, there is a transient inrush of Ca^{++} through voltage-gated calcium channels. (3) This Ca^{++} binds to calcium-releasing channel proteins [*ryanodine receptors* (RyRs)] (Capes et al. 2011) on the intracellular sarcoplasmic reticulum (SR) membranes. (4) This triggers a massive release of calcium ions from the SR through RyRs that diffuse into the myocyte's cytoplasm. (5) This locally released Ca^{++} activates the actin–myosin contractile proteins, which shorten. (6) Immediately, ATP-driven calcium ion pumps scavenge cytoplasmic Ca^{++} back into the SR, where it is stored, allowing the intracellular contractile proteins to relax. Muscle relaxation is thus an active, ATP-dependent process.

natural population growth rate: The difference between a population's birth rate and death rate.

natural selection: The process whereby *favorable traits* that are heritable propagate throughout a reproductive population. It is axiomatic that individual organisms with favorable traits are more likely to survive and reproduce than those with unfavorable traits. When these traits are the result of a specific genotype, then the number of individuals with that genotype will increase in the following generations. Over many generations, this passive process results in adaptations to a *niche,* and eventually to speciation.

neighbor of a node (in graph theory): The neighbors of node k are those nodes connected directly to it by one edge, or by one or two edges if the graph is a digraph.

network motifs (in modularity): Recurrent building block patterns seen in complex metabolic networks.

niche (in evolution and ecology): A species' *fundamental niche* is the <u>full range</u> of environmental conditions (physical and biological) under which it can exist. As the result of competition with other organisms, the species may be forced to occupy a *realized niche* smaller than the fundamental niche. G.E. Hutchinson (1958) defined a niche as an **N**-dimensional hypervolume in **N**-space. Examples of dimensions might include moisture, pH, solar radiation, salinity, algae, and so forth. A species can actively alter its niche (e.g., beavers) and generally must compete with other species for resources in their mutual niche–hypervolume. Thus, a fundamental niche can be partitioned by species' competition.

NK system (in CSs): Stuart Kauffman's NK model is used to describe the FLs of CNLSs. **N** is the number of elements (define or elaborate) that characterize the system. **K** is the average number of other elements that each element is interdependent with. "A certain degree of fitness is associated to every possible configuration of the elements of the system" (Kaufman 1995). An FL is usually shown as a 3-D surface; however, it can be M-D ($M > 2$). The fitness is shown as the surface height. In an NK system, if $\mathbf{K} \to 0$, each element contributes independently to the system's fitness, and each element must be optimized to get a maximally sharp, single fitness peak. $\mathbf{K} \gg 0$ will give a ragged, multi-peaked FL. An NK model allows us to compare model architectures characterized by a different number of elements (**N**) and/or a different level of interdependence (**K**) by changing the two parameters.

node (in graph theory): A variable or state of a system represented at a point. There are input, output, and system nodes. See also **vertex**.

node degree (in graph theory): *The degree of a node or vertex* is the number of edges incident on that node, with loops being counted twice. In directed graphs such as SFGs, the edge count for a node k can be broken down into *indegree(k) and outdegree(k)* numbering the number of directed edges ending on node k and leaving node k, respectively.

nonlinear (NL) system: One in which superposition does not apply and whose time domain outputs cannot be determined by real convolution. NL systems cannot be characterized by transfer functions or frequency response functions.

nonstationary: A system is nonstationary if its parameters change in time. A noisy system is not necessarily nonstationary.

North Atlantic oscillation (NAO): The NAO is a climatic phenomenon in the North Atlantic Ocean. A permanent low-pressure area in the atmosphere exists around Iceland, while a permanent high-pressure region occurs around the Azores. East–west oscillations in the positions of the Icelandic low and Azores high causes sea-level differences in atmospheric pressure that cause variations in the strength and direction of westerly winds in the North Atlantic, as well as affecting cyclonic storm (hurricane) tracks in the region.

Occam's razor (in CSs): A principle attributed to William of Occam (ca. 1285–1349), a friar and logician. The principle has been stated in several forms: (1) The explanation of any phenomenon should make as few assumptions as possible, eliminating or "shaving off" those that make no difference in the observable predictions of the explanatory hypothesis or theory. (2) All things being equal, the simplest solution tends to be the best one. (3) (In Latin) *Lex Parsimoniae: Entia non sunt multiplicanda praeter necessitatem. [Entities should not be multiplied beyond necessity.]* The term "razor" refers to the act of shaving away unnecessary assumptions in reaching a parsimonious, simple explanation.

Sir Francis Crick commented on the potential limitations of using Occam's razor in biology: "While Occam's Razor is a useful tool in the physical sciences, it can be a very dangerous implement in biology. It is thus very rash to use simplicity and elegance as a guide in biological research" (Science 2013). Evolution has led to complex designs!

oil sands (aka tar sands): Naturally occurring mixtures of sand clay, water, and *bitumen*—a very viscous mixture of petroleum components. Major deposits of oil sands are found in Alberta, Canada, and Venezuela. Proven reserves in Canada are approximately 1.75 trillion barrels (2.80×10^{11} m^3); reserves in Venezuela are approximately 513 billion barrels (8.16×10^{10} m^3). In the United States, eastern Utah has tar sand oil reserves of approximately 32 billion barrels (5.1×10^9 m^3).

Extraction of distillable petroleum from oil sands is energy and water intensive. About 1–1.35 GJ of energy is needed to extract a barrel of bitumen and refine it to synthetic crude oil. A barrel of oil sand syncrude has approximately 6.12 GJ of chemical energy (see Wikipedia: Oil_sands). Oil sand must be open pit–mined, or if in deep deposits, extraction wells must be installed. Vast quantities of chemically contaminated water (e.g., with among other PAH chemicals, naphthenic acids) are created during oil sand extraction and processing and must be stored (and should be recycled in the extraction process).

oil shale: A fine-grained sedimentary rock that contains significant amounts of *kerogen*. Heating oil shale to a sufficiently high temperature causes pyrolysis of the kerogen, forming vapor. The vapor is condensed to obtain *unconventional oil compounds* (e.g., shale kerosene and diesel). In some cases, kerogen-rich oil shale can be burned directly as a low-grade fuel for power generation. Estimates of recoverable oil from worldwide oil shale deposits range from 2.8 to 3.3 trillion barrels (4.50×10^{11} to 5.20×10^{11} m³) (cf. Wikipedia: Oil_shale). This exceeds the world's proven oil reserves of approximately 1.76 trillion barrels (2.1×10^{11} m³). While *oil sands* originate from the biodegradation of oil, pressure and heat have not yet transformed the kerogen in oil shale into petroleum. Oil shale oil production is energy intensive and requires water for steam and cooling. It is environmentally unfriendly because it makes large volumes of chemically polluted waste water and releases CO_2 and other air pollution (PAHs, NOx, SOx, etc.), and open-pit oil shale mining is ecologically devastating.

operating point (in CNLSs): A set of constant (average), SS parameter values around which a stable CNLS exhibits stable, piecewise-linear, input–output behavior.

opportunity cost (in economic systems): The value of a good or service forgone in order to acquire (or produce) another good or service. An opportunity cost is involved when a farmer chooses to plant corn for ethanol instead of beans. Another example: A city decides to build a parking lot on municipal vacant land. The opportunity cost is the value of *the next best thing* that might have been done with the land and funds used. The city could have built a hospital, a sports center, or a golf course or even have sold the land to reduce debt.

parametric regulation and control (PRC; in CNLSs): An NL, feedback/feedforward system in which regulation/control is effected by having one or more path gains be a function of one or more state variables (SVs). PRC is found in all living systems; for example, biochemical, molecular biological, and physiological systems PRC makes complex adaptive biological systems possible; it permits transcription regulation and *homeostasis*. It also appears in our models of economic and social systems.

Pareto distribution (in CNLSs): This power-law probability distribution was named after economist Vilfredo Pareto. It models the statistical behavior of many types of economic systems, as well as various social, geophysical, and actuarial phenomena, even the sizes of meteorites. If X is a random variable (RV) with a Pareto distribution, then the probability that X is greater than some value x is given by $\Pr(X > x) = (x/xm)^{-\gamma}$, where x_m is the minimum possible X and $\gamma > 0$. When this distribution is used to model the distribution of wealth, γ is called the *Pareto index*. By differentiation, the Pareto probability density function (PDF) is given by $p(x; \gamma; x_m) = (\gamma\, x_m^{\gamma})/x^{\gamma+1}$, for $x \geq x_m$. The expected value of the RV X following the Pareto statistics is $E(X) = \gamma\, x_m/(\gamma - 1)$. If $\gamma \leq 1$, $E(X) \to \infty$. The variance of X is $\text{var}(X) = [\gamma/(\gamma - 2)][x_m/(\gamma - 1)]^2$.

Pareto efficiency (or optimality; in CESs): Given a set of individuals with alternative *economic allocations* of goods or income, any movement from one allocation to another that can make at least one individual better off without making any of the others worse off is called a *Pareto improvement*. If economic allocation in any system is not Pareto efficient, there is a theoretical potential for a Pareto improvement, that is, an increase in Pareto efficiency. An allocation is called (strongly) Pareto optimal when no further Pareto improvements can be made.

Pascal (Pa): The SI unit of pressure: $1 \text{ Pa} \equiv 1 \text{ N/m}^2$, $1 \text{ kPa} = 0.14504 \text{ psi}$, $1 \text{ MPa} = 145 \text{ psi}$.

Path (in graph theory): A route that does not pass any node (vertex) more than once. If the path does not pass any node more than once, it is a *simple path*. See also **walk**.

Parity (in economics): When the cost of a good or service equals the cost of a competing good or service. *Grid parity* is when the cost of electricity from a renewable source such as solar PV generation reaches the cost of electrical energy supplied on the distribution grid. *Transportation parity* is when it becomes cheaper to take public transportation to go from point A to B than it is to drive one's personal automobile.

partial pressure (of gasses): In a mixture of *ideal gasses*, each gas has a *partial pressure*, which the gas would have if it occupied the volume alone. The total pressure of a gas mixture is the sum of the partial pressures of the individual gas components in the mixture. $P_{tot} = (RT/V)\sum_{k=1}^{N} n_k$

peak oil (production rate): In 1956, eminent earth scientist Marion K. Hubbert predicted that the US oil production rate would likely hit its peak sometime between 1965 and 1970. It did peak in 1970 and began a slow decline. Hubbert greatly underestimated the amount of oil that would be found and produced in the United States: oil from the North Slope in Alaska, oil from oil shale, oil from deep offshore ocean drilling (e.g., in the Gulf of Mexico), and so forth. He also did not know that future technologies would find ways of withdrawing more oil from old wells (e.g., by fracking). Demand for oil has continued to grow in the United States along with economic and population growth. According to Yergin (2011), the world has already produced approximately 1 trillion (1×10^{12}) barrels since the mid-19th century. Currently, it is thought that there is approximately 1.4 trillion barrels of oil worldwide that is technically and economically accessible as proven reserves. There may be another 3.6 trillion barrels as probable reserves. The oil that will be produced in the future will come with higher economic and environmental costs, however. The increased cost of this future oil will dampen consumption and slow production, pushing the world "peak petroleum" far into the 21st century. Yergin stated: "Based on current and prospective plans, it appears the world liquid [petroleum] production capacity should grow from about 93 million barrels per day in 2010 to about 110 mbd by 2030. This is about a 20 percent increase." The peak oil production rate *in the United States* had a discernable peak around 1970 and then a long, flat plateau with a slight downward slope. If we add in the Organization of Petroleum Exporting Countries (OPEC) and non-OPEC production rates, the total world production continues to climb briskly, as of 2012. The Earth is not running out of oil any time soon. However, certain regional oilfields such as the Alaska North Slope have reached peak production. Alaska's oil production rate peaked at approximately 2000 barrels/day in 1988 and then began to drop off.

pH: The measure of acidity or alkalinity of an aqueous solution. pH ranges from 0 < to ≤ 14. Neutral pH = 7.0. A solution with pH < 7 is acidic. In general, pH $\equiv -\log_{10}[H^+]$, where $[H^+]$ is the molar hydrogen ion concentration in the solution.

phase plane (in NL system analysis): A 2-D (or 3-D) parametric plot of two (or three) outputs of a system given certain initial conditions, or a transient input. Often, $\dot{x}(t)$ versus $x(t)$ is plotted. As an example, assume an NL second-order system is modeled by a second-order ODE of the form

$$\ddot{\mathbf{x}} + F(\dot{\mathbf{x}}, \mathbf{x}) = 0 \tag{GL30}$$

Let $\mathbf{x} = x(t)$ and $\mathbf{y} = \dot{x}(t)$. Note that the second-order NL system can also be written as two first-order ODEs *(state equations).*

$$\dot{\mathbf{y}} = -F(\mathbf{y}, \mathbf{x}) \tag{GL31}$$

$$\dot{\mathbf{x}} = \mathbf{y} \tag{GL32}$$

A plot of $\mathbf{x}(t)$ versus $\mathbf{y}(t)$ in the $\mathbf{y} = \dot{x}$ versus $\mathbf{x}$ plane for t values ≥ 0, starting at $t = 0$ at some $\mathbf{x}(0)$, $\mathbf{y}(0)$, is called a *phase plane trajectory.*

The $\mathbf{y}$, $\mathbf{x}$ plane is the phase plane. The system's phase plane trajectories provide important information about the equilibrium states, oscillations (*LCs*), and the stability of the NL system. Any Mth-order NL system modeled by a set of M NL ODEs can be written as $x_n = F_n(x_1, x_2, \ldots, x_{n-1}, x_n, x_{n+1}, \ldots, x_M)$, $n = 1, 2, \ldots, M$. Thus, M 2-D phase plane plots can be made for the $(\dot{x}_n, x_n)$ trajectories. A phase plane trajectory that converges on a closed path as $t \to 0$ gives a graphic demonstration of stable LC oscillations. Patterned, closed, SS phase plane trajectories are called *attractors.*

phase space: A multidimensional phase plane. In a system that has $N > 1$ variables, the phase space is generally a $2N$-dimensional space (the N variables plus their N first time derivatives).

photonic crystal: Periodic nanostructures composed of regularly repeating internal regions of high and low dielectric constant. Their purpose is to act as band-pass/band-reject filters for incident photon wave-lengths lying from far IR to ultraviolet (UV). This behavior gives rise to optical phenomena such as inhibition of spontaneous emission, high-reflecting omnidirectional mirrors, and low-loss waveguiding. They can also be used as collector/emitters in thermal PV (TPV) power systems to select only those wavelength bands that optimally excite the PV materials used.

Planck's blackbody radiation law: The *spectral emmittance* of an ideal, radiating blackbody at Kelvin temperature, T, is given by

$$W(\lambda,T) = \frac{c_1 \lambda^{-5}}{\exp[c_2/(\lambda T)]-1} \text{watts}/\text{m}^2 \text{per nm wavelength} \qquad \text{(GL33)}$$

where λ is the photonic wavelength in nanometers, $c_1 = 3.740 \times 10^{20}$, and $c^2 = 1.4385 \times 10^7$.

The $W(\lambda, T)$ curve is near zero for a temperature-dependent cutoff wavelength, rises to a peak value at a wavelength given by

$$\lambda_{pk} = (2.8971 \times 10^6)/T \text{ nm} \qquad \text{(GL34)}$$

and then decreases slowly with increasing λ. See Figure 4.6 for an illustration of BB radiation curves.

plankton: Any drifting organisms (plants, animals, prokaryotes—including bacteria and archaea) that inhabit the pelagic zone of oceans, seas, or freshwater lakes. More specifically, they include the following: *Phytoplanktons* obtain energy from photosynthesis. They live near the water surface where they can harvest the sun's photons. They include diatoms, cyanobacteria, dinoflagellates, and coccolithophores. Phytoplankton blooms are seen in eutrophication events. *Zooplanktons* are small eukaryotic animals (protozoans, metazoans), including the eggs and larvae of larger animals such as fish, crustaceans, and annelids. Zooplanktons include amphipods, tiny shrimp, jellyfish, copepods, siphonophora, and so forth. Zooplanktons feed on phytoplanktons and each other. They are a major food biomass for fish larvae as well as baleen whales. *Bacterioplanktons* include the bacteria and archaea. (Prokaryotic phytoplanktons are also bacterioplanktons.) All planktons have a very important and complex ecology.

plectics: A research area, named by Murray Gell-Mann, which is "...a broad transdisciplinary subject covering aspects of simplicity and complexity as well as the properties of CASs, including composite CASs consisting of many adaptive agents" (Gell-Mann 2010).

pole [in LSs considered in the frequency (Laplace) domain]: An LTIS is said to have a pole at some finite, complex frequency value, $\mathbf{s}_k$, if its transfer function magnitude, $|H(\mathbf{s}_k)| \to \infty$. That is

$$\frac{Y}{X}(\mathbf{s}_k) = H(\mathbf{s}_k) = \frac{P(\mathbf{s}_k)}{Q(\mathbf{s}_k)} = \frac{\mathbf{s}_k^m + b_{m-1}\mathbf{s}_k^{m-1} + \cdots + b_1\mathbf{s}_k + b_0}{\mathbf{s}_k^n + a_{n-1}\mathbf{s}_k^{n-1} + \cdots + a_1\mathbf{s}_k + a_0} \to \infty \qquad \text{(GL35)}$$

In general, this will occur for some finite root of the denominator polynomial, $\mathbf{s}_k = \sigma_k + j\omega_k$, such that $Q(\mathbf{s}_k) \to 0$, and $|H(\mathbf{s}_k)| \to \infty$. The kth pole is at $\mathbf{s}_k = \sigma^k + j\omega_k$.

pollination: The transfer of pollen from the stamens of a flower to the pistil. Pollination in some plants can be done by wind; in others, it can be accomplished by insects (e.g., honeybees, wild bees), hummingbirds, bats, or even humans. *Abiotic pollination* is pollination by wind or water; this form of pollination is found in grasses, most conifers, and many deciduous trees. *Biotic pollination* requires living pollinators. There are roughly 2×10^5 varieties of animal pollinators; most are insects (bees, wasps, ants, beetles, moths, butterflies, and flies). Biotic pollination is also done to a lesser degree by fruit bats, hummingbirds, honeyeaters, sunbirds, and spiderhunters. Pollination is required for fertilization and the production of seeds for all higher plants. On the surface of the pistil, pollen grains germinate and form pollen tubes that grow downward toward the ovules. During fertilization, a sperm cell in a pollen tube fuses with the egg cell of an ovule, giving rise to the plant embryo. The fertilized ovule then grows into a seed.

Bumblebees do "buzz pollination"; they vibrate flowers as they feed on nectar, releasing pollen, which reaches the pistil; tomato flowers are effectively pollinated by bumblebees.

potash: A generic term for potassium salts found in wood ashes: potassium carbonate, potassium hydroxide, potassium chloride, and various potassium sulfates. Potash is used in fertilizers.

power: The *rate of doing work:* $P = dW/dt$. The SI units of power are 1 J/s = 1 W (kilowatts, megawatts, gigawatts, and terawatts are used). The horsepower is a unit of power; 1 hp = 746 W. If an agent does 1000 J of work each second, the work done in 1 h is $3600 \times 1000 = 1$ kWh = 3.6×10^6 J. Power has the basic dimensions of ML^2T^{-3}.

precautionary principle (in the design of resilient systems): The precautionary principle considers that if an action or policy might cause severe or irreversible harm to ecosystems, the environment, or public health and welfare, restraint is called for. In the absence of verifiable evidence that harm will certainly not occur, the action or policy must not be undertaken, or alternative, less risky actions and policies should be considered, even if more expensive. Cost–benefit and cost–risk analyses need to be done including most recent data.

primary battery: An electrochemical battery that irreversibly transforms chemical energy to electrical energy. A primary battery is not rechargeable.

pulses (in botany): A generic term for peas, beans, and lentils.

Q_{10} **(in physical biochemistry):** Q_{10} (pronounced Q-ten) is a measure of a chemical reaction's *temperature sensitivity*. It is the ratio of the rate of an isothermal reaction at temperature T_0 to that of the same reaction at $T = T_0 - 10°C$. Mathematically, Q_{10} can be written more generally as

$$Q_{10} \equiv \left(\frac{R_2[10/(T_2 - T_1)]}{R_1} \right) \qquad \text{(GL36)}$$

where R_1 is the reaction rate at $T = T_1$; R_2 is the reaction rate at $T = T_2 > T_1$. Q_{10} for most biochemical reactions ≈ 2.

Another way of looking at the temperature sensitivity (S_T) of reaction rate constants is

$$S_T = \frac{\Delta R}{\Delta T} (T_1 / R_1) \qquad \text{(GL37)}$$

where $\Delta R = R_2 - R_1$, $\Delta T = T_2 - T_1$, $T_2 > T_1$, $R_2 = R$ at T_2, $R_2 > R_1$, and so forth.

quadrillion: 1 quadrillion = 10^{15}. 1 quadrillion grams = 1 petagram.

Quinoa: The edible seeds (grain) of Andean plants of the family *Amaranthaceae,* that is, *Chenopodium quinoa. C. quinoa* plants are bisexual and self-pollinating and have been raised for their nutritious, edible, gluten-free seeds in Bolivia, Peru, Ecuador, and Colombia for over 5000 years. World quinoa production in 2009 was 69,000 tonnes. Many *Chenopodium* cultivars exist. The plants are hardy and frost-proof except when blossoming. They grow best in sandy, well-drained soil. Each 100 g of uncooked quinoa has 368 calories, 52 g starch, 7 g dietary fiber, 6 g fat of which 3.3 is polyunsaturated, 14 g protein, 18 AAs (including the essential AA lysine), 13 g water, plus thiamine (vitamin B_1), riboflavin (vitamin B_2), vitamins B_6 and B_9, iron, magnesium, zinc, and phosphorus. Quinoa oil contains vitamin E and the antioxidant tocopherol, which makes the oil have a long shelf life. The UN has designated 2013 as the international year of quinoa.

rank (in matrix algebra): The rank of a matrix $\mathbf{A}$ is invariant under the interchange of two rows (or columns), the addition of a scalar multiple of a row (or column) to another row (or column), or the multiplication of any row (or column) by a nonzero scalar. For an $n \times m$ matrix $\mathbf{A}$, rank $\mathbf{A} \leq \min(n, m)$. For an $n \times n$ matrix $\mathbf{A}$, a necessary and sufficient condition for rank $\mathbf{A} = n$ is that $|\mathbf{A}| \neq 0$ (Ogata 1987).

recession (in economics): A contraction phase of the business cycle. A recession is said to occur when the real GDP is negative for two or more consecutive quarters. Recessions may include bankruptcies, deflation, foreclosures, unemployment, reduced sales, and stock market crashes.

redox reactions: Chemical, reduction–oxidation reactions. An oxidized chemical loses electrons; a reduced reactant gains electrons. An oxidizer gains electrons, that is, is reduced.

Red Queen Contest (in ecological genomics): It occurs when two, *competing*, complex biological systems interact, for example, a quasi-species (QS; bacterial, viral, or parasitic) versus the host's immune system, an antibiotic versus a bacterial population, or an insecticide versus an insect population. Red Queen Contests have been viewed as a type of evolutionary "arms race." *Red Queen Cycling* is when the QS genes that result in infectivity track host immune system gene frequencies, leading to high QS fitness and a continuing infection in a sympatric host; infectivity LCs can occur. The Red Queen Scenario, first proposed by L. van Valen (1973, A new evolutionary law, *Evol Theory* 1: 1–30), suggests that genetic diversity within a QS population is required to counter rapidly evolving pathogens, antibiotics, or toxins. A Red Queen scenario exists between pathogens and the human development and use of antibiotics. The *Red Queen Principle* can be stated: for an evolutionary system, continuing development is needed just in order to maintain its fitness relative to the systems it is coevolving with. (The Red Queen herself comes from Chapter 2 in Lewis Carroll's *Through the Looking Glass*, 1872.)

reductionist thinking (in CSs): Includes the natural human proclivity to try to manipulate a CS by changing one input state at a time. A CS model is often partitioned (i.e., reduced) by reductionists in order to simplify analysis or obtain a local solution. This approach can destroy the counterintuitive behavior of the unreduced CS. A better approach to manipulating a whole, unreduced, CS model is to use multiple inputs and observe its behavioral responses, and then choose which inputs work best.

redundancy (in biology): Redundancy is found at the molecular, cellular, physiological, and organismic levels, where multiple elements have the same functions and operate in parallel (e.g., pancreatic beta cells, hepatocytes, erythrocytes, platelets, hematopoietic stem cells, two lungs, two eyes, two ovaries, many worker bees, etc.). Redundancy is one insurance of system *robustness*. See **degeneracy**.

refractive index (in optics): A transmissive medium's refractive index **n** is the ratio of the speed of light *in vacuo* to the speed of light in the medium, that is, $\mathbf{n} = c/v$. $\mathbf{n}$ for crown glass is approximately 1.52; $\mathbf{n}$ for water is 1.333 at $\lambda = 589$ nm. $\mathbf{n}$ is a function of wavelength λ.

regulator (in control theory): A subsystem that senses the departure of certain critical system states from normal set values due to adverse external inputs or internal disturbances and then takes action to bring the SVs back to the set values. A regulator must effectively store the information about acceptable state values to compare with the perturbed SVs, either implicitly or explicitly. It must then make internal adjustments to certain system states in order to maintain constancy in the regulated parameters (e.g., states, outputs). A *parametric regulator* adjusts system branch gains to effect critical SV constancy. In general, a regulator responds to disturbances, rather than to system control inputs. However, a system controller can also act as a regulator. Regulation is generally effected by some form of negative feedback.

relaxation oscillation: Relaxation oscillation is an electronics term (Wang 1999). A simple relaxation oscillator can be made from a switch, a battery, a resistor, a capacitor, and a neon bulb. The neon bulb is placed across the capacitor, which is connected through the resistor to the battery. The capacitor is initially discharged. The switch is closed, and the capacitor charges through the resistor toward the battery voltage, V_B. When the voltage across the capacitor, V_C, reaches the ionization threshold voltage ($V_I < V_B$) for the neon bulb, the sudden increase of conductance of the neon bulb causes the capacitor to discharge rapidly through the high-conductivity neon plasma until it reaches the low extinction voltage, $V_E < V_I$, where the ionization of the neon gas $\rightarrow 0$, and the bulb's conductance $\rightarrow 0$. The capacitor now charges again through the series resistor until $V_C = V_I$, and the relaxation cycle repeats. The voltage across the capacitor, V_C, is seen to have a sawtooth shape, rising slowly from V_E until it reaches V_I and then decreasing rapidly to V_E again as the capacitor discharges through the neon bulb causing it to flash, and the cycle repeats. Many other types of relaxation oscillator exist; as a rule, they all involve threshold switching of a parameter or device.

renewable energy: Basically, energy from any source that is essentially inexhaustible. By exclusion, all FFs (coal, oil, NG) *are not* renewable energy sources. Renewable energy sources include (1) wind; (2) solar (PV, thermoelectric, solar thermal); (3) geothermal; (4) hydropower, including falling water, tidal currents, and wave energy; (5) biofuels made from growing plants and algae (e.g., biodiesel), also methane from garbage; (6) combustion of biomass (wood, wood pellets, dried cattle dung, garbage); and (7) passive solar (green buildings), including hot water heating.

resilience (in CASs): Resilience (n.) has been defined as the capacity of a system, ecosystem, community, or society to absorb disturbance and reorganize while undergoing changes so as to retain essentially the same functions, structure, identity, and feedback, that is, remain within one regime. It reflects the degree to which a CAS is capable of self-organization and the degree to which the CAS can build and sustain the capacity for learning and adaptation (Norberg and Cummings 2008). Compare with **robustness**.

road parity: When the cost of public transportation (buses, trains, subways) is less expensive per capita than personal automobile travel to a given distant destination.

robustness (in systems theory and biology): A physical or living system is said to be robust (have robustness) when it largely preserves its normal functioning in the presence of external (environmental) disturbance inputs (e.g., acceleration; temperature changes; changes in pressure, salinity, incident light flux, pO_2, etc.) and internal noise, and also shows feature persistence in the presence of internal (parametric) pathway changes or failures or subsystem (module) failures. In living systems, pathway parametric changes can result from photon energy absorption, molecular ageing, chemical damage from peroxides, free radicals, and metabolic poisons. Negative feedback can produce robustness, as can redundancy, alternate pathways, and *modularity*. See **homeostasis**. Robustness is graded. An example of a robust system is a commercial aircraft flight recorder ("black box").

route (in graph theory): A sequence of edges and nodes from one node to another. Any given edge or node might be used more than once.

scale (of CSs): The complexity of biological, economic, weather, ecosystem, political, and government systems, for example, can be studied at several levels of scale. Complex biological system behavior can be observed and modeled at molecular (biochemical), cellular, physiological, and organismic levels. An economic system's complexity can be studied at various scales, such as family, municipal, state, regional, national and global, and so forth.

In the biophysical sciences, scale is generally defined as a property of the dimensions of space and time. This view of scale has two main components: *grain* and *extent*. Grain refers to the resolution of analysis, and extent to the coverage [Cummings and Norberg: *Scale and Complex Systems*. Chapter 9 in Norberg and Cummings (2008)].

scale-free graph (in graph/network theory): A graph/network whose *degree distribution* follows a *power law* model, sometimes asymptotically. That is, $P(K) \cong ak^{-\gamma}$, where the exponent γ typically lies in the range between 2 and 3. Many observed networks appear to be scale-free. These include the World Wide Web, protein–protein interaction networks, citation networks (in scholarly papers), human sexual partners, semantic networks, and some social networks. The high-degree nodes are called *hubs*. Some properties of scale-free networks (SFNs) are as follows: (1) SFNs are more robust against random failures; the network is more likely to stay functional than a random network after removal of randomly chosen nodes. (2) SFNs are more vulnerable against targeted (nonrandom) attacks on their hubs. This means an SFN loses function rapidly when nodes are removed according to their degree. (3) SFNs have short average path lengths. The average path length L is proportional to $\log(N)/\{\log[\log(k)]\}$ (Hidalgo and Barabasi 2008).

secondary battery: A rechargeable electrochemical battery, for example, the familiar lead–acid battery used in automotive applications.

sensitivity (of complex LSs; aka gain sensitivity): Assume a *large LS* with a certain input $\mathbf{x}_i$ vectors and output vectors, $\mathbf{y}_k$. The complex gain between this input and output is defined as $\mathbf{A}_{ik} = \mathbf{Y}_k/\mathbf{X}_i$. The sensitivity of $\mathbf{A}_{ik}$, given a change in a certain internal path gain, $\mathbf{G}_{pq}$, is defined as

$$\mathbf{S}_{\mathbf{G}_{pq}}^{\mathbf{A}_{ik}} \equiv \frac{\partial \mathbf{A}_{ik}}{\partial \mathbf{G}_{pq}}(\mathbf{G}_{pq}/\mathbf{A}_{ik}) \tag{GL38}$$

Ideally, we want $\mathbf{S}_{\mathbf{G}_{pq}}^{\mathbf{A}_{ik}} \to 0$ for *robustness*. For example, consider a linear, SISO feedback system with forward gain, $\mathbf{P}$, and feedback path gain, $\mathbf{Q}$. The system's overall gain is well known: $\mathbf{A} = \mathbf{Y}/\mathbf{X} = \mathbf{P}/(1 + \mathbf{PQ})$. $\mathbf{S}_{\mathbf{P}}^{\mathbf{A}}$ is calculated for this simple system and found to be $1/(1 + \mathbf{PQ})$. That is, the effect of $\Delta\mathbf{P}$ on the gain $\mathbf{A}$ is minimized as the return difference $\mathbf{F} = (1 + \mathbf{PQ}) \to \infty$. This illustrates an important general property of negative feedback systems, that is, the reduction of gain sensitivities, hence the effect of parameter variations on overall loop gain.

For a stable, multiple-input multiple-output (MIMO) CNLS, the sensitivity for the *small-signal gain* between the ith input and kth output, $\mathbf{A}_{ik}$, as a function of a change in the nth parameter, $\mathbf{P}_n$, or another input, $\mathbf{x}_j$, can be written as

$$\mathbf{S}_{\mathbf{P}_n}^{\mathbf{A}_{ik}} = \frac{\partial \mathbf{A}_{ik}}{\partial \mathbf{P}_n}(\mathbf{P}_n/\mathbf{A}_{ik}) \text{ or } \mathbf{S}_{\mathbf{x}_j}^{\mathbf{A}_{ik}} = \frac{\partial \mathbf{A}_{jk}}{\partial \mathbf{x}_j}(\mathbf{x}_j/\mathbf{A}_{jk}) \tag{GL39}$$

These sensitivities can be approximated by

$$\mathbf{S}_{\mathbf{P}_n}^{\mathbf{A}_{ik}} \cong \frac{\Delta \mathbf{A}_{ik}}{\Delta \mathbf{P}_n}(\mathbf{P}_n/\mathbf{A}_{ik}), \text{ and } \mathbf{S}_{\mathbf{x}_j}^{\mathbf{A}_{ik}} \cong \frac{\Delta \mathbf{A}_{jk}}{\Delta \mathbf{x}_j}(\mathbf{x}_j/\mathbf{A}_{jk}) \tag{GL40}$$

For robustness, we generally desire the non-direct-path parameter sensitivities to be small or zero.

set point (in a regulated system): The desired, SS value of a regulated system's output state.

Shannon diversity index (SDI; applied to complex ecosystems): Sometimes called the Shannon–Wiener diversity index, the SDI (H') is used as an objective measure of biodiversity in complex ecosystems. [Sometimes workers use $\log_2(*)$ instead of $\ln(*)$.]

$$H' = -\sum_{i=1}^{s} p_i \ln(p_i) \tag{GL41}$$

where

n_i = number of individuals in each of a total S species; i = 1, 2, 3,...S (aka n_i = the abundance of a species).

S = the number of different species in the ecosystem (species richness).

N = the total number of *all* individuals in the ecosystem = $\sum_{i=1}^{s} n_i$.

p_i = the relative abundance of species i: $p_i \cong n_i/N$. Estimates the probability of finding species i in the ecosystem.

[It can be shown that for any given number of species, there is a maximum possible $H' = H'_{max} = \ln S$, when all S different species are present in equal numbers ($n_i = N/S$).]

Shannon's equitability index (applied to complex ecosystems): Defined as $E_H \equiv H'/H'_{max} [H'_{max} = \ln(S)]$. Equitability assumes a value between 0 and 1, with 1 being complete evenness.

signal flow graphs (SFGs): Used in describing *linear time-invariant (LTI) systems* in the frequency domain. *They are directed graphs* in which the nodes (vertices) are signal summation points [adding inputs from all incoming branches (edges)]. SFG directed edges are signal conditioning pathways that multiply (condition) the signal at a source node and input that conditioned signal additively to the sink node. *Outwardly directed branches* that leave a node do not change the signal (state) at that node. They process the signal at the source node that is the result of summing all input signals at that node. SFGs generally process signals in the frequency domain. SFGs can be used to characterize sets of linear state equations (ODEs). In NL SFGs, one or more branches operate nonlinearly on node signals. Thus, *Mason's rule* cannot be used to find the NL SFG's transfer functions.

simplicity (in systems): The absence of complexity. The property of being simple, having few states. Simple systems are easier to describe, model, explain, and understand than complicated ones. Like complexity, simplicity is a relative, graded property. Simplicity can denote beauty, purity, or clarity.

size (in graph theory): The *size* of a graph is the total number of its *edges*. In CSs, size matters.

societal collapse: A broad term for the syndrome of a downward-spiraling sociopolitical–economic system. It has many causes and manifestations. Tainter (1988) described nine attributes that a collapsing society might exhibit: "(1) a lower degree of stratification and social differentiation; (2) less economic and occupational specialization, of individuals, groups, and territories; (3) less centralized control; that is, less regulation and integration of diverse economic and political groups by elites; (4) less behavioral control and regimentation; (5) less investment in the epiphenomena of complexity, those elements that define the concept of 'civilization': monumental architecture, artistic and literary achievements, and the like; (6) less flow of information between individuals, between political and economic groups, and between a center and its periphery; (7) less sharing, trading, and redistribution of resources; (8) less overall coordination and organization of individuals and groups; (9) a smaller territory integrated within a single political unit." Collapse, once some ill-defined tipping point is reached, is rapid, taking no more than a few decades to achieve: a..."significant loss of an established level of sociopolitical complexity."

sorghum: Sorghum is a genus of numerous species of grasses. *Sorghum bicolor* is an important world crop used for a food grain, sorghum molasses, cattle fodder, a fermentation stock, and biofuels. Most varieties are drought and heat tolerant, making them an ideal "sustainability crop." Sorghum is an important food crop in Africa, Central America, and South Asia, and is the "third most important cereal crop grown in the United States, and the fifth most important cereal crop grown in the world" (Grains Council 2013). The fermentation of the sugar-rich stalks (stover) is used to make *maotai* in China and ethanol fuel in India and the United States.

specific heat capacity: Denoted by c.

$$c = \text{heat capacity}/\text{mass} = \frac{Q/\Delta T}{m} \quad \frac{\text{J}}{\text{k}_g\text{K}} \qquad \text{(GL42)}$$

where Q = heat supplied to a mass in joules, ΔT = temperature rise resulting from heat input, and m = mass of heat storage medium. $Q/\Delta T \equiv$ the *heat capacity* of the mass.

spectral response (of SC): The ratio of the current generated by the SC to the light power at wavelength λ incident on the SC. That is, amps per watt.

stability tests (for an LS): An LS is stable if its transfer function $H(s)$ has no poles (denominator roots) in the right-half s-plane. In a SISO, single-loop feedback system, the closed-loop transfer function $H(s)$ is of the form

$$H(s) = F(s)/[1 - A_L(s)] \qquad \text{(GL43)}$$

Some tests for stability examine the complex s values that make the denominator of $H(s) \to \infty$, that is, values of s that make $A_L(s) = 1 \angle 0$. This criterion has given rise to the venerable *Nyquist stability criterion*, the *Root–Locus technique*, and the Popov stability criterion.

stagflation (in economics): An economic condition in which there is simultaneous inflation in the prices of goods and economic stagnation (unemployment and low GDP).

standard of living: Various methods of measure exist: income method, consumption method and private consumption expenditure, the average personal income in a country less taxes, loans, and so forth. One generally accepted measure is a county's *average real domestic product per capita.*

standard temperature and pressure (in physical chemistry): Conditions used to compare physical properties of substances: 0°C (273 K) and 1 atm (101.325 kPa, 14.696 psi, 760 mm Hg, or 760 torr). In standard ambient temperature and pressure (SATP), the US NIST uses 20°C and 101.325 kPa.

state (in LTISs): The state of an LTIS is the smallest set of variables (called SVs) such that the knowledge of these variables at $t = t_o$, together with the input for $t \geq t_o$, completely determines the behavior of the system for any time $t \geq t_o$ (Ogata 1970).

state variable (SV; in LTISs): The SVs of an LTIS are the smallest set of variables that determine the *state* of the dynamic system. If at least n variables $x_1(t)$, $x_2(t)$,..., $x_n(t)$ are needed to completely describe the behavior of a dynamic LTIS, then such n variables $x_1(t)$, $x_2(t)$,..., $x_n(t)$ are a set of SVs (Ogata 1970). In a linear *SFG*, sums of certain SVs are found at certain nodes, and their integrals at other nodes.

stationary (in systems analysis): A *stationary signal* arises from a *stationary system* in which no parameters or branch gains are changing in time over the period that the signal is measured or calculated. Stationary also refers to noise produced from within such a system.

stem cells (in biology): There are both adult and embryonic stem cells. A stem cell is an undifferentiated cell, which, through certain physical, biochemical, and genomic signals, can be caused to differentiate to a specialized cellular form and function, such as neurons, heart muscle cells, skin cells, blood cells, and so forth. An undifferentiated stem cell can divide indefinitely in that form until signaled to differentiate. *Totipotent* stem cells are the earliest embryonic cells that can become *pluripotent* stem cells and the placenta; thus, they can give rise to a twin embryo. A *pluripotent* stem cell can generate all the structures of the developing embryo except the placenta and surrounding tissues, as well as reproducing themselves. *Committed* stem cells derive from *pluripotent* stem cells; they can differentiate into fixed categories of embryonic tissues. (For example, the *hemangioblasts* can form blood vessels, blood cells, and lymphocytes; the *mesenchymal stem cells* form connective tissues, cartilage, muscle, fat, etc.) The restriction on the potency of stem cells in embryonic development is gradual and depends on poorly understood internal and external signaling factors.

step function (in systems analysis): Used as a system input, a unit step function, $U(t)$, is 0 for $t < 0$ and 1 for $t \geq 0$. A delayed step function,

$$U(t - \tau), \text{ is 0 for } t < \tau \text{ and 1 for } t \geq \tau. \text{ Also, } U(t - \tau) = \int_0^\infty \delta(t - \tau) dt. \ \tau > 0.$$

stiff ODEs (in simulating CNLSs): Computer solutions of NL ODEs are done by numerically integrating DEs. There are many numerical integration routines that are used for these solutions (e.g., rectangular, trapezoidal, Euler, Runge–Kutta/Fehlberg, Dormand–Prince, Gear, and Adams). Use of certain integration algorithms with stiff ODEs can result in numerically unstable or noisy solutions. Such chaotic behavior can often be eliminated by making the step size extremely small and choosing an integration routine such as Gear or rectangular, which are generally well behaved with stiff ODEs. An example of a stiff, NL ODE was given by Moler (2003): An ODE that models the size of a ball of flame when a match is struck is $dr/dt = r^2 - r^3$, $r(0) = \rho$, where ρ is the initial (normalized) radius r of the flame ball; try 0.0001 to 0.01. In this IVP, we are interested in the SS solution, in this case, $r(\infty) \rightarrow 1$, and how the ODE approaches SS ($dr/dt = 0$). Moler shows, using the MATLAB ODE solver, *ode45*, that this ODE approaches its SS in a very long, noisy manner before rapidly converging on $r = 1$.

Stirling engine: A closed-cycle, reciprocating-piston heat engine, invented in 1816 by Robert Stirling, a Scottish minister/inventor. This is a Carnot-cycle, external combustion engine in which a gas working fluid operates in a closed cycle. The gas can be air, nitrogen, hydrogen, or helium. The heat source for a Stirling engine can be concentrated solar energy, heat from combustion of a fuel such as methane or NG, or waste heat from industrial processes. A solar Stirling engine/generator system typically can produce from 2 to 25 kW of electric power at 575VAC 60 Hz at 22% to 30% thermodynamic efficiency, the highest of all solar technologies. An alpha-type Stirling engine uses one or more pairs of pistons, driving a common crankshaft. The hot cylinder receives heat input from external energy (e.g., focused solar energy), and the cold cylinder allows working gas expansion and heat loss. As the cycle continues, the cool gas is compressed into the hot cylinder where it picks up solar heat and expands, doing mechanical work. Some excellent, illustrated descriptions of Stirling engines may be found in *Wikipedia*: Stirling_engine (18 pp. with many references); also see REUK (2012) and Sitingcases (2012).

stover: The leaves and stalks of corn (maize), sorghum, peanuts, or soybean plants that are left in a field after harvesting. It can be directly grazed by cattle or dried and used for fodder. Stover can also be used as biomass for fermentation to produce cellulosic ethanol or as substrate for *HTC*.

strongly connected graph (in graph theory): A *directed graph* (e.g., an SFG), directed subgraph, or module is called *strongly connected* if there is a path from each vertex (node) in the graph or subgraph to every other vertex. In particular, this means a path in each direction: a path from node **p** to node **q** and also a path from **q** to **p**.

structural gene (in genomics): A gene that codes for the structure of a protein.

subgraph (in graph theory): A subgraph of a graph G is a graph whose vertex and edge sets are subsets of the total vertices and edges of G. If a graph G is contained in graph H, H is a supergraph of G.

supercapacitor: An electric double-layer capacitor, having very high energy density storage at low voltage. See Section 4.6.6 for details.

supercritical fluid (SCF): At pressures and temperatures above a substance's *critical point* (CP), the substance exists as an SCF where distinct liquid and gas phases do not exist. An SCF can effuse through solids like a gas and dissolve materials like a liquid. Carbon dioxide has a CP at 304.1 K and 7.38 MPa (72.8 atm.). CO_2 exists as an SCF at temperatures and pressures above its CP. Do not confuse the CP with the TP where solid, liquid, and gas phases exist simultaneously. The TP for CO_2 is at $-56.6°C$ and 5.11 atm. The CP of methane (CH_4) is at 190.4 K and 4.60 MPa (45.4 atm).

sustainable development: "Improving the quality of human life while living within the *carrying capacity* of supporting ecosystems" (Leape and Humphrey 2010).

Sustainocene: Term coined by Ranganathan and Irwin (2011) to describe an epoch the Anthropocene epoch will morph into. In the *Sustainocene epoch*, all humans will have to unite to actively care for planetary ecosystems, freshwater resources, GHGs, and so forth, in order to survive. It may never occur.

swidden agriculture (also called milpa, slash-and-burn, and forest-fallow cultivation): Plots of forestland are cleared, and the cleared land is planted on for a number of years or used for grazing cattle. As yields decline and weeds encroach, the plot is abandoned, and the forest is allowed to regrow until fully established. This fallow period may be as long as 25 years before the forest is again cut and burned (Tainter 1988). The charcoal from the burning enriches the soil as *terra preta*.

synchronization (of N oscillating systems): Frequency synchronization causes two or more oscillators to oscillate at exactly the same frequency. Phase-lock synchronization causes two or more oscillators to oscillate with the same frequency, in phase.

syngas (synthesis gas): A mixture of carbon monoxide, hydrogen, and carbon dioxide (CO, H_2, CO_2). (Syngas can also contain methane, depending on the process that created it.) It can be used to synthesize methanol or ethanol by the process of catalytic hydrogenation of CO_2 with H_2 and to produce various alkanes (leading to diesel fuel) by the *Fischer–Tropsch process*. One means of producing syngas is by pyrolysis of organic materials (including household garbage) with electric plasma torches. *Geoplasma*, a startup firm in Atlanta, GA, is a leader of the plasma torch pyrolysis-syngas technology. Syngas can be used as a feedstock in a number of non-FF (NFF) chemical synthesis processes.

synthetic biology: An emerging area of *biological engineering*. The name *synthetic biology* was coined in 1974 by the Polish geneticist Waclaw Szybalski. It was used again in 1978 by Szybalski and Skalka. One of the aims of synthetic biology is to create artificial gene circuits that perform designated functions outside the normal capabilities of a host organism (e.g., a bacterium). Synthetic biological gene systems have been designed to be "orthogonal," that is, to be substantially independent of the gene systems of the host organism. That is, orthogonal messenger RNA (*o*-mRNA) and *o*-ribosome pairs are used to make new proteins [and consequently, genetically modified organisms (GMOs)] that have utility in various fields of human endeavor (e.g., medicine, agriculture, materials, etc.). *o*-Ribosomes generally do not recognize endogenous mRNAs.

system: A group of interacting, interrelated, or interdependent *elements* (also: *parts, entities, states*) forming or regarded as forming a collective entity. There are many definitions of *system*, which are generally context dependent. (This definition is very broad and generally acceptable.)

system boundary: The notional dividing line between a system and its system environment (and other systems).

system environment: The system's niche. Defined as: That set of *entities* outside the *system boundary*, the state of which set is affected by the system or which affects the state of the system itself. Think arctic ice cap and polar bears.

temporary energy storage: Energy production methods such as by SC, wind turbines, tidal turbines, and wave generation are not constant. To be effective, their peak energy outputs must be stored and the stored energy released into the power grid in off-peak times.

One of the oldest means of energy storage is pumped hydroelectric storage (PHS), where water is pumped by electric pumps to a reservoir above the river, giving the water in the reservoir potential energy. The water is released during peak electric demand times and drives turbogenerators connected to the distribution network. PHS is expensive to implement but relatively cheap to operate. Typically, up to 2 GW can be stored.

A second means of short-term energy storage is by electrochemical batteries. Local solar and wind systems generally use batteries. DC from the batteries must be converted to line frequency AC when the batteries are supplying energy to the local grid. Electrochemical batteries are relatively expensive, require maintenance, and have finite lives.

A third means of short-term storage is by compressed air energy storage. Caverns produced by dissolved mineral mining (e.g., salt mines) are generally airtight and can be filled with compressed air by pumps driven by excess energy from solar and wind sources, as well as conventional sources during off-peak times. Compressed air storage has the potential to store 50–300 MW. Stored air pressure drives turbogenerators during peak-power-demand periods.

A fourth means of short-term storage is by heat. In solar–thermal power systems, excess solar energy can be stored as heat in water, certain (molten) salts, and rocks.

A fifth means of short-term energy storage is by the KE of a spinning flywheel.

Yet another proposed means of storing energy is in chemical form: for example, excess DC power from solar and wind sources can be used to separate H_2 and O_2 gasses from water by electrolysis. They can be used in fuel cells or in direct combustion to produce heat for turbogeneration.

terra preta (literally, "black earth" in Portuguese): A type of very dark, fertile anthropogenic soil originally found in the Amazon Basin. It was made by adding ground charcoal, bone, and manure to the otherwise relatively infertile soil there. Terra preta has 13%–14% organic matter. Transforming nutrient-poor soil to terra preta requires mixing in charcoal derived from biochar (low-temperature charcoal), or from HTC, plus other organic nutrients.

thermophotovoltaic (TPV) energy conversion: A direct energy conversion process that converts a thermal differential to near infrared (NIR) photoelectrons, producing electric power. A TPV system consists basically of an *absorber–emitter material* that absorbs thermal radiation (e.g., BB radiation from the sun) and converts it to IR photons that power a *PV power converter* (PPC). Most TPV systems also include additional components such as concentrators, filters, and reflectors.

A thin slab of an emitter material [e.g., polycrystalline silicon carbide, tungsten, rare earth oxides (ceria), photonic crystals, etc.] traps source long-wave IR radiation (LIR) energy and rises to an equilibrium operating temperature of approximately 900°C to 1300°C. The absorber–emitter can be heated by sunlight or combustion. By etching or depositing structures on the surfaces of the emitter, it can be given the property of *controlled incandescence* (Greffet 2011), where it radiates photon energy to the PPC mostly in the NIR band of wavelengths where it has maximum PV sensitivity. A wide variety of PPCs can be made from such materials as crystalline silicon, germanium, gallium antimonide, indium gallium arsenide antimonide, indium gallium arsenide, and indium phosphide arsenide antimonide.

time constant: If a quantity x decays exponentially from an initial value, X_o, according to the equation

$$x(t) = X_o e^{-t/\tau}$$

τ is the time constant; the time it takes $x(\tau) = 0.36788X_o = X_o e^{-1}$.

time constant form (of a rational polynomial in s): For example, the transfer function $H(s)$ in Laplace form is written as

$$H(s) = \frac{K(s+a)}{s^2 + 2\xi\omega_n s + \omega_n^2} \tag{GL44}$$

In *time constant form*, $H(s)$ is written as

$$H(s) = \frac{\left(Ka/\omega_n^2\right)(s\tau + 1)}{s^2/\omega_n^2 + s2\xi/\omega_n + 1} \tag{GL45}$$

where $\tau = 1/a$.

The time constant form is used for Bode plotting; the Laplace form is used to find the output time function by using the inverse Laplace transform.

tipping point (in CNLSs): The tipping point of a CNLS occurs when a critical set of initial conditions and/or inputs is reached such that the system rapidly changes its overall input/output dynamic behavior. In the *phase plane,* this change is indicated by a switch from one *attractor* to another. *One example of a tipping point* is in global warming climate change. The slow buildup of manmade atmospheric GHGs (e.g., CO_2, CH_4, etc.) has caused a slow increase of global temperature from trapped solar IR radiation. This temperature increase causes the slow melting of arctic permafrost, the arctic ice cover, and glaciers. Melting permafrost releases more methane and carbon dioxide into the atmosphere; melting ice sheets and glaciers expose more IR-absorbing ground (causing a decrease in arctic albedo), so more heat is absorbed, accelerating the melting process. These events can be viewed as local positive feedback (autocatalytic) processes, which accelerate the process of global warming, hence global temperature rise. A tipping point for these processes occurs at some critical atmospheric CO_2 concentration. Of course, many other parameters affect global climate, including atmospheric dust and SO_3 gas from volcanism, solar energy output, alteration of heat-carrying ocean currents by fresh melt water, and so forth. See also **Hopf bifurcation**.

Another ecological tipping point has recently occurred off the coast of Namibia. Namibia once had a large and profitable sardine fishery. Sardines eat planktons. The Namibian sardine population fell abruptly from overfishing, leading to a plankton bloom. The planktons died, and their bodies settled on the sea floor. Soon, bacterial decomposition of the dead planktons led to the release of an extensive volume of methane gas and also poison hydrogen sulfide gas (H_2S). These gasses dissolved in seawater create a hostile environment for sardines *and* planktons, so the sardine population was prevented from recovering. If the sardine catch had been regulated, there would have been a sustainable catch and no plankton superbloom.

ton (aka short ton): 2000 lb. or 907.18474 kg.

tonne (aka metric ton): A unit of mass equal to 1000 kg = 2204.6 lb. 1 tonne of CO_2 contains $12/44 = 0.2727$ tonnes of carbon.

Tragedy of the Commons: The title of a prescient paper written in 1968 by Garrett Hardin. The paper was about finite resources and population control and used the analogy of "tragedy of freedom in a commons." A commons was a large pasture open to all farmers in a village, common in England in the 17th–19th centuries. A number of herdsmen graze their flocks (of cattle, horses, sheep, goats, etc.) on the commons. It is expected that each herdsman tries to keep as many animals as possible on the commons. Hardin points out that such an arrangement may work reasonably satisfactorily for centuries because wars, poaching, famine, and disease keep the numbers of both men and animals well below the carrying capacity of the land. Finally, social stability leads to tragedy in the following way: As a rational, self-interested person, each herdsman seeks to maximize his gain. He (and his fellow herdsmen) ask: "What is the utility *to me* of adding one more animal to my herd?" This utility has a negative and a positive component. The positive component is a function of the benefits from one animal; since the herdsman receives all the proceeds generated by the new animal, the positive utility approaches +1. The negative component is a function of the additional overgrazing by the animal. Since these negative effects are shared by all the herdsmen, the negative utility for the animal's

owner is only a fraction of –1. Adding up the component partial utilities, the herdsman and his fellows all pursue the strategy of adding
another animal to their herds, then another, and so forth. Quoting
Hardin: "Each man is locked into a system that compels him to
increase his herd without limit—in a world that is limited." This is
the tragedy—"Freedom in a commons brings ruin [unsustainability]
to all."

transmittance: In absorption spectrophotometry, a beam of monochromatic
light at wavelength λ with optical power P_{in} is passed through a
sample cell of length L containing the analyte. The emergent beam
has optical power P_{out}. The transmittance is defined simply as $T(\lambda) \equiv$
P_{out}/P_{in}. The percent transmittance is defined as $\%T(\lambda) = 100\, P_{out}/P_{in}$.
See also **absorbance**.

transpiration (in plants): The process of photosynthesis requires carbon
dioxide and water. Plants transport the water their roots take up
from the soil up their stems in xylem to the leaves where photosynthesis takes place. An excess of water is transported to the leaves;
the portion not involved in photosynthesis is lost from the leaves
through regulatory pores called *stomata*. This water loss, which
includes the phase change of liquid water to vapor, is called transpiration. "Stomata control not only the amount of water lost by
leaves but also the amount of carbon gained. Thus stomatal function
often represents a compromise between two conflicting demands"
(Schulze et al. 1987).

An acre of corn (maize) transpires approximately 11,400–15,100
L/day, and a large oak tree can transpire approximately 151,000 L/
year (Perlman 2011). The stomata in the leaves of most plants close
at night, preventing transpiration (water loss from the plants) and
turning off CO_2 flux into the leaves. Plants do not photosynthesize at
night.

transport lag (TL; in systems; aka delay, or dead time): A TL is when a train
enters a tunnel and emerges τ minutes later, unchanged. As a physiological example, a TL can be assigned to the time it takes a nerve
impulse to propagate down an axon or the time it takes a hormone
molecule released into the blood by the adrenal gland to reach the
heart. A TL can be modeled in the time domain as $\mathbf{y}(t) = \mathbf{x}(t - \tau)$. That
is, the TL operator output $\mathbf{y}$ is the input $\mathbf{x}$ delayed by τ seconds.

tree (in graph theory): A connected, acyclic, *simple graph*. A tree has no path
loops.

trillion: 1 trillion = 10^{12}.

tropopause: The top of the Earth's atmospheric troposphere, approximately
12 km above the surface.

total fertility rate (TFR): The average number of children that *would be born
to a woman over her lifetime* if (1) she were to experience the exact current age-specific fertility rates (ASFRs) through her lifetime and (2)

she were to survive from birth through the end of her reproductive life. It is obtained by summing the single-year age-specific rates at a given time. TFR is also called the *fertility rate, period TFR* (PTFR), or *total period fertility rate* (TPFR).

ultrastable system (complex interacting systems): A term coined by W. R. Ashby. Two CSs of continuous variables interact (one is an *environment*, or *niche*, and the other is an *organism* or *reacting part*). The interaction is in the form of two feedback modalities—the first is a rapid, *primary feedback* (through complex sensory and motor channels). The second feedback works intermittently at a slower speed from the environment to certain continuous variables in the organism. These continuous variables affect the values of *step mechanisms* in the organism only when their values fall outside set threshold limits (upper or lower thresholds or "windows"). The changes in the step mechanisms determine how the organism reacts to its environment. If environmental conditions become adverse (but not fatal), an ultrastable organism may modify its environment or move out of it to a more suitable environment. See **complex adaptive system**.

unit impulse (a system input): Used to characterize linear and NL systems. Mathematically, a unit impulse, $\delta(t - \tau)$, occurs at $t = \tau$ and is zero elsewhere. As a rectangular pulse, its height is $1/\varepsilon$, and its width is ε, giving it unity area. In the limit as $\varepsilon \to 0$, the pulse becomes infinitely tall and has 0 width, and its area is unity: $\int_0^\infty \delta(t-\tau)dt \equiv 1.0 U(t-\tau)$, beginning at $t \geq \tau$, else 0. A *linear* system's response to a unit impulse (called its weighting function) completely characterizes its dynamics.

utility (in economics): The condition or quality of being useful; usefulness.

van der Pol equation: An NL, second-order differential equation with oscillatory (periodic) solutions as an IVP: It is written as

$$\ddot{x} + \dot{x}\mu(x^2 - 1) + x = 0, \quad \text{given ICs on } \dot{x} \text{ and } x.$$

In SV form, it is

$$\dot{x} = \mu[y - f(x)], \quad \dot{y} = -x/\mu, \quad f(x) = (x^3/3 - x)$$

See Northrop (2011), Section 3.4.3.

vertex (in graph theory; the same as a node): A fundamental unit out of which graphs are formed. The vertices or nodes in a graph are connected by edges (branches), which can be directed or undirected. In *SFGs*, the nodes have values representing the *states* of the system that the graph models. For example, they can represent chemical concentrations in a biochemical system or voltages and currents in an electronic system. A *source vertex* is a vertex with only directed branches leaving it. A *sink vertex* has no directed branches leaving it.

vertex distance (in graph theory): See **distance**.

volt-ampere: AC power measure. A volt-ampere is a vector quantity, having a real and imaginary part. VAR stands for volt-ampere reactive, the imaginary component of a generator's load, for instance. For maximum generator efficiency, it is desired to minimize a generator's VAR output. The *real power* supplied from a generator is $P = VI \cos\varphi$ watts, where VI is the output volt-amperes, and φ is the phase angle between output voltage V and current I. P is obviously maximum when V and I are in phase ($\varphi = 0$).

walk (in graph theory): A walk from node i to node j is an alternating sequence of nodes and edges (a sequence of adjacent nodes) that begins with i and ends with j. The *length l of a walk* is the number of edges in the sequence. A walk is *closed* if its first and last vertices are the same, and *open* if they are different. For an open walk, $l = N - 1$, where N is the number of vertices visited (a vertex is counted each time it is visited). $l = N$ for a closed walk. A *cycle* is a closed walk of at least three nodes in which no edge is repeated. (Compare **walk** with **SFG path**.)

water footprint, annual: The total volume of water in a year used globally to produce the goods and services consumed by its inhabitants (or volume per capita). The *external water footprint* is the fraction of the water footprint due to the consumption of imported goods (i.e., the water used in the country that exported the goods to produce them).

watt: SI unit of power, the rate of doing work. 1 W = 1 J/s.

Wiener index, W (a complexity measure in graph theory): The sum of distances over all pairs of vertices in an undirected graph:

$$W \equiv 1/2 \sum_{ij=1}^{N} d_{ij} = d_{11} + d_{12} + d_{13} + \cdots + d_{1N} + d_{21} + d_{22} + d_{23} + \cdots \quad \text{(GL46)}$$
$$+ d_{2N} + \cdots + d_{N1} + d_{N2} + \cdots d_{NN}$$

where $d_{kk} \equiv 0$, and $d_{jk} = d_{kj}$; $1 \le k,j \le N$

work (in physics): *Mechanical work* is done when a force moves a mass through a distance. This may be expressed by the differential vector equation $dW = \mathbf{F} \cdot d\mathbf{x}$. So $W = \int dW = \int_{x_1}^{x_2} F \cos\theta \, dx$ for a 2-D vector space. θ is the angle between the path of the mass and the force vector, $\mathbf{F}$. The SI units of work are newton-meters, or joules. Electrical work W_e is done when a DC source supplies a current I_L at a constant electromotive force (EMF) E_s to a load over a time T. $W_e = I_L E_s T$ watt-seconds = joules. Several other units of work are used: 1 J = 0.7376 ft.-lb. 1 kWh = 60,000 J. In centimeter-gram-second (CGS) units, 1 dyne-cm = 1 erg = 10^{-7} J. The basic dimensions of work are ML^2T^{-2}.

xerophytes: Plants that can thrive under low water conditions. Examples include aloe, cactus, jade plant, yucca, and so forth.

xylan: A linear polymer formed by covalent, oxygen-linked D-xylose sugar molecules (a glycan). It has the formula $[C_5H_8O_4]_N$. Many strands of xylan are cross-linked by hydrogen bonding, similar to cellulose. Xylan is found in the walls of plant cells.

Younger Dryas: A geologically brief (ca. 1300 years) period of cold climatic conditions between 12,800 and 11,500 years before present. Its cause may have been the shutdown of the North Atlantic Conveyor (NAC) current that brings warm tropical water northward (the Gulf Stream is part of the NAC system). The NAC shutdown may have been due to cold freshwater from glacier and ice cap melting interrupting the thermohaline circulation.

Zooxanthellae (aka Symbiodinium): A genus of unicellular algae from the phylum *Dinoflagellata*. These symbiotic unicellular algae commonly reside in the endoderm of tropical cnidarians such as corals, sea anemones, and various jellyfish species. They also can be found in sponges, flatworms, giant clams, *foraminifera*, and some ciliates. They pass on their products of photosynthesis (O_2, sugars) to their host and, in turn, receive inorganic nutrients from metabolic waste of their host (e.g., CO_2, NH_4^+). They are the most abundant eukaryotic microbes found in coral reef ecosystems. Coral "bleaching" is due to a loss of these pigmented algae from adverse environmental conditions (high water temperature, UV radiation, low salinity, etc.).

Bibliography and Recommended Reading

Abelard. 2003. Global warming: A briefing document. 16 pp. Web paper. Available at: www.abelard.org/briefings/global_warming.htm, accessed December 8, 2004.

ABM Software. 2010. Comparison of agent-based modeling software. Available at: en.wikipedia.org/wiki/Comparison_of_agent-based_modeling software, accessed February 11, 2010.

Abta, A., A. Kaddar & H.T. Alaoui. 2008. Stability of limit cycle in a delayed IS-LM business cycle model. *Appl. Math. Sci.* 2(50): 2459–2471.

AEBIOM. 2007. Pellets for small-scale domestic heating systems. *Sixth Framework Programme.* 16 pp. Available at: www.erec-renewables.org/fileadmin/erec_docs/Project_Documents/RESTMAC/Pellets_small_ scale_heat.pdf, accessed July 5, 2012.

AFDC. 2012. *Hydrogen Basics.* US DoE Alternative Fuels Data Center. Available at: www.afdc.energy.gov/fuels/hydrogen_basics.html, accessed January 27, 2013.

Agency for Toxic Substances & Disease Registry. 2011. Nitrate/nitrite toxicity: What are the physiological effects of exposure to nitrates/nitrites? Web published by the U.S. Centers for Disease Control & Prevention. Available at: www.atsdr.cdc.gov/csem/nitrate/no3physiologic_effects.html, accessed February 19, 2011.

Aksglaede, L., K. Sørensen, J. Petersen, N. Skakkebæk & A. Juul. 2009. Recent decline in age at breast development: The Copenhagen Puberty Study. *Pediatrics* 123(5): e932–e939. doi:10.1542/peds.2008-2491.

Albert, R. 2005. Scale-free networks in cell biology. *J. Cell Sci.* 118: 4947–4957.

Alexopoulos, C.J., C.W. Mims & M. Blackwell. 1996. *Introductory Mycology.* New York, NY, USA: John Wiley & Sons. ISBN: 0-471-52229-5.

Algae. 2011. Keeping heads above water. Web news brief, 25 July 2011. Available at: news.nationalgeographic.com/news/2011/07/pictures/110725-algae-china-beaches-qingdao-swimming-science-environment-world/, accessed August 2, 2011.

Algae Oil. 2011. Algae oil extraction. 3 pp. Web article. Available at: www.oilgae.com/algae/oil/extract/extract.html, accessed October 25, 2011.

Ali, A. 2011. High bacteria levels found in meat. *Reuters*, 15 April 2011. Available at: www.reuters.com/article/2011/04/15/us-usa-meat-bacteria-idUSTRE73E80D20110415, accessed April 18, 2011.

Al-Kaisi, M. 2001. Soil erosion and crop productivity: Topsoil thickness. *Integr. Crop Manag.* 486(1): 11. Web published by Iowa State University. Available at: www.ipm.iastate.edu/ipm/icm/2001/1-29-2001/topsoilerosion.html, accessed February 19, 2011.

Allesina, S., A. Bodini & M. Pascual. 2009. Functional links and robustness in food webs. *Phil. Trans. Royal Soc. B* 364: 1701–1709.

Allsopp, M.H., W. J. de Lange & R. Veldtman. 2008. Valuing insect pollination services with cost of replacement. *Plos ONE* 3(9): September 2008. e3128. 8 pp.

Alon, U. 2003. Biological networks: The tinkerer as an engineer. *Science* 301: 1866–1867.

Alon, U. 2007. *An Introduction to Systems Biology*, Boca Raton, FL: Chapman & Hall/ CRC. ISBN 1-58488-624-0.

Andreev, V.M., V.P. Khvostikov, O.A. Khvostikova, A.S. Vlasov, P.Y. Gazaryan, N.A. Sadchikov & V.D. Rumyantsev. 2005. Solar thermophotovoltaic system with high temperature tungsten emitter. *Proc. 31st IEEE Photovoltaic Specialists Conference and Exhibition, 2005*. St. Petersburg, FL. 4 pp. Available at: ieeexplore.org/xpls/ abs_all.jsp?arnumber=1488220&tag=1, accessed January 16, 2012.

Angus-Monash. 2011. Why is economics not an evolutionary science? 34 pp. Web presentation. Available at: users.monash.edu.au/~sangus/cgi-bin/moinres. cgi/sangus?action=AttachFile&do=get&target=Angus-MonashEconomics-ResearchSeminary.pdf, accessed October 7, 2011.

Antonietti, M. & M.M. Titirici. 2008. Hydrothermal carbonization of biomass: Black carbon with refined structure without charring. 30 pp. Slide show. Available at: www.biochar-international.org/images/Antonietti_HTChouston2008.pdf, accessed January 31, 2011.

Antúnez, K., R. Martín-Hernández, L. Prieto, A. Meana, P. Zunino & M. Higes. 2009. Immune suppression in honey bee *(Apis mellifera)* following infection by *Nosema ceranae* (Microsporidia). *Environ. Microbiol.* doi:10.1111/j.1462-2920.2009.01953.x.

Argonne National Laboratory. 2005. *Nitrate and Nitrite: Human Health Fact Sheet*. Web published by the Environmental Sciences Division of the U.S. Department of Energy, August 2005. Available at: www.ead.anl.gov/pub/doc/nitrate-ite.pdf, accessed February 19, 2011.

Arnoux, R. 2011. Spring is blooming, construction is booming. Online news release, 5 April 2011. Available at: www.iter.org/newsline/171/668, accessed April 8, 2011.

Arora, N. 2011. Gluten intolerance: Why so many, why so common? 3pp web article. Available at: www.naturopathicliving.com/cms/gluten-intolerance-why-so-many-why-so-common/, accessed June 5, 2012.

Arthur, W.B. 1999. Complexity and the economy. *Science* 284: 107–109.

Arthur, W.B., S.N. Durlauf & D.A. Lane. 1997. *The Economy as an Evolving Complex System II*. Addison-Wesley & The Santa Fe Institute, Reading, MA.

Ashby, W.R. 1958. Requisite variety and its implications for the control of complex systems. *Cybernetica* 1(2): 83–89. Available at: pespmc1.vub.ac.be/Books/ AshbyReqVar.pdf, accessed November 20, 2009.

ATMI. 2012. ATMI Begins commercialization of new carbon dioxide capture system co-developed with SRI International. Press release, 28 June 2012. Available at: www.globenewswire.com/newsroom/news.html?d=260604, accessed July 2, 2012.

Ausra. 2007. An introduction to solar thermal electric power. 11 pp. Web tutorial paper. Available at: www.ausra.com/pdfs/SolarThermal101_final.pdf, accessed April 20, 2011.

Axelrod, R. & L. Tesfatsion. 2009. On-line guide for newcomers to agent-based modeling in the social sciences. 16 pp. Available at: www.econ.iastate.edu/tesfatsi/ abmread.htm, accessed March 15, 2010.

Ay, N., E. Olbrich, N. Bertschinger & J. Jost. 2006. A unifying framework for complexity measures of finite systems. Available at: sbs-net.sbs.ox.ac.uk/complexity_ PDFs/ECCS06/Conference_Procedings/PDF/p202.pdf, accessed December 3, 2009.

Bade Shrestha, S.O., G. Narayanan & G. Narayanan. 2008. Landfill gas with hydrogen addition as a fuel for SI engines. *Fuel* 87(17/18): 3616–3626.

Badgley, C., J. Moghtadera, E. Quinteroa, E. Zakema, M.J. Chappella, K. Avilés-Vázqueza, A. Samulona & I. Perfectoa. 2007. Organic agriculture and the global food supply. *Renewable Agric. Food Syst.* 22(2): 86–108.

Bahcall, R. 2000. How the sun shines. Online paper on the official Nobel Prize Web site. Available at: www.nobelprize.org/nobel_prizes/physics/articles/fusion/, accessed August 17, 2011.

Balboa, B. 2012. Cellulosic ethanol on verge of production. 3 pp. Online news bulletin. Available at: www.feedandgrain.com/news/10656675/cellulosic-ethanol-on-verge-of-production, accessed May 2, 2012.

Banerjee, D., G.A. Strobel et al. 2010. An endophytic *Myrothecium inundatum* producing volatile organic compounds. Web published 30 October 2010 by Mycosphere. Available at: www.mycosphere.org/pdfs/MC1_3_No6.pdf, accessed September 2, 2011.

Banerjee, N. & R.D. White. 2011. Turmoil in OPEC nation drives oil prices up sharply. *Los Angeles Times.* 22 Feb. 2011. Available at: articles.latimes.com/2011/feb/22/business/la-fi-oil-20110222, accessed January 23, 2013.

Bannerjee, A. & E. Duflo. 2011. More than 1 billion people are hungry in the world*. *But what if the experts are wrong? *Foreign Policy* May–June: 67–72.

Barnes, F.S. & J.G. Levine. 2011. *Large Energy Storage Systems Handbook.* CRC Press, Boca Raton, FL. ISBN: 9781420086003.

Barnes. 1983. *Handbook of Infrared Radiation Measurements.* 81 pp. Barnes Engineering Company, Stamford, CT.

Barnett, W.A., C. Deissenberg & G. Feichtinger. 2004. *Economic Complexity: Non-Linear Dynamics, Multi-Agents Economics, and Learning.* 492 pp. Amsterdam, Nethderlands: Elsevier Science. ISBN: 9780444514332.

Barrett, J. 2005. Airborne bacteria in CAFOs: Transfer of resistance from animals to humans. *Environ. Health Perspect.* 113(2): February. Published on the NIH Web site. Available at: www.ncbi.nlm.nih.gov/pmc/articles/PMC1277892/pdf/ehp0113-a0116b.pdf, accessed January 17, 2011.

Barrett-Connor, E., C. Garland, J.B. McPhillips, K.T. Khaw & D.L. Wingard. 1990. A prospective, population-based study of androstenedione, estrogens, and prostatic cancer. *Cancer Res.* 50: 169.

Barriopedro, D., E.M. Fischer, J. Luterbacher, R.M. Trigo & R. Garcia-Herrera. 2011. The hot summer of 2010: Redrawing the temperature record map of Europe. 32 pp. Published 17 March 2011 on *Science Express.* doi:10.1126/science.1201224. Available at: www.sciencemag.org/cgi/content/full/science.1201224/DCI, accessed March 13, 2012.

Bartlett, A.A. 1997. Is there a population problem? *Wild Earth* 7(3): 88–90. Available at: www.ecofuture.org/pop/rpts/bartlett_pop_prob.html, accessed December 29, 2010.

Bartlett, A.A. 1998. Reflections on sustainability, population growth and the environment. 25 pp. Web paper. Available at: www.populationress.org/essays/essay-bartlett-pop.html, accessed December 8, 2010.

Basin. 2011. Ivanpah solar electric generating system: Our comments of the draft environmental impact statement/final staff assessment. 9 pp. Web critique. Available at: www.basinandrangewatch.org/IvanpahFSA-reliability.html, accessed April 20, 2011.

Bauer, D.M. & I.S. Wing. 2010. Economic consequences of pollinator declines: a synthesis. *Agric. Resour. Econ. Rev.* 39(3) (October 2010): 368–383.

BBC. 2010. Satellite spies vast algal bloom in Baltic Sea. BBC Web site. Available at: www.bbc.co.uk/news/science-environment-10740097, accessed July 23, 2010.

Beckham, R. 2011. Raising and feeding mealworms. 3 pp. Web article. Available at: www.efinch.com/, accessed May 13, 2011.

Beckie, H.J., K.N. Harker, A. Légère, M.J. Morrison, G. Séguin-Swartz & K.C. Falk. 2011. GM canola: the Canadian experience. *Farm Policy Journal* 8(1): 43–49.

Bedi, E. & H. Falk. 2000. Solar thermal power production. 13 pp. Web tutorial paper. Available at: energy.saving.nu/solarenergy/thermal.shtml, accessed April 20, 2011.

Beespotter. 2007. The economic importance of bees. 5 pp. Web article. Available at: beespotter.mste.illinois.edu/topics/economics/, accessed July 5, 2012.

Benjaminson, M.A., J.A. Gilchrest & M. Lorenz. 2002. In vitro edible muscle protein production system (MPPS): Stage 1, fish. *Acta Astronaut.* 51(12): 879–889.

Bennett, J.E. 1996. Chapter 9. Antifungal agents. In: Hardman, J.G., Limbird, L.E., Goodman, L.S. & Gilman, A., eds. 1996. *Goodman & Gilman's the Pharmacological Basis of Therapeutics*, 9th ed. pp. 1175–1190. New York, New York, USA: McGraw-Hill.

Benson, S.M. & D.R. Cole. 2008. CO_2 sequestration in deep sedimentary formations. *Elements* 4: 325–331. Available at: 171.66.125.216/cgi/reprint/4/5/325, accessed June 27, 2011.

Berge, N.D. et al. 2011. Hydrothermal carbonization of municipal waste streams. *Environ. Sci. Technol.* 45: 5696–5703.

Bermel, P. et al. 2011. Tailoring photonic metamaterial resonances for thermal radiation. *Nanoscale Res. Lett.*, 6: 549. 5 pp. Available at: www.nanoscalereslett.com/content/pdf/1556-276X-6-549.pdf, accessed January 12, 2012.

Bielo, D. 2008. Fertilizer runoff overwhelms streams and rivers—Creating vast "dead zones." *Scientific American*, 14 March 2008, p. 12.

Biello, D. 2010a. What the frack? Natural gas from subterranean shale promises U.S. energy independence—With environmental costs. *Scientific American*, March 30, 2010, 3 pp. Available at: www.scientificamerican.com/article.cfm?id=shale-gas-and-hydraulic-fracturing, accessed January 3, 2011.

Biello, D. 2010b. What causes the North Atlantic plankton bloom? *Scientific American*, 14 April 2010. 2 pp. Available at: www.scientificamerican.co/article.cfm?id = north-atlantic-plankton-bloom, accessed May 17, 2011.

Biobest. 2011. Bumblebees. Web article and ad. Available at: www.intertechserv.com/pollination.htm, accessed March 22, 2011.

Biodiesel. 2012a. National Biodiesel Board (NBB) home page with numerous hyperlinks. Available at: www.biodiesel.org/, accessed June 1, 2012.

Biodiesel. 2012b. Biodiesel FAQ's. 3 pp. Web article. Available at: www.biodiesel.org/what-is-biodiesel/biodiesel-faq's, accessed June 1, 2012.

Biodiesel. 2013. Production statistics. Available at: www.biodiesel.org/production/production-statistics, accessed January 23, 2013.

Bioenergie. 2007. Bioenergie: Grosses Potentzial für Pellets-Märkte in Europa und weltweit. Available at: www.solarserver.de/news/news-7242.html, accessed July 5, 2012.

Bioenergy. 2011. Q & A about Miscanthus. 3 pp. Available at: /bioenergy.ornl.gov/papers/miscanthus/miscanthus.html, accessed November 30, 2011.

Biofuels. 2007. The facts about biofuels: Ethanol from cellulose. 3 pp. FAQ Web document. Available at: www.energyfuturecoalition.org/biofuels/fact_ethanol_cellulose.htm, accessed September 2, 2011.

Birkett, J. & J. Lester. 2003. *Endocrine Disruptors in Wastewater and Sludge Treatment Processes*. Lewis Publishers, CRC Press, London.

Blasing, T.J. 2010, updated 2012. Recent greenhouse gas concentrations. 5 pp. Web paper. Available at: cdiac.ornl.gov/pns/current_ghg.html, accessed December 8, 2010.

Blazer, V.S., L.R. Iwanowicz, D.D. Iwanowicz, D.R. Smith, J.A. Young, J.D. Hedrick, S.W. Foster & S.J. Reeser. 2007. Intersex (testicular oocytes) in smallmouth bass from the potomac river and selected nearby drainages. *J. Aquat. Anim. Health* 19: 242–253. Available at: http://antietamflyanglers.org/docs/jaahintersex.pdf, accessed January 19, 2013.

Blehert, D.S., J.M. Lorch, A.E. Ballmann, P.M. Cryan & C.U. Meteyer. 2011. Bat White-Nose Syndrome in North America. *Microbe* 6(6): 267–273. Available at: http://www.microbemagazine.org/images/stories/images/june_2011/znw00611000267.pdf, accessed January 19, 2013.

Bloom, D.E. 2011. 7 billion and counting. *Science* 333: 462–569.

BLS. 2010. Consumer price index—All urban consumers. US Bureau of Labor Statistics. Available at: data.bls.gov/PDQ/servlet/SurveyOutputServlet, accessed February 19, 2010.

Blue whale. 2008. Blue whale population. 2 pp. Available at: www.blue-whale.info/Blue_Whale_Population.html, accessed January 3, 2011.

Boa, E. 2004. Wild edible fungi: A global overview of their use and importance to people. UN FAO document. Available at: www.fao.org/docrep/007/y5489e/y5489e05.htm#P138_20454, accessed September 20, 2008.

Bobleter, O. 1994. Hydrothermal degradation of polymers derived from plants. *Prog. Polym. Sci.* 19(5): 797–841.

Boccaletti, S., V. Latora, Y. Moreno, M. Chavez & D.-U. Hwang. 2006. Complex networks: Structure and dynamics. *Phys. Rep.* 424: 175–308. Available at: ftp://ftp.elet.polimi.it/users/Carlo.Piccardi/VarieDsc/BoccalettiEtAl2006.pdf, accessed January 19, 2013.

BOM. 2012. Monitoring the weather: Air movement along isobars. Web article by the Australian Bureau of Meteorology—Forecasting the weather. Available at: www.bom.gov.au/info/ftweather/page_14.shtml, accessed May 18, 2012.

Bonabeau, E. 2002. Agent-based modeling: Methods and techniques for simulating human systems. *PNAS*. 99(suppl. 3): 7280–7287.

Bonchev, D. 2004. Complexity analysis of yeast proteome network. *Chem. Biodivers.* 1: 312–326.

Borch, T. 2010. Steroid hormone runoff from an agricultural field applied with biosolids. Presentation published online by Colorado State University. Available at: www.cwi.colostate.edu/(S(dgtvi045gveiv24513j1ak45))/Workshops/WaterScienceDay/Presentations/Borch.pdf, accessed January 29, 2011.

Boucher, O., P. Friedlingstein, B. Collins & K. Shine. 2009. The indirect global warming potential and global temperature change potential due to methane oxidation. *Environ. Res. Lett.* 4: 044007 (5pp). Available at: http://iopscience.iop.org/1748-9326/4/4/044007/pdf/1748-9326_4_4_044007.pdf, accessed January 19, 2013.

Bourne, J.K. Jr. & R. Clark. 2007. Green dreams. *National Geographic,* October 2007, p. 41.

Bourouni, K., M.T. Chaibi & A.A. Taee. 2001. Water desalination by humidification and dehumidification of air. *Desalination.* 137(1–3): 167–176.

Bourzac, K. 2011. The rare earth crisis. *Technol. Rev.* 114(3, Special Issue on Emerging Technologies): 58–63.

Bowles, S. & H. Gintis. 2002. *Homo reciprocans. Nature* 415: 125–128.

Bowman, B., W. Knotts & K. Swope. 2011. Solar thermal electric power plants. 11 pp. Student survey paper. Available at: me1065.wikidot.com/solar-thermal-electric-power-plants, accessed February 6, 2012.

BP. 2011. *BP Energy Outlook 2030.* 80 pp. Online booklet. BP, London, UK. Available at: www.bp.com/liveassets/bp_internet/globalbp/globalbp_uk_english/reports_and_publications/statistical_energy_review_2011/STAGING/local_assets/pdf/2030_energy_outlook_booklet.pdf, accessed January 27, 2012.

Bradsher, K. 2011. A radical kind of reactor. 4 pp. Web news reprint. *The New York Times,* 25 March 2011. Available at: www.nytimes.com/2011/03/25/business/energy-environment/25chinanuke.html?pagewanted=all, accessed October 25, 2011.

Brain, M. & R. Lamb. 2000, 2011. How nuclear power works. Web article. Available at: www.science.howstuffworks.com/nuclear-power.htm, accessed April 8, 2011.

Brandt, A.R. 2008. Converting oil shale to liquid fuels: Energy inputs and greenhouse gas emissions of the Shell in situ conversion process. *Environ. Sci. Technol.* 42: 7489–7495.

Brauer, F. & C. Castillo-Chávez. 2001. *Mathematical Models in Population Biology and Epidemiology.* New York: Springer.

Breifing. 2011. World: Automotive outlook. 13 pp. Web article from the Economist Intelligence Unit, 7 January 2011. Available at: www.eiu.com/index.asp?layout=ib3PrintArticle&article_id=1547747539&printer=printer, accessed July 6, 2011.

Brennan, S.R. & J. Withgott. 2007. Chapter 6. Environmental systems: Connections, cycles, and feedback loops. In: *Environment: The Science Behind the Stories,* 6th ed. 641 pp. Benjamin Cummings. ISBN: 0805344276. Available at: www.aw-bc.com/scp/brennan/assetts/downloads/ch06.pdf, accessed August 6, 2012.

Briefing. 2011. World: Automotive outlook. 13pp web article from the Economist Intelligence Unit.7 January 2011. Available at: www.eiu.com/index.asp?layout=ib3PrintArticle&article_id=1547747539&printer=printer, accessed July 6, 2011.

British Broadcasting Corporation. 2010. Satellite spies vast algal bloom in Baltic Sea. Published on BBC Web site, 23 July 2010. Available at: www.bbc.co.uk/news/science-environment-10740097.

Brock, W.A. & A.G. Malliaris. 1992. *Differential Equations, Stability and Chaos in Dynamic Economics,* Ist repr. Amsterdam, Netherlands: Elsevier Science, B.V.

Brodie, C. 2004. Geometry and patterns in nature 1: Exploring the shapes of diatom frustules with Johan Gielis' superformula. 9 pp. *Micscape Magazine,* April 2004. Available at: www.microscopy-ik.org.uk/mag/artapr04/cbiatom2.html, accessed January 9, 2012.

Bromenshenk, J.J., C. Henderson, C. Wick, M. Stanford, A. Zulich, R. Jabbour, S. Deshpande, P. McCubbin, R. Seccomb, P. Welch, T. Williams, D. Firth, E. Skowronski, M. Lehmann, S. Bilimoria, J. Gress, K. Wanner & R. Cramer. 2010. Iridovirus and microsporidian linked to honey bee colony decline. *PLoS ONE* 5(10): e13181. 11 pp. Available at: plosone.org/article/info%3Adoi%2F10.1371%2Fjournal.pone.0013181, accessed March 28, 2011.

Brown, L.R. 2011. The new geopolitics of food. *Foreign Policy,* May–June 2011, pp. 54–62. Available at: www.foreignpolicy.com/articles/2011/04/25/the_new_geopolitics_of_food, accessed March 25, 2012.

Browne, M.W. 1989. Physicists debunk claim of a new kind of fusion. *New York Times,* 3 May 1989. 5 pp. Available at: partners.nytimes.com/library/national/science/050399sci-cold-fusion.html, accessed August 2, 2012.

Buchanan, M. 2009. Meltdown modelling: Could agent-based computer models prevent another financial crisis? *Nature* 460(7256): 680–682.

Bullis, K. 2008. Ethanol from garbage and old tires. 2 pp. Web article. Available at: http:/pages.uoregon.edu/recycle/events_topics_Ethanol_text.htm, accessed January 24, 2011.

Bumblebee. 2011. Bees are responsible for pollinating plants that provide much of our food. 2 pp. Web paper. Available at: www.bumblebee.org/economic.htm, accessed March 22, 2011.

Burgermeister, J. 2009. Austria flexes its bioenergy muscles. *Renewable Energy World*, 20 April 2009. Available at: www.renewableenergyworld.com/rea/news/article/2009/04/austria-flexes-its-bioenergy-muscles, accessed July 5, 2012.

Burkhardt, U. & B. Kärcher. 2011. Global radiative forcing from contrail cirrus. *Nat. Clim. Chang.* 1: 54–58. doi:10.1038/NCLIMATE1068.

Burlington. 2012. Joseph C. McNeil Generating Station. 4pp web information bulletin, v. 6/25/12. Available at: www.burlingtonelectric.com/page.php?pid=75&name=mcneil, accessed June 26, 2012.

Business Bestseller. 2000. Review of Frederic Vester's book: *The Art of Networked Thinking—Ideas and Tools for a New Dealing with Complexity.* Book Review in English from *Business Bestseller,* No. 10/00. 12 pp. Available at: www.frederic-vester.de/eng/books/complete-review/, accessed March 14, 2011.

Buyer's Guide. 2009. Buyer's guide to pellet- and wood-burning stoves: The pros, cons and costs vs. natural gas, oil, and coal. *Consumer Reports,* August 2009. Available at: www.consumerreports.org/cro/appliances/heating-cooling-and-air/wood-stoves/buyers-guide-to-pellet-and-wood-burning-stoves-1-07/overview/0701_pellet-stove.htm, accessed December 8, 2011.

Cai, Z., R. Clarke, B. Glowacki, W. Nuttall & N. Ward. 2010. Ongoing ascent to the helium production plateau—insights from system dynamics. *Resour. Policy* 35: 77–89.

Caldwell, C. 1995. Graph theory glossary. On-line tutorial. Available at: www.utm.edu/departments/math/graph/glossary.html, accessed September 7, 2009.

California. 2010a. Ivanpah solar electric generating system. 5 pp. Online posting of Hearing Docket No. 07-AFC-5C. Available at: www.energy.ca.gov/sitingcases/ivanpah/index.html, accessed April 19, 2011.

California. 2010b. Large solar energy projects. 5 pp. Information bulletin of California Department of Energy. Available at: www.energy.ca.gov/siting/solar/index.html, accessed April 20, 2011.

California Environmental Protection Agency. 2010. Proposition 65: Request for relevant information on chemicals being considered for listing by the authoritative bodies mechanism. November 30, 2010. Available at: oehha.ca.gov/prop65/CRNR_notices/admin_listing/requests_info/dci_abpkg43_112610.html, accessed January 29, 2011.

Cameron, S.A., J.D. Lozier, J.P. Strange, J.B. Koch, N. Cordes, L.F. Solter & T.L. Griswold. 2010. Patterns of widespread decline in North American bumble bees. 6 pp. *PNAS Early Edition.* Available at: www.pnas.org/cgi/doi/10.1073/pnas.1014743108/, accessed March 23, 2011.

Campbell, D.T. 1976. Assessing the impact of planned social change. Paper #8, *Occasional Paper Series,* Dartmouth College Public Affairs Center, Hanover, NH. 70 pp. Available at: www.wmich.edu/evalctr/pubs/ops/ops08.pdf, accessed February 17, 2010.

Cannon, R.H. Jr. 1967. *Dynamics of Physical Systems*. New York, New York, USA: McGraw-Hill Book Co.

Capes, E.M., R. Loaiza & H. Valdivia. 2011. Ryanodyne receptors. *Skelet. Muscle* 1: 18. 13 pp.

CAPP. 2008. Environmental challenges and progress in Canada's oil sands. Canadian Association of Petroleum Producers. Available at: www.capp.ca/getdoc. aspx?DocID = 135721, accessed May 9, 2011.

Carafa, R., S. Dueri & J.-M. Zaldívar. 2007. Linking terrestrial and aquatic ecosystems: Complexity, persistence and biodiversity in European food webs. *European Commission, Joint Research Centre Report*. ISSN: 1018-5593, ISBN: 978-92-79-06931-4. 40 pp. Available at: publications.jrc.ec.europa.eu/repository/bitstream/111111111/10719/1/eur_eu_ecosystems_final.pdf, accessed October 4, 2011.

CCC. 2010. Civilian Conservation Corps (CCC), 1933–1941. Web essay. Available at: www.u-s-history.com/pages/h1586.html, accessed January 4, 2010.

CDC. 2011. Measuring radiation. 3 pp. Available at: emergency.cdc.gov/radiation/measurement.asp, accessed April 11, 2011.

CDC. 2012. Trends in foodborne illness in the United States, 1996–2010. Available at: www.cdc.gov/foodborneburden/trends-in-foodborne-illness.html (page last updated 5/22/12), accessed June 9, 2012.

CDIAC. 2011a. Frequently asked global change questions. 15 pp. FAQ Web document from the Carbon Dioxide Information Analysis Center (CDIAC) at ORNL. Available at: cdiac.ornl.gov/pns/faq.html, accessed June 30, 2011.

CDIAC. 2011b. Recent greenhouse gas concentrations. 5 pp. Web data. Available at: cdiac.ornl.gov/pns/current_ghg.html, accessed June 30, 2011.

CDIAC. 2011c. Global CO_2 emissions from fossil-fuel burning, cement manufacture, and gas flaring. Data tables, June 10, 2011. Available at: cdiac.ornl.gov/ftp/ndp030/global.1751_2008.ems, accessed May 30, 2012.

Cellstrom. 2010. Vanadium redox flow battery. 2 pp. Web article. Available at: www.cellstrom.com/Technology.7.0.html?&L=1, accessed March 18, 2011.

Centre. 2012. Biomass energy in Canada. Online report. Available at: www.centreforenergy.com/AboutEnergy/Biomass/Overview.asp?page=6, accessed July 7, 2012.

Chakravarty, S.R. & A. Majumder. 2005. Measuring human poverty: A generalized index and an application using basic dimensions of life and some anthropometric indicators. *J. Hum. Dev.* 6(3): 283. Available at: biblioteca.hegoa.ehu.es/system/ebooks/15362/original/Measuring_Human_Poverty._A_Generalized_Index. pdf, accessed October 17, 2011.

Chambers, W. & R. 1896. *Chambers's Encyclopedia: A Dictionary of Universal Knowledge*. pp. 252–253. Available at: books.google.com/books?id=4pwMAAAAYAAJ&pg=PA252, accessed July 2, 2012.

Chandler, D.L. 2011. A novel way to concentrate sun's heat. *MIT News*. Available at: web.mit.edu/newsoffice/2011/thermo-photovoltaics-1202.html, accessed December 31, 2011.

Charles, D. 2011. A tale of two seed farmers: Organic vs. engineered. Published online by National Public Radio, 25 January 2011. Available at: www.npr.org/2011/01/25/133178893/a-tale-of-two-seed-farmers-organic-vs-engineered, accessed February 24, 2011.

ChemistryViews. 2010. Stop evaporation losses. 1 p. Web bulletin. Available at: www. chemistryviews.org/details/news/861957/Stop_Evaporation_Losses.html, accessed September 3, 2012.

ChemistryViews. 2011. New drought-tolerant crops. 1 p. Web bulletin. Available at: www.chemistryviews.org/details/news/1308511/New_Drought-Tolerant_ Crops.html, accessed September 3, 2012.

Cheung, W.-H. & H. Kung. 1994. *Methanol Production and Use.* New York, New York, USA: Marcel Dekker, Inc. ISBN-13: 978-0824792237.

Cheze, B., P. Gastineau & J. Chevallier. 2011. Forecasting world and regional aviation jet fuel demands to the mid-term (2025). *Energy Policy.* doi:10.1016/ j.enpol.2011.05.049. 13 pp.

Christie, E. 2009. Carbon capture and storage: Selected economic and institutional aspects. The Vienna Institute for International Economic Studies, Research Report No. 360, December 2009. 46 pp. Available at: www.wiiw.ac.at/mod- Publ/download.php?publ=RR360, accessed July 4, 2011.

Chu, J. 2009. Cellulosic ethanol on the cheap. 2 pp. Web article. Available at: www.tech nologyreview.com/printer_friendly_article.aspx?id=22673, accessed January 24, 2011.

CIA. 2012. Country comparison, total fertility rate, *World Fact Book.* Available at: www.cia.gov/library/publications/the-world-factbook/rankorder/2127rank. html, accessed June 23, 2012.

Clark, J. 2004. Phase diagrams of pure substances. 13 pp. Web tutorial. Available at: www. chemguide.co.uk/physical/phaseeqia/phasediags.html, accessed July 14, 2011.

Clark, J. 2002. The Haber process. 6 pp. Web tutorial. Available at: www.chemguide. co.uk/physical/equilibria/haber.html, accessed July 28, 2011.

Cleveland, C. & A. Roman. 2007. Orinoco Heavy Oil Belt, Venezuela. 2 pp. Web article from the *Encyclopedia of Earth.* Available at: www.eoearth.org/article/Orinoco_ Heavy_Oil_Belt,_Venezuela, accessed March 16, 2011.

Club of Rome. 2011. Rapid acceleration of methane release emissions come as a shock. 2 pp. News release 12/21/11. Available at: www.clubofrome.org/?p= 3401&print=1, accessed January 16, 2012.

Cohen-Kupiec, R. & I. Chet. 1998. The molecular biology of chitin digestion. *Curr. Opin. Biotechnol.* 9: 270–277.

Coliban. 2006. Coliban water evaporation reduction trial using Watersavr. A case study. 9 pp. Available at: www.flexiblesolutions.com/products/watersavr/documents/ TheColbanTrial-WaterSavrCaseStudy.pdf, accessed September 3, 2012.

Competing. 2012. Competing schools of economic thought. Available at: www.scribd. com/doc/47720473/Competing-Schools-of-Economic-Thought, accessed July 5, 2012.

Condon, T. 2008. The Dust Bowl: A cautionary tale. Op ed article in the *Hartford Courant*, 24 August 2008. p C4.

Consumer Reports. 2010. How safe is that chicken? *Consumer Reports Magazine*, January 2010. Available at: www.consumerreports.org/cro/magazine- archive/2010/january/food/chicken-safety/overview/chicken-safety-ov.htm, accessed January 17, 2011.

Cook, E.R., R. Seager, R.R. Heim Jr., R.R., Vose, R. S., Herweijer, C. & Woodhouse, C. 2009. Megadroughts in North America: Placing IPCC projections of hydroclimatic change in a long-term paleoclimate context. *J. Quat. Sci.* 25: 48–61. ISSN: 0267-8179.

Cook, J. 2011. From May 1958 to May 2011, atmospheric CO_2 up 24.1% (a record high). 1 p. Web article. Available at: irregulartimes.com/index.php/archives/2011/06/14/from-may-1958-to-may-2011-atmospheric-co2-up-24-1-a-record-high/, accessed June 27, 2011.

Corning, P.A. 1998. Complexity is just a word. Web paper. Available at: www.complexsystems.org/commentaries/jan98.html, accessed November 16, 2009.

Corps of Engineers. 2011. Water budget in the United States. 4 pp. Web article. Available at: www.nwrfc.noaa.gov/info/water_cycle/hydrology.cgi, accessed July 18, 2011.

Costa, S. (Inventor). 2009. Whey-based fungal microfactory technology for enhanced biological pest management using fungi. *Non-Confidential Invention Disclosure* (patent pending). University of Vermont Office of Technology Commercialization. Available at: www.uvm.edu/~uvmpr/theview/article.php?id=2238, accessed May 5, 2012.

Cowan, L. 2008. How heat pumps Work. 7 pp. Web paper. Available at: ome.howstuffworks.com/home-improvement/heating-and-cooling/heat-pump.htm/printable, accessed February 14, 2011.

Crawford, D. 1992. Modularization and McCabe's cyclomatic complexity. *Commun. ACM*. 35(12): 17–19.

Crickets. 2011. How to raise your own crickets. 4 pp. Web article. Available at: www.wikihow.com/Raise-Your-Own-Crickets, accessed May 13, 2011.

Crowe, B. 1969. The tragedy of the commons revisited. (Reprinted in *Managing the Commons*, by G. Hardin & J. Baden. W.H. Freeman, 1977; ISBN: 0-7167-0476-5). Available at: dieoff.org/page95.htm/, accessed January 3, 2011.

CSCC. 2012. Spain is the world leader in solar thermal energy as of year-end 2011. Published online, 11 January 2012, by the California-Spain Chamber of Commerce. Available at: www.californiaspainchamber.org/index.php?option=com_k2&view=item&id=105:spain-is-the-world-leader-in-solar-thermal-energy-as-of-year-end-2011&Itemid=46&lang=en, accessed July 4, 2012.

CSE Group. 2008. Complexity measures. 9 pp. Web paper. Available at: cse.ucdavis.edu/~cmg/Group/group_documents/MeasuresofComplexity.pdf, accessed December 1, 2009.

CSPI. 2011. 3-page letter + 14 anecdotal examples of Quorn allergic reactions. To the Deputy Commissioner for Foods, USFDA, from M.F. Jacobson, Executive Director, CSPI, 16 Nov. 2011. Available at: cspinet.org/new/pdf/quorn-letter-to-fda-nov-15-2011.pdf, accessed June 4, 2012.

Cuaresma, J.C., T. Palokangas & A. Tarasyev, eds. 2009. *Dynamic Systems, Economic Growth, and the Environment (Dynamic Modeling and Econometrics in Economics and Finance)*. 307 pp. New York, New York, USA: Springer. ISBN-13: 978-3642021312.

Cui, X., M. Antonietti & S.-H. Yu. 2006. Structural effects of iron oxide nanoparticles and iron ions on the hydrothermal carbonization of starch and rice carbohydrates. *Small* 2(6): 756–759.

Dai, A. 2010. Drought under global warming: A review. 21 pp. Web paper. Available at: onlinelibrary.wiley.com/doi/10.1002/wcc.81/pdf, accessed October 22, 2010.

Dailymail. 2011. North Sea 'closes' to avert cod shortage. 1 p. Press release. Available at: www.dailymail.co.uk/news/article-17695/North-Sea-closes-avert-cod-shortage.html, accessed August 15, 2011.

Das, T. & J.D. McCalley. 2012. Compressed air energy storage. 35 pp. Online technical paper. Available at: home engineering.iastate.edu/~jdm/wind/ Compressed%20Air%20Energy%20Storage_Chapter_TRISHNA%20DAS.pdf, accessed August 1, 2012.

Day, R.H. 1994. *Complex Economic Dynamics: An Introduction to Dynamical Systems and Market Mechanisms.* Vol. I. MIT Press, Cambridge, MA.

Deacon, J. 2005. *Fungal Biology.* Blackwell Publishers, Cambridge, MA. ISBN: 1-4051-3066-0.

DeFoliart, G.R. 1992. A concise summary of the general nutritional value of insects. Reprinted from *Crop Prot.* 11: 395–399. G.R. DeFoliart. Insects as human food... 1992. Available at: www.food-insects.com/Insects%20as%20Human%20Food. htm, accessed January 9, 2012.

DeFoliart, G.R. 1999. Insects as food: Why the western attitude is important. *Annu. Rev. Entomol.* 44: 21–50.

DeJong-Hughes, J., J.F. Moncrief, W.B. Voorhees & J.B. Swan. 2001. *Soil Compaction: Causes, Effects and Control.* Published online by the University of Minnesota Extension. Available at: www.extension.umn.edu/distribution/cropsystems/ dc3115.html, accessed February 19, 2011.

Demirbas, A. 2007. Importance of biodiesel as transportation fuel. *Energy Policy* 35: 4661–4670.

DEPweb. 2004. World Population Growth. Chapter III in: *Beyond Economic Growth,* 2nd ed. The World Bank Group. Online book. Available at: www.worldbank. org/depweb/english/beyond/global/chapter3.html, accessed February 8, 2012.

Deshpande, M.V. 1999. Mycopesticide production by fermentation: Potential and challenges. *Crit. Rev Microbiol.* 25(3): 229–243.

Diamond, J. 2005. *Collapse.* USA: Penguin, New York. ISBN: 0-14-30.3655-6.

Dittrick, P. *Oil Gas J.* 2012. DuPont lets engineering contract for Iowa cellulosic ethanol plant. 1p web article. Available at: www.ogj.com/articles/2012/07/dupont-lets-engineering-contract-for-iowa-cellulosic-ethanol-project._printArticle.html, accessed January 27, 2013.

Doak, T. 2012. *Symbiodinium* sp. 1 p. Online description. Available at: genome.wustl. edu/genomes/view/symbiodinium_sp/, accessed August 2, 2012.

DOE. 2012a. Biodiesel. Home page with numerous hyperlinks, v. 6/01/12. Available at: www.fueleconomy.gov/feg/biodiesel.shtml, accessed June 1, 2012.

DOE. 2012b. Hydrogen. Home page with numerous hyperlinks, v. 6/01/12. Available at: www.fueleconomy.gov/feg/hydrogen.shtml, accessed June 1, 2012.

DOE. 2012c. How fuel cells work. Home page with numerous hyperlinks, v. 6/01/12. Available at: www.fueleconomy.gov/feg/fcv_PEM.shtml, accessed June 1, 2012.

DOE. 2012d. Ethanol production and distribution. 2 pp. Information bulletin from US DOE Alternate Fuels Data Center. Available at: www.afdc.energy.gov/afdc/ fuels/ethanol_production, accessed September 10, 2012.

Donnelly, J. 2011. The Irish famine. *BBC History.* Available at: www.bbc.co.uk/his tory/british/victorians/famine_01.shtml, accessed June 23, 2012.

Doty Wind Fuels. 2009. Compressed air energy storage (CAES)—utilities and cars. 10 pp. Web paper (updated 12/08/09) by Doty Wind Fuels. Available at: dotyenergy.com/Markets/CAES.htm, accessed May 14, 2012.

Doty, F.D. 2004. A realistic look at hydrogen price projections. 12 pp. Web paper. Available at: www.dotynmr.com/PDF/Doty_H2Price.pdf, accessed June 9, 2011.

Doumit, M.E., D.R. Cook & R.A. Merkel. 1993. Fibroblast growth factor, epidermal growth factor, insulin-like growth factors, and platelet-derived growth factor-BB stimulate proliferation of clonally derived porcine myogenic satellite cells. *J. Cell. Physiol.* 157(2): 326–332.

Dörner, D. 1997. *The Logic of Failure: Recognizing and Avoiding Error in Complex Situations.* Basic Books. 240 pp. ISBN-10: 0201479486.

Dresser-Rand. 2010. Compressed air energy storage (CAES). 7 pp. Web article. Available at: www.dresser-rand.com/literature/general/85164-10-CAES.pdf, accessed May 14, 2012.

Duff, B. & M. Davila. 2005. Presence/absence of estrogen in the raccoon river watershed and Des Moines water works treatment process. Report published online by the Des Moines Waterworks Laboratory, August 2005. Available at: www.dmww.com/Laboratory/EstrogenReport.pdf, accessed February 6, 2011.

Dunmore, C. 2011. EU states can ban GM crops for public order. Reuters Web article, 3 February 2011. Available at: www.reuters.com/article/2011/02/03/us-eu-gmo-bans-idUSTRE71267M20110203/, accessed April 6, 2011.

Durlauf, S.N. 1997. What should policymakers know about economic complexity? Online paper. Available at: www.santafe.edu/research/publications/working papers/97-10-080.pdf, accessed January 19, 2013.

Dyni, J.R. 2010. Oil shale. In: Clarke, A.W. & Trinnaman, J.A. eds. 2010. *Survey of Energy Resources,* 22nd ed. pp 93–123. ISBN: 978-0-946121-02-1. Available at: www.worldenergy.org/documents/ser_2010_report.pdf, accessed January 19, 2013.

Easley, D. & J. Kleinberg. 2010. *Networks, Crowds and Markets: Reasoning about a Highly Connected World.* Cambridge University Press. Cambridge, UK.

Ebner, P. 2007. *CAFOs and Public Health: The Issue of Antibiotic Resistance.* Concentrated Animal Feeding Operations Series. Available at: www.extension.purdue.edu/extmedia/ID/cafo/ID-349.pdf, accessed January 17, 2011.

Econfaculty. 2010. Walter E. Williams. 2 pp. Biography. Available at: econfaculty.gmu.edu/wew/, accessed February 20, 2011.

Economic Complexity Observatory. 2011. Homepage. Available at: macroconnections.media.mit.edu/featured/economic-complexity-observatory/, accessed October 13, 2011.

Economic Situation. 2010. Global Outlook, Ch 1 in *World Economic Situation and Prospects 2011.* 46 pp. Pre-release of Chapter 1 of the World Economic Situation and Prospects 2011, issued 1 December 2010 in New York. UN Dept. of Economic and Social Affairs (DESA). Available at: www.un.org/en/development/desa/policy/wesp/wesp_current/2011wesp_prerelease1.pdf, accessed October 17, 2011.

Economist. 2007. Hands to the pump. Figure illustrating petrol consumption by country in 2003; appeared in *The Economist,* 3 July 2007. Available at: www.economist.com/images/ga/2007w27/Petrol.jpg, accessed July 6, 2011.

Economist. 2011a. Who rules the waves? *The Economist,* 12–18 February 2011. p. 90.

Economist. 2011b. Climate change in black and white. *The Economist,* 19–25 February 2011. pp. 89–91.

Economist. 2011c. The 2011 oil shock. *The Economist,* 5–11 March 2011. p. 13.

Economist. 2011d. The most surprising demographic crisis. *The Economist,* 7–13 May 2011. pp. 43–44.

Economist. 2011e. A man-made world. *The Economist,* 28 May 3–June 2011. pp. 81–83.

Economist. 2011f. A painful eclipse. *The Economist,* 15–21 October 2011. pp. 75–76.

Economist. 2011g. The thirsty road ahead. *The Economist,* 12–18 November 2011. pp. 40–41.

Economist. 2011h. Unquenchable thirst. *The Economist,* 19–25 November 2011. pp. 27–29.

Economist. 2011i. Reinventing the wheel. *The Economist,* 3–9 December 2011. pp. 24, 26.

Economist. 2011j. Revenge of the petrolheads. *The Economist,* 10–16 December 2011. p. 73.

Economist. 2011k. Christina the alchemist. *The Economist,* 5–11 November 2011. p. 98.

Economist. 2012a. Building a better suntrap. *The Economist,* 31 December 2011–6 January 2012. 2 pp. Available at: www.economist.com/node/21542157, accessed March 13, 2012.

Economist. 2012b. Flower power. *The Economist,* 21–27 January 2012. p. 91.

Economist. 2012c. Packing some power. *The Economist,* 3–9 March 2012. pp. 15–16.

Economist. 2012d. In a hole? *The Economist,* 17–23 March 2012. p. 90.

Economist. 2012e. A shiny new pipe dream. *The Economist,* 12–18 May 2012. pp. 84–55.

Economist. 2012f. Economic and financial indicators. *The Economist,* 11–17 August 2012. pp. 80–81.

Economist. 2012g. Special report: Natural gas. *The Economist,* 14–20 July 2012. 18 pp.

Economist. 2012h. Special report: The Arctic. by James Astill. *The Economist,* 16–22 June 2012. pp. 3–16.

Economy, E.C. 2010. *The River Runs Black,* 2nd ed. Cornell University Press, Ithaca, NY.

Edmonds, B. 1999a. Syntactic measures of complexity. PhD dissertation, University of Manchester, UK.

Edmonds, B. 1999b. What is complexity? The philosophy of complexity *per se* with application to some examples in evolution. In: Heylighen, F., Aerts, D., eds. *The Evolution of Complexity.* Kluwer, Dordrecht, Netherlands. Available at: cfpm. org/~bruce/evolcomp/, accessed December 3, 2009.

EFSA. 2009. Marine biotoxins in shellfish—Saxitoxin group. *EFSA J.* 1019: 1–76. Available at: www.efsa.europa.eu/en/efsajournal/doc/1019.pdf, accessed January 10, 2012.

Ehrlich, P.R. 1991. *The Population Explosion.* New York, New York, USA: Touchstone Books. ISBN-13: 978-0671732943.

Ehrlich, P.R. 1968. *The Population Bomb.* New York, New York, USA: Ballantine Books. ISBN-10: 1568495870.

Ehrlich, P.R. 2010. The MAHB, the culture gap, and some really inconvenient truths. *PLoS Biol.* 8(4): 3 pp. Available at: www.ncbi.nlm.nih.gov/pmc/articles/ PMC2850377/pdf/pbio.1000330.pdf/, accessed January 1, 6, 2011.

Ehrlich, P.R. & A.H. Ehrlich. 2009. The population bomb revisited. *Electronic J. Sustain. Dev.* 1(3): 63–71. Available at: 173-45-244-96.slicehost.net/docs/The_ Population_Bomb_Revisited.pdf, accessed May 18, 2012.

Ehrlich, P.R. & A.H. Ehrlich. 2010. The culture gap and its needed closures. *Int. J. Environ. Stud.* 67(4): 481–492.

Ehrlich, P.R. & A.H. Ehrlich. 2012. The population explosion. 3 pp. Web article. Available at: www.2think.org/tpe.shtml, accessed May 18, 2012.

Ehrlich, P.R. & J. Holdren. 1971. The impact of population growth. *Science* 171: 1212–1217.

EIA. 2011. *International Energy Outlook 2011.* US Energy Information Administration. Report No. DOE/EIA-0484(2011). Available at: www.eia.gov/forecasts/ieo/index.cfm, accessed January 26, 2012.

EIA. 2012. *Annual Energy Outlook 2012: With Projections to 2035.* ix + 252pp. Available at: www.eia.gov/f...o/pdf/038(2012).pdf, accessed January 25, 2013.

Electric Fuel. 2012. The zinc-air fuel cell system for electric vehicles. 3 pp. Web information brochure. Available at: www.electric-fuel.com/EV, accessed June 5, 2012.

Elert, G., ed. 2008. Volume of Earth's oceans. 4 pp. Fact sheet. Available at: hypertext book.com/2001/SyedQadri.shtml, accessed May 14, 2012.

Elkin, W.B. 1902. An inquiry into the causes of the decrease of the Hawaiian people. *Am. J. Sociol.* 8(3): 398–411.

Elsis, M.R. 2000. We have passed our sustainability. 3 pp + 3 tables. Web paper, 1 May 2000. Available at: www.overpopulation.net/, accessed May 30, 2012.

Emonet, T., C.M. Macal, M.J. North, C.E. Wickersham & P. Cluzel. 2005. Agent cell: A digital single-cell assay for bacterial chemotaxis. *Bioinformatics* 21(11): 2714–2721. Available at: http://emonet.biology.yale.edu/papers/2005_bioinformatics.pdf, accessed January 19, 2013.

Engineering ToolBox. 2012. Rechargeable batteries. Online table. Available at: www.engineeringtoolbox.com/rechargeable-batteries-d_1219.html, accessed August 1, 2012.

EPA. 2005. Emission facts: Average carbon dioxide emissions resulting from gasoline and diesel fuel. EPA420-F-05-001. 2 pp. Web article. Available at: www.epa.gov/otaq/greenhousegases.htm, accessed July 6, 2011.

EPA. 2008. Chapter 2. Landfill gas basics. In: *ATSDR 2008. Landfill Gas Primer—An Overview for Environmental Health Professionals.* pp. 3–14. Available at: www.eps.gov/lmop/documents/pdfs/pdh_chapter2.pdf, accessed June 15, 2011.

EPA. 2010. Methane and nitrous oxide emissions from natural sources. 19 pp. EPA report, EPA 430-R-10-001, April 2010.

EPA. 2011a. Landfill Methane Outreach Program: Basic information. 3 pp. Web paper. Available at: www.epa.gov/Imop/basic-info/index.html, accessed June 13, 2011.

EPA. 2011b. Investigation of ground water contamination near Pavillion, Wyoming. EPA Draft Report EPA 600/R-00/000. 121 pp. Available at: www.epa.gov/ord, accessed December 12, 2011.

EPA. 2012. Investigation of ground water contamination near Pavillion, Wyoming: Phase V sampling event: Summary of methods and results. Available at: www.epa.gov/region8/superfund/wy/pavillion/phase5/PavillionSeptember 2012Narrative.pdf, accessed January 24, 2013, accessed January 25, 2013.

Estabrook, B. 2012. Organic can feed the world. *The Atlantic* online. 3 pp. Article. Available at: www.theatlantic.com/health/archive/2011/12/organic-can-feed-the-world/249348/, accessed August 30, 2012.

EurActiv. 2010. Carbon capture and storage. 12 pp. Web paper. Available at: www.euractiv.com/en/climate-change/carbon-capture-storage/article-157806, accessed July 4, 2011.

European Union's Scientific Committee on Veterinary Measures Relating to Public Health. 1999. *Assessment of Potential Risks to Human Health from Hormone Residues in Bovine Meat and Meat Products.* 30 April 1999. Available at: ec.europa.eu/food/fs/sc/scv/out21_en.pdf, accessed January 29, 2011.

Evans, J.D., K. Aronstein, Y.P. Chen, C. Hetru, J.-L. Imler, H. Jiang, M. Kanost, G.J. Thompson, Z. Zou & D. Hultmark. 2006. Immune pathways and defence mechanisms in honey bees *Apis mellifera*. *Insect Mol. Biol.* 15(5): 645–656.

Exmoor Pharma Concepts. 2008. *The In Vitro Meat Consortium Preliminary Economics Study*, Project 29071 V5. Available at: http://www.newharvest.org/img/files/culturedmeatecon.pdf, accessed January 21, 2013.

Fairless. 2012. Fairless Hills Steam Generating Station. 2 pp. Web article. Available at: www.exeloncorp.com/powerplants/fairlesshills/Pages/profile.aspx, accessed August 27, 2012. Also see Using landfill gas to generate electricity. 3 pp. Article. Available at: www.exeloncorp.com/energy/generation/landfillgas.aspx, accessed August 27, 2012.

Fairley, P. 2011. Alberta's oil sands heat up. *Technol. Rev.* 114(6): 52–59.

Faiman, D. 2012. Solar energy in Israel. 5pp web article. Available at: www.jewishvirtuallibrary.org/jsource/Environment/Solar.html, June 4, 2012.

Fan, W.-Q. 2007. Atrazine-induced aromatase expression is SF-1 dependent: Implications for endocrine disruption in wildlife and reproductive cancers in humans. *Fam. Environ. Health Perspect.* 115(5): 720–727. doi:10.1289/ehp.9758.

FAO. 2010. Algae-based biofuels: Applications and co-products. 117 pp. Environment and Natural Resources Management Working Paper No. 44. Published online by the Food and Agriculture Organization of the United Nations. Available at: www.fao.org/docrep/012/i1704e/i1704e00.pdf, accessed July 12, 2012.

FAO. 2012. *Wood Energy*. Published online by the Food and Agriculture Organization of the United Nations. Available at: www.fao.org/forestry/energy/en/, accessed July 7, 2012.

Farber, S. 2000. Sweden to examine health risks of burning biomass. 1 p. Web press release. Available at: health.phys.iit.edu/extended_archive/0012/msg00062.html, accessed February 14, 2011.

Fargione, J., J. Hill, D. Tilman, S. Polasky & P. Hawthorne. 2008. Land clearing and the biofuel carbon debt. *Science* 319(5867): 1235–1238. Available at: www.sciencemag.org/cgi/reprint/319/5876/1235.pdf, accessed November 19, 2009.

Farmer, J.D. & D. Foley. 2009. The economy needs agent-based modelling. *Nature* 460(7256): 685–686.

Farrell, J. 2012. *Solar Grid Parity 101*. Available at: www.renewableenergyworld.com/rea/blog/post/2012/01/solar-grid-parity-101, accessed June 25, 2013.

Farwell, J. 2010. Pioneering tidal energy technology unveiled. 1 p. Press release, 19 February 2010. Available at: www.mainebiz.biz/news45921.html, accessed May 19, 2011.

Faupel, K. & A. Kurki. 2002. Biodiesel: A brief overview (ATTRA document). Available at: www.attra.ncat.org/attra-pub/PDF/biodiesel.pdf, accessed November 4, 2009.

FCBasics. 2011. Fuel cell basics. 2 pp. Web paper. Available at: www.fctec.com/fctec_types_pem.asp, accessed January 24, 2011.

Featherman, A. 2011. "Yuck" or "yum"?: Pondering lab-grown meat. 3 pp. Web article. Available at: greenanswers.com/blob/241176/%E2%80%9Cyuck%E2%80%9D-or-%E2%80%9Cyum%E2%80&9D-pondering-lab-grown-meat#ixzz1NckuxX8s, accessed December 1, 2011.

Federer, H.M., B.J. Johnson, S. O'Connell, E.D. Shapiro, A.C. Steere, G.P. Wormser, W.A. Agger, H. Artsob, P. Auwaerter, J.S. Dumler, J.S. Bakken, L.K. Bockenstedt, J. Green, R.J. Dattwyler, J. Munoz, R.B. Nadelman, I. Schwartz, T. Draper, E. McSweegan, J.J. Halperin, M.S. Klempner, P.J. Krause, P. Mead, M. Morshed,

R. Porwancher, J.D. Radolf, R.P. Smith, S. Sood, A. Weinstein, S.J. Wong & L. Zemel. 2007. A critical appraisal of "chronic Lyme disease." *N. Engl. J. Med.* 345: 1422–1430.

Feed. 2011. How to feed the world in 2050. 35 pp. UN FAO Report. Available at: www. fao.org/fileadmin/templates/wsfs/docs/expert_paper/How_to_Feed_the_ World_in_2050.pdf, accessed October 17, 2011.

Feldman, D.P. 2008. Theory and applications of complex networks. Class One, College of the Atlantic. 21 pp. Web presentation. Available at: hornacek.coa. edu/Teaching/Networks.08/feldman.lec.01.2008.pdf, accessed October 7, 2011.

Feldman, S. 2010. Fracking chemicals will be disclosed, drilling companies say. 4 pp. Web paper. Available at: www.truth-out.org/fracking-chemicals-will-be-dis closed-drilling-companies-say73217, accessed December 29, 2010.

Fernández-del Castillo, C., G. Robles-Diaz, V. Diaz-Sanchez & A. Altamirano. 1990. Pancreatic cancer and androgen metabolism: high androstenedione and low testosterone serum levels. *Pancreas* 5(5): 515–518.

Ferreira, M.L. Jr. 2008. CrossFire fusion reactor. 6 pp. Disclosure of patent pending PCT/IB2008/054254. Available at: www.crossfirefusion.com/nuclear-fusion-reactor/overview.html, accessed April 8, 2011.

Ferrentino, J.M., I.H. Farag, & L.S. Jahnke. 2006. Microalgal oil extraction and in-situ transesterification. 7 pp. Web report. Available at: www.ntnu.no/users/skoge/prost/ proceedings/aiche-2006/data/papers/P69332.pdf, accessed October 25, 2011.

Fertguide. 2011. Chapter 3. Managing algal productivity. In *Fert Guide.* pp. 16–35. Available at: pdacrsp.oregonstate.edu/pubs/fertguide_PDF/chapter_3_of_ fert_guide.pdf, accessed October 25, 2011.

Ffestiniog. 2012. Ffestiniog Power Station. 4pp web article. Available at: www.fhc. co.uk/ffestiniog.htm, accessed June 8, 2012.

Fischer, D. 2011. A detailed look at America's poisonous mushrooms. Available at: americanmushrooms.com/toxicms.htm, accessed July 8, 2012.

Fitzgerald, A.E. & C. Kingsley Jr. 1952. *Electric Machinery.* New York, New York, USA: McGraw-Hill Book Co.

Flory, A.R., Kumar, S., Stohlgren, T.J. & Cryan, P.M. 2012. Environmental conditions associated with bat white-nose syndrome mortality in the north-eastern United States. *J. Appl. Ecol.* 49(3): 680–689. doi:10.1111/j.1365-2664.2012.02129.x. 10 pp.

Flottum, K. 2010. The grove gamble: Will there be enough bees to pollinate this spring? 2 pp. Web article. Available at: www.thedailygreen.com/environmental-news/ blogs/bees/2011-almond-crop-0705, accessed March 13, 2012.

Food Insecurity. 2010. The state of food insecurity in the world. 60 pp. UN FAO Report. ISBN: 978-92-5-106610-2. Available at: www.fao.org/docrep/013/ i1683e/i1683e.pdf, accessed October 17, 2011.

FoodWeb3D. 2004. Images of ecosystem webs. Available at: peacelab.cloudapp.net/ gallery_index.html, accessed October 3, 2011.

Forrester, J.W. 2003. Dynamic models of economic systems and industrial organizations. *Syst. Dyn. Rev.* 19(4): 331–345. (Reprinted in *SDR* from the 1956 manuscript of JWF.) doi:10.1002/sdr284.

Forster, J., A.G. Hirst & D. Atkinson. 2011. How do organisms change size with changing temperature? The importance of reproductive method and ontogenic timing. *Funct. Ecol.* 25: 1024–1031. Available at: onlinelibrary.wiley.com/ doi/10.1111/j.1365-2435.2011.01852.x/pdf, accessed October 20, 2011.

Forster, P.M., V. Ramaswami, P. Artaxo, T. Berntsen, R. Betts, J. Haywood & R. Van Dorland. 2007. Changes in atmospheric constituents and in radiative forcing. In: *Climate Change 2007: The Physical Science Basis*. Contribution of Working Group I to the Fourth Assessment Report of the Intergovernmental Panel on Climate Change (IPCC). [Solomon, S. et al. eds.] Cambridge University Press, NY and UK. 106 pp. Available at: www.ipcc.ch/pdf/assessment-report/ar4-wg1-chapter2.pdf, accessed June 19, 2012.

Foster, J. 2004. From simplistic to complex systems in economics. Discussion Paper No. 335, October 2004, School of Economics, the University of Queensland, St. Lucia, Australia.

Foster, L.J. 2011. Interpretation of data underlying the link between CCD and an invertebrate iridescent virus. *Mol. Cell. Physiol. MCP Papers in Press*. Published 4 January 2011 as MS # O110.0066387. Available at: www.mcponline.org/content/early/2011/01/04/mcp.O110.006387.full.pdf, accessed June 15, 2011.

Franken, M. 2006. Germany's giant. *New Energy: Magazine for Renewable Energy*, May 2006, p. 24. Available at: www..newenergy.info/index.php?id=1266, accessed July 4, 2012.

Friedman, T.L. & M. Mandelbaum. 2011. America really was that great (but that doesn't mean we are now). *Foreign Policy*, November 2011, pp. 76–78.

Friedrich, K. & P. Bissoli. 2011. Analysis of temperatures and precipitation recorded at stations in Eastern Europe during the heat wave in summer 2010. 11 pp. Online report by *Deutscher Wetterdienst*. Available at: www.dwd.de/bvbw/generator/DWDWWW/Content/Oeffentlichkeit/KU/KU2/KU23/rcc-cm/products/SWE/European/20110124_Hitzwelle_Russland, accessed March 13, 2012.

Fuel Cells. 2010. What automakers are saying about fuel cell vehicles... 3 pp. Web article. Available at: www.fuelcells.org/wp-content/uploads/2012/02/automaker_quotes.pdf, accessed May 16, 2012.

Fuel Cell Today. 2011. *The Fuel Cell Today Industry Review, 2011*. Report published online. Available at: www.fuelcelltoday.com/media/1351623/the_industry_review_2011.pdf, accessed July 7, 2012.

Fulmer, M. 2000. Taco Bell recalls shells that used bioengineered corn. *Los Angeles Times*, 23 September 2000.

Funes, P. 2008. Complexity measures for complex systems and complex objects. (On-line class notes.) 12 pp. Available at: www.cs.brandeis.edu/~pablo/complex.maker.html, accessed November 16, 2009.

Funke, A. & F. Zeigler. 2011. Heat of reaction measurements for hydrothermal carbonization of biomass. *Biores. Technol.* 102: 7595–7598.

Gaia, K. 2010. Population: The last taboo. *Mother Jones Magazine*, May/June 2010. 11 pp. Available at: www.overpopulation.org/solutions.html, accessed December 8, 2010.

Galbraith, K. 2010. How bad is the Ogallala Aquifer's decline in Texas? 3 pp. Web article in the *Texas Tribune*, 17 June 2010. Available at: www.texastribune.org/texas-environmental-news/water-supply/how-bad-is-the-ogallala-aquifer-decline-in-texas/, accessed March 14, 2011.

GAO. 2010. Energy–water nexus: A better and coordinated understanding of water resources could help mitigate the impacts of potential oil shale development. US Government Accountability Office Report GAO-11-35, October 2010.

Gardner, J.L., A. Peters, M.R. Kearney, L. Joseph & R. Heinsohn. 2011. Declining body size: a third universal response to warming? *Trends Ecol. Evol.* 26: 285–291.

gas2.org. 2009. Complete list of cellulosic ethanol plants operating or under-construction in the US. 3pp. Available at: http://gas2.org/2009/03/05/complete-list-of-cellulosic-ethanol-plants-operating-or-under-condtruction-in-the-us/, accessed January 27, 2013.

Gasland. 2010. Hydraulic fracturing FAQs. 2 pp. Available at: gaslandthemovie.com/whats-fracking, accessed December 29, 2010.

Geballe, T. 2011. IR transmission spectra. Gemini Observatory WV data. Available at: www.gemini.edu/?q=node/10789, accessed August 4, 2011.

Gee, D. 2002. The Pebble Bed Modular Reactor. 12 pp. *EEE 460 Web Project,* Spring 2002. Available at: www.eas.asu.edu/~holbert/eee460/dfg/index.html, accessed October 28, 2011.

Gehrman, E. 2007. The unintended consequences of holding people accountable. Web essay. Available at: www.iq.harvard.edu/news/unintended_consequences_holding_people_accountable, accessed February 17, 2010.

Gell-Mann, M. 2010. Transformations of the twenty-first century: Transitions to greater sustainability. Ch 1 in: Schellnhuber, H. J., Molina, M., Stern, N., Huber, V. & Kadner, S. (eds.) (2010). *Global Sustainability: A Nobel Cause.* Cambridge, England: Cambridge University Press.

Georgescu-Roegen, N. 1971. *The Entropy Law and the Economic Process.* Harvard University Press, Cambridge, MA.

Geothermal. 2009. Geothermal economics 101: Economics of a 35 MW binary cycle geothermal plant. Available at: www.glacierpartnerscorp.com/geothermal.php, accessed October 17, 2009.

GFDL. 2011. Climate impact of quadrupling CO_2: An overview of Gfdl climate model results. 8 pp. Web paper. Available at: www.gfdl.noaa.gov/climate-impact-of-quadrrupling-co2, accessed June 30, 2011.

GHD. 2003. Methods for reducing evaporation from storages used for urban water supplies. Australian Department of Natural Resources and Mines: Final Report. March 2003. Available at: ncea-linux.usq.edu.au/farmdammanagement/images/stories/publications/DNRM_2003.pdf, accessed September 4, 2012.

Gilbert, N. 2007. *Agent-Based Models.* Sage Publications, London. ISBN: 978-1-4129-4964-4.

Gilbert, N. & P. Terna. 1999. How to build and use agent-based models in social science. Web paper. Available at: web.econ.unito.it/terna/deposito/gil_ter.pdf, accessed December 10, 2009.

Gillam, C. 2008. U.S. organic food industry fears GMO contamination. Published online by *Reuters,* March 12, 2008. Available at: www.reuters.com/article/2008/03/12/us-biotech-crops-contamination-idUSN1216250820080312, accessed February 24, 2011.

Gladwell, M. 2002. *The Tipping Point.* New York: Little, Brown & Co.

Glantz, M., ed. 1989. The Ogallala Aquifer depletion. 4 pp. Article from *Forecasting by Analogy: Societal Responses to Regional Climatic Change.* Summary Report of Environmental and Societal Impacts Group NCAR. Available at: www.iitap.iastate.edu/gccourse/issues/society/ogallala/ogallala.html, accessed March 14, 2011.

Gliński, Z. & K. Buczek. 2003. Response of the *Apoida* to fungal infections. *Apiacta.* 38: 183–189. Available at: www.apimondia.org/apiacta/articles/2003/glinski_1.pdf, accessed March 23, 2011.

Global. 2011. The global carbon cycle. 4 pp. Web article. Available at: www.global-greenhouse-warming.com/global-carbon-cycle.htm, accessed June 30, 2011.

Global Carbon Emissions. 2010. Global carbon emissions. 3 pp. Data sheet. Available at: co2now.org/Current-CO2/CO2-Now/global-carbon-emissions.html, accessed May 30, 2012.

Global Crop Diversity Trust. 2011a. Svalbard Global Seed Vault. Published online. Available at: www.croptrust.org/main/arcticseedvault.php?itemid=211, accessed February 24, 2011.

Global Crop Diversity Trust. 2011b. Svalbard Global Seed Vault: Resources. Published online. Available at: www.croptrust.org/main/resources.php, accessed February 24, 2011.

Global Envision 2011. Test-tube meat: Could it feed the world one day? Global Envision. Available at: www.globalenvision.org/topics/agriculture?page=1, accessed February 3, 2012.

Globe. 2011. An introduction to the global carbon cycle. 12 pp. Web paper. Available at: globecarboncycle.unh.edu/CarbonCycleBackground.pdf, accessed June 30, 2011.

Godfrey, K. 1983. *Compartmental Models and Their Application.* Academic Press, London.

Gomez, N. & M. Belda. 2010. Demand for lignocellulose biomass in Europe. Report published online by Elobio, a project of the EU. Available at: www.elobio.eu/fileadmin/elobio/user/docs/Additional_D_to_WP3-Lignocellulosic_biomass.pdf, accessed July 7, 2012.

Goodland, R. & J. Anhang. 2009. Livestock and climate change. 10 pp. Paper in the November/December 2009 online journal, *World Watch.* Available at: www.worldwatch.org/files/pdf/Livestock%20and%20Climate%20Change.pdf, accessed August 27, 2012.

Goodscience. 2011. Nutritional value of mopane worms. 4 pp. Web paper. Available at: informalscientist.com/tag/nutritional-value-of-mopane-worms/, accessed May 13, 2011.

Goodyear, D. 2011. Grub. *The New Yorker*, 15 & 22 August 2011. pp. 38–46.

Goulson, D. 2003. *Bumblebees: Their Behaviour and Ecology.* New York, New York, USA: Oxford University Press. ISBN: 0-9552211-0-2.

Govindarajan, L., N. Raut & A. Alsaeed. 2009. Novel solvent extraction for extraction of oil from algae biomass grown in desalination reject stream. *J. Algal Biomass Util.* 1(1): 18–28.

Grains Council. 2013. Sorghum handbook. 14 pp on-line information bulletin. Available at: www.grains.org/images/stories/technical_publications/Sorghum_Handbook.pdf, accessed January 24, 2013.

Greco, L. 1995. Why so many intolerant to gluten? 6 pp. Web paper. Available at: www.celiac.com/articles/76/1/WhySoManyIntolerantToGluten?, accessed June 5, 2012.

Greenpeace Canada. 2007. Water depletion. 25 October 2007. Available at: www.greenpeace.org/canada/en/campaigns/tarsands/threats/water-depletion, accessed February 12, 2012.

Green Car Congress. 2012a. DuPont breaks ground on commercial-scale cellulosic biorefinery in Iowa. 2 pp. Web article. Available at: www.greencarcongress.com/2012/11/dupont-20121130.html#more, accessed January 27, 2013.

Green Car Congress. 2012b. GraalBio to build $145M cellulosic ethanol plant in Brazil; first in the Southern Hemisphere. 2 pp. Web article. Available at: www.greencarcongress.com/2012/05/graalbio-20120523.html, accessed January 27, 2013.

Green World. 2011. List of major biomass power plants in the world—scale increasing. *Green World Investor,* 9 March 2011. Available at: www.green worldinvestor.com/2011/03/09/list-of-major-biomass-power-plants-in-the-world-scale-increasing/, accessed July 7, 2012.

Greffet, J.-J. 2011. Controlled incandescence. *Nature* 478: 191–192.

Gristmill. 2004. Available at: uow.academia.edu/AdamLucas/Papers/951334/_Pre-Modern_Grist_Review_of_John_Langdon_Mills_in_the_Medieval_Economy_England_1300_-1540_2004.

Grogg, K. 2005. Harvesting the wind: The physics of wind turbines. 41 pp. Web paper. Carleton College Physics and Astronomy Department, Physics and Astronomy Comps Papers. Available at: digitalcommons.carleton.edu/pacp/7, accessed February 14, 2011.

Grotelüschen, F. 2010. Nuclear fusion reactor faces delays, budget woes. Web press release. Available at: www.dw-world.de/dw/article/0,,5841888,00.html, accessed April 8, 2011.

Guardian. 2012. UK aid helps to fund forced sterilization of India's poor. 2 pp. News article by G. Chamberlin. Available at: www.guardian.co.uk/world/2012/apr/15/uk-aid-forced-sterilisation-india, accessed July 17, 2012.

Gurian-Sherman, D. 2011. Monsanto corn unlikely to help drought-stricken farmers. Statement by D. Gurian-Sherman, Senior Scientist, UCS Food & Environment Program, 22 December 2011. Available at: www.ucsusa.org/news/press_release/monsanto-drought-corn-1363.html, accessed January 28, 2012.

Gurian-Sherman, D. 2012. *High and Dry: Why Genetic Engineering Is Not Solving Agriculture's Drought Problem in a Thirsty World.* Report published by the Union of Concerned Scientists, Cambridge, MA, June 2012. Available at: http://www.ucsusa.org/assets/documents/food_and_agriculture/high-and-dry-report.pdf. Accessed January 19, 2013.

Guy, A.B. 2011. Wild pollinators worth up to $2.4 billion to farmers, study finds. 3 pp. Web news bulletin. Available at: newscenter.berkeley.edu/2011/06/20/wild-pollinators-worth-billions-to-farmers/, accessed July 5, 2012.

Guyton, A. 1991. *Textbook of Medical Physiology.* W.B. Saunders Co., Philadelphia, PA.

Gylfe, A., S. Bergström, J. Lundström & B. Olsen. 2000. Reactivation of *Borrelia* infection in birds. *Nature* 403: 724–725.

Haar, D. 2012. Jobless rate falls to 8%. *Hartford Courant,* 13 March 2012. p. 1A.

Habal, R. & J. Martinez. 2011. Mushroom toxicity. Published online by Medscape, 27 July 2011. Available at: emedicine.medscape.com/article/167398-overview#a0104, accessed July 7, 2012.

Halder, G., P. Callerets & W.J. Gehring. 1995. Induction of ectopic eyes by targeted expression of the eyeless gene in *Drosophila. Science* 267: 1788–1792.

Hallberg, M.C. 2003. Historical perspective on adjustment in the food and agriculture sector. Pennsylvania State University, October. Available at: agadjust.aers.psu.edu/Workshop_files/Hallberg.pdf, accessed January 29, 2011.

Hallinan, J. & P.T. Jackway. 2005. Network motifs, feedback loops and the dynamics of genetic regulatory networks. *Proc. 2005 IEEE Symposium on Computational Intelligence in Bioinformatics and Computational Biology,* 11–15 November 2005. Available at: ieeexplore.ieee.org/Stamp/Stamp.jsp?tp=&arnumber=1594903&isnumber=33563, accessed November 16, 2009.

Hallinan, J. & J. Wiles. 2004. Evolving genetic regulatory networks using an artificial genome. Paper presented at the 2nd Asia-Pacific Bioinformatics Conference (APBC2004), Dunedin, New Zealand. *Conferences in Research and Practice in Information Technology,* Vol. 29. Yi-Ping Phoebe Chen, ed. pp. 291–286.

Hallinan, J. 2003. Self-organization leads to hierarchical modularity in an Internet community. In: *Lecture Notes in Computer Science,* Vol. 2773/2003. Springer, Berlin, Germany. pp. 914–920. ISBN: 978-3-540-40803-1. Available at: www.springerlink. com/content/82r8ha86g3jluhdl/fulltext.pdf, accessed November 24, 2009.

Hallinan, J. 2004a. Gene duplication and hierarchical modularity in intracellular interaction networks. *BioSystems* 74: 51–62.

Hallinan, J. 2004b. Cluster analysis of the p53 genetic regulatory network: Topology and biology. *Proc. 2004 IEEE Symposium on Computational Intelligence in Bioinformatics and Computational Biology,* La Jolla, CA, 7–8 October 2004. 8 pp. Available at: www.staff.ncl.ac.uk/j.s.hallinan/pubs/CIBCB2004.pdf, accessed November 16, 2009.

Hallinan, J. & G. Smith. 2002. Iterative vector diffusion for the detection of modularity in large networks. *Int. J. Complex Syst. B.* 9 pp. Available at: research.imb. uq.edu.au/~j.hallinan/ICCS2002.htm, accessed November 12, 2009.

Hammons, T.J. 2009. Chapter 2. Tidal energy technologies: Currents, wave and offshore wind power in the United Kingdom, Europe and North America. In: Hammons, T.J., ed., *Renewable Energy.* pp. 463–504. New York, New York, USA: InTech. ISBN: 978-953-7619-52-7.

Handel, A. & D.E. Rozen. 2009. The impact of population size on the evolution of asexual microbes on smooth versus rugged fitness landscapes. *BMC Evol. Biol.* 9: 236. Available at: www.biomedcentral.com/1471-2148/9/236, accessed January 23, 2011.

Hansen, M, J.M. Halloran, E. Groth & L. Lefferts. 1997. *Potential Public Health Impacts of the Use of Recombinant Bovine Somatotropin in Dairy Production.* Prepared for a Scientific Review by the Joint Expert Committee on Food Additives. September 1997. Published online by Consumer's Union. Available at: www. consumersunion.org/pub/core_food_safety/002272.html, accessed January 29, 2011.

Hardin, G. 1968. The tragedy of the commons. *Science* 162(3859): 1243–1248.

Hardin, G. 1972. *Exploring New Ethics for Survival: The Voyage of the Spaceship Beagle.* New York, New York, USA: Viking Press, 273 pp.

Hardin, G. 1977. The ethical implications of carrying capacity. 12 pp. Web essay. Available at: jayhanson.us/page96.htm, accessed June 1, 2011.

Hardman, J.G., L.E. Limbird, L.S. Goodman & A. Gilman. eds. 1996. *Goodman and Gilman's the Pharmacological Basis of Therapeutics,* 9th ed. New York, New York, USA: McGraw-Hill.

Harford, T., C.A. Hidalgo & A. Simoes. 2011. The art of economic complexity. *New York Times Magazine,* 15 May 2011. pp. 40–41.

Hariri, S.S. & R. Chou. 2008. Method of estimating biogenic methane production at fill sites. 7 pp. Web paper. Available at: www.sccaepa.org/docs/onlineJv1no12nd article.pdf, accessed June 10, 2011.

Hasunuma, T., E. Fukusaki & A. Kobayashi. 2003. Methanol production is enhanced by expression of an *Aspergillus niger* pectin methylesterase in tobacco cells. *J. Biotechnol.* 106: 45–52.

Hawaii Data Book. 2010. Available at: hawaii.gov/dbedt/info/economic/databook/2010-individual/17/170710.pdf, accessed at July 4, 2012.

Hayes, B. 2012. Computation and the human predicament. *Am. Sci.* 100(3): 186–191.

Hayes, M.J. 2010. What is drought? Drought and climate change. 4 pp. Web paper. Available at: www.drought.unl.edu/whatis/cchange.htm, accessed October 21, 2010.

Hayutin, A. 2007. Global aging: The new new thing. The big picture of population change. 4 pp. Web paper from the Stanford Center on Longevity. Available at: longevity1.stanford.edu/files2/GlobalAgingTheNewThing.pdf, accessed July 17, 2012.

Heart, S.F. 2008. Albert Einstein quotes. Available at: www.sfheart.com/einstein.html, accessed October 22, 2008.

Heilmann, S.M., H.T. Davis, L.R. Jader, P.A. Lefebvre, M.J. Sadowsky, F.J. Schendel, M.G. von Keitz & K.J. Valentas. 2010. Hydrothermal carbonization of micro-algae. *Biomass Bioenergy.* doi:10.1016/j.biombioe.2010.01032. 8 pp. Available at: www.elsevier.com/locate/biombioe, accessed December 23, 2010.

Heinz Center. 2010. Carbon capture and storage development in China. 43 pp. Report by the H. John Heinz Center for Science, Economics and the Environment, Washington, DC. Available at: www.heinzctr.org/publications/PDF/Carbon_Capture_in_China.pdf, accessed July 4, 2011.

Helms, M., P. Vastrup, P. Gerner-Smidt & K. Mølbak. 2002. Excess mortality associated with antimicrobial drug-resistant *Salmonella typhimurium*. Published online by the Centers for Disease Control and Prevention. Available at: www.cdc.gov/ncidod/eid/vol8no5/01-0267.htm, accessed January 17, 2011.

Hemschemeier, A., A. Melis & T. Happe. 2009. Analytical approaches to hydrogen production in unicellular green algae. *Photosynth. Res.* 102: 523–540.

Heppenstall, A., A.J. Evans & M.H. Birkin. 2005. A hybrid multi-agent/spatial interaction model system for petrol price setting. *Trans. GIS.* 9(1): 35–51.

Herman-Giddens, M., E.J. Slora, R.C. Wasserman, C.J. Bourdony, M.V. Bhapkar, G.G. Koch & C.M. Hasemeier. 1996. Secondary sexual characteristics and menses in young girls seen in office practice: A study from the Pediatric Research in Office Settings Network. *Pediatrics* 99(4): 505–512. Available at: pediatrics.aappublications.org/cgi/content/abstract/99/4/505?ijkey=LbPAOIXHCOVeU, accessed February 5, 2011.

Herzog, H. 2003. Assessing the feasibility of capturing CO_2 from the air. 15 pp. Report, MIT LFEE 2003-002 WP, from MIT Laboratory for Energy and the Environment. Available at: step.berkeley.edu/Journal_Club/paper1_02092010.pdf, accessed July 7, 2011.

Heylighen, F. & C. Joslyn. 2001. The law of requisite variety. Web paper. Available at: http://pespmc1.vub.ac.be/REQVAR.html, accessed August 16, 2009.

Hidalgo, C.A. 2009. The dynamics of economic complexity and the product space over a 42 year period. CID Working Paper No. 189, December 2009. 20 pp. Harvard University Center for International Development. Available at: www.hks.harvard.edu/var/ezp_site/storage/fckeditor/file/pdfs/centers-programs/centers.cid/publications/faculty/wp/189.pdf, accessed October 10, 2011.

Hidalgo, C.A. & A.-L. Barabasi. 2008. Scale-free networks. *Scholarpedia* 3(1): 1716. Available at: www.scholarpedia.org/article/Albert-Barabasi, accessed January 25, 2013.

Hidalgo, C.A. & R. Hausmann. 2008. A network view of economic development. Developing alternatives. 10 pp. Available at: www.chidalgo.com/Papers/ HidalgoHausmann_DAI_2008.pdf, accessed October 10, 2011.

Hidalgo, C.A. & R. Hausmann. 2009. The building blocks of economic complexity. *PNAS.* 106(26): 10570–10575 + 42 pp supplementary Material. Available at: www.childalgo.com/Papers/HidalgoHausmann_PNAS_2009_PaperAndSM. pdf, accessed August 15, 2011.

Hintze, A. & C. Adami. 2008. Evolution of complex modular biological networks. *PLoS Comput. Biol.* 4(2): 1–12.

History Learning. 2012. The Great Famine of 1845. 3 pp. Online encyclopedia article. Available at: www.historylearningsite.co.uk/ireland_great_famine_of_1845. htm, accessed July 17, 2012.

Holland, J.H. 1992. *Adaptation in Natural and Artificial Systems: An Introductory Analysis with Applications to Biology, Control and Artificial Intelligence.* MIT Press, Cambridge, MA. ISBN: 0-262-58111-6.

Holland, J.H. 1995. *Hidden Order: How Adaptation Builds Complexity.* Cambridge, Massachusetts, USA: Helix Books/Perseus Book Group. ISBN-13: 978-0201442304.

Holland, J.H. & J.H. Miller. 1991. Artificial adaptive agents in economic theory. *AEA Papers and Proceedings,* May 1991. pp. 365–370. Available at: zia.hss.cmu.edu/ miller/papers/aaa.pdf, accessed December 1, 2009.

Holm, A. 2010. Geothermal energy: International market update. Available at: www. geo-energy.org/pdf/reports/GEA_International_Market_Report_Final_ May_2010.pdf, accessed May 24, 2010.

Hoover. 2010. Thomas Sowell. 1 p. Biography. Available at: www.hoover.org/fel lows/9767, accessed February 10, 2011.

Hoppe, H. & N.S. Sariciftci. 2004. Organic solar cells: An overview. *J. Mater. Res.* 19(7): 1924–1940.

Höppener, J.W., S. Mosselman, P.J. Roholl, C. Lambrechts, R.J. Slebos, P. de Pagter-Holthuizen, C.J. Lips, H.S. Jansz & J.S. Sussenbach. 1988. Expression of insulin-like growth factor-I and II genes in human smooth muscle tumours. *EMBO J.* 7: 1379–1385.

Horgan, J. 20 May 2008. Can chaoplexy save economics? *The Scientific Curmudgeon* Web blog. Available at: www.stevens.edu/csw/cgi-bin/blogs/csw/?p=150, accessed November 16, 2009.

Horrigan, L., R.S. Lawrence & P. Walker. 2002. How sustainable agriculture can address the environmental and human health harms of industrial agriculture. *Environ. Health. Perspect.* 110(5). doi:10.1289/ehp.02110445. Graph available at: ehp03.niehs.nih.gov/article/slideshow.action?uri=info:doi/10.1289/ehp.02110 445&imageURI=info:doi/10.1289/ehp.02110445.t001#, accessed February 6, 2012.

Howard, P.H. 2009. Visualizing consolidation in the global seed industry: 1996–2008. *Sustainability 2009,* Vol. 1, 1266–1287; doi:10.3390/su1041266.

HRH the Prince of Wales. 2012. *On the Future of Food.* New York, New York, USA: Rodale Books. ISBN-13: 978-1-60961-471-3.

Hubbert, M.K. 1982. *Techniques of Prediction as Applied to Production of Oil and Gas.* US Department of Commerce, NBS Special Publication 631, May 1982.

Huffington. 2012. New England tar sands pipeline plan opposed by environmental groups. Posted 6/19/12, updated 6/20/12. 2 pp. Online article. Available at: www.huffingtonpost.com/2012/06/19/new-england-tar-sands-pipeline_n_1609174.htm, accessed August 2, 2012.

Huffstutter, P.J. 2011. As soybeans die, a theory blooms. *Hartford Courant*, 14 April 2011. p. A5.

Huijgen, W.J.J. 2007. Carbon dioxide sequestration by mineral carbonization. PhD dissertation, van Wageningen Universiteit, Prof. Dr. M.J. Kropff, adviser. Available at: edepot.wur.nl/121870, accessed May 17, 2011.

Hutchings, J. 1996. Spatial and temporal variation in the density of northern cod: A review of hypotheses for the stock's collapse. *Can. J. Aquat. Sci.* 53: 943–952.

Hutchins, S. 2001. *Potential of Concentrated Animal Feed Operations (CAFOs) to Contribute Estrogens to the Environment*. Published online by the EPA, 19 September 2001. Available at: www.epa.gov/ORD/NRMRL/EDC/pdf/hutchins_09192001.txt, accessed January 29, 2011.

Hutchins, S. 2007. Assessing potential for ground and surface water impacts from hormones in CAFOs. Presentation by the Ground Water & Ecosystems Restoration Division of the Environmental Protection Agency at the EPA Office of Research & Development CAFO workshop in Chicago, IL, 21 August 2007. Available at: www.epa.gov/ncer/publications/workshop/pdf/hutchins_ord82007.pdf, accessed February 6, 2011.

Hutter, M. 2008. Algorithmic complexity. (Web article v. 19 April 2008.) Available at: www.scholarpedia.org/article/Algorithmic_complexity, accessed November 20, 2009.

IBI. 2011. International Biochar Initiative. Homepages. Available at: www.biochar international.org/, accessed January 31, 2011.

ICIMOD. 2012. Honeybees and pollination. 2 pp. Web article. Available at: www.ici mod.org/?q=1231, accessed March 13, 2012.

ICIS pricing. 2010. Biodiesel (USA): Domestic prices, 7 April 2011. Available at: www. icispricing.com/il_shared/Samples/SubPage10100131.asp, accessed October 13, 2011.

Identifying. 2011. Identifying harmful marine dinoflagellates. Available at: botany.si. edu/references/dinoflag/Taxa/Gpolygramma.htm, accessed May 19, 2011.

IEA. 2011. World energy outlook 2011 factsheet. 6 pp. Web article. International Energy Agency, Paris, France. Available at: www.iea.org, accessed January 26, 2012.

IEA. 2012. Did you know? 2 pp. Web fact sheet. International Energy Agency, Paris, France. Available at: www.iea.org/journalists/fastfacts.asp, accessed January 26, 2012.

Ikweiri, F.S., H. Gabril, M. Jahawi & Y. Almatrdi. 2008. Evaluating the evaporation water loss from the Omar Muktar open water reservoir. *Proc. 12th International Water Technology Conference, IWTC12 2008*, Alexandria, Egypt. pp. 893–899. Available at: www.iwtc.info/2008_pdf/10-4.PDF, accessed September 3, 2012.

Ince, P.J. 1979. How to estimate recoverable heat energy in wood or bark fuels. USDA Forest Service, Forest Products Laboratory. *General Technical Report FPL 29*. pp. ii + 7.

Incidence. 2011. Lyme disease incidence map. Available at: www.aldf.com/usmap. shtml, accessed July 28, 2011.

IndexMundi. 2011a. Hydroelectric power consumption by country. Online data. Available at: www.indexmundi.com/energy.aspx?product=hydro&graph=con sumption-growth-rate, accessed May 30, 2012.

IndexMundi. 2011b. World hydroelectric power production by year (billion kWh). Online data. Available at: www.indexmundi.com/energy.aspx?product=hydro &graph=production-by-year, accessed May 30, 2012.

IPCC [Intergovernmental Panel on Climate Change]. 2001. Trace gasses: Current observations, trends and budgets. Ch 4 in: *Climate Change 2001: The Scientific Basis*. Available at: www.grida.no/climate/ipcc_tar/wg1/134.htm, accessed June 14, 2011.

IPCC. 2007a. Summary for policymakers. In: *Climate Change 2007: Synthesis Report*. 22 pp. Available at: www.ipcc.ch/pdf/assessment-report/ae4/syr/ar4-syr-spm.pdf, accessed November 30, 2011.

IPCC. 2007b. Summary for policymakers. In: Solomon, S., Qin, D., Manning, M., Chen, Z., Marquis, M., Averyt, K.B., Tignor, M. & Miller, H.L. (eds.). *Climate Change 2007: The Physical Science Basis. Contribution of Working Group I to the Fourth Assessment Report of the Intergovernmental Panel on Climate Change*. New York, New York, USA: Cambridge University Press. Available at: http://www.ipcc.ch/publications_and_data/publications_ipcc_fourth_assessment_report_wg1_report_the_physical_science_basis.htm, accessed January 19, 2013.

IPCC. 2007c. Summary for policymakers. In: Parry, M.L., Canziani, O.F., Palutikof, J.P., van der Linden, P.J., & Hanson, C.E., eds., *Climate Change 2007: Impacts, Adaptation and Vulnerability. Contribution of Working Group II to the Fourth Assessment Report of the Intergovernmental Panel on Climate Change*. Cambridge, England: Cambridge. Available at: www.ipcc.ch/pdf/assessment-report/ar4/wg2/ar4-wg2-spm.pdf, accessed November 30, 2011.

IPCC. 2007d. Summary for policymakers. In: Metz, B., Davidson, O.R., Bosch, P.R., Dave, R. & Meyer, L.A., eds. *Climate Change 2007: Mitigations of Climate Change. Contribution of Working Group III to the Fourth Assessment Report of the Intergovernmental Panel on Climate Change*. Cambridge, England: Cambridge University Press. Available at: http://www.ipcc.ch/publications_and_data/ar4/wg3/en/contents.html, accessed January 19, 2013.

IRIN. 2005. Fears of witchcraft lead to widespread infanticide in remote north. 2 pp. Online article. Available at: www.irinnews.org/printreport.aspx?reportid=55489, accessed August 9, 2012.

Isidore, C. 2011. 2 economists: Peril exists for economy. *Hartford Courant*, 4 August 2011. p. A12.

ISSHA. 2010. Paralytic shellfish poisoning, saxitoxins and organisms. 2 pp. Web hot links list by the International Society for the Study of Harmful Algae. Available at: www.issha.org/Welcome-to-ISSHA/Harmful-Algae-Links/Phycotoxins/Saxitoxins-PSP, accessed January 10, 2012.

Jackson, T. 2011. *Prosperity without Growth: Economics for a Finite Planet*. Paperback, 288 pp. Washington, DC, USA: Earthscan. ISBN-10: 1849713235.

James, A., J.W. Pitchford & J. Brindley. 2003. The relationship between plankton blooms, the hatching of fish larvae, and recruitment. *Ecol. Model*. 160: 77–90.

Jana, B.K., S. Biswas, M. Majumder, P. Roy & A. Mazumdar. 2010. Estimation of carbon dioxide emission contributing GHG level in ambient air of a metro city: A case study for Kolkata. Ch 1 in: Jana, B.K., Majumder, M., eds. *Impact of Climate Change on Natural Resource Management*. Section 1.5.3: CO_2 Emission from Human Respiration. New York, New York, USA: Springer. ISBN: 978-90-481-3580-6.

Japan Times. 2012. Preparatory drilling for methane hydrate off Aichi coast set to start. 2 pp. Press release, Wednesday, 15 February 2012. Available at: www.japantimes.co.jp/print/nb20120215a4.html, accessed May 18, 2012.

Jennings, D.H. & G. Lysek. 1996. *Fungal Biology: Understanding the Fungal Lifestyle.* Bios Scientific Publishers Ltd., Guildford, UK. ISBN: 978-1-85996-150-6.

Jennings, G.P. 1933. *Greens Farms Connecticut: The Old West Parish of Fairfield.* Greens Farms, Connecticut, USA: The Congregational Society of Greens Farms.

Jeol.com. 2007. A 30 Wh/kg supercapacitor for solar energy and a new battery. 3 October 2007. Available at: www.jeol.com/NEWSEVENTS/PressReleases/tabid/521/articleType/ArticleView/articleId/112/A-30-Whkg-Supercapacitor-for-solar-Energy-and-a-New-Battery.aspx, accessed September 13, 2011.

Jha, A. 2011. Bees in freefall as study shows sharp US decline. *The Guardian,* 3 January 2011. Available at: www.guardian.co.uk/environment/2011/jan/03/bumble bees-study-us-decline, accessed June 24, 2012.

Johnson, C. 2009. Feds: Bomb could have downed plane. *Hartford Courant,* 29 December 2009.

Johnson, I. 2011. The road to 2050 (Part I). 3 pp. Web essay. Available at: www.theglobalist.com/printStoryId.aspx?StoryId=9396, accessed January 16, 2012.

Johnson, K. 2010. Scientists and soldiers solve a bee mystery. *New York Times,* 6 October 2010. Available at: www.nytimes.com/2010/10/07/science/07bees.html, accessed May 18, 2012.

Junkins, C. 2010. What's in the water?: Fracking chemicals under microscope. 3 pp. Web paper. Available at: theintelligencer.net/page/content.detail/id/549992/What-s-In-The-Water-Fracking-Chemicals-Under-Microscope.html?nav=515, accessed December 29, 2010.

Kadak, A.C. 2007. MIT pebble bed reactor project. *Nucl. Eng. Technol.* 39(2): 95–102. Available at: web.mit.edu/pebble-bed/papers1_files/MIT_PBR.pdf, accessed October 28, 2011.

Kafri, O. 2008. Sociological and economic inequality and the second law. Web paper. Available at: mpra.ub.uni-muenchen.de/9175, accessed September 3, 2008.

Kajiwara, Y. & M. Imahori. 2009. Effects of estrone on both embryonic and extra-embryonic regions in mouse embryos during pre-organogenesis in vitro. *Congenit. Anom. (Kyoto)* 49(1): 42–45.

Kanda, H. & T. Mimaki. 2010. Successful extraction of "green crude oil" from blue-green algae: High yield extraction at room temperature without drying nor pul-verizing process. 8 pp. Press release, 17 March 2010, NEDO & CRIEPI. Available at: criepi.denken.or.jp/en/e_publication/pdf/den445.pdf, accessed October 25, 2011.

Kandel, E.R., J.H. Schwartz & T.M. Jessell. 1991. *Principles of Neural Science,* 3rd ed. Appleton & Lange, Norwalk, CT.

Kaplan, M. 2009. Breeding and raising the house cricket. 5 pp. Web paper. Available at: www.anapsid.org/crickets.html, accessed May 13, 2011.

Kaplan, R.M. 2009. These ain't your father's parasites: Dewormer resistance and new strategies for parasite control in horses. 10 pp. Web paper. Available at: www.animal.ufl.edu/extension/equine/documents/2009equineinstit/kaplanpara sites.pdf, accessed June 17, 2012.

Kaplan, R.M. & A.N. Vidyashankar. 2012. An inconvenient truth: Global worming and anthelmintic resistance. *Vet. Parasitol.,* 186(1–2): 70–78.

Kaplan, R.M. & A. Wolstenholme. 2010. Monsters inside our animals, 34th Annual Report, University of Georgia Veterinary Medical Experiment Station, College of Veterinary Medicine. 24 pp. Web article. Available at: www.vet.uga.edu/research/vmes/annreports/VMES2010-smaller.pdf, accessed June 17, 2012.

Karabin, S. 2007. Infanticide, abortion responsible for 60 million girls missing in Asia. 5 pp. News release. Available at: www.foxnews.com/story/0,2933,281722,00. html, accessed August 9, 2012.

Kaufman, S. 1955. *At Home in the Universe: The Search for the Laws of Self-Organization and Complexity.* New York, New York, USA: Oxford University Press.

Kavlock, R.J., G.P. Daston, C. DeRosa, P. Fenner-Crisp, L.E. Gray, S. Kaattari, G. Lucier, M. Luster, M.J. Mac, C. Maczka, R. Miller, J. Moore, R. Rolland, G. Scott, D.M. Sheehan, T. Sinks & H.A. Tilson. 1996. Research needs for the risk assessment of health and environmental effects of endocrine disruptors: A report of the U.S. EPA–sponsored workshop. *Environ. Health Perspect.* 104(suppl. 4): 715–740.

Keating, J.E. 2011. How food explains the world. *Foreign Policy*, May–June 2011. pp. 73–75.

Keeling, R.F. & C.D. Whorf. 2008. Atmospheric CO_2 records from sites in the SIO air sampling network. In: *Trends: A Compendium of Data on Global Change.* Oak Ridge National Laboratory Carbon Dioxide Analysis Center, Oak Ridge, TN. Available at: http://cdiac.ornl.gov/trends/trends.htm, accessed January 21, 2013.

Keleman, A. & H.G. Rano. 2008. Biofuels and rising food prices: Mexico's 2007 tortilla crisis. Presentation at the Food Security and Environmental Change Conference, 2–4 April 2008, Oxford University, UK. Available at: www.gecafs.org/docu ments/PP12Keleman_000.pdf, accessed March 25, 2012.

Kennedy, C.W. 2008. Fly ash flood covers acres. Web news release. Available at: www.knoxnews.com/news/2008/dec/23/fly-ash-flood-covers-acres/, Accessed January 6, 2011.

Khan, J.R., J.F. Klausner, D.P. Ziegler & S.S. Garimella. 2010. Diffusion driven desali-nation for simultaneous fresh water production and desulfurization. *J. Therm. Sci. Eng. Appl.* 2: 031006-1–031006-14.

Kim, J. & T. Wilhelm. 2008. What is a complex graph? *Physica A* 387(11): 2637–2652.

Kimball, J. 2011. Checks on population growth. 11 pp. Web essay. Available at: users. rcn.com/jkimball.ma.ultranet/BiologyPages/P/Populations2.html, accessed May 18, 2012.

Kinealy, C. 1995. *The Great Calamity: The Irish Famine 1845–1952.* Gill & Macmillan, London, UK. ISBN: 1-57098-034-1.

Klausner, J.F., Y. Li, M. Darwish & R. Mei. 2004. Innovative diffusion driven desalina-tion process. *ASME J. Energy Resour. Technol.* 126(3): 219–225.

Klein, W. 2000. Chapter 7. *Westport Connecticut.* Greenwood Press, Westport, CT.

Kolbert, E. 2011. Enter the age of man. *Natl. Geogr. Mag.* 219(3): 60–85.

Kolodziej, E.P. 2010. Faculty research summary. Assistant Professor, Department of Civil and Environmental Engineering, University of Nevada, Reno, NV. Available at: www.unr.edu/cee/homepages/kolodziej/research.html, accessed January 28, 2011.

Kolpin, D.W., E. Furlong, M. Meyer, E.M. Thurman, S. Zaugg, L. Barber & H. Buxton. Pharmaceuticals, hormones, and other organic wastewater contaminants in U.S. streams, 1999–2000: A national reconnaissance. *Environ. Sci. Technol.* 36(6): 1202–1211. Available at: http://digitalcommons.unl.edu/cgi/viewcontent. cgi?article=1064&context=usgsstaffpub, accessed January 21, 2013.

Kolstad, C. & D. Young. 2010. Cost analysis of carbon capture and storage for the Latrobe Valley. Bren School of Environmental Science and Management, UCSB, CA. 35 pp. Available at: www.calera.com/uploads/files/ccs_costs_latrobe-final-2.pdf, accessed July 4, 2011.

Konarka. 2011. Konarka Power Plastic reaches 8.3% efficiency. Web news blurb. Available at: www.pv-tech.org/news/nrel_validates_konarkas_8.3_power_plastic.pv-tech.org, accessed July 5, 2011.

KPL. 2011. Farmers learn farming edible insects. KPL Lao News Agency news release. 30 March 2011. Available at: laovoices.com/2011/03/29/farmers-learn-farming-edible-insects, accessed May 11, 2011.

Krieger, R. 2001. *Handbook of Pesticide Toxicology*, 2nd ed. Academic Press, San Diego, CA.

Krill. 2012. Krill oil decline reasons. 4 pp. Available at: www.squidoo.com/krill-Oil-decline, accessed January 9, 2012.

Kristoff, N. 2011. When food kills. *The New York Times*, 11 June 2011.

Krivit, S.B. 2007. The mistakes of Pons and Fleischmann and why their discovery was initially thought to be a mistake. *New Energy Times*, 23 March 2007. 2 pp. Web article. Available at: newenergytimes.com/v2/reports/MistakesOfFleischmannAndPons.shtml, accessed August 2, 2012.

Krugman, P. 1994. Complex landscapes in economic geography. *Am. Econ. Rev.* 84(2): 412–416.

Kundert, K. 2008. Modeling dielectric absorption in capacitors. Version 2d, June 2008. 19 pp. Web paper. Available at: www.designers-guide.org/Modeling/da.pdf, accessed August 4, 2011.

Kunzig, R. 2009. The Canadian oil boom. 7 pp. *National Geographic* article. Available at: ngm.nationalgeographic.com/2009/03/canadian-oil-sands/kunzig-text/8, accessed March 14, 2011.

Kuo, B.C. 1982. *Automatic Control Systems*, 4th ed. Prentice-Hall, Inc., Englewood Cliffs, NJ.

Kurlansky, M. 1997. *Cod*. Penguin Books, Ltd., London, UK.

Kurtzman, R.H., Jr. 2005. Mushrooms: Sources for modern western medicine. *Micol. Aplicada Int.* 17(2): 21–33. Available at: www.oystermushrooms.net/kurtzman.pdf, accessed August 26, 2011.

Labossiere, R. 2009. Yale scientist's quest: Search for fuel's gold. *The Hartford Courant*, 30 November 2009.

Lacey, L.A., D.R. Horton & D.C. Jones. 2008. The effect of temperature and duration of exposure of potato tuber moth (*Lepidoptera gelechiidae*) in infested tubers to the biofumigant fungus *Muscador albus*. *J. Invertebrate Pathol.* 97: 159–164.

Lackner, K.S. 2003. A guide to CO_2 sequestration. *Science* 300: 1677–1678.

Laessig, S. & D. Snow. 2011. Special symposium: Hormones in the Environment at the SETAC North America 2010 Annual Meeting. *SETAC Globe*, 21 January 2011, Volume, 12 Issue 1. Published online by the Society of Environmental Toxicology and Chemistry. Available at: www.setac.org/globe/2011/january/hormones.html, accessed January 29, 2011.

Lagi, M., Y. Bar-Yam, K.Z. Bertrand & Y. Bar-Yam. 2011. The food crises: A quantitative model of food prices including speculators and ethanol conversion. arXiv: 1109.4859v1,[q-fin.GN] 2011. 5 pp. Web paper. Available at: arxiv.org/abs/1109.4859, accessed March 25, 2012.

Lagi, M., Y. Bar-Yam, K.Z. Bertrand & Y. Bar-Yam. 2012. Update, February 2012—The food crises: Predictive validation of a qualitative model of food prices including speculators and ethanol conversion. arXiv: 1203.1313, 6 March 2012. 5 pp. Web paper. Available at: necsi.edu/research/social/foodprices/update/food_prices_update.pdf, accessed March 25, 2012.

Lansing, J.S. 2003. Complex adaptive systems. *Ann. Rev. Anthropol.* 32: 183–204.

Larkum, A.W.D. 2010. Limitations and prospects of natural photosynthesis for bio-energy production. *Curr. Opin. Biotech.* 21: 271–276. DOI 10.1016/j.copbio.2010.03.004.

Larsen, M.C. 2009. *Statement of the Associate Director for Water, U.S. Geological Survey, U.S. Department of the Interior, Before the Committee on Natural Resources, Subcommittee on Insular Affairs, Oceans and Wildlife*, June 9, 2009. Published online by the USGS. Available at: www.usgs.gov/congressional/hearings/docs/larsen_09june09.doc, accessed February 5, 2011.

Larson, K. et al. 2011. Methicillin-resistant *Staphylococcus aureus* in pork production shower facilities. *Appl. Environ. Microbiol.* 77(2): 696–698. doi:10.1128/AEM.01128-10.

Larson, E.D. & H. Yang. 2004. Dimethyl ether (DME) from coal as a household cooking fuel in China. *Energy Sustain. Dev.* VIII(3): 115–126.

LaSalle, T. & P. Hepperly. 2008. *Regenerative Organic Farming: A Solution to Global Warming.* Published online by the Rodale Institute, July 2008. Available at: www.scribd.com/doc/17134792/Rodale-Research-Paper073008, accessed February 22, 2011.

LaSalle, T., P. Hepperly, & A. Diop. 2008. *The Organic Green Revolution.* Published online by the Rodale Institute. Available at: www.rodaleinstitute.org/files/GreenRevUP.pdf, accessed February 24, 2010.

Lawrie, J. & J. Hearne. 2007. Reducing model complexity via output sensitivity. *Ecol. Model.* 207: 137–144.

Leakey, R. & R. Lewin. 1996. *The Sixth Extinction: Patterns of Life and the Future of Humankind.* New York, New York, USA: Anchor Books. ISBN: 0-385-46809-1.

Leal, W.S., J.J. Bromenshenk, C.B. Henderson, C.H. Wick, M.F. Stanford, A.W. Zulich, R.E. Jabbour, S.V. Deshpande, P.E. McCubbin, R.A. Seccomb, P.M. Welch, T. Williams, D.R. Firth, E. Skowronski, M.M. Lehmann, S.L. Bilimoria, J. Gress, K.W. Wanner & R.A. Cramer. 2010. Iridovirus and microsporidian linked to honey bee colony decline. *PLoS ONE* 5(10): e13181. Available at: http://www.ncbi.nlm.nih.gov/pubmed/20949138, accessed January 21, 2013.

Leape, J.P. & S. Humphrey. 2010. Towards a sustainable future. Ch 5 in: Schellnhuber, H.J., Molina, M., Stern, N., Huber, V., Kadner, S., eds. *Global Sustainability.* New York, New York, USA: Cambridge University Press. pp. 49–64. Available at: www.nobel-cause-de/book/NobelCauseBook_chapter5.pdf, accessed December 7, 2010.

Lee, R. 2011. The outlook for population growth. *Science* 333: 669–573.

Levin, D.B., L. Pitt & M. Love. 2004. Biohydrogen production: prospects and limitations to practical application. *Int. J. Hydrogen Energy* 29: 173–185.

Li, X., A. Ahlman, X. Yan, H. Lindgren & L.-H. Zhu. 2009. Genetic transformation of the oilseed crop *Crambe abyssinica. Plant Cell Tiss. Organ Cult.* Online paper. Available at: http://icon.slu.se/ICON/Documents/Publications/GeneticTransfCrambeLiHua.PDF, accessed January 21, 2013.

Li, J. & X. Zou. 2009. Generalization of the Kermack-McKendrick SIR model to a patchy environment for a disease with latency. *Math. Model. Nat. Phenom.* 4(2): 92–118.

Liebhardt, W. 2001. Get the facts straight: Organic agriculture yields are good. *Organic Farming Research Foundation Information Bulletin* 10. Summer, 2001.

Lim, Y.S. & S.L. Koh. 2009. Chapter 12. Marine tidal current electric power generation: State of art and current status. In: Hammons, T.J., ed. *Renewable Energy*. New York, New York, USA: InTech. ISBN: 978-953-7619-52-7. 211-226. Available at: http://cdn.intechweb.org/pdfs/9330.pdf, accessed January 21, 2013.

Lindner, S., S. Peterson & W. Windhorst. 2009. An economic and environmental assessment of carbon capture and storage (CCS) power plants—A case study for the City of Kiel. Kiel Institute for the World Economy Working Paper No. 1527, June 2009. 24 pp. Available at: ideas.repec.org/p/kie/kieliw/1527.html, accessed July 4, 2011.

Lipman, T. 2011. An overview of hydrogen production and storage systems with renewable hydrogen case studies. Published online by the Clean Energy States Alliance. Available at: www.cleanenergystates.org/assets/2011-Files/Hydrogen-and-Fuel-Cells/CESA-Lipman-H2-prod-strorage-050311.pdf, accessed July 5, 2012.

Lippman, M., M. E. Monaco & G. Bolan. 1977. Effects of estrone, estradiol, and estriol on hormone-responsive human breast cancer in long-term tissue culture. *Cancer Res*. 37: 1901.

Lipson, H., J.B. Pollack & N.P. Suh. 2002. On the origin of modular variation. *Evolution* 56(8): 1549–1556.

Litvak, A. 2010. Marcellus shale well blowout prompts second DEP suspension. *Pittsburgh Business Times*. Available at: pittsburgh.bizjournals.com/pittsburgh/stories/2010/06/07/daily32.html, accessed June 9, 2010.

Liu, B. & M. Pop. 2009. ARDB—Antibiotic resistance genes database. *Nucleic Acids Res*. 37(database issue): D443–D447.

Liu, J. & J.A. Curry. 2010. Accelerated warming of the Southern Ocean and its impacts on the hydrological cycle and sea ice. *PNAS Early Edition*. pp. 14989–14993. Available at: www.pnas.org/cgi/dci/10.1073/pnas.1003336107/

Ljundgren, D. & R. Palmer. 2011. Canada to quit Kyoto climate change treaty. *The Hartford Courant*, 13 December 2011. p. A6. Reuters.

Lloyd, S. 2001. Measures of complexity: A nonexhaustive list. *IEEE Control Systems Magazine*. August 2001. Web paper. Available at: web.mit.edu/esd83/www/notebook/Complexity.PDF, accessed November 12, 2009.

Lopes, J.C., J. Dias & J. Ferreira do Amaral. 2008. Assessing economic complexity in some OECD countries with input–output based measures. Paper presented at the ECOMOD2008-International Conference on Policy Making, 2–4 July 2008, Berlin, Germany. 25 pp. Available at: www.ecomod.org/files/papers/746.pdf, accessed November 18, 2009.

Lopez, T. 2011. Oil and gas in the South China Sea. *Manila Times*, 21 July 2011. Available at: www.manilatimes.net/index.php/opinion/2441-oil-and-gas-in-the-south-china-sea, accessed November 18, 2011.

Lorenz, E.N. 1963. Deterministic non-periodic flow. *Journal of the Atmospheric Sciences*. 20: 130–141.

Lorenz, E.N. 1993. *The Essence of Chaos*. Seattle, Washington, USA: University of Washington Press.

Lovgren, S. 2007. Mystery bee disappearances sweeping U.S. *National Geographic News*. Available at: news.nationalgeographic.com/news/2007/02/070223-bees.html, accessed March 23, 2011.

Lucas, Adam. 2005. Pre-modern grist: Review of John Langdon, mills in the medieval economy: England 1300-1540. *Metascience* 14: 447–451.

Lucas, C. 2006. Quantifying complexity theory. Web paper, v. 4.83. Available at: www.calresco.org/lucas/quantify.htm, accessed November 4, 2009.

Lustgarten, A. 2009. EPA: Chemicals found in Wyo. drinking water might be from fracking. *Pro Publica*. Available at: www.propublica.org/article/epa-chemicals-found-in-wyo.-drinking-water-might-be-from-fracking-825, accessed December 29, 2010.

Lustgarten, A. & N. Kusnetz. 2011. Feds link water contamination to fracking for the first time. 3 pp. Online article, 8 December 2011. Available at: www.propublica.org/article/feds-link-water-contamination-to-fracking-for-the-first-time, accessed December 11, 2011.

Lustig, K. 2010. Map of China carbon capture and storage (CCS) projects. 1 p. Available at: www.earthtrendsdelivered.org/map_of_china_carbon_capture_and_storage_ccs_projects, accessed July 4, 2011.

Lyme Cases. 2009. Total Lyme cases reported by CDC 1990–2008. Map. Available at: module.lymediseseassociation.net/Maps/images/hires/90-08map_300dpi.gif, accessed July 28, 2011.

Lyons, G. 1999. Endocrine disrupting pesticides. *Pesticides News* 46: 16–19. Reproduced on the Pesticide Action Network UK Web site. Available at: www.panuk.org/pestnews/Actives/endocrin.htm, accessed February 6, 2011.

Macal, C. & M. North. 2006a. Tutorial on agent-based modeling and simulation. Part 2: How to model with agents. In: Perrone, L.F., Lawson, B., Liu, J. & Wieland, F.P., eds. *Proc. 2006 Winter Simulation Conference*. Monterey, CA, 3–6 December 2006. pp. 73–83. ISBN 1-4244-0501-7. Available at: www.informs-sim.org/wsc06papers/008.pdf, accessed December 10, 2009.

Macal, C. & M. North. 2006b. Introduction to agent-based modeling and simulation (37 presentation slides). *MCS LANS Informal Seminar*. Argonne National Laboratory. Available at: www.cas.anl.gov/, accessed March 15, 2010.

MacDonald, G.J. 1990. Role of methane clathrates in past and future climates. *Climatic Change*. 16: 247–281. Available at: http://biodav.atmos.colostate.edu/kraus/Papers/MethaneClathrate/MacDonald-Clathrate.pdf, accessed June 19, 2012.

Mader, E., M. Vaughan, M. Shepherd & S.H. Black. 2010. Alternative pollinators: Native bees. 26 pp. Web paper. Available at: attra.ncat.org/attra-pub/PDF/nativebee.pdf, accessed March 22, 2011.

Magiawala, K.R., J.L. Sollee, M.G. Wickham & S.W. Fornaca. 2007. Methane extraction method and apparatus using high-energy diode lasers or diode-pumped solid state lasers. US Patent Application Pub. No.: US 2007/0267220 A1. Pub. date: November 22, 2007.

Makoka, D. & M. Kaplan. 2005. Poverty and vulnerability. Term paper, November 2005. Center for Development Research, University of Bonn. 31 pp. Available at: www.zef.de/fileadmin/downloads/forum/docprog/Termpapers/2005_2b_Kaplan_Makoka.pdf, accessed October 17, 2011.

Malik mzsg. 2011. Sensitivity Model Prof. Vester®. The computerized system tools for a new management of complex problems. 12 pp. Available at: www.malik-mzsg.ch, or info@frederic-vester.de, accessed March 31, 2011.

Marketwatch. 2012. Ivanpah solar project reaches halfway mark and peak of construction employment. Available at: www.marketwatch.com/story/ivanpah-solar-project-reaches-halfway-mark-and-peak of-construction-employment-2012-08-07, accessed January 25, 2013.

Mascoma. 2009. Mascoma announces major cellulosic biofuel technology break-through. 3 pp. Press release. Available at: www.mascoma.com/download/ TechnologyAdvancesRelease-050709-Final.pdf, accessed July 9, 2012.

Mascoma. 2011. Mascoma awarded $80 million from the DOE for construction of commercial-scale hardwood cellulosic ethanol facility in Kinross, Michigan. 2 pp. Mascoma press release. Available at: www.mascoma.com/Mascoma_DOE Press Release FINAL.pdf, accessed July 9, 2012.

Mason, S.J. 1953. Feedback theory—Some properties of signal flow graphs. *Proc. IRE* 41(9): 1144–1156.

Mason, S.J. 1956. Feedback theory—Further properties of signal flow graphs. *Proc. IRE*. 44(7): 920–926.

Mathworld. 2011. Kermack–McKendrick model. Available at: mathworld.wolfram. com/Kermack-McKendrickModel.html, accessed August 15, 2011.

Mauro, R. 2003. Opinion: Our hydrogen energy future: Do we have one? 3 pp. Web editorial, *NHA News*. Available at: www.hydrogenus.com/newsletter/ad92h-2future.asp, accessed June 9, 2011.

Mauseth, J.D. 2008. *Botany: An Introduction to Plant Biology.* Sudbury, Massachusetts, USA: Jones & Bartlett. ISBN: 978-0-7637-5345-0.

Max, M.D. & W.P. Dillon. 1998. Oceanic methane hydrate: The character of the Blake Ridge hydrate stability zone, and the potential for methane extraction. *J. Pet. Geol.* 21(3): 343–357.

Mburu, J., L.G. Hein, B. Gemmill & L. Collettte. 2006. Economic valuation of pollination services: Review of methods. UNFAO Report, June 2006. 43 pp. Section 2.3. Economic valuation of the pollination service. pp. 14–20. Section 3. Experiences with the valuation of the pollination service. pp. 21–33. Available at: www.fao. org/fileadmin/templates/agphome/documents/Biodiversity-pollination/ econvaluepoll1.pdf, accessed July 9, 2012.

McBride, J. 2010. Helium prices jump to curb debt. 2 pp. News release. Available at: amarillo.com/news/local-news/2010-09-13/helium-prices-jump-curb-debt, accessed November 7, 2011.

McCabe, T. 1976. *A Complexity Measure.* Available at: www.literateprogramming. com/mccabe.pdf.

McCauley, J.L. & C.M. Küffner. 2004. Economic system dynamics. *Discrete Dyn. Nat. Soc.* 1: 213–220. Available at: mrpa.ub.uni-muenchen.de/2158/1/MRPA_ paper_2158.pdf MPRA- Munich Personal RePEc Archive, accessed November 18, 2009.

McGowan, E. 2010. Secrecy of fracking chemicals takes beltway spotlight. 4 pp. Web paper. Available at: solveclimatenews.com/news/20101201/secrecy-fracking-chemicals-takes-beltway-spotlight?page=3, accessed December 29, 2010.

McJannet, D., F. Cook, J. Knight & S. Burn. 2008. Evaporation reduction by mono-layers: Overview, modelling and effectiveness. Urban Water Security Research Alliance: Technical Report No. 6. 25 pp. Available at: www.urbanwateralliance. org.au/publications/UWSRA-tr6.pdf, accessed September 3, 2012.

McLean, R. et al. 2010. The evolution of antibiotic resistance: Insight into the roles of molecular mechanisms of resistance and treatment context. *Discovery Medicine,* 4 August 2010. Available at: www.discoverymedicine.com/R-Craig-MacLean/2010/08/04/the-evolution-of-antibiotic-resistance-insight-into-the-roles-of-molecular-mechanisms-of-resistance-and-treatment-context/, accessed January 23, 2011.

Mead, P.S., L. Slutsker, V. Dietz, L.F. McCaig, J.S. Bresee, C. Shapiro, P.M. Griffin & R.V. Tauxe. 1999. Food-related illness and death in the United States. *Emerging Infectious Diseases.* 5(5): 607–625. Available at: http://wwwnc.cdc.gov/eid/article/5/5/99-0502_article.htm, accessed January 21, 2013.

Meadows, D.H., D.L. Meadows, J. Randers & W.W. Behrens III. 1972. *The Limits to Growth: A Report for the Club of Rome's Project on the Predicament of Mankind.* New York, New York, USA: Universe Books.

Meadows, D.H., J. Randers & D.L. Meadows. 2004. *Limits to Growth: The 30-Year Update.* Chelsea Green Publishing, White River Junction, VT.

Meeuse, B.J.D. 2011. Pollination. *Britannica Online Encyclopedia.* 6 pp. Available at: www.britannica.com/EBchecked/topic/467948/pollination/, accessed June 1, 2012.

Mekonnen, M. & A. Hoekstra. 2012. A global assessment of the water footprint of farm animal products. *Ecosystems* 15: 401–415. doi:10.1007/s10021-011-9517-8.

Melián, C.J. & J. Bascompte. 2004. Food web cohesion. *Ecology* 85(2): 352–358.

Mellgard, P. 2010. Trouble brews in the South China Sea. 2 pp. Web article from the *Foreign Policy Association,* 22 June 2010. Available at: foreignpolicyblogs.com/2010/06/22/trouble-brews-in-the-south-china-sea/, accessed November 18, 2011.

Mellis, A., L. Zhang, M. Forestier, M.L. Ghirardi & M. Seibert. 2000. Sustained photobiological hydrogen gas production upon reversible inactivation of oxygen evolution in the green alga *Chlamydomonas reinhardtii. Plant Physio.* 122: 127–135.

Melzer, E.J. 2010. Michigan considers chemical disclosure rules for fracking. 3 pp. Press release. Available at: michiganmessenger.com/44943/michigan-considers-chemical-disclosure-rules-for-fracking, accessed December 29, 2010.

Mercier, J. & J.I. Jiménez. 2007. Potential of the volatile-producing fungus *Muscadore albus* for control of building molds. *Can. J. Microbiol.* 53: 404–410.

Methane. 2011. Methanol as a fuel: From bugs to cars and cows. 4 pp. Web paper. Available at: www.wiley.com/college/boyer/0470003790/cutting_edge/methanol/methanol.htm, accessed January 24, 2011.

Mick, J. 2008. Production Honda FCX fuel cell cars hit American streets. *Daily Tech,* 17 June 2008. Available at: www.dailytech.com/Production+Honda+FCX+Fuel+Cell+Cars+Hit+American+Streets/article12101.htm, accessed July 5, 2012.

Miller, J. 2003. *Review of Water Resources and Desalination Technologies.* Sandia National Laboratories. Available at: prod.sandia.gov/techlib/access-control.cgi/2003/030800.pdf, accessed July 4, 2012.

Miller, J.H. & S.E. Page. 2008. *Complex Adaptive Systems: An Introduction to Computational Models of Social Life.* Princeton, New Jersey, USA: Princeton University Press. ISBN-10: 0691127026.

Milloy, S. 2002. Quorn & CSPI: The other fake meat. *Fox News* release. Available at: www.undueinfluence.com/milloy.htm, accessed May 30, 2012.

Milner, L.S. 1998. A brief history of infanticide. 3 pp. Web paper. Society for the Prevention of Infanticide. Available at: www. infanticide.org/history.htm, accessed January 17, 2011.

Mineral County. 2012. Yucca Mountain Oversite Program. 6 pp. Online article. Available at: mcnucprojects.com/yuccahistory.htm, accessed August 2, 2012.

Mirjafari, P., K. Asghari & N. Mahinpey. 2007. Investigating the application of enzyme carbonic anhydrase for CO_2 sequestration purposes. *Ind. Eng. Chem. Res.* 46: 921–926.

MIT Media Lab. 2011. The Economic Complexity Observatory. 2 pp. Announcement. Available at: macroconnections.media.mit.edu/featured/economic-complexity-observatory, accessed May 24, 2011.

Mitchell, C., D. Delaney, & K. Balkcom. 2008. A historical summary of Alabama's Old Rotation (circa 1886): The world's oldest, continuous cotton experiment. *Agron J.* 100: 1493–1498.

Mitchell, C. & J.A. Entry. 1998. Soil C, N and crop yields in Alabama's long-term 'Old Rotation' cotton experiment. *Soil Tillage Res.* 47: 331–338.

Mithen, S. 2003. *After the Ice: A Global Human History 20,000-5,000 BC.* Harvard University Press, Cambridge, MA.

Miyamoto, K. 2012. Energy conversion by photosynthetic organisms. *Renewable Biological Systems for Alternative Sustainable Energy Production.* (Report produced by FAO Agriculture and Consumer Protection.) 12 pp. Web document. Available at: www.fao.org/docrep/w7241e/w7241e06.htm#TopOfPage, accessed June 4, 2012.

Mohanty, R.I. 2012. Trash bin babies: India's female infanticide crisis. 3 pp. Online *Atlantic* article. Available at: www.theatlantic.com/international/print/2012/05/trash-bin-babies-indias-female-infanticide-crisis/257672/, accessed August 9, 2012.

Molento, M.B. 2009. Parasite control in the age of drug resistance and changing agricultural practices. *Vet. Parasitol.* 163: 229–334. Available at: cnia.inta.gov.ar/helminto/confe09/paper-waavp09v163.pdf, accessed June 17, 2012.

Moler, C. 2003. *Stiff Differential Equations.* Matlab News & Notes. May, 2003. Available at: www.mathworks.com/company/newsletters/news_notes/clevescorner/may03_cleve.html, accessed November 13, 2009.

Monolayers. 2012. A list of water evaporation-inhibiting chemicals and covers published online. Available at: ncea-linux.usq.edu.au/farmdammanagement/images/stories/publications/Product_Review.pdf, accessed September 3, 2012.

Monsanto. 2012. Monsanto to introduce genuity droughtgard hybrids in Western Great Plains in 2013. Press release issued by Monsanto on September 11, 2012. Available at: http://monsanto.mediaroom.com/genuity-droughtgard-hybrids-2013, accessed January 19, 2013.

Montague, P. 2009. GMOs contaminate crops worldwide: Organic farmers penalized. Published online April 17, 2009 by Food Freedom. Available at: foodfreedom.wordpress.com/2009/06/26/gmos-contaminate-crops-worldwide-organic-farmers-penalized/, accessed February 24, 2011.

Morris, W., ed. 1973. *The American Heritage Dictionary of the English Language.* American Heritage Pub. Co. & Houghton Mifflin Co., Boston, MA.

Morse, R.A. & N.W. Calderone. 2000. The value of honey bees as pollinators of US crops in 2000. Report. Cornell University, Ithaca, NY.

Morton, O. 2011. Cooling the Earth. *The Economist,* November 22, 2010. Available at: http://www.economist.com/node/17492961, accessed January 21, 2013.

Mosier, N.S. 2006. Cellulosic ethanol—Biofuel beyond corn. Online Purdue Extension Bulletin ID-335. Available at: www.ces.purdue.edu/extmedia/ID/ID-335.pdf, accessed August 2, 2011.

Mulchandani, E. 2006. Aircraft—Biggest CO_2 contributor by 2020. 2 pp. Web article. Available at: www.4ecotips.com/eco/article_show.php?aid=671&id=287, accessed July 11, 2011.

Mulholland, P.J. et al. 2008. Stream denitrification across biomes and its response to anthropogenic nitrate loading. *Nature* 452(7184): 13; 202–205.

Muska, D.D. 2011. Will eco-obstructionists block domestic oil? Op ed commentary in the 9 July 2011 *Chronicle* newspaper. p. 5. Willimantic, CT.

Mycoprotein. 2012. Course: FNH200/2011w Team05 Mycoprotein. 10 pp. Article on Quorn from the University of British Columbia, Course FNH200. Available at: wiki.ubc.ca/Course:FNH200/2011w_Team05_Mycoprotein, accessed May 30, 2012.

Naish, C., I. McCubbin, O. Edberg & M. Harfoot. 2007. Outlook of energy storage technologies. 57 pp. Technical Report No. IP/A/ITRE/ST/2007-07 to the European Parliament's Committee on Industry, Research and Energy (ITRE). Available at: www.europarl.europa.eu/document/activities/cont/201109/20110906ATT26009EN.pdf, accessed May 18, 2012.

NAMA. 2012. Mushroom poisoning syndromes. Published online by the North American Mycological Association. Available at: www.namyco.org/toxicology/poison_syndromes.html, accessed July 7, 2012.

Nan, F. & D. Adjeroh. 2004. On complexity measures for biological sequences. *Proc. 2004 IEEE Computational Systems Bioinformatics Conference* (CSB 2004). 5 pp.

NASA. 2012a. Satellite image of eutrophic dead zone off the Mississippi River delta. Available at: www.nasaimages.org/luna/servlet/detail/NSVS~3~3~7277~107277:Mississippi-Dead-Zone, accessed June 10, 2012.

NASA. 2012b. Satellite image of algal bloom in the Baltic Sea. Available at: earthobservatory.nasa.gov/images/2001184110526.LIA_HROM_lrg.jpg, accessed June 10, 2012.

NASA. 2012c. Watching your heat budget. 7 pp. Web article. Available at: http://education.gsfc.nasa.gov/experimental/all98invProject.Site/Pages/trl/inv2-1.abstract.html, accessed June 8, 2012.

National Academy of Sciences. 2003. *Frontiers in Agricultural Research: Food, Health, Environment, and Communities*. Washington, DC.

National Biological Information Infrastructure. 2011a. Agriculture and genetic diversity. Web published by the U.S. Geological Survey. Available at: www.nbii.gov/portal/server.pt/community/agriculture_genetic_diversity/406, accessed February 19, 2011.

National Biological Information Infrastructure. 2011b. Sustainable agriculture. Published online. Available at: www.nbii.gov/portal/server.pt/community/agriculture_genetic_diversity/406/sustainable_agriculture/589, accessed February 24, 2011.

National Cancer Institute. 2006. National Cancer Institute fact sheet: Estimating breast cancer risk. Available at: www.cancer.gov/cancertopics/factsheet/risk/estimating-breast-cancer-risk, accessed February 5, 2011.

Natural gas. 2012. Natural gas gross withdrawals and production. US EIA data. Available at: 205.245.135.7/dnav/ng/ng_prod_sum_dcu_NUS_a.htm, accessed July 11, 2012.

Nature News. 2010. GM crop escapes into the American wild. Available at: www.nature.com/news/2010/100806/full/news.2010.393.html, accessed August 24, 2011.

Nature's Dry. 2009. Nutrient values in dried/dehydrated mushrooms. 4 pp. Web paper. Available at: www.naturesdry.com/dried-mushrooms-nutrition-values.html, accessed August 26, 2011.

Neck, R., ed. 2003. *Modeling and Control of Economic Systems*. Oxford, England: Elsevier Science. 442 pp.

Neill, F. 2010. Puberty blues. *Intelligent Life Magazine*, 29 June 2010. Available at: more intelligentlife.com/print/2815, accessed February 5, 2011.

NETL1. 2011. Energy resource potential of methane hydrate. 23 pp. Online Report of the US National Energy Technology Laboratory, DOE. Available at: www. netl.doe.gov/technologies/oil-gas/publications/Hydrates/2011Reports/MH_ Primer.pdf, accessed April 11, 2011.

NETL2. 2011. Fire in the ice. 28 pp. January 2011 methane hydrate newsletter. Available at: www.netl.doe.gov/technologies/oil-gas/publications/Hydrates/ Newsletter/MHNews_2011_01.pdf, accessed April 11, 2011.

Newsletter. 1996. Raising mealworms. *The Food Insects Newsletter* 9(1): 6 pp. Available at: www.hollowtop.com/finl_html/mealworms.htm, accessed May 13, 2011.

NG. 2011. Natural gas. 4 pp. Web paper. Available at: www.natural.gas.org/overview/ background.asp/, accessed January 10, 2011.

NHDES (New Hampshire Department of Environmental Services). 2006. *Nitrate and Nitrite: Health Information Summary*. Web published. Available at: des.nh.gov/ organization/commissioner/pip/factsheets/ard/documents/ard-ehp-16.pdf, accessed February 19, 2011.

Nicol, S. & Y. Endo. 1997. Krill fisheries of the world. On-line FAO Fisheries Technical Paper 367. 100 pp. ISBN: 92-5-104012-5. Available at: www.fao.org/DOCREP/003/ W5911/w5911e00.htm#Contents, accessed January 10, 2012.

Nikulshina, V., C. Gebald & A. Steinfeld. 2009. CO_2 capture from atmospheric air via consecutive CaO-carbonization and $CaCO_3$-calcination cycles in a fluidized-bed solar reactor. *Chem. Eng. J.* 146: 244–248.

Njiti, V.N., O. Myers Jr., D. Schroeder & D.A. Lightfoot. 2003. Roundup ready soybean: Glyphosate effects on *Fusarium solani* root colonization and sudden death syndrome. *Agron. J.* 95: 1140–1145.

Nisbet, R.E.R., R. Fisher, R.H. Nimmo, D.S. Bendall, P.M. Crill, A.V. Gallego-Sala, E.R.C. Hornibrook, E. López-Juez, D. Lowry, P.B.R. Nisbet, E.F. Shuckburgh, S. Sriskantharajah, C.J. Howe & E.G. Nisbet. 2009. Emission of methane from plants. *Proc. Royal. Soc. B* 276: 1347–1354.

NIST. 2006. Kolmogorov complexity by (CRC-A). Available at: www.nist.gov/dads/ HTML/kolmogorov.html, accessed November 18, 2009.

NOAA. 2008a. U.S. drought. State of the Science Fact Sheet, May 2008. 2 pp. Available at: nrc.noaa.gov/plans_docs/2008/US_drought_FINAL.pdf, accessed November 14, 2011.

NOAA. 2008b. Ocean acidification. State of the Science Fact Sheet, May 2008. 2 pp. Available at: nrc.noaa.gov/plans_docs/2008/Ocean_acidification_FINAL.pdf, accessed November 14, 2011.

NOAA. 2010. Greenhouse gases: Frequently asked questions. National Climatic Data Center. 4 pp. Available at: www.ncdc.npaa.gov/oa/climate/gases.html, accessed November 14, 2011.

NOAA. 2011. Natural variability main culprit of deadly Russian heat wave that killed thousands. 2 pp. *NOAA news* online. Available at: www.noaanews.noaa.gov/ stories2011/20110309_russianheatwave.html, accessed March 13, 2012.

NOAA. 2012. Recent Mauna Loa CO_2. Graphs. Available at: www.esrl.noaa.gov/ gmd/ccgg/trends/, accessed May 30, 2012.

NOAA graph. 2009. Global distribution of atmospheric methane (plot of [CH_4] vs. time and latitude). Available at: en.wikipedia.org/wiki/File:Ch4_surface_color. svg, accessed June 18, 2012.

Noone, C.J., M. Torrilhon & A. Mitsos. 2011. Heliostat field optimization: A new computationally efficient model and biomimetric layout. *Solar Energy* 86: 792–803.

Norberg, J. & G.S. Cummings, eds. 2008. *Complexity Theory for a Sustainable Future.* Columbia University Press, New York.

Nordling, L. 2010. Pebble-bed nuclear reactor gets pulled: South Africa cuts funding for energy technology project. 2 pp. News article from *Nature* 463: 1008–1009. Available at: www.nature.com/news/2010/100223/full/4631008b.html, accessed October 25, 2011.

Northfield. 2011. Northfield Mountain Station. 1 p. Web article. Available at: www. firstlightpower.com/generation/north.asp, accessed March 14, 2011.

Northrop, R.B. 1967. Electrofishing. *IEEE Trans. Biomed. Eng.* 14(3): 91–200.

Northrop, R.B. 1981. The use of underwater electric fields to control the movements of migratory fish. Report No. 1 in *Progress Report on the Use of Underwater Electric Fields to Control the Movements of Migratory Fish at Holyoke, Massachusetts.* R&D Report Submitted to Northeast Utilities Service Company.

Northrop, R.B. 2000. *Exogenous and Endogenous Regulation and Control of Physiological Systems.* Chapman & Hall/CRC Press, Boca Raton, FL.

Northrop, R.B. 2001. *Dynamic Modeling of Neuro-Sensory Systems.* CRC Press, Boca Raton, FL.

Northrop, R.B. 2005. *Introduction to Instrumentation and Measurements,* 2nd ed. CRC Press, Boca Raton, FL.

Northrop, R.B. 2010. *Signals and Systems Analysis in Biomedical Engineering,* 2nd ed. CRC Press, Boca Raton, FL. ISBN: 978-1-4398-1251-8.

Northrop, R.B. 2011. *Introduction to Complexity and Complex Systems.* 531 pp. CRC Press. Boca Raton, FL. ISBN: 978-1-4398-3901-0.

Northrop, R.B. 2012. *Analysis and Application of Analog Electronic Circuits to Biomedical Instrumentation,* 2nd ed. CRC Press, Boca Raton, FL. ISBN: 978-1-4398-6669-6.

Northrop, R.B. & A.N. Connor. 2009. *Introduction to Molecular Biology, Genomics and Proteomics for Biomedical Engineers.* CRC Press, Boca Raton, FL. ISBN: 978-1-4200-6119-2.

NREL. 2007a. Cellulosic ethanol. 8 pp. Online paper, NREL/BR-510-40742, March 2007. Available at: www.nrel.gov/biomass/pdfs/40742.pdf, accessed May 5, 2012.

NREL. 2007b. Validation of hydrogen fuel cell vehicle and infrastructure technology. NREL online Report NREL/FS-560-42284, October 2007. Available at: www. afdc.energy.gov/afdc/pdfs/42284.pdf, accessed May 16, 2012.

NREL. 2012. Learning about renewable energy: Geothermal energy basics. NREL's homepage for geothermal energy with hot links. Available at: www.nrel.gov/ learning/re_geothermal.html?print, accessed May 30, 2012.

NREL. 2012b. Ivanpah Solar Electric Generating Station. Available at: www.nrel.gov/ csp/solarplaces/project_detail.cfm/projectID=62?print, accessed January 25, 2013.

Nuclear. 2012. Nuclear power reactors. 25 pp. Online tutorial paper. Available at: www.world-nuclear.org/info/inf32.htm, accessed August 2, 2012.

Nuclear power history. 2007. History of the global nuclear power industry. A graph. Available at: http://en.wikipedia.org/wiki/File:Nuclear_Power_History.png, accessed June 26, 2012. (Data from the IAEA.)

Nus, J. 1993. Drought resistance. *Golf Course Management,* July 1991. pp. 20–26, 102–103. Available at: archive.lib.msu.edu/tic/acman/article/1993jul20.pdf, accessed July 18, 2011.

Nutrition. 2000. Nutritional value of various insects per 100 grams. Table. Available at: www.ent.iastate.edu/misc/insectnutrition.html, accessed May 13, 2011.

Nutrition Facts. 2012. Quinoa, cooked. Complete nutritional data. Available at: quick-nutritionfacts.com/nutritional-value/20137/quinoa-cooked, accessed July 17, 2012.

Oberg, E., F.D. Jones & H.L. Horton. 1976. Flywheels. In: *Machinery's Handbook,* 20th ed. New York, New York, USA: Industrial Press, Inc. pp. 341–345.

Obert, R. & B.C. Dave. 1999. Enzymatic conversion of carbon dioxide to methanol: Enhanced methanol production in silica gel matrices. *J. Am. Chem. Soc.* 121(51): 12192–12193.

Ogata, K. 1970. *Modern Control Engineering.* Prentice-Hall Inc., Englewood Cliffs, NJ.

Ogata, K. 1987. *Discrete-Time Control Systems.* Prentice-Hall Inc., Englewood Cliffs, NJ. ISBN 0-13-216102-8.

Oil sands. 2008. Canadian oil sands vs Colorado's oil shale. 1 p. Web article. Available at: canadianoilsand.org/canadian-oil-sands-versus-colorados-oil-shale/, accessed February 18, 2011.

Olah, G.A. 2005. Beyond oil and gas: The methanol economy. *Angew. Chem. Int. Ed. Engl.* 44(18): 2636–2639.

Olenev, N. 2007. A normative balance dynamic model of regional economy for study economic integrations. Paper presented at International Conference on Economic Integration Competition and Cooperation. Session 6. 19–20 April 2007. Available at: mpra.ub.uni-muenchen.de/7823/, accessed November 18, 2009.

Olsen, O.W. 1974. *Animal Parasites: Their Life Cycles and Ecology,* 3rd ed. University Park Press, Baltimore, MD. ISBN: 0-8391-0643-2.

Orcutt, M. 2012. The carbon capture conundrum. *Technology Review.* September/October. p. 21.

Ormerod, P., H. Johns & L. Smith. 2001. An agent-based model of the extinction patterns of capitalism's largest firms. Web paper. 15 pp. Available at: www.complexity-society.com/papers/model_of_capitalism.pdf, accessed December 10, 2009.

ORNL. 2002. Methane extraction and carbon sequestration. 6 pp. *ORNL Rev.* 35(2). Available at: www.ornl.gov/info/ornlreview/v35_2_02/methane.shtml, accessed May 18, 2012.

Owen, J. 2003. Environmentalists fight plans to farm cod in Scotland. 3 pp. *National Geographic News.* 22 July 2003. Available at: news.nationalgeographic.com/news/pf/30422362.html, accessed March 14, 2011.

Pacinst. 2012. *Hydraulic Fracturing and Water Resources: Separating the Frack from the Fiction.* Cooley, H., Donnelly, K., authors; Ross, N., Luu, P., eds. 34 pp. Pacific Institute report. ISBN: 1-893790-40-1. Available at: www.pacinst.org/reports/fracking/full_report.pdf, accessed July 20, 2012.

Palaeos. 2008. Fungi phylogeny. 7 pp. Web paper. Available at: www.palaeos.org/Fungi_phylogeny, accessed August 22, 2011.

Paoletti, M.G., L. Norberto, E. Cozzarini & S. Musumeci. 2009. Role of chitinases in human stomach for chitin digestion: AMCase in the gastric digestion of chitin and chit in gastric pathologies. Ch 20 in: *Binomium Chitin-Chitinase: Recent Issues.* ISBN: 978-1-60692-339-9.

Papua. 2008. Papua New Guinea women performing infanticide to end tribal war. 3 pp. Online news release. Available at: www.foxnews.com/story/0,2933,460166,00. html, accessed August 9, 2012.

Paralytic. 2012. Paralytic shellfish poisoning. 2 pp. Web article. Available at: www. nwfsc.noaa.gov/hab/habs_toxins/marine_biotoxins/psp/index.html, accessed January 10, 2012.

Parker, R. 2006. Early human diet—Insects as food. 17 pp. Web paper. Available at: www.coconutstudio.com/Insects.htm, accessed May 6, 2011.

Parker, R.C. (with J.F. Morgan). 1950. *Methods of Tissue Culture,* 2nd ed. New York, New York, USA: Paul B. Hoeber, Inc., 294 pp.

Parks Canada. 2012. Fundy National Park of Canada: Tides. Available at: www.pc.gc. ca/pn-np/nb/fundy/visit/marees-tides.aspx, accessed July 5, 2012.

Parliament. 2010. Where's the hydrogen economy? Library of Parliament (Canada), Background Paper. Publication 2010-16-E. Available at: www.parl.gc.ca/ Content/LOP/ResearchPublications/2010-16-e.pdf, accessed January 27, 2013.

Partap, U. 2002. Case Study No. 10. Cash crop farming in the Himalayas: The importance of pollinator management and managed pollination. *Biodiversity and the Ecosystem Approach in Agriculture, Forestry and Fisheries.* Available at: www.fao.org/DOCREP/005/Y4586/y4586e11.htm, accessed March 28, 2011.

Pearce, M. 2012. Oil spill reshapes Michigan Town. *Hartford Courant,* 31 July 2012. p. A4.

Peer, R.L. 1992. Development of an empirical model of methane emissions from landfills. US EPA Project Summary No. EPA/600/SR-92/037. April 1992. 5 pp.

Pellets. 2011. What are pellets? 3 pp. Web article. Available at: pelletheat.org/pellets/ what_are_pellets/, accessed May 14, 2012.

Perlman, H. 2011. The water cycle: Transpiration. 3 pp. Web article. Available at: ga.water.usgs.gov/edu/watercycletranspiration.html, accessed July 18, 2011.

Petrosky, H. 2009. Infrastructure. *Am Sci.* 97(5): 370–375.

PewClimate. 2009. Carbon capture and storage (CCS). 8 pp. Online Climate TechBook. Pew Center for Global Climate Change. Available at: www.pewclimate.org/ docUploads/CCS-Fact-Sheet1_0.pdf, accessed July 4, 2011.

Physicians for Social Responsibility. 2011. U.S. meat production. Published online. Available at: www.psr.org/chapters/oregon/safe-food/industrial-meat-system. html, accessed January 17, 2011.

Pidwirny, M. 2010. Atmospheric composition. 5 pp. Web paper. Available at: www. eoearth.org/article/Atmospheric_composition, accessed July 11, 2011.

Pidwirny, M. 2011. Introduction to the hydrosphere: (b) The hydrological cycle. 2 pp. Available at: www.physicalgeography.net/fundamentals/8b.html, accessed August 22, 2011.

Pienkos, P.T., L. Laurens & A. Aden. 2011. Making biofuel from microalgae. *Am Sci.* 99(6): 474–481.

Pimentel, D. 2005. Environmental and economic costs of the application of pesticides primarily in the United States. *Environ. Dev. Sustain.* 7: 229–252.

Pimentel, D., P. Hepperly, J. Hanson, R. Seidel & D. Douds. 2005. Organic and conventional farming systems: Environmental and economic issues. Report 05-1. Online paper published by the Rodale Institute, Kutztown, Pennsylvania, USA. Available at: http://dspace.library.cornell.edu/bitstream/1813/2101/1/pimentel_report_05-1.pdf, accessed January 21, 2013.

Pimentel, D., C. Harvey, P. Resosudarmo, K. Sinclair, D. Kurz, M. McNair, S. Crist, L. Shpritz, L. Fitton, R. Saffouri & R. Blair. 1995. Environmental and economic costs of soil erosion and conservation. *Science* 267: 1117–1123.

Pimm, S.L., G.J. Russell, J.L. Gittelman & T.M. Brooks. 1995. The future of biodiversity. *Science* 269: 347–350.

Piqueira, J.R.C. 2008. A mathematical view of biological complexity. *Commun. Nonlinear Sci. Numer. Simul.* 14(6): 2581–2586.

Plasmans, J.E.J.K., J. Engwerda, B. van Aarle, G. di Bartolomeo & T. Michalak. 2010. *Dynamic Modeling of Monetary and Fiscal Cooperation among Nations* (Dynamic Modeling and Econometrics in Economics and Finance). 335 pp. New York, New York, USA: Springer. ISBN-13: 978-1441939104.

Poloni, C.R., M.A. Ramírez, R. Morejón, J.M. Dell´Amico, D. Morales, L. Alfonso & O. Ledea. 2009. Disminución de la evaporación del agua desde superficies libres utilizando como retardador el alcohol estearílico [Decreased water evaporation from free surfaces using stearyl alcohol as a retardant]. *Cultrop [Tropical Crops]*, vol.30, n.1. ISSN 1819-4087. Available at: http://scielo.sld.cu/scielo.php?script=sci_arttext&pid=S0258-59362009000100002&lng=es&nrm=iso&tlng=es, accessed January 21, 2013.

Potts, S.G., J.C. Biesmeijer, C. Kremen, P. Neumann, O. Schweiger & William E. Kunin. 2010. Global pollinator declines: Trends, impacts and drivers. *Trends Ecol. Evol.* 25: 345–353. Available at: http://nature.berkeley.edu/kremenlab/Articles/Global%20pollinator%20declines.pdf, accessed January 21, 2013.

Power Partners. 2009. Power Partners[SM] resource guide: Landfill methane. 6 pp. Web article. Available at: www.uspowerpartners.org/Topics/SECTION6Topic-LandfillMethane, accessed June 13, 2011.

Powers-Fraites, M.J., R.L. Cooper, A. Buckalew, S. Jayaraman, L. Mills & S.C. Laws. 2009. Characterization of the hypothalamic–pituitary–adrenal axis response to atrazine and metabolites in the female rat. *Toxicol. Sci.* 112(1): 88–99.

PowerStream. 2003. PowerStream battery chemistry FAQ. 26 pp. Web paper. Available at: www.powerstream.com/BatteryFAQ.html, accessed June 13, 2011.

Pretty, J.N., A.D. Noble, D. Bossio, J. Dixon, R.E. Hine, F.W. Penning de Vries & J.I. Morison. 2006. Resource-conserving agriculture increases yields in developing countries. *Environ. Sci. Technol.* 40(4): 1114–1119.

Prosek, J. 2010. *Eels.* New York, New York, USA: HarperCollins. ISBN: 978-0-06-056611-1.

Provost, C. 2012. Global land grab could trigger conflict, report says. *The Guardian,* 2 February 2012. Available at: www.guardian.co.uk/global-development/2012/feb/02/global-land-grab-trigger-conflict-report, accessed March 25, 2012.

PRWeb. 2010. GRT's future commercial air-capture products will address CO_2 market needs in sectors ranging from agriculture to industry. 2 pp. Press release. Available at: www.prweb.com/printer/545763.htm, accessed July 7, 2011.

Pryor, F.L. 1996. *Economic Evolution and Structure: The Impact of Complexity on the US Economic System.* Cambridge University Press, NY. ISBN: 0-521-55097-1.

PSP. 2012. Paralytic shellfish poisoning (PSP). 4 pp. Web article. Available at: yyy.rsmas.miami.edu/groups/ohh/science/psp.htm, accessed January 10, 2012.

Puechmaille, S.J., G. Wibbelt, V. Korn, H. Fuller, F. Forget, K. Mühldorfer, A. Kurth, W. Bogdanowicz, C. Borel, T. Bosch, T. Cherezy, M. Drebet, T. Görföl, A-J Haarsma, F. Herhaus, G. Hallart, M. Hammer, C. Jungmann, Y. Le Bris, L. Lutsar, M. Masing, B. Mulkens, K. Passior, M. Starrach, A. Wojtaszewski, U. Zöphel & E. Teeling. 2011. Pan-European distribution of White-Nose Syndrome fungus *(Geomyces destructans)* not associated with mass mortality. *PLoS ONE.* 6(4): e19167. 11 pp.

Quastler, H. 1958. A primer on information theory. In: Yockey, H.P., Platzman, R.L., Quastler, H., eds. *Symposium on Information Theory in Biology.* New York, New York, USA: Pergamon Press. pp. 3–49.

Quayle, S. 2011. SI radiation measurement units: conversion factors. 2 pp. Available at: www.stevequayle.com/ARAN/rad.conversion.html, accessed April 11, 2011.

Quijano, R., L. Panganiban & N. Cortes-Maramba. 1993. Time to blow the whistle; dangers of toxic chemicals. *World Health* 46(5): 26–27.

Quorn. 2011a. Our products. 2 pp. Web blurb. Available at: www.quorn.us/Products/, accessed August 22, 2011.

Quorn. 2011b. Quorn. 4 pp. SwissPedia article. Available at: www.swisscorner.com/wiki.php?title=Quorn#cite_note-28, accessed May 30, 2012.

Ragheb, M. 2010. Modern wind generators. Published online by the University of Illinois. Available at: netfiles.uiuc.edu/mragheb/www/NPRE%20745%20Wind%20Power%20Systems/Modern%20Wind%20Generators.pdf, accessed July 4, 2012.

Railsback, S.F., S.L. Lytinen & S.K. Jackson. 2006. Agent-based simulation platforms: Review and development recommendations. *Simulation* 82(9): 609–623. Available at: sim.sagepub.com/cgi/reprint/82/9/609, accessed February 11, 2010.

Raine, A., J. Foster & J. Potts. 2006. The new entropy law and the economic process. *Ecol. Complex.* 3: 354–360.

Rajapakse, A., D. Muthumuni & N. Perera. 2009. Chapter 7. Grid integration of renewable energy systems. In: Hammons, T.J., ed., *Renewable Energy.* pp. 463–504. New York, New York, USA: InTech. ISBN: 978-953-7619-52-7.

Raloff, J. 2002. Hormones: Here's the beef: environmental concerns reemerge over steroids given to livestock. *Sci. News* 161(1): 10.

Ramanan, R., K. Kannan, S.D. Sivanesan, S. Mudliar, S. Kaur, A.K. Tripathi & T. Chakrabarti. 2009. Bio-sequestration of carbon dioxide using carbonic anhydrase enzyme purified from *Citrobacter freundii.* *World J. Microbiol. Biotechnol.* 25(6): 981–987.

Ranganathan, J. & F. Irwin. 2011. Letter to the editors of *The Economist,* 11–17 June 2011. p. 22 (concerning the age of man).

Rapal, B. 2002. Chrysler Natrium: The power of Na. *The Car Connection,* 5 May 2002. Available at: www.thecarconnection.com/tips-article/1003728_chrysler-natrium-the-power-of-na, accessed July 5, 2012.

Ratliff, E. 2007. One molecule could cure our addiction to oil. *Wired Magazine* 15(10): 7 pp. Available at: www.wired.com/print/science/planetearth/magazine/15-10/ff_plant, accessed October 10, 2011.

Reactor. 2011. Reactor could produce fuel from sunlight. Press release. Available at: www.theengineer.co.uk/news/reactor-could-produce-fuel-fromsunlight/1006723.article/, accessed January 10, 2011.

REN21. 2011. Renewables 2011 global status report. Published online by the Renewable Energy Policy Network for the 21st Century. Available at: www.ren21.net/Portals/97/documents/GSR/REN21_GSR2011.pdf, accessed July 8, 2012.

renewableenergyworld. 2011. Advent of ultracapacitors signals change in wind turbine capabilities. Available at: www.renewableenergyworld.com/rea/news/article/2011/03/advent-of-ultracapacitors-signals-change-in-wind-turbine-capabilities/March 2100, accessed September 13, 2011.

Repetto, R. & S.S. Baliga. 1996. *Pesticides and the Immune System: The Public Health Risks*. World Resources Institute, Washington, DC. Available at: pdf.wri.org/pesticidesandimmunesystem_bw.pdf, accessed February 6, 2011.

REUK. 2008. La Rance Tidal Power Plant. 2 pp. Web article. Available at: www.reuk.co.uk/print.php?article=La-Rance-Tidal-Power-Plant, accessed May 24, 2011.

REUK. 2012. Stirling engine solar power. 2 pp. Web article. Available at: www.reuk.co.uk/print.php?article=Stirling-Engine-Solar-Power.htm, accessed February 10, 2012.

Reuters. 2012. Up to 5000,000 new refugees could flee to S. Sudan: WFP. By Holland, H., 30 January 2012. News article on Sudanese refugees. Available at: af.reuters.com/article/topNews/idAFJOE80T08220120130, accessed June 23, 2012.

Rhinefrank, K., E.B. Agamloh, A. von Jouanne, A.K. Wallace, J. Prudell, K. Kimble, J. Aills, E. Schmidt, P. Chan, B. Sweeny & A.A. Schacher. 2006. Novel ocean energy permanent magnet linear generator buoy. *Renewable Energy*. 31: 1279–1298.

Ridley. 2006. Future of the electric car. 39 pp. Presentation by Ridley Engineering. Available at: www.switchingpowermagazine.com/downloads/Future of the Electric Car.pdf, accessed January 23, 2012.

Riebeek, H. 2011. The carbon cycle. Online paper. Available at: earthobservatory.nasa.gov/Features/CarbonCycle/printall.php, accessed August 3, 2012.

Riga, E., L.A. Lacey & N. Guerra. 2008. *Muscador albus*, a potential biocontrol agent against plant-parasitic nematodes of economically important vegetable crops in Washington State, USA. *Biol. Control* 45: 380–385.

Ripley, J. et al. 2008. Utilization of protein expression profiles as indicators of environmental impairment of smallmouth bass (*Micropterus dolomieu*) from the Shenandoah River, Virginia, USA. *Environ. Toxicol. Chem.* 27(8): 1756–1767.

Robbins, C., M. Whiteman, S. Hillis, K. Curtis, J. McDonald, P. Wingo, A. Kulkarni & P. Marchbanks. 2009. Influence of reproductive factors on mortality after epithelial ovarian cancer diagnosis. *Cancer Epidemiol Biomarkers Prev*. July. 18: 2035. doi: 10.1158/1055-9965.EPI-09-0156.

Robinson, M. 2009. 20% Wind by 2030: Technology & science challenges. Presentation at the Cornell University Workshop on Large-Scale Wind Power, June 13, 2009. Available at: cfd.mae.cornell.edu/~caughey/WindPower_09/Presentations/Robinson.pdf, accessed July 4, 2012.

Robinson, M.C. 2006. Renewable energy technologies for use on the outer continental shelf. 34 pp. PowerPoint slide presentation. Ocean Energy Technology Conf. National Renewable Energy Lab. 6 June 2006. Available at: ocsenergy.anl.gov/documents/docs/NREL_Scoping_6_06_2006_web.pdf, accessed August 18, 2011.

Rogers, I., K. Northstone, D. Dunger, A. Cooper, A. Ness & P. Emmett. 2010. Diet throughout childhood and age at menarche in a contemporary cohort of British girls. *Public Health Nutr.* 13: 2052–2063. doi:10.1017/S1368980010001461.

Rogers, L. 2009. Scientists grow pork meat in a laboratory. 2 pp. Web article from *The Sunday Times*, (London) 29 November 2009. Available at: voidmanufacturing.wordpress.com/2009/12/16/scientists-grow-pork-meat-in-a-laboratory, accessed December 6, 2011.

Rosen, N., D. Yee, M.E. Lippman, S. Paik & J.J. Cullen. 1991. Insulin-like growth factors in human breast cancer. *Breast Cancer Res. Treat.* 18: S55–S62.

Rosser, J.B., Jr. 2008. Econophysics and economic complexity. 27 pp. Web essay. Available at: cob.jmu.edu/rosserjb/ECONOPHYSICS%20AND%20ECONOMIC%20COMPLEXITY.doc, accessed June 1, 2012.

Rothwell, J. 1997. Cold fusion and the future: Part 1—Revolutionary technology. *Infinite Energy Magazine*, Issue 12. Available at: www.infinite-energy.com/iemagazine/issue12/coldfusion.html, accessed April 6, 2011.

Rothwell, J. & E. Storms. 2011. Does cold fusion exist? Viewpoint: Yes,... 5 pp. Web essay. Available at: www.scienceclarified.com/dispute/Vol-2/Does-cold-fusion-exist, accessed April 6, 2011.

Rotman, D. 2012. King natural gas. *Technology Review*, September/October 2012, pp. 77–79.

Rudoph, J.C. 2011. Physicist group's study raises doubts on capturing carbon dioxide from air. *New York Times,* 9 May 2011. Available at: www.nytimes.com/2011/05/10/science/earth/10carbon.html, accessed July 7, 2011.

Runge, C.F. & B. Senauer. 2007. How biofuels could starve the poor. *Foreign Affairs,* May–June 2007. Available at: www.foreignaffairs.com/articles/62609/c-ford-runge-and-benjamin-senauer/how-biofuels-could-starve-the-poor, accessed March 25, 2012.

Ruth, M. & B. Hannon. 1997. *Modeling Dynamic Economic Systems.* New York, New York, USA: Springer. 339 pp.

Ryan, D. 2004. *Biodiesel—A Primer.* ATTRA document. Available at: attra.ncat.org/attra-pub/PDF/biodiesel.pdf, accessed November 4, 2009.

Rycroft, C.H., G. Grest, J. Landry & M. Bazant. 2006. Analysis of granular flow in a pebble-bed nuclear reactor. *Phys Rev E Stat Nonlin Soft Matter Phys.* 2006 Aug;74(2 Pt 1):021306. Epub 2006 Aug 24.

Sample, I. 2006. Rice fungus's killer gene discovered. 1 p. Web press release. Available at: www.guardian.co.uk/science/2006/mar/23/food.gm, accessed August 26, 2011.

Samuels, C.A. 2011. USA: cheating scandals intensify focus on test pressures. 4 pp. Web essay. Available at: jorgewerthein.blogspot.com/2011/08/usa-cheating-scandals-intensify-focus.html, accessed August 15, 2011.

Samuelson, P.A. & W.D. Nordhaus. 2001. *Economics*, 17th ed. New York, New York, USA: McGraw-Hill, p. 157.

Sanogo, S., X.B. Yang & H. Scherm, 2000. Effects of herbicides on *Fusarium solani* f. sp. glycines and development of sudden death syndrome in glyphosate-tolerant soybean. *Phytopathology* 90: 57–66.

Santner, S.J., B. Ohlsson-Wilhelm & R.J. Santen. 1993. Estrone sulfate promotes human breast cancer cell replication and nuclear uptake of estradiol in MCF-7 cell cultures. *Int. J. Cancer* 54(1): 119–124.

Sapkota, P. & H. Kim. 2009. Zinc-air fuel cell, a potential candidate for alternative energy. *J. of Industrial and Engineering Chemistry.* 15: 450–455.

SARA. 2011. MHD wave energy conversion. 2 pp. Homepage. Available at: www.sara.com/RAE/ocean_ wave.html, accessed May 11, 2011.

Saslow, W.M. 1999. An economic analogy to thermodynamics. *Am. J. Phys.* 67(12): 1239–1247.

Sauser, B. 2007. Ethanol demand threatens food prices. *Technology Review,* 13 February 2007. Available at: www.technologyreview.com/Energy/18173/?a=f, accessed March 25, 2012.

Savinell, R.F. 2011. Flow batteries. 6 pp. *Electrochemistry Encyclopedia*. Available at: electrochem.cwru.edu/encycl/art-b03-flow-batt.htm, accessed August 1, 2012.

Saxton. 2009. Electric vehicle efficiency analysis. 3 pp. Web article. T. & C Saxton. Available at: www.saxton.org/EV/efficiency.php, accessed January 23, 2012.

Scardaci, S.C. 2003. A new disease in California. Agronomy Fact Sheet Series 1997-2. U. California-Davis. Available at: www.plantsciences.ucdavis.edu/uccerice/afs/agfs0297.htm, accessed October 20, 2010.

Scheffer, M. 2009. *Critical Transitions in Nature and Society*. Princeton University Press, Princeton, NJ. ISBN: 978-0-691-12204-5.

Schellnhuber, H.J., M. Molina, N. Stern, V. Huber & S. Kadner, eds. 2010. *Global Sustainability: A Nobel Cause*. Cambridge University Press, Cambridge, MA. ISBN: 978-0-521-76934-1.

Schirber, M. 2007. Why desalination doesn't work (yet). 3 pp. Web article. Available at: www.livescience.com/environment/070625_desalination_membranes.html/, accessed December 7, 2010.

Schneider, E. & J. Kay. 1994. Life as a manifestation of the second law of thermodynamics. *Math. Comput. Model.* 19(6–8): 25–48.

Schoof, R. 2011. For military branches, next foe is fossil fuels. *Hartford Courant*, 26 June 2011.

Schulze, E.-D., R.H. Robichaux, J. Grace, P.W. Rundel & J.R. Ehleringer. 1987. Plant water balance. *BioScience* 37(1): 30–37.

Schuster, H.G. 2005. *Complex Adaptive Systems: An Introduction*. Scator Verlag, Saarbrücken, Germany. ISBN: 3-9807936-0-5. Available at: www.theo-physik.uni kiel.de/theo-physik/schuster/cas.html, accessed June 1, 2012.

Science. 2013. Science quotes by Francis Crick. Available at: http://todayinsci.com/C/Crick_Francis/CrickFRancis-Quotations.htm, accessed January 25, 2013.

ScienceDaily. 2008a. Economic value of insect pollination worldwide estimated at U.S. $217 billion. ScienceDaily news release. 2 pp. Available at: www.sciencedaily.com/releases/2008/09/080915122725.htm, accessed July 5, 2012.

ScienceDaily. 2008b. Solar power: New world record for solar-to-grid conversion efficiency set. Available at: www.sciencedaily.com/releases/2008/02/080213172955.htm, accessed July 4, 2012.

scienceofdoom. 2011. Evaluating and explaining climate science: Radiative forcing and the surface energy balance. 10 pp. Web paper. Available at: scienceofdoom.com/2011/09/02/radiative-forcing-and-the-surface-energy-balance/, accessed November 21, 2011.

Scionix. 2011. Scintillation crystals. 6 pp. Web paper. Available at: www.helgesona.es/m02/firmas/SCIONIX/pages/navbar/sci_cry, accessed April 11, 2011.

Searchinger, T. et al. 2008. Use of U.S. croplands for biofuels increases greenhouse gasses through emissions from land use change. *Science* 319(5867): 1238–1240.

Sears, F.W. 1949. *Optics*. Addison-Wesley Press, Inc., Cambridge, MA.

Sears, F.W. 1950. *Mechanics, Heat and Sound*, 2nd ed. Addison-Wesley Press, Inc., Cambridge, MA.

Seat61. 2010. CO_2 emissions% global warming: Trains versus planes. 3 pp. Web article. Available at: www.seat61.com/CO2flights.htm, accessed July 11, 2011.

Secretariat of the Rotterdam Convention on the Prior Informed Consent Procedure for Certain Hazardous Chemicals and Pesticides in International Trade. 2008. Draft decision guidance document: Alachlor. Available at: www.pic.int/en/DGDs/Alachlor/Alalchlor%20DGD%20after%20CRC5.pdf, accessed February 6, 2011.

Secretary. 2009. The Secretary of Energy talks with *Technology Review* about the future of nuclear power post Yucca Mountain. *Q & A: Steven Chu.* 14 May 2009. Available at: www.technologyreview.com/printer_friendly_article.aspx?id=22651, accessed August 1, 2012.

Sedlak, D. 2011. The fate of hormones in the aquatic environment. Published online by the University of California at Berkeley. Available at: www.ce.berkeley. edu/~sedlak/research_1.html, accessed January 29, 2011.

Seed Savers Exchange. 2011. About us. Published online. Available at: www.seedsavers. org/Content.aspx?src=aboutus.htm, accessed February 22, 2011.

Seekingalpha. 2012. BP gives up on cellulosic ethanol in the US. 2 pp. Web article. October 26, 2012. Available at: http://seekingalpha.com/article/954581-bp-gives-up-on-cellulosic-ethanol-in-the-u-s, accessed January 27, 2013.

Seetharaman, D. 2012. High gas prices help Detroit. *The Hartford Courant,* 16 March 21012. p. A12.

Seligsohn, D. et al. 2010. CCS in China: Toward an environmental, health, and safety regulatory framework. World Resources Institute Report. Washington, DC. Available at: www.wri.org/publication/ccs-in-china, accessed July 4, 2011.

Selvam, J.N., N. Kumaravadivel, A. Gopikrishnan, B.K. Kumar, R. Ravikesavan, R. & M.N. Boopathi. 2009. Identification of a novel drought tolerance gene in *Gossypium hirsutum* L. cv KC3. *Commun. Biometry Crop Sci.* 4(1): 9–13.

Semelsberger, T.A., R.L. Borup & H.L. Greene. 2006. Dimethyl ether (DME) as an alternative fuel. *J. of Power Sources* 156: 497–511.

Semuels, A. 2011. Crime growing like a weed on California farms. *Hartford Courant* newspaper, 16 January 2011. p. A6.

Senöz, E. & R.P. Wool. 2009. Hydrogen storage on carbonized chicken feather fibers. Paper 14 at the 13th Annual Green Chemistry & Engineering Conference, Tuesday, 23 June 2009. University of Maryland.

SERC. 2011. Earth exploration toolbox: Part 1—What causes a phytoplankton bloom in the Gulf of Maine? 5 pp. Web paper. Available at: serc.carleton.edu/eet/phy toplankton/primer.html, accessed May 17, 2011.

Sesma, A. & A.E. Osbourn. 2004. The rice leaf blast pathogen undergoes developmental processes typical of root-infecting fungi. *Nature* 431: 582–586 (Letters to Nature).

Sevilla, M. & A.B. Fuertes. 2009. The production of carbon materials by hydrothermal carbonization of cellulose. *Carbon* 47: 2281–2289.

Sevilla, M., J.A. Maciá-Agulló & A.B. Fuertes. 2011. Hydrothermal carbonization of biomass as a route for the sequestration of CO_2: Chemical and structural properties of the carbonized products. *Biomass Energy* 35: 3152–3159.

Shah, A. 2011. PV solar cells—Guide to solar cell costs and efficiency, buying solar cells, major cell manufacturers. 2 pp. Web article. Available at: greenworld-investor.com2011/04/05/pv-solar-cells-guide-to-solar-cell-costs-and-efficiency buying-solar-cells-major-cell-manufacturers/, accessed April 18, 2011.

Shannon, C. 1948. A mathematical theory of communication. *Bell Syst. Tech. J.* 7: 535–563.

Shannon, C. & W. Weaver. 1949. *Mathematical Theory of Communications.* University of Illinois Press, Urbana, IL.

Shekar, C. 2006. Methanol: The new hydrogen. 2 pp. Available at: www.technology review.com/printer_friendly_article.aspx?id=16629, accessed January 24, 2011.

Shenhua. 2010. Shenhua launches China's first carbon capture and storage program. 1 p. Press release. *People's Daily Online*. Available at: english.peopledaily.com. cn/90001/90778/90860/7011478.html, accessed July 4, 2011.

Sheridan, J.A. & D. Bickford. 2011. Shrinking body size as an ecological response to climate change. *Nat. Clim. Chang.* doi:10.1038/nclimate1259. Published online 16 October 2011. Available at: www.nature.com/nclimate/journal/vaop/ncur rent/full/nclimate1259.html, accessed October 20, 2011.

Shiklomanov, I.A. 2000. Appraisal and assessment of world water resources. *Water Int.* 25(1): 11–32.

Shore, L. & A. Pruden. 2009. *Hormones and Pharmaceuticals Generated by Concentrated Animal Feeding Operations: Transport in Water and Soil* (Emerging Topics in Ecotoxicology). Springer, New York.

Sialis. 2011. Raising mealworms: Everything you always wanted to know (and more). 13 pp. Web article. Available at: sialis.com/raisingmealworms.htm, accessed May 13, 2011.

Siddiqui, O. & R. Bedard. 2005. Feasibility assessment of offshore wave and tidal current power production. *Proc. IEEE PES 05 GM,* paper 05GM0538, San Francisco, CA, June 2005. pp. 1–6.

Sigma Scan. 2011. Plight of the bumble bee: Decline in pollinator populations? 2 pp. Web article. Available at: www.sigmascan.org/Live/Issue/ViewIssue/151/5/plight-of-the-bumble-bee, accessed February 14, 2012.

Simon, J.L. 1998. *The Ultimate Resource 2*. Princeton, New Jersey: Princeton University Press. ISBN-10: 0691003815.

Simpkins, D.M. 2012. Malthus, Thomas Robert. 7 pp. Web encyclopedia biography and bibliography. Available at: www.encyclopedia.com/topic/Thomas_Robert_Malthus.aspx, accessed June 4, 2012.

Simpson, J., C. McConnell & Y. Matsuda. 2011. *Economic Assessment of Carbon Capture and Storage Technologies—2011 Update*. 57 pp. Global CCS Institute Report Online. Available at: info@globalccsinstitute.com, accessed July 4, 2011.

Singh, S.K., G.A. Strobel, B. Knighton, B. Geary, J. Sears & D. Ezra. 2011. An endophytic *Phomopsis* sp. possessing bioactivity and fuel potential with its volatile organic compounds. *Microb. Ecol.* 61: 729–739. doi:10.1007/s00348-011-9818-7.

Sitingcases. 2012. Appendix B: Solar stirling engine. 14 pp. Available at: www.energy. ca.gov/sitingcases/solartwo/documents/applicant/afc/volume_02+03/ MASTER_AppendixB.pdf, accessed February 10, 2012.

Smil, V. 2001. *Enriching the Earth: Fritz Haber, Carl Bosch, and the Transformation of World Food Production*. MIT Press, Cambridge, MA. ISBN: 0-262-69313-5.

Smil, V. 2006. *Energy: A Beginner's Guide*. Oneworld Pub. Co., London, UK. ISBN: 1851684522.

Smil, V. 2008. *Energy in Nature and Society: General Energetics of Complex Systems*. MIT Press, Cambridge, MA.

Smil, V. 2011a. Global energy: The latest infatuation. *Am. Sci.* 99(3): 212–219.

Smil, V. 2011b. Gluttony. *Foreign Policy*, November 2011. p. 67.

Smith, D.J. 2004. Systems thinking: The knowledge structures and the cognitive process (online essay). 33 pp. Available at: www.smithsrisca.co.uk/systems-think ing.html, accessed June 9, 2011.

Smith, E.M., T.W. Goodwin & J. Schillinger. 2003. Challenges to the worldwide supply of helium in the next decade. *Adv. Cryog. Eng.* 49 A (710): 119–138. doi:10.1063/1.1774674.

Smith, E.N. & R.T. Taylor. 1982. Acute toxicity of methanol in the folate-deficient acatalasemic mouse. *Toxicology* 25(4): 271–287.

Smith, T. 2011. UK to invest $4 million(US) in three wave and tidal energy technologies. 1 p. Press release. Available at: www.oceanpowermagazine.net/2011/02/07/uk-to-invest-4-millionus-in-three-wave-and-tidal-energy-technologies/, accessed May 24, 2011.

Soerensen, H.C. & A. Weinstein. 2008. Ocean energy: Position paper for IPCC. Keynote paper for the *IPCC Scoping Conference on Renewable Energy*, Lübeck, Germany, January 2008. Available at: www.eu-oea.com/euoea/files/ccLibraryFiles/Filename/000000000400/Ocean_Energy_IPCC_final.pdf, accessed May 19, 2011.

Solarbenzin. 2011. ETH macht aus Sonnenenergie Benzin. Press release. Available at: www.20min.ch/wissen/news/story/ETH-macht-aus-Sonnenenergie-Benzin-14440579/, accessed January 10, 2011.

Solé, R.V. & B. Luque. 1999. Statistical measures of complexity for strongly interacting systems. (Submitted to *Physical Review E*, 27 August 1999). 12 pp. Available at: arxiv.org/PS_cache/adap-org/pdf/9909/9909002v1.pdf, accessed November 30, 2009.

Sornette, D. 2003. *Why Stock Markets Crash*. Princeton University Press, Princeton, NJ. ISBN: 0-691-09630-9.

Soto, A.M. & C. Sonnenschein. 2010. Environmental causes of cancer: Endocrine disruptors as carcinogens. *Nat. Rev. Endocrinol.* 6: 363–370.

Soystats. 2011. US biodiesel production 1999–2010. Available at: www.soystats.com/2011/page_24.htm, accessed October 13, 2011.

Spath, P.L. & M.K. Mann. 2004. Biomass power and conventional fossil systems with and without CO_2 sequestration—Comparing the energy balance, greenhouse gas emissions and economics. 28 pp. NREL technical paper. NREL/TP-510-32575. January 2004. Available at: www.osti.gov/bridge, accessed May 18, 2012.

Specter, M. 2011. Test-tube burgers. *New Yorker* magazine, 23 May 2011. pp. 32–34.

Spelter, H. & D. Toth. 2009. North America's wood pellet sector. 21 pp. USDA research paper FPL-RP-656. Available at: www.fpl.fs.fed.us/documnts/fplrp/fpl_rp656.pdf, accessed May 14, 2012.

Spenser, J. 2004. Fuel cells in the air. *Boeing Frontiers Online*. Available at: www.boeing.com/news/fronteirs/archive/2004/july/ts_sfta.htm, accessed January 25, 2013.

Spruce Budworm. 2011. Balsam fir/spruce budworm foodweb v2 (3-D dynamic view of network). Available at: www.youtube.com/watch?v=zSggb5WmSYO, accessed October 7, 2011.

Stamp, J.W. 2003. Plant and mushroom growth medium. US Patent No.: US 6.609,331 B1. Pub. date: 26 August 2003.

Standard. 2003. How do we measure "standard of living"? 8 pp. Web paper. Available at: www.bos.frb.org/education/ledger/ledger03/winter/measure.pdf, accessed October 13, 2011.

Steering Committee. 2009. Colony collapse disorder progress report. 45 pp. CCD Steering Committee. Available at: www.extension.org/mediawiki/files/c/c7/CCDReport2009.pdf, accessed March 22, 2011.

Steering Committee. 2010. Colony collapse disorder progress report. 7 pp + 31 pp Appendix. CCD Steering Committee. Available at: www.ars.usda.gov/is/br/ccd/ccdprogressreport2010.pdf, accessed March 29, 2011.

Steinberg, M. 1997. The Carnol process system for CO_2 mitigation and methanol production. *Energy* 22(2/3): 143–149.

Stelzl, U. & E.E. Wanker. 2006. The value of high quality protein–protein interaction networks for systems biology. (ScienceDirect.com.) *Curr. Opin. Chem Biol.* 10: 551–558.

Stelzl, U., U. Worm, M. Lalowski, C. Haenig, F.H. Brembeck, H. Goehler, M. Stroedicke, M. Zenkner, A. Schoenherr, S. Koeppen, J. Timm, S. Mintzlaff, C. Abraham, N. Bock, S. Kietzmann, A. Goedde, E. Toksöz, A. Droege, S. Krobitsch, B. Korn, W. Birchmeier, H. Lehrach & E. Wanker. 2005. A human protein–protein interaction network: A resource for annotating the proteome. *Cell* 122: 957–968.

Stern, D. 2004. The energy of the sun. Online paper by NASA. Available at: www-istp. gsfc.nasa.gov/stargaze/Sun7enrg.htm, accessed August 17, 2011.

Stokstad, E. 2007. Deadly wheat fungus threatens world's breadbaskets. *Science* 315: 1786–1787.

Stolaroff, J.K., D.W. Keith & G.V. Lowry. 2008. Carbon dioxide capture from atmospheric air using sodium hydroxide spray. *Environ. Sci. Technol.* 42(8): 2728–2735. Available at: pubs.ac.org/doi/pdf/10.1021/es702607w, accessed July 7, 2011.

Strachan, H. 2003. *The First World War.* Penguin Books, London, UK. ISBN: 978-0-14-303518-3.

Strobel, G.A., Strobel, G.A., B. Knighton, K. Kluck, Y. Ren, T. Livinghouse, M. Griffin, D. Spakowicz & J. Sears. 2010. *Corrigendum.* The production of myco-diesel hydrocarbons and their derivatives by the endophytic fungus *Gliocladium roseum* (NRRL 50072). *Microbiology* 156: 3830–3833.

Succar, S. & R.H. Williams. 2008. Compressed air energy storage: Theory, resources, and applications for wind power. *Report of the Princeton Environmental Institute.* 81 pp. 8 April 2008. Available at: www.princeton.edu/pei/energy/publica tions/texts/SuccarWilliams_PEI_CAES_2008April8.pdf, accessed April 11, 2011.

Sun. 2004. (S-7) The energy of the sun. NASA online tutorial. Available at: www-spof. gsfc.nasa.gov/stargaze/Sun7enrg.htm, accessed July 20, 2012.

SunPower. 2011. Corporate homepage. Available at: investors.sunpowercorp.com/ governance.cfm?sh_print=yes&, accessed October 13, 2011.

Suontama, J. 2006. Macrozooplankton as a feed source for farmed fish—growth, product quality and safety. PhD dissertation at the University of Bergen, Norway. 110 pp. Available at: bora.uib.no/bitstream/1956/2070/17/Main_Thesis.pdf, accessed January 10, 2012.

Sussmann, R., F. Forster, M. Rettinger & P. Bousquet. 2012. Renewed methane increase for five years (2007–2011) observed by solar FTIR spectrometry. *Atmos. Chem. Phys.* 12: 4885–4891. doi:10.5194/acp-12-4885-2012.

Sustainable Table. 2011a. Artificial hormones. Published online. Available at: www. sustainabletable.org/issues/hormones/, accessed January 29, 2011.

Sustainable Table. 2011b. Biodiversity. Published online. Available at: www.sustain abletable.org/issues/biodiversity/, accessed February 22, 2011.

Sustainable Table. 2011c. Heritage and heirloom foods. Published online. Available at: www.sustainabletable.org/issues/heritage/, accessed February 22, 2011.

Sustainable Table. 2011d. Sustainable vs. industrial: A comparison. Published online. Available at: www.sustainabletable.org/intro/comparison/, accessed February 24, 2011.

Sutton, P. 2000. Sustainability: What does it mean? 7 pp. Web paper. Available at: www.green-innovations.asn.au/sustblty.htm, accessed May 7, 2012.

Suzawa, M., H.A. Ingraham & A. Iwaniuk. 2008. The herbicide atrazine activates endocrine gene networks via non-steroidal NR5A nuclear receptors in fish and mammalian cells. *PLoS ONE* 3(5): e2117. doi:10.1371/journal.pone.0002117.

Svoboda, E. 2010. The hard facts about fracking. *Popular Mechanics,* 13 December 2010. 3 pp. Available at: www.popularmechanics.com/science/energy/coal-oil-gas/the-hard-facts-about-fracking, accessed May 30, 2012.

Swamy, P.K. 1974. Plankton as a source of human food. *Seafood Export J.* 64(2): 2–26. Available at: eprints.cmfri.org.in/7444/, accessed May 11, 2011.

Swayze, V.W. 1995. Frontal leukotomy and related psychosurgical procedures in the era before antipsychotics (1935–1954). *Am. J. Psychiatry* 152(4): 505–515.

Tabarrok, A. 2008. The law of unintended consequences. Blog. Available at: www.marginalrevolution.com/marginalrevolution/2008/01/the-law-of-unin.html, accessed November 19, 2009.

Taheripour, F. & W. Tyner. 2007. Ethanol subsidies, who gets the benefits? 18 pp. Paper at the *Bio-Fuels, Food and Feed Tradeoffs Conference.* 12–13 April, St. Louis, MO. Available at: www.farmfoundation.org/projects/documents/TaheripourandTyner_St_Louis.pdf, accessed November 14, 2011.

Tainter, J.A. 1988. *The Collapse of Complex Societies.* Cambridge, England: Cambridge University Press. ISBN: 978-0-521-38673-9.

Talbot, D. 2012. The great German energy experiment. *Technol. Rev.* 115(4): 41–55.

Tang, Y., J.-S. Xie & K. Chen. 2002. Hand pollination of pears and its implications for biodiversity conservation and environmental protection: A case study from Hanyuan County, Sichuan Province, China. 22 pp. Web paper. Available at: ftp://ftp.cgiar.org/ifpri/WeiZhang/Literature/Pollination/Yaetal.pdf, accessed March 28, 2011.

Tar Sands. 2011. About tar sands. Web article. Available at: osteis.anl.gov/guide/tarsands/index.cfm?printversion=true/, accessed March 14, 2011.

Tesfatsion, L. 2002. Agent-based computational economics: Growing economies from the bottom up. *Artif. Life* 8: 55–82.

Tesfatsion, L. 2005a. Syllabus of readings for complex adaptive systems and agent-based computational economics. 4. Biological evolution. Web paper. Available at: www.econ.iastate.edu/tesfatsi/bioevol.htm, accessed November 19, 2009.

Tesfatsion, L. 2005b. Agent-based computational economics: a constructive approach to economic theory. pp. 1–55. (Preprint of a chapter that appeared in Tesfatsion, L. & K.L. Judd, eds. 2006. *Handbook of Computational Economics, Volume 2: Agent-Based Computational Economics.* Amsterdam, Netherlands: Elsevier/North Holland. Available at: www.econ.iastate.edu/tesfatsi/hbintlt.pdf, accessed December 10, 2009).

Tesfatsion, L. 2009. Agent-Based Computational Economics (ACE) home page. Available at: www.econ.iastate.edu/tesfatsi/ace.htm, accessed December 10, 2009.

Tesfatsion, L. & K.L. Judd, eds. 2006. *Handbook of Computational Economics, Volume 2: Agent-Based Computational Economics.* Elsevier/North Holland. (Preface, topics, contributors & chapter abstracts.) Available at: www.econ.iastate.edu/tesfatsi/hbace.htm, accessed December 11, 2009.

Thermoanalytics. 2012. Battery types. Available at: www.thermoanalytics.com/support/publications/batterytypesdoc.html, accessed July 5, 2012.

Thomas, G. & J. Keller. 2003. Hydrogen storage—Overview. Presentation at Sandia National Laboratories' Hydrogen Delivery and Information Workshop, 7–8

May 2003. Available at: www1.eere.energy.gov/hydrogenandfuelcells/pdfs/
bulk_hydrogen_stor_pres_sandia.pdf, accessed July 5, 2012.

Thomasnet. 2012. Is cellulosic ethanol the Loch Ness monster of biofuel? 4 pp. Web
article. Available at: http://news.thomasnet.com/green_clean/2012/01/31/
is-cellulosic-ethanol-the-loch-ness-monster-of-biofuel/, accessed January 27, 2013.

Times. 2011. Relying on hard and soft sells, India pushes sterilization. *New York Times*,
22 June 2011.

Titirici, M.M., D. Murach & M. Antonietti. 2010. Chapter 27. Opportunities for tech-
nological transformations: From climate change to climate management? In:
Schellnhuber, H.J. et al., eds. *Global Sustainability: A Nobel Cause*. New York, New
York, USA: Cambridge University Press. pp. 320–330.

Titirici, M.M., A. Thomas & M. Antonietti. 2007. Back in the black: Hydrothermal
carbonization of plant material as an efficient chemical process to treat the CO_2
problem? *New J. Chem.* 31(6): 787–789.

Titman, S. 2010. Oil sands hold promise as energy resource. 5 pp. Web article published 16 December 2010 on Texas Enterprise. Available at: texasenterprise.org/
print/569, accessed March 16, 2011.

Tokuda, N., T. Kanno, T. Hara, T. Shigematsu, Y. Tsutsui, A. Ikeuchi, T. Itou & T.
Kumamoto. 2000. Development of a redox flow battery system. *SEI Technical
Rev.* 50: 88–94.

Tononi, G., O. Sporns & G.M. Edelman. 1994. A measure for brain complexity:
Relating functional segregation and integration in the nervous system. *PNAS*
91: 5033–5037.

Tononi, G., O. Sporns & G.M. Edelman. 1999. Measures of degeneracy and redundancy in biological networks. *PNAS* 96: 3257–3262.

Tools. 2010. Tools for agent-based modelling. *SwarmWiki*. Available at: www.swarm.
org/index. php/Tools_for_Agent_Based_Modeling/, accessed March 15, 2010.

Torp, T.A. & J. Gale. 2003. Demonstrating storage of CO_2 in geological reservoirs:
The Sleipner and SACS projects. In: Gale, J., Kaya, Y., eds. *Proc. 6th International
Conf. on Greenhouse Gas Control Technologies*. Paper B1-1. Pergamon, Amsterdam,
Netherlands. pp. 311–316. Available at: smartpipe.com/project/IK23430000%20
SACS/Publications/Tprp_and_Gale_SACS_overview.pdf, accessed July 6, 2011.

Torres Galvis, H.M., J.H. Bitter, C.B. Khare, M. Ruitenbeek, A.I. Dugulan & K.P. de
Jong. 2012. Supported iron nanoparticles as catalysts for sustainable production of lower olefins. *Science* 335: 835–838. doi:10.1126/science.1215614.

TransBiodiesel. 2009. Overview: TransBiodiesel: A fast-growing company. 7 pp.
Web overview. Available at: www.transbiodiesel.com/General/Questions-
Answers/qaa/, accessed June 27, 2011.

Treehugger. 2009. Crambe: One more plant to be turned into biofuels. Web paper.
Available at: www.treehugger.com/files/2009/02/crambe-one-more-plant-
turned-into-biofuels.php, accessed January 3, 2011.

Tricoli, J.V., L.B. Rall, C.P. Karakousis, L. Herrera, N.J. Petrelli, G.I. Bell & T.B. Shows.
1986. Enhanced levels of insulin-like growth factor messenger RNA in human
colon carcinomas and liposarcomas. *Cancer Res.* 46: 6169–6173.

Truxal, J.G. 1955. *Automatic Feedback Control System Synthesis*. New York, New York,
USA: McGraw-Hill Book Co.

Tulloch, D. 2011. Does cold fusion exist? Viewpoint: No... 4 pp. Web paper. Available at:
www.scienceclarified.com/Vol-2/Does-cold-fusion-exist.html, accessed April
6, 2011.

Turnbaugh, B. 2009. EPA finds secret fracking chemicals in drinking water. 3 pp. Web paper. Available at: www.ombwatch.org/node/10353, accessed December 29, 2010.

Turner, D. 2011. Run like the mob: US school cheating scandal details emerge. 3 pp. News release. Available at: www.msnbc.msn.com/id/43779246/ns/us_news-life/t/run-mob-us-school-cheating-scandal-details-emerge/, accessed August 15, 2011.

TVA. 2010. From the new deal to a new century: A short history of TVA. Web article. Available at: www.tva.gov/about/history.htm, accessed January 4, 2010.

Tyner, W.E. 2007. Biofuels, energy-security and global warming policy interactions. 12 pp. Online paper, presented at the National Agricultural Biotechnology Conference, at South Dakota State University, Brookings, SD, 22–24 May 2007. Available at: www.agecon.purdue.edu/papers/biofuels/S_Dakota_paper_May07_Tyner.pdf, accessed May 30, 2012.

UCAR (University Corporation for Atmospheric Research). 2009. Global warming: Cuts in greenhouse gas emissions would save Arctic ice, reduce sea level rise. 4 pp. Press release. Available at: www.ucar.edu/news/releases/2009/greenhousecuts.jsp, accessed October 26, 2010.

UCAR. 2010. Climate change: Drought may threaten much of globe within decades. 5 pp. Web paper by Aiguo Dai. Available at: www2.ucar.edu/news/climate-change-drought-may-threaten-much-globe-within-decades, accessed October 22, 2010.

Ulrich, W. 2005. Can nature teach us good research practice? A critical look at Frederic Vester's bio-cybernetic systems approach. *J. Res. Pract.* 1(1): Article R2. 10 pp. Available at: jrp.icaap.org/index.php/jrp/article/view/1/1, accessed March 14, 2011.

UN (United Nations). 2010. UN data: Total fertility rate (children per woman). Available at: esa.un.org/unpd/wpp/p2k0data.asp/, accessed November 14, 2011.

UN. 2011. World population prospects: The 2010 revision. United Nations, New York, New York, USA. Available at: esa.un.org/unpd/wpp/index.htm, accessed November 14, 2011.

Underhill, R. 2009. Bees make cleansing flights. 1 p. Web article. Available at: peacebeefarm.blogspot.com/2009/12/bees-make-cleansing-flights.html, accessed March 23, 2011.

UNEP (United Nations Environmental Programme). 2010. UNEP emerging issues: Global honey bee colony disorder and other threats to insect pollinators. 12 pp. Online United Nations Environmental Programme Report. Available at: www.unep.org/dewa/Portals/67/pdf/Global_Bee_Colony_Disorder_and_Threats_insect_pollinators.pdf, accessed July 17, 2012.

UNEP. 2012. State of the environment and policy retrospective: 1972–2002. Online report. Available at: www.unep.org/geo/geo3/english/pdfs/chapter2-2_land.pdf, accessed June 23, 2012.

UNEP (United Nations Environmental Programme) & UNCTAD (United Nations Conference on Trade and Development). 2008. Organic agriculture and food security in Africa. Report released October 2008, pp. 1–61.

Uniongas. 2011. Chemical composition of natural gas. 2 pp. Web table. Available at: www.uniongas.com/aboutus/aboutng/composition.asp/, accessed January 10, 2011.

United Nations Convention on Biological Diversity. 2008. *Biodiversity for food and nutrition*. Web published. Available at: www.cbd.int/agro/food-nutrition/, accessed February 19, 2011.

US Census Bureau. 2012. Current population clock. Available at: www.census.gov/ main/www/popclock.html, accessed June 23, 2012.

USDA (United States Department of Agriculture). 1979. *How to estimate recoverable heat energy in wood or bark fuels.* Available at: www.fpl.fs.fed.us/documents/fplgtr/ fplgtr29.pdf, accessed July 2, 2012.

USDA (United States Department of Agriculture). 2006. Transitioning to organic production. Published online by USDA Sustainable Agriculture Research and Education. Available at: www.sare.org/publications/organic/organic01.htm, accessed February 24, 2011.

USDA (United States Department of Agriculture). 2007. Organic agriculture overview. Published online by the USDA, Cooperative State Research, Education, and Extension Service (CSREES). Available at: www.csrees.usda.gov/ ProgViewOverview.cfm?prnum=16643, accessed January 24, 2011.

USDA (United States Department of Agriculture). 2010. Agricultural biotechnology: Adoption of biotechnology and its production impacts. Published online by the Economic Research Service of the United States Department of Agriculture, July 1, 2010. Available at: http://www.ers.usda.gov/data-products/adoption-of-genetically-engineered-crops-in-the-us/recent-trends-in-ge-adoption.aspx, accessed January 19, 2013.

USDA (United States Department of Agriculture). 2011. Agricultural concentration. Report published online by the United States Department of Agriculture. Available at: www.usda.gov/documents/Agricultural_Concentrationd.doc, accessed January 17, 2011.

US DOE (United States Department of Energy). 2004. China supports plan to build world's largest tidal power plant. *EERE News,* 14 November 2004. Available at: apps1.eere.energy.gov/news/news_detail.cfm/news_id=8286, accessed July 5, 2012.

US DOE. 2008. Comparison of fuel cell technologies. Table. Available at: www.hydro gen.energy.gov/, accessed January 20, 2011.

US DOE. 2011. U.S. Billion ton update: Biomass supply for a bioenergy and bioproducts industry. R.D. Perlak & B.J. Stokes (leads), ORNL/TM-2011/224. Oak Ridge National Laboratory, Oak Ridge, TN. 227 pp. Available at: www1.eere.energy .gov/biomass/pdfs/billion_ton_update.pdf, accessed October 25, 2011.

US DOE. 2012. *Heating fuel comparison calculator.* (Last updated May 2012.) Available at: www.eia.doe.gov/neic/experts/heatcalc.xls, accessed July 4, 2012.

USEIA. 2010. Monthly biodiesel production report. December 2009. DOE/ EIA0642(2009/12). Available at: www.eia.gov/fuelrenewable.html, accessed October 13, 2011.

USEIA. 2011a. Table 5.6b. Average retail price of electricity to ultimate customers by end-use sector, by state, year-to-date. US Energy Information Administration: Electric Power Monthly. Available at: www.eia.doe.gov/electricity/monthly/, accessed January 30, 2012.

USEIA. 2011b. Petroleum & other liquids. Table and graph of Weekly All Countries Spot Price FOB Weighted by Estimated Export Volume. Available at: www.eia. doe.gov/dnav/pet/hist/LeafHandler.ashx?n=PET&s=WTOTWORLD&f=W, accessed April 6, 2011.

USEIA. 2012. How much of our electricity is generated from renewable energy? Published online by the US Energy Information Agency. Available at: 205.254.135.7/energy_in_brief/renewable_electricity.cfm, accessed July 8, 2012.

USFDA (US Food and Drug Administration). 2002a. Letter from M.F. Jacobson, Executive Director, CSPI, to J. Levitt, Director, Center for Food Safety and Applied Nutrition, USFDA. Available at: cspinet.org/new/quornltr.pdf, accessed May 30, 2012.

USFDA (US Food and Drug Administration). 2002b. The use of steroid hormones for growth promotion in food-producing animals. *FDA Veterinarian Newsletter* 16(5): 2002.

USFDA (US Food and Drug Administration). 2010. 2009 summary report on antimicrobials sold or distributed for use in food-producing animals. Report released December 2010. Available at: www.fda.gov/downloads/ForIndustry/UserFees/AnimalDrugUserFeeActADUFA/UCM231851.pdf, accessed January 17, 2011.

USGS (United States Geological Survey). 2008. 90 billion barrels of oil and 1,670 trillion cubic feet of natural gas assessed in the Arctic. USGS, 27 July 2008. Available at: www.usgs.gov/newsroom/article.asp?ID=1980&from=rss_home#.T-eGg StYv9M, accessed May 9, 2012.

USGS. 2009. An estimate of recoverable heavy oil resources of the Orinoco oil belt, Venezuela. Fact Sheet 2009–3028, October 2009. Available at: pubs.usgs.gov/fs/2009/3028, accessed January 27, 2013.

USGS. 2010a. Water science for schools. 3 pp. Web article. Available at: ga.water.usgs.gov/edu/drinkseawater.html/, accessed December 7, 2010.

USGS. 2010b. white-nose syndrome threatens the survival of hibernating bats in North America. 10 pp. Web article. Available at: www.fort.usgs.gov/wns/, accessed August 19, 2010.

USGS. 2011. The water cycle: Transpiration. 3 pp. Web article. Available at: ga.water.usgs.gov/edu/watercycletranspiration.html, accessed July 18, 2011.

Utzinger, J.D., W.B. Brooks & R.D. Touse. 2011. Effect of wood ashes on garden soil. Cornell University Cooperative Extension of Schenectady County. Fact Sheet. Available at: http://counties.cce.cornell.edu/schenectady/new/pdf/ag%20fact%20sheets/soil/Wood%20Ashes%20on%20Garden%20Soil.pdf, accessed January 21, 2013.

Vajda, A., L. Barber, J. Gray, E. Lopez, J. Woodling & D. Norris. 2008. Reproductive disruption in fish downstream from an estrogenic wastewater effluent. *Environ. Sci. Technol.* 42(9): 3407–3414. Available at: pubs.acs.org/doi/abs/10.1021/es0720661, accessed February 5, 2011.

Vanderklippe, N. 2011. EnCana slams EPA water contamination report. 2 pp. Press release. *Calgary Globe and Mail.* Available at: www.theglobeandmail.com/report-on-business/industry-news/encana-slams-epa-water-contamination-report, accessed December 11, 2011.

van der Werf, G.R., D.C. Morton, R.S. DeFries, J.G.J. Olivier, P.S. Kasibhatla, R.B. Jackson, G.J. Collatz & J.T. Randerson. 2009. CO_2 emissions from forest loss. *Nature Geosci.* 2: 737–738.

van Engelsdorp, D., J. Evans, C. Saegerman, C. Mullin, E. Haubruge, B.K. Nguyen, M. Frazier, J. Frazier, D. Cox-Foster, Y. Chen, R. Underwood, D. Tarpy & J. Pettis. 2009. Colony collapse disorder: A descriptive study. *PLoS ONE* 4(8): e6481. 17 pp. Available at: plosone.org/article/info:doi/10.1371/journal.pone.0006481, accessed March 28, 2011.

van Noorden, R. 2007. Air, can we have our carbon back? 2 pp. Web news bulletin. Available at: www.rsc.org/chemistryworld/News/2007/October/05100701.asp, accessed July 7, 2011.

Vasectomy. 2008. The history of vasectomy. 10 pp. Web article. Available at: www. vasectomy-information.com/moreinfo/histoty.htm, accessed May 14, 2012.

Vast. 2012. Vast methane "plumes" seen in Arctic Ocean as sea ice retreats. 2 pp. News release. Available at: www.independent.co.uk/news/science/vast-methane-plumes-seen-in-arctic-ocean-as-sea-ice-retreats-6276278.html, accessed January 16, 2012.

Vatican. 2009. Fighting poverty to build peace. Message of His Holiness Pope Benedict XVI for the Celebration of the World Day of Peace, 1 January 2009. Available at: www.vatican.va/holy_father_benedict_xvi/messages/peace/documrnts/hf_ben-xvi_mes_20081208_xlii-world-day-peace_en.html, accessed February 10, 2012.

Veazey, M. 2006. Alberta's oil sands: Not just for caulking canoes. 5 pp. Web article from *Rigzone*. Available at: www.rigzone.com/training/heavyoil/insight. asp?i_id=186, accessed March 16, 2011.

Vegan. 2011. Oyster mushroom nutrition. 4 pp. Web article. Available at: www.bestve ganguide.com/oystermushroom-nutrition.html, accessed August 26, 2011.

Verhoest, C. & Y. Rickman. 2012. Industrial wood pellets report. Published online 22 March 2012 by Enplus Pellets. Available at: www.enplus-pellets.eu/wp-con tent/uploads/2012/04/Industrial-pellets-report_PellCert_2012_secured.pdf, accessed July 7, 2012.

Vester, F. 1999. *Die Kunst vernetzt zu Denken: Ideen und Werkzeuge für neuen Umgang mit Komplexität [The Art of Networked Thinking: Ideas and Tools for a New Way of Dealing with Complexity]*, 6th ed. (in German). Deutsche Verlags-Anstalt. Stuttgart, Germany. ISBN: 3-421-05308-1. (Amazon.de offers an English translation of this book published by MCB Verlag. November 2007. ISBN-10: 3939314056.) Abstract available at: www.frederic-vester.de/eng/books/preface-of-the-book/, accessed March 14, 2011.

Vester, F. 2004. Sensitivity Model/Sensitivitäts modell Prof. Vester®. v. SMW 5.0e. Commercial software package in English, German or Spanish language. (Original version 1991.) For Windows 95, 98, NT, 2000 & XP. Munich, Germany. Frederic Vester GmbH. Available at: www.frederic-vester.de/eng/sensitivity-model/, accessed March 14, 2011.

Victor, D.G. 2011. *Global Warming Gridlock: Creating More Effective Strategies for Protecting the Planet*. Cambridge, England: Cambridge University Press. 358 pp. ISBN: 978-0-521-86501-2.

Vine, M. 2011. Why fuel-cell vehicles may replace battery vehicles as the cleantech poster boy. 3 pp. Blog. Available at: seekingalpha.com/article/311290-why-fuel-cell-vehicles-mas-replace=battery-vehicles-as-the-cleantech-poster-boy, accessed May 16, 2012.

Vining, J. 2005. Ocean wave energy conversion. Advanced independent study report for ECE 699. December 2005. University of Wisconsin, Madison. 37 pp. Available at: homepages.cae.wisc.edu/~vining/JVining-WaveEnergyConversion. pdf, accessed May 11, 2011.

Vogel, H. 1979. A better way to construct the sunflower head. *Math. Biosci.* 44(44): 179–189.

Wagner, A. 2005. Distributed robustness versus redundancy as causes of mutational robustness. *BioEssays* 27(2): 176–188.

Wagner, G.P., M. Pavlicev & J.M. Cheverud. 2007. The road to modularity. *Nat. Rev. Genet.* 8: 921–931.

Wallinga, D. & M. Mellon. 2008. Factory farms feeding antibiotic crisis. OP-ed essay in the *Hartford (CT) Courant*, 14 July 2008.

Walter, K. 2011. Methane hydrate: A surprising compound. 6 pp. Online article. Available at: www.llnl.gov/str/Durham.html, accessed April 11, 2011.

Wang, D. 1999. Relaxation oscillators and networks. In: Webser, J.G., ed. *Wiley Encyclopedia of Electrical and Electronics Engineering*, Volume 18. New York, New York, USA: J. Wiley & Sons.

Wang, H., Z. Xu, L. Gao & B. Hao. 2009. A fungal phylogeny based on 82 complete genomes using the composition vector method. *BMC Evol. Biol.* 9: 195. doi:10.1186/1471-2148-9-195.

Wargo, J. 1996. *Our Children's Toxic Legacy: How Science and Law Fail to Protect Us from Pesticides*. Yale University Press, New Haven, CT.

Warner, M. 2005. Lawsuit challenges a meat substitute. 2 pp. *New York Times* press release. Available at: www.nytimes.com/2005/05/03/business/03food.html, accessed May 30, 2012.

Water Cycle. 2011. Dr. Art's guide to planet earth. The water cycle. Web article. Available at: www.planetguide.net/book/chapter_2/water_cycle.html, accessed August 22, 2011.

WaterSavr. 2012. Conserve water and reduce drought problems. 4 pp. Online bulletin. Available at: www.flexiblesolutions.com/products/watersavr/documents/WS_09_lowres.pdf, accessed September 3, 2012.

Watson, D. & M. Adams. 2011. *Design for Flooding: Architecture, Landscape and Urban Design for Resilience to Flooding and Climate Change*. New York, New York, USA: John Wiley & Sons, Inc. ISBN: 978-0-470-47565-5.

Wave. 2006. Technology white paper on wave energy potential on the U.S. Outer Continental Shelf. US Minerals Management Service, US Department of the Interior. 11 pp. Available at: ocsenergy.anl.gov/documents/docs/OCS_EIS_WhitePaper_Wave.pdf, accessed April 28, 2011.

Weber, L. 2010. *Demographic Change and Economic Growth: Simulations on Growth Models* (Contributions to Economics). 289 pp. Heidelberg, Germany: Physica-Verlag. ISBN-13: 978-3790825893.

Webster: krill. 2012. Extended definition: krill. 11 pp. *Webster's Online Dictionary*. Available at: www.websters-online-dictionary.org/definitions/krill?cx=partner-pub-0939450753529744%3Av0qd01-tdlq&cof=FORID%3A9&ie=UTF-8&q=krill&sa=Search#906, accessed January 10, 2012.

Webster: krill fishery. 2012. Krill fishery. 1 p. *Webster's Online Dictionary*. Available at: www.websters-online-dictionary.org/definitions/Krill+fishery, accessed January 10, 2012.

Webster, K.D. et al. 2012. Using open-path laser measurement of atmospheric methane concentration along a major shear zone in western Greenland as an analogue for exploration on Mars. 2 pp. Abstract of paper presented at the 43rd Lunar and Planetary Science Conference, The Woodlands, TX, 19–23 March. Available at: www.lpi.usra.edu/meetings/lpsc2012/pdf/1514.pdf, accessed June 26, 2012.

Weier, J. 1999. Changing currents color the Bering Sea a new shade of blue. 7 pp. Web paper. Available at: earthobservatory.nasa.gov/Features/Coccoliths/printall.php, accessed January 9, 2012.

Weldon, J. 2006. The Atlantic cod: The potential for farming in Shetland. 7 pp. Web paper published by the North Atlantic Fisheries College. Available at: www.thefishsite.com/articles/163/the-atlantic-cod-the-potential-for-farming-in-shetland/, accessed March 14, 2011.

White. 2006. Technology white paper on wave energy on the U.S. Outer Continental Shelf. US Minerals Management Service, US Department of the Interior. May 2006. 11 pp. Available at: ocsenergy.anl.gov, accessed May 19, 2011.

White, R.D. & D. Lee. 2011. Quake could roil energy costs. (Tribune newspapers.) *Hartford Courant*, 12 March 2011.

Wholesale solar. 2012. Solar panels on sale. 11 pp. Available at: www.wholesalesolar.com/solar-panels.html, accessed January 25, 2013.

Wibbelt, G., A. Kurth, D. Hellmann, M. Weishaar, A. Barlow, M. Veith, J. Prüger, T. Görföl, L. Grosche, F. Bontadina, U. Zöphel, H.-P Seidl, P. Cryan & D. Blehert. 2010. White-nose syndrome fungus *(Geomyces destructans)* in bats, Europe. *Emerg. Infect. Dis.* 16(8): 1237–1242.

Wikipedia. 2011. Fusion power. 20 pp. (Tokamak table.) Available at: en.wikipedia.org/wiki/Fusion_power, accessed April 8, 2011.

Wikipedia. 2013. Tokamaks. Available at: http://en.wikipedia.org/wiki/Tokamaks, accessed January 25, 2013.

Williams, E. 1991. *Dinorwig – The Electric Mountain*. A National Grid (UK) Publication.

Williams, D. 2009. The evolution of the pebble bed reactor. 4 pp. Web article. Available at: nuclearstreet.com/nuclear_power_industry_news/b/nuclear_power_news/archive/2009/11/27/under_the_hood_with_duncan_williams-the evolution-of-the-pebble-bed-reactor-11271.aspx, accessed October 25, 2011.

Wilson, E.O. 2002. *The Future of Life*. New York, New York, USA: Vintage Books. ISBN: 0-679-76811-4.

Wiltsee, G. 2000. Lessons learned from existing biomass power plants. 143 pp. Technical Report NREL/SR-570-26946.

National Renewable Energy Laboratory, Golden, CO. Available at: www.doe.gov/bridge, accessed January 17, 2011.

Windholz, M., ed. 1976. *The Merck Index*, 9th ed. Merck & Co., Rahway, NJ.

Wisconsin DNR (Department of Natural Resources). 2010. Manure management and water quality. Web published, 1 November 2010. Available at: dnr.wi.gov/runoff/ag/waterquality.htm, accessed February 19, 2011.

wisteme. 2011. What are the blue whale population trends? 2 pp. Web paper. Available at: www.wisteme.com/question.view?targetAction=viewQuestionTab&id=8441, accessed March 14, 2011.

WNS. 2012. What we do/white-nose syndrome. *Bat Conservation International*, 30 April 2012. (Article has detailed incidence map of WNS in eastern US.) Available at: www.batcon.org/index.php/what-we-do/white-nose-syndrome.html, accessed April 30, 2012.

Woo, J. 2001. Enzyme reactors making methanol from CO_2. Biochemistry 462B Honors Presentation, The University of Arizona. Available at: www.biochem.arizona.edu/classes/bioc462/462bh2008/462honorsprojects/, accessed January 24, 2011.

Wool, R.P., A. Campanella, K. Danner, E. Senoz, J. Stanzione III, C. Watson & M. Zhan. 2010. Chicken feather fibers for hydrogen storage. 3 pp. Final Report on EPA Grant No. SU834324. Available at: cfpub.epa.gov/ncer_abstracts/index.cfm/fuseaction/display.abstractDetail/abstract/9002/report/F, accessed February 4, 2011.

Wooldridge, M. 2002. *An Introduction to Multi-Agent Systems*. New York, New York, USA: Wiley & Sons. ISBN: 0-471-49691-X.

World Briefing. 2011. Texas-based firm touts natural gas find in Alaska. World briefing. *Hartford Courant*, 7 November 2011. p. 4.

World Energy. 2010. *Survey of Energy Resources.* London, England: World Energy Council. 618 pp. Available at: www.worldenergy.org/documents/ser_2010_report_1.pdf, accessed July 9, 2012.

World Health Organization. 2003. Impacts of antimicrobial growth promoter termination in Denmark. Report published on the WHO Web site. Available at: www.who.int/gfn/en/Expertsreportgrowthpromoterdenmark.pdf, accessed January 17, 2011.

Wright, D.A. & P. Welborn, 2002. *Environmental Toxicology.* Cambridge University Press, London. 630 pp.

Xu, P., M., Shi, & X.-X. Chen. 2009. Antimicrobial peptide evolution in the Asiatic honey bee *Apis cerana. PloS ONE* 4(1): e4239. 9 pp.

Yang, Y.-Y., J.L. Gray, E.T. Furlong, J. G. Davis, R.C. ReVello & T. Borch. 2012. Steroid Hormone Runoff from Agricultural Test Plots Applied with Municipal Biosolids, *Environmental Science & Technology* (Volume 46, No. 5, January 30, 2012), Pages 2746–2754.

Yasuo, M. 2003. Association of serum estrone levels with estrogen receptor-positive breast cancer risk in postmenopausal Japanese women. *Clin. Cancer Res.* 9(6): 2229–2233.

Yergin, D. 2011. *The Quest: Energy, Security, and the Remaking of the Modern World.* New York, New York, USA: Penguin Press. ISBN: 978-1-59420-283-4. 803 pp.

Yokayo. 2011. Yokayo Biofuels price schedule, effective 18 April 2011. Yokayo Biofuels, Ukiah, CA. Available at: www.ybiofuels.org/biodiesel/distribution.php, accessed October 13, 2011.

Yoon, I., R.J. Williams, E. Levine, S. Yoon, J. Dunne & N. Martinez. 2004. Webs on the Web: 3D visualization of ecological networks on the WWW for collaborative research and education. *Proc. IS&T/SPIE Symposium on Electronic Imaging, Visualization and Data Analysis* 5295: 124–132. Available at: http://spiedigitalli brary.org/data/Conferences/SPIEP/22247/124_1.pdf, accessed January 21, 2013.

Yvkoff, L. 2011. Fuel cell vehicles you can drive now (if you qualify). *CNET Reviews,* 17 March 2011. Available at: reviews.cnet.com/8301-13746_7-20043071-48.html, accessed July 7, 2012.

Zadeh, L.A., G.J. Klir & B. Yuan, eds. 1996. *Fuzzy Sets, Fuzzy Logic, and Fuzzy Systems: Selected Papers by Lofti A. Zadeh.* World Scientific, Singapore.

Zhang, N., K. O'Donnell, D.A. Sutton, F.A. Nalim, R.C. Summerbell, A.A. Padhye & D.M. Geiser. 2006. Members of the *Fusarium solani* species complex that cause infections in both humans and plants are common in the environment. *J. Clin. Microbiol.* 44(6): 2186–2190.

Zhang, W.-B. 2005. Chapter 1. Differential equations in economics. In: *Differential Equations, Bifurcations and Chaos in Economics.* Singapore, Singapore: World Scientific Pub. Co., Inc. ISBN-13: 978-9812563330.

Zheng, J.P. 2002. The limitation of energy density for battery/double layer capacitor hybrid cell. In: Brodd, R.J., ed. *Proc. Symposium on Advances in Electrochemical Capacitors & Hybrid Power Systems. Electrochemical Society Proceedings.* 1 p. Available at: www.electrochem.org/dl/ma/201/pdfs/0228.pdf, accessed December 14, 2011.

Zhu, Y., H. Chen, J. Fan, Y. Wang, Y. Li, J. Chen, J.X. Fan, S. Yang, L. Hu, H. Leung, T.W. Mew, P.S. Teng, Z. Wang & C. Mundt. 2000. Genetic diversity and disease control in rice. *Nature* 406: 718–722. doi:10.1038/35021046.

Zhu, L., R. Gao, L. Liu, Y. Wang & Y.Y. Wang. 2011. Study on catalytic experiments of methanol synthesis from cornstalk syngas. *Proc. Third International IEEE Conference on Measuring Technology and Mechatronics Automation.* pp. 423–425. doi:10.1109/ICMTMA.2011.677.

Zimmer, C. 1999. Complex systems: Life after chaos. *Science* 284(5411): 83–86. doi:10.1126/science.284.5411.83.

Index

Page numbers followed by *f* and *t* indicate figures and tables, repectively.

Printed and bound by CPI Group (UK) Ltd, Croydon, CR0 4YY

18/10/2024

01776270-0014